SUPERSYMMETRY:
A Decade of Development

SUPERSYMMETRY: A Decade of Development

Edited by **Peter C West**, King's College, London

Adam Hilger, Bristol and Boston

British Library Cataloguing in Publication Data

Supersymmetry: a decade of development.
1. Supersymmetry
I. West, P. C.
530.1′43 QC174.17.S9

ISBN 0-85274-572-9

Consultant Editor: **Professor R F Streater**, King's College London

Published under the Adam Hilger imprint by IOP Publishing Limited
Techno House, Redcliffe Way, Bristol BS1 6NX, England
PO Box 230, Accord, MA 02018, USA

Printed in Great Britain by J W Arrowsmith Ltd, Bristol

CONTENTS

LIST OF CONTRIBUTORS

L ALVAREZ-GAUMÉ
Department of Physics
Harvard University
Cambridge
Massachusetts 02138
USA

E CREMMER
Laboratoire de Physique Théorique
Ecole Normale Supérieure
24 Rue Lhomond
F-75231 Paris Cedex 05
France

S FERRARA
CERN
CH-1211 Geneva 23
Switzerland

present address

Department of Physics
University of California, Los Angeles
California 90024
USA

YU A GOLFAND
Lebedev Physical Institute
Academy of Sciences of the USSR
Leninsky Prospekt 53
117 924 Moscow
USSR

M B GREEN
Department of Physics
Queen Mary College
University of London
Mile End Road
London E1 4NS
UK

M T GRISARU
Department of Physics
Brandeis University
Waltham
Massachusetts 02254
USA

P S HOWE
Department of Mathematics
King's College
University of London
Strand
London WC2R 2LS
UK

E P LIKHTMAN
Lebedev Physical Institute
Academy of Sciences of the USSR
Leninsky Prospekt 53
117 924 Moscow
USSR

P VAN NIEUWENHUIZEN
Institute of Theoretical Physics
State University of New York
Stony Brook
New York 11794
USA

S RABY
Los Alamos National Laboratory
Theory Division, T-8
MS B285
Los Alamos
New Mexico 87545
USA

G G ROSS
Department of Theoretical Physics
University of Oxford
1 Keble Road
Oxford OX1 3NP
UK

J H SCHWARZ
Lauritsen Laboratory of Physics
California Institute of Technology
Pasadena
California 91125
USA

W SIEGEL
Department of Physics
University of California, Berkeley
California 94720
USA

present address

Department of Physics and Astronomy
University of Maryland
College Park
Maryland 20742
USA

P C WEST
Department of Mathematics
King's College
University of London
Strand
London WC2R 2LS
UK

D ZANON
Department of Physics
Brandeis University
Waltham
Massachusetts 02254
USA

PREFACE

It is now more than a decade since supersymmetry was discovered. Although only a few people were concerned with its original development, after a few years a substantial part of the theoretical physics community was working on this subject. Although there was and still is no experimental indication that supersymmetric theories are relevant to nature, interest in the subject was sustained by the rich structure of these theories that enabled physicists to make a series of new developments that provoked further interest.

This began with the discovery of the theories of rigid supersymmetry including those with more than one supersymmetry. With the construction of the supergravity theories, the couplings between supergravity and matter were then found. One of the most remarkable properties of supersymmetric theories is their ultraviolet behaviour in the sense that it was found that there were dramatic cancellations between infinite quantities resulting in fewer than expected infinities. Indeed it was this development that led to a technical solution of the hierarchy problem and gave some indication of how supersymmetric models may be relevant to nature. However, the most dramatic result of this type was the finiteness (or superconformal invariance) of a large class of the rigid supersymmetric theories. While these developments were taking place, realistic models of supersymmetry were constructed. These models had the surprising property that they related phenomena at the weak scale to that at around the Planck scale.

On a different front, supersymmetry was used to give simplified proofs of mathematical theorems, such as the positivity of the energy in general relativity and the Atiyah–Singer index theorem. Finally, the one development which is now viewed with much greater importance than before was the further development of string theories.

As such supersymmetry has become part of the culture of modern physics and the aim behind this book is that it should give a balanced view of this decade of development. Although there are a number of books on this subject, they naturally represent the viewpoints of the authors. In this book, a number of people closely connected with some of these significant developments have given a pedagogical account of their subject with the benefit of hindsight. An attempt has been made to enlist the help of people, who, although they have made major contributions, have not for one reason or another written a review type of article. We have also avoided including topics about which substantial reviews are currently being written.

Recently, there has been a considerable shift of effort away from the traditional supersymmetry approach towards that of string theory. It is perhaps not a coincidence that this occurred when it seemed as if most of the more tractable properties of supersymmetric theories had been found. In particular, it became clear that point supersymmetry field theories were unlikely to resolve the dilemma between gravity and quantum mechanics. Even if one assumes that string theories are relevant to nature, it is not clear how the 'traditional', i.e. point field theory approach to supersymmetric theories, will fit into this picture. It may be hoped that this book will be helpful in preserving some of the more important aspects of supersymmetry theories as well as the mentality with which they were viewed.

I would like to thank my colleagues for suggesting to me many of the contributors to this book.

P C West

ON $N=1$ SUPERSYMMETRY ALGEBRA AND SIMPLE MODELS

YU A GOLFAND AND E P LIKHTMAN

INTRODUCTION

When fifteen years ago we were thinking on supersymmetry (the word would arise later) our goal was rather modest. Somewhat earlier one of the authors (YuAG) considered spinorial extensions of Poincaré group wishing to come across new no-go theorems. After several kinds of such extensions were formulated we decided to construct a field theory subject to the new invariance principle. We chose the most simple algebra (henceforth denoted by R). Curiously, in our realization (we used Dirac spinors) the algebra R seemed to be noninvariant under space reflections. Hence we thought that supersymmetry provided a natural way to build interactions with parity nonconservation. Later we understood that the question of parity conservation in the framework of supersymmetry was far from clear.

To construct the field-theoretical model we used Hamilton's method because it was not clear to us then how to formulate Lagrange's approach consistently with our invariance principle. On the contrary Hamilton's approach appeared to be very simple because the Hamiltonian of the system was among the generators of the basic algebra. It became very simple to build a representation of the algebra in terms of free field operators. Starting with this we developed a method to determine the form of interaction. At this point it was found that the basic invariance principle had to be completed with a special locality principle (Likhtman, 1973). Only after that did we obtain a nonlinear set of equations for the possible interaction terms. The method seemed complicated enough and we succeeded only in constructing several simple models. The Lagrangean method developed later is much simpler and leads to the same results.

After the illustrious works of Wess and Zumino (1974) the theory of supersymmetry started to develop in an explosive manner. What are the reasons for this phenomenon? Frankly speaking we cannot give any arguments to answer the question. Probably supersymmetry contains a grain of truth. But we think that to gain the whole truth (we mean serious progress in this theory) at least one more very significant step has to be done. At the present time, in our opinion, the main problem is to investigate and reconsider along with the development of the principal topics of supersymmetry, its deeper structure.

This goal determines the character of the present article. The first section is of a mathematical character (on a "physical" level). The basic concepts of supermathematics are described very superficially and then the matrix method of superalgebra investigation (a very powerful one, from our point of view) is illustrated by several simple examples. In the second section a certain universal superalgebra is built. It contains all the useful representations of the algebra R and other quantities necessary to construct a supersymmetric theory. Then superfields are introduced in such a way that they together with the universal algebra form a broader superalgebra. This technique is illustrated by the simplest example of Wess-Zumino model and then a nonstandard version of supersymmetric electrodynamics (without photino) is proposed.

1. SUPERALGEBRAS AND SUPERGROUPS

The supersymmetry of a physical system is based on an extension of the concept of classical group. This extension is connected with introduction in the theory, besides usual (real or complex) numbers, of some new quantities - the so-called Grassmann numbers - which anticommute with each other. As a matter of fact anticommuting quantities occurred in theoretical physics rather long ago. Such quantities are necessary for the description of fermions in the second-quantized theory or in some versions of the functional method in the field theory. But at that stage only the simplest mathematical properties of these quantities were needed for the construction of the theory and naturally theorists avoided dipping into the jungle of mathematical details.

In the early seventies when physical supersymmetry appeared many aspects of the theory were obscure. The exposition of the theory in the original articles (Golfand and Likhtman, 1971, 1972; Volkov and Akulov, 1973a, b, c; Wess and Zumino, 1974a, b) had a handicraft character and suffered from inexactitudes. Now the situation is better. As had already been the case several times, mathematicians developed appropriate sections of mathematics which appeared to be adequate for the new physical problems. In our case we deal with the section of mathematics concerning algebra and analysis with anti-commuting variables. Somewhat tentatively it can be termed supermathematics. The development of supermathematics to a great extent is connected with the name of Felix Berezin (1966, 1970, 1980, 1983). At the present time supermathematics forms a reliable foundation for supersymmetry. It is not complicated and should serve as a working tool in theoretical physics.

This section is of a mathematical character. We begin with a very brief survey of basic concepts of the supermathematics limiting ourselves by the notions most necessary for our purpose.

1.1 Graded linear space

A linear space L is called Z_2-graded if there exists in it a subspace ${}^0L$ vectors of which have the property of being even and another subspace ${}^1L$ consisting of odd vectors. The whole space L has to be a direct sum:

$$L = {}^0L \oplus {}^1L. \tag{1.1}$$

According to (1.1) each vector of L admits unique representation as a sum of even and odd vectors. Even and odd vectors are both called homogeneous. For them the evenness function

$\alpha(x)$ can be defined:

$$\alpha(x) = \begin{matrix} 0, & x \in {}^0L, \\ 1, & x \in {}^1L. \end{matrix} \quad (1.2)$$

Dimensionality of graded space, dimL=(p,q), is determined by a pair of numbers p,q which are the dimensionalities of the subspaces ${}^0L$ and ${}^1L$ respectively. We shall designate graded space of dimensionality (p,q) as K^{pq} where K=R in the case of real space and K=C for complex space. The even subspace of K^{pq} will be denoted as K^{po}, the odd subspace as K^{oq}. It is convenient to introduce in K^{pq}, the standard basis e_i in such a way that for $1 \leqslant i \leqslant p$ and for $p+1 \leqslant i \leqslant p+q$ e_i are bases of the subspaces K^{po} and K^{oq} respectively. In the standard basis, vectors of K^{pq} are determined by their components. We shall denote them as a column $\begin{pmatrix} x \\ \xi \end{pmatrix}$ where x and ξ symbolize the even and odd components of the vector.

1.2 Linear operators in K^{pq}

A linear operator in the space K^{pq} can be represented by a matrix referred to the standard basis e_i. If the basis is fixed the linear operators are in one-to-one correspondence with matrices. We shall not consider here simple questions concerning transformation from one standard basis to another and in the following we shall not differentiate between an operator and the corresponding matrix unless it is misleading. The set of all matrices forms an associative algebra with respect to matrix multiplication. We shall denote this algebra as Mat(p,q). The graded property of the space K^{pq} transfers naturally to the matrix algebra Mat(p,q). The decomposition into even and odd parts (according to (1.1)) of any vector referred to the standard basis is given by:

$$\begin{pmatrix} x \\ \xi \end{pmatrix} = \begin{pmatrix} x \\ 0 \end{pmatrix} + \begin{pmatrix} 0 \\ \xi \end{pmatrix}. \quad (1.3)$$

According to (1.3) we represent a matrix from Mat(p,q) in block form:

$$a = \begin{pmatrix} a_{11} & a_{12} \\ a_{21} & a_{22} \end{pmatrix} = \begin{pmatrix} a_{11} & 0 \\ 0 & a_{22} \end{pmatrix} + \begin{pmatrix} 0 & a_{12} \\ a_{21} & 0 \end{pmatrix} \quad (1.4)$$

where a_{11} is p x p-matrix, a_{12} is p x q-matrix etc. The decomposition (1.4) of a matrix $a \in Mat(p,q)$ into even and odd parts is induced by the decomposition of a vector (1.3). Under such a definition of evenness the matrix algebra Mat(p,q) becomes a graded algebra. That means that the evenness function (1.2) obeys the relation (for homogeneous a,b):

$$\alpha(ab) = \alpha(a) + \alpha(b) \qquad (\text{mod } 2). \tag{1.5}$$

1.3 Grassmann algebras

Grassmann algebra Λ_n is associative algebra with unity and n generating elements ξ_i $(1<i<n)$. The quantities ξ_i are supposed to be anticommuting with each other:

$$\xi_i \xi_j + \xi_j \xi_i = 0 \tag{1.6}$$

and it is required that (1.6) be a complete set of relations i.e. that any relation between ξ_i would be a consequence of the relations (1.6). The general form of an element $f \in \Lambda_n$ is:

$$f = f(\xi) = f_o + \sum_{k>0} \sum_{i_1 \dots i_k} f_{i_1 \dots i_k} \xi_{i_1} \dots \xi_{i_k} . \tag{1.7}$$

The numerical coefficients $f_{i_1 \dots i}$ (real or complex) are antisymmetric in the indices i_j. (1.7) terminates at k>n due to the relations (1.6). The dimension of the algebra Λ_n is 2^n. Grassmann algebra is the simplest example of graded algebra. An element of the form (1.7) is regarded as even if all the coefficients $f_{i \dots i_k} = 0$ for odd k and as odd if $f_{i \dots i_k} = 0$ for even k. All odd elements of the algebra Λ_n anticommute with each other and commute with even elements.

1.4 Grassmann envelope

The concept of Grassmann envelope is useful because it enables us to associate a superalgebra with any Lie algebra and in this way to construct the corresponding group. The Grassmann envelope is built as follows. We take the graded linear space K^{pq} and Grassmann algebra Λ_n. No connection between the numbers p,q and n is supposed. The only condition on the algebra Λ_n is to be real if K=R and complex if K=C. The Grassmann envelope of the space K^{pq} (denoted by $K^{pq}(\Lambda)$) is a set of all quantities of the form:

$$X = \sum_{1}^{p} c_i e_i + \sum_{p+1}^{p+q} \gamma_j e_j \tag{1.8}$$

where e_i is the standard basis of K^{pq}, c_i are even and γ_j odd elements of algebra Λ. The definition apparently involves a concrete form of the standard basis. But in fact the structure of the space $K^{pq}(\Lambda)$ does not depend on the choice of the standard basis.

When the space K^{pq} has any complementary structure (e.g. it can be an associative algebra or a superalgebra) the commutation rules between Grassmann coefficients in (1.8) and basis elements e_i should be added. That can be done in two ways:

$$\begin{array}{lll} \text{(i)} & c_i e_i = e_i c_i, & \gamma_j e_j = e_j \gamma_j, \\ \text{(ii)} & c_i e_i = e_i c_i, & \gamma_j e_j = -e_j \gamma_j. \end{array} \tag{1.9}$$

The resulting Grassmann envelope is called Grassmann envelope of the first and of the second kind respectively. In the column representation of type (1.3) an element of $K^{pq}(\Lambda)$ has as x even and as ξ odd elements of Λ. In particular the Grassmann envelope of the matrix algebra Mat(p,q), denoted by $\mathrm{Mat}(p,q|\Lambda)$, consists of matrices of the form (1.4) but matrix elements of blocks a_{11} and a_{22} are even while those of blocks a_{12} and a_{21} are odd elements of Λ. $\mathrm{Mat}(p,q|\Lambda)$ is the Grassmann envelope of the first kind. Some examples of Grassmann envelopes of the second kind will be considered later.

1.5 Superalgebra

The consideration in previous subsections was of preparatory character. Now we come to the main topic, that is, superalgebra. Supersymmetry of a physical system means invariance with respect to any superalgebra and "ordinary" symmetry means invariance with respect to any Lie algebra. Superalgebra is a generalization of the customary concept of Lie algebra to the case of graded space. The formal definition is the following:

A superalgebra g is graded linear space K^{pq} on which a bilinear operation $[x,y]$ called a commutator (just as in the case of ordinary Lie algebra) is specified. For homogeneous elements x,y and z the commutator obeys the relations:

$$\alpha([x,y]) = \alpha(x) + \alpha(y) \quad (\text{mod } 2), \tag{1.10}$$

$$[x,y] = [y,x](-1)^{\alpha(x)\alpha(y)+1}, \tag{1.11}$$

$$[x,[y,z]](-1)^{\alpha(x)\alpha(z)} + [z,[x,y]](-1)^{\alpha(z)\alpha(y)} + [y,[z,x]](-1)^{\alpha(y)\alpha(x)} = 0. \tag{1.12}$$

An arbitrary element of the superalgebra is the sum of its even and odd elements. Hence the definition of commutator can be extended by means of bilinearity to the case of arbitrary x and y.

The relations (1.11-12) resemble corresponding relations for Lie algebra. The only difference is in signs depending on evenness character of elements x, y, z. The formula (1.12) is the generalization of Jacobi identity to the case of superalgebra. The even subspace K^{po} forms an ordinary Lie algebra. Instead of the above axioms, superalgebra can be formulated by means of structure constants. The commutators of two elements of the standard basis are determined by the relations:

$$[e_i,e_j] = C^k_{ij}e_k. \tag{1.13}$$

From equalities (1.10-12) the conditions on C^k_{ij} follow:

$$\alpha(i) + \alpha(j) + \alpha(k) = 0 \quad (\text{mod } 2), \tag{1.14}$$

$$C^k_{ij} = C^k_{ji}(-1)^{\alpha(i)\alpha(j)}, \tag{1.15}$$

$$C^s_{it}C^t_{kl}(-1)^{\alpha(i)\alpha(l)} + C^s_{kt}C^t_{li}(-1)^{\alpha(k)\alpha(i)} + C^s_{lt}C^t_{ik}(-1)^{\alpha(l)\alpha(k)} = 0 \tag{1.16}$$

where for brevity we set $\alpha(i) = \alpha(e_i)$ etc.

A graded associative algebra A can be turned into a superalgebra if we define the commutator (for homogeneous x and y) by the formula:

$$[x,y] = xy - yx(-1)^{\alpha(x)\alpha(y)} \tag{1.17}$$

or by the formula:

$$[x,y] = xy(-1)^{\alpha(x)\alpha(y)} - yx. \qquad (1.18)$$

From this formula we see that the commutator in the case of superalgebra is defined in the usual way unless both elements x and y are odd. In the latter case it is replaced by the anticommutator (with the opposite sign for (1.18)). The Jacobi identity is fulfilled automatically as consequence of associativeness of multiplication in the basic algebra A. Both resulting superalgebras are isomorphic to each other in the case of complex algebra A. For real A these superalgebras are two real forms of the corresponding complex superalgebra.

One can associate a superalgebra with any Lie algebra. For this purpose we consider a graded associative algebra A and its Grassmann envelope $A(\Lambda)$. An arbitrary element of $A(\Lambda)$ is the sum of monomials ax where $a \in \Lambda$ and x is a homogeneous element of A. Therefore it is enough to define commutators for monomials in the algebra $A(\Lambda)$. Because all elements of $A(\Lambda)$ are even by construction we set:

$$[ax,by] = axby - byax = ab[x,y]. \qquad (1.19)$$

Here the commutator [x,y] is defined by formula (1.17) if $A(\Lambda)$ is the Grassmann envelope of the first kind and by formula (1.18) in the case of the envelope of the second kind. In both cases $A(\Lambda)$ turns into Lie algebra closely connected to the superalgebra defined on A. Having a Lie algebra we can construct by means of the exponential formula a group corresponding to the given superalgebra. The only difference compared to the case of ordinary Lie group is that group parameters are not numbers but elements of a Grassmann algebra. We consider now a very simple example of superalgebra which in spite of (but may be owing to) its simplicity plays a basic role in the majority of supersymmetric theories. We denote this superalgebra as R. Generators of the algebra R consist of generators of the Poincaré algebra $M_{\mu\nu}$ and P_μ complemented by two spinorial generators Q and $\bar{Q}$. Poincaré generators form the even subalgebra ${}^0R$. They obey the usual commutation relations:

$$[M_{\mu\nu},M_{\lambda\sigma}]_- = i(g_{\mu\lambda}M_{\nu\sigma} + g_{\nu\sigma}M_{\mu\lambda} - g_{\mu\sigma}M_{\nu\lambda} - g_{\nu\lambda}M_{\mu\sigma}),$$

$$[M_{\mu\nu},P_\lambda]_- = i(g_{\mu\lambda}P_\nu - g_{\nu\lambda}P_\mu), \tag{1.20}$$

$$[P_\mu,P_\nu]_- = 0.$$

The two-component spinors Q and $\bar{Q}$ form the basis of the odd subspace ^1R of the graded space R. Under the group SL(2,C) they transform according to representations $(\frac{1}{2},0)$ and $(0,\frac{1}{2})$ respectively. This defines the commutators between Poincaré and spinorial generators:

$$[M_{\mu\nu},Q]_- = \tfrac{1}{2}\sigma_{\mu\nu}Q, \qquad [M_{\mu\nu},\bar{Q}]_- = \tfrac{1}{2}\tilde{\sigma}_{\mu\nu}\bar{Q},$$

$$[P_\mu,Q]_- = [P_\mu,\bar{Q}]_- = 0. \tag{1.21}$$

The σ- matrices are defined below after (1.38). To complete the commutation relations anticommutators between spinor generators are necessary. By conjecture they are:

$$[Q,\bar{Q}]_+ = 2\sigma_\mu P_\mu, \qquad [Q,Q]_+ = [\bar{Q},\bar{Q}]_+ = 0. \tag{1.22}$$

The commutation relations (1.20-22) determine the superalgebra R. It can be proved that R is the only superalgebra being the extension of the Poincaré algebra by two two-component spinor generators. We shall not check the conditions (1.10-12). A little later we consider another algorithmic method of superalgebra construction in which the conditions (1.10-12) are fulfilled automatically. The significant part of the superalgebra R is its subalgebra $\tilde{t}$ generated by translations P_μ and spinor Q,$\bar{Q}$ (we call them generators of spinor translations). The subalgebra $\tilde{t}$ we call generalized translations algebra. It is convenient to divide the commutation relations (1.20-22) into two parts. The first part includes the relations between generators of the superalgebra $\tilde{t}$, the second part includes all the rest. Only the first part is nontrivial. The commutation relations from the second part are immediate consequences of transformation law of generators P_μ, Q, $\bar{Q}$ with respect to Lorentz group. Hence by specifying the superalgebra $\tilde{t}$ we determine the whole superalgebra R. In the next subsection some other examples of generalized translation superalgebras will be given.

$\tilde{t}$ is an invariant subalgebra of R. That means that the commutator [x,y], where x is an arbitrary element of R and y is an element of $\tilde{t}$, belongs to the subalgebra $\tilde{t}$. Such a subalgebra is called an ideal. The factor-algebra $R/\tilde{t}$ is the Lorentz algebra L. The whole superalgebra R is a semidirect sum of its subalgebras $\tilde{t}$ and L.

All these concepts transfer naturally to the corresponding group. We illustrate this for the example of the Poincaré group. The Poincaré algebra is the semidirect sum of ideal $\tilde{t}$ generated by translations P_μ and Lorentz algebra L generated by momenta $M_{\mu\nu}$. Accordingly the Poincaré group P is the semidirect product of the translations group T and Lorentz group L. T is an invariant subgroup of P and the factorgroup P/T is L. In more detail. Every element G of the Poincaré group admits unique representation in the form:

$$G = (a)L \tag{1.23}$$

where (a) is translation by the fourvector a_μ and L is a Lorentz transformation. The composition law is given by the formula:

$$(a_1)L_1(a_2)L_2 = (a_1 + L_1a_2)L_1L_2. \tag{1.24}$$

This representation allows us to reduce consideration of the Poincaré group to the simpler group T and L.

Quite similar is the situation with the superalgebra R and corresponding supergroup. This supergroup is the semidirect product of the supergroup of generalized translations $\tilde{T}$ and the Lorentz group L. In establishing the explicit form of supergroup R we can limit ourselves by the consideration of generalized translations of the group $\tilde{T}$. We drop some delicate points concerning the fact that in the case of supergroup parameters, the Lorentz transformation are even elements of the Grassmann algebra. In the following it is enough to consider ordinary Lorentz transformations.

We form the Grassmann envelope $\tilde{t}(\Lambda)$ of the superalgebra $\tilde{t}$. Elements of $\tilde{t}(\Lambda)$ are written as $a_\mu P^\mu + \zeta Q + \bar{\zeta}\bar{Q}$ where a_μ is even and $\zeta, \bar{\zeta}$ are odd elements of Λ. These quantities play the role of group parameters. Elements of the supergroup $\tilde{T}$ are of the form:

$$G(a,\zeta,\bar{\zeta}) = \exp i(a_\mu P^\mu + \zeta Q + \bar{\zeta}\bar{Q}). \tag{1.25}$$

To derive the composition law we remark that when in any algebra all double commutators vanish, and that is the case, then according to the Campbell-Hausdorf formula:

$$\exp a \exp b = \exp(a + b + [a,b]). \tag{1.26}$$

Using (1.26) we obtain:

$$\begin{aligned} &G(a_1,\zeta_1,\bar{\zeta}_1)\, G(a_2,\zeta_2,\bar{\zeta}_2) = G(a,\zeta,\bar{\zeta}), \\ &a_\mu = a_{1\mu} + a_{2\mu} + i(\zeta_1\sigma_\mu\bar{\zeta}_2 - \zeta_2\sigma_\mu\bar{\zeta}_1), \\ &\zeta = \zeta_1 + \zeta_2, \qquad \bar{\zeta}_1 = \bar{\zeta}_1 + \bar{\zeta}_2. \end{aligned} \tag{1.27}$$

The relations (1.27) represent the multiplication law in the generalized translations group.

1.6 Matrix method

A very powerful method of investigation of superalgebras is to consider them as subalgebras of a matrix superalgebra. In this way a great deal of interesting results have been achieved. We shall not enter into details, referring the reader to original works (Berezin and Tolstoy, 1980; Berezin, 1983). Here we limit ourselves to several illustrations. The matrix algebra Mat(p,q) graded according to (1.4) turns into a superalgebra if one defines commutators by the formula (1.17). This superalgebra will be denoted by mat(p,q). We obtain examples of superalgebras by considering subalgebras of mat(p,q).

The first very simple example is the subalgebra of mat(1,1) consisting of all matrices of the form:

$$\left(\begin{array}{c|c} a & \alpha \\ \hline \alpha^* & a \end{array}\right) \tag{1.28}$$

where a is real and α is a complex number. Asterisk means complex conjugation. We explain here the style of designations to avoid repetitions in the following more complicated examples. Matrices are broken down by lines into blocks to separate their even and odd parts. We can introduce a basis formed from matrices:

$$H = \left(\begin{array}{c|c} 1 & 0 \\ \hline 0 & 1 \end{array}\right), \qquad \Theta = \left(\begin{array}{c|c} 0 & 1 \\ \hline 0 & 0 \end{array}\right), \qquad \Theta^+ = \left(\begin{array}{c|c} 0 & 0 \\ \hline 1 & 0 \end{array}\right) \tag{1.29}$$

and write down the matrix (1.28) as a linear combination:

$$\left(\begin{array}{c|c} a & \alpha \\ \hline \alpha^* & a \end{array}\right) = aH + \alpha\Theta + \alpha^*\Theta^+ \tag{1.30}$$

It is easily seen that the set of matrices (1.28) forms the superalgebra. This superalgebra can be also represented by commutation relations between basis elements (1.29):

$$\begin{aligned} &[H,\Theta]_- = [H,\Theta^+]_- = 0, \\ &[\Theta,\Theta^+]_+ = H, \qquad [\Theta,\Theta]_+ = [\Theta^+,\Theta^+]_+ = 0 \end{aligned} \tag{1.31}$$

We hope that a detailed consideration of the simplest superalgebra (1.31) will be enough for the easy understanding of the following examples.

The superalgebra (1.31) has been applied in physics for constructing supersymmetric models of nonrelativistic quantum mechanics (Witten, 1981; Salomonson, 1982).

The next example concerns the superconformal algebra. This algebra was first introduced by Wess and Zumino (1974a) and by Ferrara (1974) and then considered by many authors in various aspects. The matrix representation of the superconformal algebra is of the form:

$$\left(\begin{array}{cc|c} a & b & \alpha \\ c & -a^+ & \beta \\ \hline \beta^+ & -\alpha^+ & 0 \end{array}\right) \tag{1.32}$$

where a,b,c are complex 2x2-matrices, α, β are two-component spinors, α^+, β^+ are their Hermitian conjugate. The matrices b and c are Hermitean ($b^+=b$, $c^+=c$). The superconformal algebra (1.32) is interesting for us because it gives indications how to construct different extensions of the Poincaré algebra. In our works (1971, 1972) extensions of Poincaré algebra by bispinor generators were constructed. We rewrite the commutation relations of these superalgebras somewhat altering the notation. We break down bispinors W and $\bar{W}$ into two-component constituents:

$$W = \begin{pmatrix} Q \\ S \end{pmatrix}, \qquad \bar{W} = (\bar{S}, \bar{Q}). \tag{1.33}$$

All the superalgebras are semidirect sums of the ideal of $\tilde{t}$-generalized translations and the Lorentz algebra L. Therefore it would be enough to write down commutation relations for superalgebra $\tilde{t}$ only as has been explained in consideration of the superalgebra R.

We consider consequently the three superalgebras given in our work (1972). First we write down their commutation relations using the notation (1.33) and then exhibit the form of their matrix representation. For all the following algebras the commutation relations:

$$[Q,Q]_+ = [Q,S]_+ = [S,S]_+ = 0 \tag{1.34}$$

are common. We shall not write them down bearing in mind (1.34). Some commutation relations (e.g. for $[P_\mu, Q]$) immediately follow from the others as a result of Hermitian conjugation. For the sake of brevity we do not indicate such relations. The first superalgebra $\tilde{t}_1$:

$$\begin{aligned} &[P_\mu, Q]_- = [P_\mu, S]_- = 0, \\ &[Q, \bar{Q}]_+ = \sigma_\mu P^\mu, \\ &[Q, \bar{S}]_+ = [S, \bar{S}]_+ = 0. \end{aligned} \tag{1.35}$$

We see that the commutation relations between generators P_μ, Q and $\bar{Q}$ are the same as in (1.21-22) (apart from the nonsignificant coefficient 2). Therefore the superalgebra $\tilde{t}_1$ contains a superalgebra isomorphic to $\tilde{t}$. The whole t_1 is the direct sum of $\tilde{t}$ and commutative algebra generated by S and $\bar{S}$. Hence we can drop the generators S and $\bar{S}$ and take the first algebra as the superalgebra $\tilde{t}$ considered earlier. Now we give the matrix representation of the superalgebra R being semidirect sum of ideal $\tilde{t}$ and Lorentz algebra L. The representation consists of matrices of the form:

$$\left(\begin{array}{cc|c} a & b & \alpha \\ 0 & -a & 0 \\ \hline 0 & -\alpha & 0 \end{array}\right) \tag{1.36}$$

where a and b are complex 2x2-matrices satisfying the conditions:

$$\mathrm{tr}\ a = 0, \qquad b^{+} = b \tag{1.37}$$

α and $\bar{\alpha}$ are two-component spinors. The even part of the matrices (1.36) forms representation of the Poincaré algebra. The matrices a and b can be rewritten in a more customary form:

$$a = \omega^{\mu\nu}\sigma_{\mu\nu}, \qquad b = b^{\mu}\sigma_{\mu} \tag{1.38}$$

where σ_i are Pauli matrices, $\sigma_{ij} = \varepsilon_{ijk}\sigma_k$, $\sigma_{ok} = i\sigma_k$, $\sigma_\mu = (1,\vec{\sigma})$, $w^{\mu\nu}$ and b^μ are real numbers; coefficients of the algebra. The matrices $\sigma_{\mu\nu}$ form basis of the Lorentz algebra (to be more precise the algebra sl(2,C). The matrices σ_μ form the basis of the irreducible representation $(\frac{1}{2},\frac{1}{2})$.

The second superalgebra $\tilde{t}_2$:

$$[P_\mu,Q]_- = [P_\mu,S]_- = 0,$$

$$[Q,\bar{Q}]_+ = \sigma_\mu P^\mu, \qquad [S,\bar{S}]_+ = \tilde{\sigma}_\mu P^\mu, \tag{1.39}$$

$$[Q,\bar{S}]_+ = 0.$$

Here $\tilde{\sigma}_\mu = (1,-\vec{\sigma})$. This superalgebra is similar to the previous one. Its matrix representation is of the form:

$$\left(\begin{array}{cc|cc} a & b & \alpha & \bar{\beta} \\ 0 & -a^{+} & 0 & 0 \\ \hline 0 & -\bar{\alpha} & 0 & 0 \\ 0 & -\beta & 0 & 0 \end{array}\right) \tag{1.40}$$

where a and b are matrices (1.38), α and β are two-component spinors, $\bar{\alpha}$ and $\bar{\beta}$ are their complex conjugate.

It is relevant to point out here that extended superalgebra involving an internal symmetry admits a matrix representation analogous to (1.40). It can be written as:

$$\left(\begin{array}{cc|c} a & b & \alpha \\ 0 & -a^{+} & 0 \\ \hline 0 & -\alpha^{+} & d \end{array}\right). \tag{1.41}$$

In (1.41) a and b are matrices (1.38), α is a complex 2xN-matrix, α^+ is its Hermitian conjugate, d is a Hermitian NxN-matrix representing algebra of internal symmetry. The third superalgebra $\tilde{t}_3$:

$$\begin{aligned} &[P_\mu, Q]_- = \sigma_\mu S, \qquad [P_\mu, S]_- = 0, \\ &[Q, \bar{Q}]_+ = [Q, \bar{S}]_+ = [S, \bar{S}]_+ = 0. \end{aligned} \tag{1.42}$$

This superalgebra is of a somewhat different character than the previous two. The main difference consists in noncommutativity of linear momentum with spinor generator Q. The matrix representation is of the form:

$$\left(\begin{array}{cc|c:ccc} a & b & \bar{\beta} & & & \\ 0 & -a^+ & -\bar{\alpha} & & & \\ \hline 0 & 0 & 0 & & & \\ \hdashline & & & a & b & 0 \\ & & & 0 & -a^+ & 0 \\ \cline{4-6} & & & \alpha & -\beta & 0 \end{array}\right). \tag{1.43}$$

The designations are the same as in the previous examples. the matrices (1.43) for convenience are divided into diagonal blocks. Within each block the subdivision in even and odd parts is performed.

We considered these examples in detail because we think that the matrix method is a good framework for investigation of superalgebras. Possibly it could help us to find other superalgebras applicable in physics.

In the next section one more representation of superalgebras is considered. This representation is convenient for applications to supersymmetric theories. It is also suitable for any kind of generalization of the basic superalgebra.

2. REPRESENTATIONS AND MODELS

In the section 1 various mathematical aspects of the theory of superalgebras were treated. Now we consider applications of supersymmetry to field-theoretical problems. First of all we have built a suitable representation of superalgebra. We restrict ourselves to the consideration of the superalgebra R defined by the relations (1.20-22) and any superalgebra related to it.

2.1 <u>The universal superalgebra Σ</u>

We begin with the construction of a rather simple superalgebra

Σ among representations of which there are all interesting representations of the superalgebra R. The other quantity (such as covariant derivatives) necessary for building supersymmetric field-theoretical models are also elements of Σ. As building material for Σ serve two-component spinors θ^α, $\bar\theta^{\dot\alpha}$ and ∂_α, $\bar\partial_{\dot\alpha}$ which together form Clifford algebra:

$$[\theta^\alpha,\theta^\beta]_+ = [\theta^\alpha,\bar\theta^{\dot\beta}]_+ = [\bar\theta^{\dot\alpha},\bar\theta^{\dot\beta}]_+ = 0,$$
$$[\partial_\alpha,\partial_\beta]_+ = [\partial_\alpha,\bar\partial_{\dot\beta}]_+ = [\bar\partial_{\dot\alpha},\bar\partial_{\dot\beta}]_+ = 0,$$
$$[\partial_\alpha,\theta^\beta]_+ = \delta^\beta_\alpha, \qquad [\partial_\alpha, \bar\theta^{\dot\beta}]_+ = 0,$$
$$[\bar\partial_{\dot\alpha},\theta^\beta]_+ = 0, \qquad [\bar\partial_{\dot\alpha},\bar\theta^{\dot\beta}]_+ = \delta^{\dot\beta}_{\dot\alpha}.$$

One could consider symbols ∂ as derivatives with respect to θ but it is not necessary since in the following we express all relations in terms of commutators and that is why the relations (2.1) are enough.

Some remarks on operations with spinors. The spinors with dotted and undotted indices transform under the Lorentz group (more precisely the group SL(2,C)) according to complex conjugate representations. Indices are raised and lowered with matrices $\varepsilon^{\alpha\beta}$ and $\varepsilon_{\alpha\beta}$:

$$\theta^\alpha = \varepsilon^{\alpha\beta}\theta_\beta, \qquad \theta_\alpha = \varepsilon_{\alpha\beta}\theta^\beta. \tag{2.2}$$

The matrices ε obey the relations:

$$\varepsilon_{\alpha\beta} = -\varepsilon_{\beta\alpha}, \qquad \varepsilon^{\alpha\beta} = -\varepsilon^{\beta\alpha}, \tag{2.3}$$
$$\varepsilon^{\alpha\gamma}\varepsilon_{\gamma\beta} = \delta^\alpha_\beta.$$

The same relations hold for dotted indices. From two spinors with similar indices a Lorentz invariant can be formed:

$$\theta\theta = \theta^\alpha\theta_\alpha, \qquad \bar\theta\bar\theta = \bar\theta^{\dot\alpha}\bar\theta_{\dot\alpha} \tag{2.4}$$

and only one (due to anticommutativity) tensor of the second rank:

$$\theta^\alpha\theta_\beta = \tfrac{1}{2}\delta^\alpha_\beta\theta\theta, \qquad \bar\theta^{\dot\alpha}\bar\theta_{\dot\beta} = \tfrac{1}{2}\delta^{\dot\alpha}_{\dot\beta}\bar\theta\bar\theta. \tag{2.5}$$

The tensor of the second rank made up of two spinors with indices of different kind is equivalent to vector. The correspondence between spinor and ordinary vector indexation is stated by formulae:

$$q_\mu = \tfrac{1}{2}(\sigma_\mu)^{\alpha\dot{\alpha}} q_{\alpha\dot{\alpha}}, \qquad q_{\alpha\dot{\alpha}} = (\sigma_\mu q^\mu)_{\alpha\dot{\alpha}} \tag{2.6}$$

where $\sigma_\mu = (1, \vec{\sigma})$. Sometimes to use spinor indices is more convenient than vector ones.

The significant constituent of the superalgebra Σ is the quantity:

$$P = \Theta^\alpha P_{\alpha\dot{\alpha}} \bar{\Theta}^{\dot{\alpha}}. \tag{2.7}$$

Here $P_{\alpha\dot{\alpha}}$ is the operator of linear momentum (in spinor indexation). The superalgebra Σ is generated by elements P, ∂_α and $\bar{\partial}_{\dot{\alpha}}$. The element P by definition is even while ∂_α and $\bar{\partial}_{\dot{\alpha}}$ are odd. We can easily construct all commutation relations of the superalgebra Σ using formulae (2.1) and taking into account that momentum P_μ commutes with all elements of Clifford algebra (2.1). The basis of Σ consists of the following elements:

$$P,\ P_{\alpha\dot{\alpha}} \quad \text{(even)}, \qquad \partial_\alpha,\ \bar{\partial}_{\dot{\alpha}}, \quad \eta_\alpha = P_{\alpha\dot{\alpha}}\bar{\Theta}^{\dot{\alpha}},\ \bar{\eta}_{\dot{\alpha}} = \Theta^\alpha P_{\alpha\dot{\alpha}} \text{(odd)} \tag{2.8}$$

The commutation relations between quantities(2.8) are (only nonzero commutators are written down):

$$[P,\partial_\alpha]_- = -\eta_\alpha, \qquad [P,\bar{\partial}_{\dot{\alpha}}]_- = \bar{\eta}_{\dot{\alpha}}, \qquad [\partial_\alpha,\bar{\eta}_{\dot{\alpha}}]_+ = [\bar{\partial}_{\dot{\alpha}},\eta_\alpha]_+ = P_{\alpha\dot{\alpha}}. \tag{2.9}$$

A curious feature of the algebra Σ is that its basis elements (2.8) are dimensional quantities. We require that P be dimensionless and because $P_{\alpha\dot{\alpha}}$ has the dimension L^{-1} the other quantities acquire dimensions: $\Theta \sim L^{\frac{1}{2}}$, $\partial \sim L^{-\frac{1}{2}}$, $\eta \sim L^{-\frac{1}{2}}$.

Consider the following sets of elements of the superalgebra Σ:

$$Q_\alpha = \partial_\alpha + \eta_\alpha, \qquad \bar{Q}_{\dot{\alpha}} = \bar{\partial}_{\dot{\alpha}} + \bar{\eta}_{\dot{\alpha}}, \tag{2.10}$$

$$Q^L_\alpha = \partial_\alpha, \qquad \bar{Q}^L_{\dot{\alpha}} = \bar{\partial}_{\dot{\alpha}} + 2\bar{\eta}_{\dot{\alpha}}, \tag{2.11}$$

$$Q_\alpha = \partial_\alpha + 2\eta_\alpha, \qquad \bar{Q}^R_{\dot{\alpha}} = \bar{\partial}_{\dot{\alpha}}. \tag{2.12}$$

It follows immediately from (2.9) that each set of quantities (2.10-12) obeys the commutation relations (1.22) and hence generates the superalgebra R. (It should be noted that here and henceforth speaking of the superalgebra R we can restrict ourselves to the consideration of its subalgebra of generalized translation $\tilde{t}$. All relations connected with angular momenta will be fulfilled automatically as was explained in Section 1). We have different representations of the superalgebra R realized by different subalgebras of Σ. The representation (2.10) is called symmetrical, the representations (2.11) and (2.12) are called chiral left-hand and right-hand respectively. The sense of these terms will become clearer later.

The generators (2.10-12) can be represented in another form:

$$Q_\alpha = e^{-P}\partial_\alpha e^{P}, \qquad \bar{Q}_{\dot\alpha} = e^{P}\bar\partial_{\dot\alpha}e^{-P}, \tag{2.10'}$$

$$Q^L_\alpha = \partial_\alpha, \qquad \bar{Q}^L_{\dot\alpha} = e^{2P}\bar\partial_{\dot\alpha}e^{-2P}, \tag{2.11'}$$

$$Q^R_\alpha = e^{-2P}\partial_\alpha e^{2P}, \qquad \bar{Q}^R_{\dot\alpha} = \bar\partial_{\dot\alpha}. \tag{2.12'}$$

One can see from these relations that the subalgebras (2.10-12) are conjugate within the algebra Σ:

$$\{Q^L_\alpha,\bar{Q}^L_{\dot\alpha}\} = e^{P}\{Q_\alpha,\bar{Q}_{\dot\alpha}\}e^{-P} = e^{2P}\{Q^R_\alpha,\bar{Q}^R_{\dot\alpha}\}e^{-2P}. \tag{2.13}$$

The symbol {....} denotes subalgebra generated by the elements inside the braces.

In an analogous way one can find within the Σ covariant derivatives (using the terminology common for supersymmetry). In the representations (2.10-12) they are expressed in the form:

$$\begin{aligned} D_\alpha &= e^{P}\partial_\alpha e^{-P} = \partial_\alpha - \eta_\alpha, \\ \bar{D}_{\dot\alpha} &= e^{-P}\bar\partial_{\dot\alpha}e^{P} = \bar\partial_{\dot\alpha} - \bar\eta_{\dot\alpha}, \end{aligned} \tag{2.14}$$

$$\begin{aligned} D^L_\alpha &= e^{2P}\partial_\alpha e^{-2P} = \partial_\alpha - 2\eta_\alpha, \\ \bar{D}^L_{\dot\alpha} &= \bar\partial_{\dot\alpha}, \end{aligned} \tag{2.15}$$

$$\begin{aligned} D^R_\alpha &= \partial_\alpha, \\ \bar{D}^R_{\dot\alpha} &= e^{-2P}\bar\partial_{\dot\alpha}e^{2P} = \bar\partial_{\dot\alpha} - 2\bar\eta_{\dot\alpha}. \end{aligned} \tag{2.16}$$

One can easily verify be means of relations (2.9) that quantities (2.14-16) anticommute with corresponding generators (2.10-12).

2.2 Representation of field multiplets

In the previous subsection all necessary operators connected with the superalgebra R were constructed in terms of the algebra Σ. Now a special representation of this algebra is built. We remark that according to (2.1) the quantities Θ^{α}, $\bar{\Theta}^{\dot{\alpha}}$ generate the Grassmann algebra Λ_4. The Grassmann envelope of the Grassmann algebra Λ_4 will be denoted by Π. Obviously Π is a commutative algebra. (We do not specify what kind of Grassmann envelope the algebra is. The envelope of the second kind was used in original works. The subsequent results are valid irrespective of the kind of the envelope). We shall deal with functions $\phi(x)$ depending on space-time point x with values in the algebra Π. Such a function is exactly the superfield (up to any nuances of treatment) introduced by Salam and Strathdee (1974a). We retain the term superfield. The superfields form a commutative algebra $\Pi(x)$. The most significant thing is that the commutator of any element of the superalgebra Σ with a superfield $\phi(x)$ is a superfield again. Hence we obtain (due to Jacobi identities) the representation of the superalgebra Σ by linear transformations of the algebra $\Pi(x)$. This representation coincides with the one found by Wess and Zumino in component form and verified later by superfield techniques.

One can obtain the general form of the superfield $\phi(x)$ by rules (1.7) and (1.8) but it is convenient to slightly modify the form of basis of the Π algebra using the relation (2.5) for anticommuting spinors Θ^{α}, $\bar{\Theta}^{\dot{\alpha}}$. As a result we express the superfield in the form:

$$\begin{aligned}\phi(x) &= C(x) + \Theta\Theta M(x) + \bar{\Theta}\bar{\Theta}N(x) + \Theta\Theta\bar{\Theta}\bar{\Theta}D(x) \\ &+ \Theta^{\alpha}\bar{\Theta}^{\dot{\alpha}}V_{\alpha\dot{\alpha}}(x) + \Theta\chi(x) + \bar{\Theta}\bar{\chi}(x) \\ &+ \bar{\Theta}\bar{\Theta}\Theta\lambda(x) + \Theta\Theta\bar{\Theta}\bar{\lambda}(x).\end{aligned} \tag{2.17}$$

Some remarks should be added. The coefficients in (2.17) are ordinary fields. They are supplied with spinorial indices and the expression (2.17) is made up in such a way that $\phi(x)$ is Lorentz scalar. Generally the coefficients can have an extra Lorentz index and then $\phi(x)$ transforms under the Lorentz group in a definite way. In any case superalgebra unites in one representation a finite number of ordinary fields with integer and half-integer spins. These fields form a supermultiplet. The supermultiplet (2.17) consists of four scalars, one vector and four two-component spinors.

By construction of the Grassmann envelope the fields of integer spin are even while spinors are odd elements of the Grassmann algebra. (It should be noted that in the case of a superfield with an extra spinor index some modification in the construction of superfield is required.) Thus a supermultiplet gives the correct relationship between spin and statistics. That is in our opinion one of the most attractive features of supersymmetry.

The superfield of general form (2.17) comprises irreducible representation of the superalgebra Σ but this representation is reducible with respect to its subalgebra R. We are not concerned with the procedure of reduction (this question is illuminated in detail in numerous reviews) but consider only a very simple example of irreducible representation of the subalgebra R. It is convenient to start with left-hand chiral representation (2.11). Under operators Q^L_α and $\bar{Q}^L_{\dot\alpha}$ not only superfields of the general form (2.17) but much more narrow set of superfields of the form:

$$\phi^L(x) = \tfrac{1}{2}A(x) + \Theta\psi(x) + \tfrac{1}{2}\Theta\Theta F(x) \tag{2.18}$$

is invariant. The superfield (2.18) is called left-hand chiral because it contains only left-hand Weil spinor. Similary one can build a right-hand chiral superfield invariant under operators (2.12):

$$\phi^R(x) = \tfrac{1}{2}\bar{A}(x) - \bar{\Theta}\bar{\psi}(x) - \tfrac{1}{2}\bar{\Theta}\bar{\Theta}\bar{F}(x). \tag{2.19}$$

Coefficients $\frac{1}{2}$ are introduced for convenience by consideration of the expression for action. The minus signs in (2.19) are introduced to present the superfield ϕ^R in the form consistent to be Hermitian conjugate of ϕ^L. (We remark that the definition of Hermitian conjugation has not been formulated. The form of superfields (2.18-19) is justified by the correct form of the final expression of action).

The superfields ϕ^L and ϕ^R transform under different representations of the superalgebra R. However these representations are equivalent within the algebra Σ according to formulae (2.13). Owing to this chiral superfields can be transformed to symmetrical form:

$$\tilde{\phi}^L = e^{-P}\phi^L e^P, \qquad \tilde{\phi}^R = e^P \phi^R e^{-P} \tag{2.20}$$

when both of them transform according to representation (2.10).

2.3 Superinvariants

To build the action of a supersymmetric field theory we need expressions composed of superfields and invariant under transformations by elements of the superalgebra Σ (or any subalgebra of it). Such expressions are called superinvariants. The way to construct superinvariants which is used in the supersymmetric field theory is the following. Infinitesimal transformation of the superfield $\phi(x)$ corresponding to an element a of the superalgebra Σ is of the form:

$$\delta\phi = [a\zeta,\phi]. \tag{2.21}$$

Here ζ is an infinitesimal parameter of transformation (it is odd Grassmann number for odd a). It is significant that the relation (2.21) gives a representation of the superalgebra Σ in the algebra of superfields $\Pi(x)$. That follows immediately from the Jacobi identity: two transformations $a\zeta$ and $b\eta$ obey the relation:

$$[[a\zeta,b\eta],\phi] = [a\zeta,[b\eta,\phi]] - [b\eta,[a\zeta,\phi]].$$

The transformation (2.21) for the product of two superfields is of the form:

$$\delta(\phi_1\phi_2) = \delta\phi_1\phi_2 + \phi_1\delta\phi_2 = [a\zeta,\phi_1\phi_2].$$

Hence the form of the transformation (2.21) is valid for the product of several superfields.

The first superinvariant can be written in the form:

$$I = \int d^4x\ \Delta\bar{\Delta}\Psi(x). \tag{2.22}$$

Here operations Δ and $\bar{\Delta}$ are defined by the equalities:

$$\Delta\Psi(x) = -\tfrac{1}{4}[\partial^{\alpha}, [\partial_{\alpha}, \Psi(x)]],$$
$$\bar{\Delta}\Psi(x) = -\tfrac{1}{4}[\bar{\partial}^{\alpha}, [\bar{\partial}_{\dot{\alpha}}, \Psi(x)]]. \qquad (2.23)$$

$\Psi(x)$ in (2.22) is a superfield of the general form (2.17). It can be composed of products and sums of "elementary" superfields representing particles. We use sometimes the notations of the type $I(\Psi)$ to indicate explicitly which superfield-argument the invariant depends on.

Invariance of the quantity (2.22) under transformation (2.21) (a can be any element of Σ) is stated very simply. The invariance under ∂_{α} and $\bar{\partial}_{\dot{\alpha}}$ follows from anticommutativity of ∂-symbols and the structure of the expressions (2.33). The rest of the elements of Σ contain derivatives. The commutator of such an element with a superfield is the divergence of some superfields. It vanishes after integration. Usually in works on supersymmetry the invariant (2.22) is expressed by covariant derivatives or by integration over Grassmann coordinates Θ. Our way is exactly equivalent to this and is in some sense simpler. We emphasize once more that the expression (2.22) is invariant under the whole superalgebra Σ.

The chiral superfields (2.18) and (2.19) form chiral algebras $\sqcap^{L}(x)$ and $\sqcap^{R}(x)$. We can build for left-hand superfields the chiral invariant:

$$I^{L} = \int d^{4}x\Delta\Psi^{L}(x) \qquad (2.24)$$

and an analogous invariant I^{R} for right-hand superfields. It is interesting that the expression (2.24) is invariant in the above-mentioned sense not only under the subalgebra R (in chiral representation) but under the whole superalgebra Σ. We point out once more the kind of invariants which has not been found in works on supersymmetry. The operation M is defined by the equality:

$$M\phi(x) = C(x) \qquad (2.25)$$

where $\phi(x)$ is a superfield of the general form (2.17) and $C(x)$ is its first coefficient. The operation M possesses the properties:

$$M(\phi_1+\phi_2) = M\phi_1+M\phi_2, \qquad M(\phi_1\phi_2) = M\phi_1 M\phi_2. \tag{2.26}$$

The superfield ϕ^S obeying the connection:

$$\phi^S(x) = M\phi^S(x) \tag{2.27}$$

we call a singlet superfield. It is clear that all singlet superfields form a subalgebra $\Pi^S(x)$ of the algebra $\Pi(x)$. The operation M according to (2.26) produces a homomorphic mapping of the algebra $\Pi(x)$ onto the algebra $\Pi^S(x)$. The invariant I^S is defined by the relation:

$$I^S = \int d^4 x\, M\Psi^S(x). \tag{2.28}$$

In spite of the fact that the invariant I^S is defined only for singlet superfields the expression (2.28) is invariant with respect to the whole superalgebra Σ, and moreover, this occurs for the other superinvariants defined earlier. The difference between these superinvariants consists in a different type of domain on which the invariant is defined. The invariant I is defined on the algebra $\Pi(x)$. This algebra according to (2.21) forms a representation of the superalgebra Σ. The invariant I^L is defined only on the algebra of chiral superfields $\Pi^L(x)$ (expression (2.24) is noninvariant for superfields not belonging to Π^L). The algebra $\Pi^L(x)$ does not form representation of superalgebra Σ. It forms the representation of subalgebra of Σ generated by elements (2.11). Finally the superinvariant I^S is defined only on algebra $\Pi^S(x)$. This algebra forms a representation only of the subalgebra of Σgenerated by $P_{\alpha\dot\alpha}$. This subalgebra is nothing but the ordinary translation algebra.

2.4 The Wess-Zumino model

The simplest supersymmetry field-theoretical model describes the selfinteraction of chiral supermultiplets. The minimal number of the supermultiplets is two: one left-hand (2.18) and one right-hand (2.19). The action can be written in terms of superinvariants (2.22) and (2.24) in the form:

$$\begin{aligned} S = &-2I(\tilde\phi^R\tilde\phi^L) + I^L(m(\phi^L)^2 + \tfrac{4}{3}g(\phi^L)^3) \\ &-I^R(m(\phi^R)^2) + \tfrac{4}{3}g(\phi^R)^3). \end{aligned} \tag{2.29}$$

Here m is the mass common for the supermultiplet, g is dimensionless coupling constant. $\mathcal{Z}^L$ and $\mathcal{Z}^R$ are symmetrical forms of chiral superfields given by formulae (2.20). By means of simple calculations we can express the action (2.29) in terms of fields A, F, Ψ and their complex conjugates:

$$S = \tfrac{1}{2}\int d^4x\{\partial_\mu A\partial^\mu \bar{A} - i\bar{\varphi}\gamma_\mu\partial^\mu\varphi + F\bar{F} + m(AF + \bar{A}\bar{F} - \bar{\varphi}\varphi) + g(A^2F + \bar{A}^2\bar{F} - (A + \bar{A})\bar{\varphi}\varphi - (A - \bar{A})\bar{\varphi}\gamma_5\varphi\}. \quad (2.30)$$

We introduced one Majorana spinor $\varphi = \begin{pmatrix}\Psi_\alpha \\ \bar{\Psi}_{\dot{\alpha}}\end{pmatrix}$ in (2.30) instead of two Weil spinors Ψ and $\bar{\Psi}$. The field F enters in action (2.30) without derivatives. Its equation of motion is of the algebraic form:

$$F + m\bar{A} + g\bar{A}^2 = 0. \quad (2.31)$$

Eliminating the field F by means of (2.31) we obtain the ultimate form of the action:

$$S = \int d^4x\{\tfrac{1}{2}[\partial_\mu a\partial^\mu a + \partial_\mu b\partial^\mu b - m^2(a^2 + b^2) - i(\bar{\varphi}\gamma_\mu\partial^\mu\varphi - m\bar{\varphi}\varphi)] - mga(a^2 + b^2) - \tfrac{1}{2}g^2(a^2 + b^2)^2 - \bar{\varphi}(a + ib\gamma_5)\varphi\}. \quad (2.32)$$

Here a and b are real and imaginary parts of the complex field A in formula (2.30). The action (2.32) contains three interacting fields: real scalar a, pseudoscalar b and Majorana spinor φ. Their masses and coupling constants are strictly connected by the condition of supersymmetry. As a result the field-theoretical model (2.32) appears to be even less divergent than one could conclude from naive counting of indices of diagrams. The model (2.32) was constructed by Wess and Zumino (1974b).

The field theory (2.32) cannot be considered as realistic not only in view of equality of all masses (the mass difference could arise from symmetry breaking) but because it contains only neutral fields and especially the Majorana spinor φ, which makes it impossible to describe the conservation of fermion number. In the next subsection we try to overcome this point.

2.5 Supergauge theories

If we want to consider more complicated supermultiplets containing a vector field we must deal with some kind of gauge theory. The first example of such theory was given in our work (1971). The model describes the interaction of a massive vector supermultiplet (in modern terminology with two mutually conjugate chiral supermultiplets. The Lagrangean contains parity violating term which is a characteristic feature for our representation of the superalgebra R. The supersymmetric versions of gauge field theory were constructed by Wess and Zumino (1974c), by Salam and Strathdee (1974b) and by Ferrara and Zumino (1974). In these theories a vector supermultiplet of the type (2.17) is considered as a gauge field and parameters of the gauge transformation are chiral superfields. As a result we have a very nonlinear theory which simplifies only in special choice of the gauge.

A very interesting (and considerably simpler) version of super Yang-Mills was recently worked out by Ogievetski and his collaborators (Galperin et al, 1984) for N=2 and N=3 supersymmetry. Regretably these topics are outside the field of the present article.

We present here a new approach to constructing supergauge theories. The resulting theory differs from the known ones in its physical content. But the main interest, in our opinion, is that the new method can shed light on the structure of supersymmetric theories and indicate the point in which its foundations should be changed. The starting point is the replacement of the superalgebra Σ defined by the relations (2.7-9) to a more complicated superalgebra Σ^A. The superalgebra Σ^A is built analogously to Σ. The only difference is that instead of the generator P defined by (2.7) we take the generator D:

$$D = P - \hat{e}A, \qquad A = \theta^\alpha A_{\alpha\dot{\alpha}}\bar{\theta}^{\dot{\alpha}}. \tag{2.33}$$

Here $A_{\alpha\dot{\alpha}}$ (or A_μ in vector indexation) is the operator of multiplication on the vector-function $A_\mu(x)$ (in x-representation). We confine ourselves to the case of Abelian gauge since we want to demonstrate the idea without redundant complications. $\hat{e}$ is a charge operator. Introduction of the charge operator is necessary in our formalism because all quantities are expressed in terms of commutators within the algebra Σ^A as well as in the transformation of superfields. The charge operator can be introduced in the most natural way in a quantized field formalism. We give here a somewhat simplified version introducing two quantities b_+ and b_- which discriminate between positive and negative charged fields. The charge operator $\grave{e}$ obeys the relations:

$$[\hat{e},b_\pm] = \pm e b_\pm . \tag{2.34}$$

Henceforth we shall multiply positive and negative superfields by b_+ and b_- respectively. In this way we ensure the correct form of the covariant derivatives. •

One can get a concrete realízation of the relations (2.34):

$$\hat{e} = e\lambda\,\partial/\partial\lambda, \qquad b_+ = \lambda, \qquad b_- = \lambda^{-1}. \tag{2.35}$$

Here λ is an auxiliary parameter. This form is convenient for technical reasons. Firstly b_+ and b_- in (2.35) commute with each other and secondly, any neutral product of charged superfields does not depend on λ and hence it commutes with the charge operator.

The main distinguishing feature of the superalgebra Σ^A compared to Σ is that for charged superfields the ordinary derivative ∂_μ is replaced by the covariant derivative $\nabla_\mu = \partial_\mu - i\hat{e}A_\mu$. That gives rise to consistency between gauge transformations and Σ^A-transformations of superfields. The gauge transformations are of the form:

$$\phi(x) \to \exp(\hat{e}\beta(x))\phi(x)\exp(-\hat{e}\beta(x)). \tag{2.36}$$

Here $\beta(x)$ is a numerical function. According to condition (2.34) the transformation (2.36) acts in the right way on positive, negative and neutral superfields. The corresponding relation for the quantity D is:

$$\begin{aligned} &\exp(-\hat{e}\beta(x))D\exp(\hat{e}\beta(x)) = D' = P - \hat{e}A', \\ &A'_\mu = A_\mu - \partial_\mu\beta. \end{aligned} \tag{2.37}$$

The change in D produced by a gauge transformation can be compensated by an appropriate transformation of the field A_μ. The replacement of the ordinary derivatives by covariant ones in the superalgebra R was considered in the de Witt and Freedman work (1975). These authors derived the known form of super Yang-Mills action by a new method. Our way is somewhat different mainly in the treatment of the gauge field as a supermultiplet. As a result we obtain another physical theory.

The structure of the superalgebra Σ^A can be easily revealed because we have an explicit form of its generating elements:

$$\Sigma^A = \{D, \partial_\alpha, \bar{\partial}_{\dot{\alpha}}\}.$$

The first commutators have a form similar to (2.9);

$$[D,\partial_\alpha]_- = -\eta_\alpha, \qquad [D,\bar{\partial}_{\dot{\alpha}}]_- = \bar{\eta}_{\dot{\alpha}}, \qquad [\partial_\alpha,\bar{\eta}_{\dot{\alpha}}]_+ = [\bar{\partial}_{\dot{\alpha}},\eta_\alpha]_+ = D_{\alpha\dot{\alpha}} \tag{2.38}$$

where

$$\eta_\alpha = D_{\alpha\dot{\alpha}}\bar{\theta}^{\dot{\alpha}}, \qquad \bar{\eta}_{\dot{\alpha}} = \theta^\alpha D_{\alpha\dot{\alpha}}, \qquad D_{\alpha\dot{\alpha}} = P_{\alpha\dot{\alpha}} - \hat{e}A_{\alpha\dot{\alpha}}. \tag{2.39}$$

But subsequent commutators do not vanish. The algebra Σ^A appears to be infinite in contrast to the case of Σ. The higher commutators can be evaluated without any difficulty. For instance the commutator:

$$[\eta_\alpha,\eta_\beta]_+ = C_{\alpha\beta}\cdot\bar{\theta}\bar{\theta} \tag{2.40}$$

where $C_{\alpha\beta}$ is expressed through electrical and magnetic fields in the form:

$$C_{\alpha\beta} = (\varepsilon\vec{\sigma}(\vec{E}+i\vec{H}))_{\alpha\beta}. \tag{2.41}$$

Generally the commutators in the algebra Σ^A whose analog in the case of algebra Σ vanishes depend on the electromagnetic field tensor $F_{\mu\nu}$ or on its derivatives. We denote by Σ^F the set of elements of the algebra Σ^A which vanish when $F_{\mu\nu} = 0$. Obviously Σ^F is a subalgebra and what is more an ideal of the algebra Σ. The factoralgebra Σ^A/Σ^F is isomorphic to the algebra Σ. This fact is very important because it permits us to build representations of the algebra Σ^A up to elements of Σ^F. These representations coincide with representations of the algebra Σ. Another important point is that if transformations from the algebra Σ^A act on a neutral superfield the vector-potential in (2.33) vanishes and the algebra Σ^A reduces to the algebra Σ. These remarks will be used in construction of a super and gauge invariant action.

We consider the interaction of charged chiral superfields with the electromagnetic field. The fermion is described by a Dirac spinor $\Psi = \begin{pmatrix}\varphi_\alpha \\ \bar{\chi}^{\dot{\alpha}}\end{pmatrix}$ Let Ψ be negative charged and its conjugate positive charged. We distribute the Weil constituents of Ψ and $\bar{\Psi}$ among different chiral superfields. To complete the action we need the following supermultiplets:

$$\begin{aligned} \phi^L_- &= b_-(B+\Theta\varphi+\Theta\Theta G), & \phi^R_+ &= b_+(\bar{B}-\bar{\Theta}\bar{\varphi}-\bar{\Theta}\bar{\Theta}\bar{G}), \\ \phi^L_+ &= b_+(C+\Theta\bar{\chi}+\Theta\Theta H), & \phi^R_- &= b_-(\bar{C}-\bar{\Theta}\chi-\bar{\Theta}\bar{\Theta}\bar{H}). \end{aligned} \qquad (2.42)$$

Besides the spinor field the superfield (2.42) contains two complex scalar fields B and C and two auxiliary fields G and H. The chiral supermultiplets (2.42) are considered as a representation of chiral subalgebras of the factoralgebra Σ^A/Σ^F in the manner mentioned below.

We take the action in the form:

$$\begin{aligned} S = &-\tfrac{1}{4}I^S(F_{\mu\nu}F^{\mu\nu}) - I(\tilde{\phi}^R_+\tilde{\phi}^L_- + \tilde{\phi}^L_+\tilde{\phi}^R_-) \\ &+ mI^L(\phi^L_+\phi^L_-) - mI^R(\phi^R_+\phi^R_-). \end{aligned} \qquad (2.43)$$

Here all superinvariants (2.22), (2.24) and (2.28) are utilized The superfields with tilde form a symmetrical realization of chiral superfields. They are defined by formulae (2.20) with the only difference that instead of P the quantity D defined in (2.33) is put in the exponent. The expression (2.43) is neutral and hence it is invariant under the whole superalgebra Σ^A. After rather simple calculations and elimination of the auxiliary fields G and H the action (2.43) assumes the conventional form:

$$\begin{aligned} S = \int d^4x\{&-\tfrac{1}{4}F_{\mu\nu}F^{\mu\nu} + (\nabla_\mu B\nabla^\mu\bar{B}-m^2B\bar{B}) \\ &+ (\nabla_\mu C\nabla^\mu\bar{C}-m^2\bar{C}C) + \tfrac{1}{2}(-i\bar{\Psi}\gamma_\mu\nabla^\mu\Psi-m\bar{\Psi}\Psi)\}. \end{aligned} \qquad (2.44)$$

This is the ordinary form of the elctromagnetic interaction of the set of field B, C, Ψ.

The most unusual feature of the action (2.43) is its invariance "up to F-terms". Due to that it becomes possible to consider the free Lagrangian of the electromagnetic field as a super-singlet and to obtain the theory without the photon's super-partners. On the other hand we have lost exact supersymmetry. Relations following from supersymmetry now are fulfilled partially, up to terms proportional to $F_{\mu\nu}$ or its derivatives. Of course that is a difficult point of the theory. In no way do we pretend to get the ultimate form of the super-electrodynamics action. We consider this example only as an illustration of the method. Instead of treating the known version of super-gauge theory we thought that it would be more interesting to look at a rather nontrivial toy.

CONCLUDING REMARKS

We have not been concerned with many important topics of supersymmetric science such as extended supersymmetry and supergravity. On these lines intensive work is now in progress and many remarkable results have been obtained. But in spite of that the theory is far from complete. Looking at the supersymmetry in total we should say that in its present state it does not look like a candidate to a genuine physical theory. Surely supersymmetry has some features which seem to be adequate to reality. That is first of all the idea of super-symmetry itself - the unification of fermions and bosons within one multiplet. What is more the supersymmetry demonstrates wonderful examples of divergences cancellation. A particular version of supersymmetric field theory (with definite internal symmetry) appears as finite.

But it is unclear to what extent these results arise due to the foundations of present supersymmetric theories. We mean the invariance under the superalgebra R or something allied to it. Can it be possible that the attractive features of super-symmetry are related with some deep properties of super-mathematical structures? For instance the fact that an irreducible supermultiplet contains a finite number of fields is tightly correlated with the Grassmanian character of the quantities involved. The cancellation of divergences seems to be caused by the same reason. We think therefore that it would be extremely interesting to reconsider critically the foundations of supersymmetry and its intrinsic mechanisms.

The authors are indebted to Peter West for this kind invitation to contribute to the book on supersymmetry owing to which they got the opportunity to give their point of view on the subject.

REFERENCES

Berezin F A 1966 The method of 2nd quantization, Academic press
Berezin F A 1980 Group theoretical methods in physics, Vol 1, Proceedings of the international seminar, Zvenigorod, 28-30 November, Publishing Haus "Nauka", Moscow pp 10-12
Berezin F A and Tolstoy V N 1980, ibid., pp 12-21
Berezin F A and Katz G I 1970, Mathemat. Sbornik (USSR) 82 343
Berezin F A 1983, Introduction in algebra and analysis with anticommuting variables, Univ. Press, Moscow
Ferrara S 1974, Nucl. Phys. B77 73
Ferrara S and Zumino B 1974, Nucl. Phys. B79 413
Galperin A, Ivanov E, Kalitzin S, Ogievetsky V, Sokatchev E 1984a, Dubna preprint E2-84-441
Galperin A, Ivanov E, Kalitzin S, Ogievetsky V, Sokatchev E 1984b, Triest preprint IC/84/43
Golfand Yu A and Likhtman E P 1971, JETP Letters 13 452
Golfand Yu A and Likhtman E P 1972 Problems of theoretical physics, A memorial volume to Igor E Tamm, Publishing Haus "Nauka", Moscow pp 37-44
Haag R, Lopuszansky J T and Sohnius M 1975, Nucl. Phys. B88 257
Likhtman E P 1973, Theor. Math. Phys. 15 142
Salam A and Strathdee J 1974a, Nucl. Phys. B76 477
Salam A and Strathdee J, 1974b, Phys. Lett. 51B 353
Salam A and Strathdee J, 1975, Phys. Rev. D11 1521
Salomonson P and van Holten J W, 1982, Nucl. Phys. B196 509
Volkov D V and Akulov V P 1973a, JEPT Lett. 16 621
Volkov D V and Akulov V P 1973b, JEPT Lett. 17 367
Volkov D V and Akulov V P 1973c, Phys. Lett. B46 109
Volkov D V and Akulov V P 1974, Theor. Math. Phys. 18 39
Wess J and Zumino B, 1974a Nucl. Phys. B70 39
Wess J and Zumino B, 1974b Phys. Lett. B49 52
Wess J and Zumino B, 1974c Nucl. Phys. B78 1
de Witt B and Freedman D Z 1975, Phys. Rev. D12 2286
Witten E 1981, Nucl. Phys. B188 513

Reviews

Ogievetsky V I and Mezincescu L 1975, Symmetries between bosons and fermions and superfields, Uspekhi Phys. Nauk. 117 No 4 pp 637-683

Ferrara S and Fayet P 1977, Supersymmetry, Phys. Rep. 32 No 5 249

SUPERGRAPHS

M T GRISARU AND D ZANON

1. INTRODUCTION

From the early days of supersymmetry it has been clear that quantum properties of supersymmetric theories are best investigated and exhibited by using superfield perturbation theory and supergraphs. Much of the simplicity of superfield calculations as compared with a component approach, occurs because one deals with objects having fewer Lorents indices. Moreover the results of calculations are manifestly supersymmetric, cancellations due to supersymmetry are automatic and the calculations themselves are generally simpler since one evaluates simultaneously bosonic and fermionic contributions.

The original superfield Feynman rules (Salam and Strathdee, 1975; Ferrara and Piguet, 1975) were not in optimal form. Although simpler than component calculations, the formalism was not manifestly supersymmetric and it was necessary to manipulate complicated expressions involving explicit spinor variables to finally get a simple result. An improved version (Grisaru, Siegel and Rocek, 1979), along with better calculational techniques was presented in 1979, just preceding the renewed interest in supersymmetry. Using these improved methods for supergraphs, explicit supersymmetry is maintained at every stage of the calculations, power counting is straightforward and higher loop calculations are drastically simplified.

Since then we have made rapid advances in developing the formalism for both global and local supersymmetry. In this review we shall attempt to illustrate the methods and summarize the results: In Section 2 we give a concise description of the superspace formalism. Ordinary supergraph techniques are described in Section 3. In Section 4 we review the quantization of supersymmetric Yang-Mills theories in the background field method and the derivation of the corresponding Feynman rules. In Section 5 we describe covariant D-algebra and covariant supergraph techniques and, as an example, we compute two-loop self-energy corrections in supersymmetric QED. Supergraphs in locally supersymmetric theories are described in Section 6.

2. REVIEW OF SUPERSPACE

Our notation and convention are the same as in "Superspace" (Gates, Grisaru, Rocek and Siegel, 1983). We use two component anticommuting spinors with dotted and undotted indices ψ_α, $\bar{\chi}_{\dot\alpha}$ and four-vectors $A_a = A_{\alpha\dot\alpha}$. Spinor indices are raised and lowered with the antisymmetric symbol

$$C_{\alpha\beta} = C^{\beta\alpha} = \begin{pmatrix} 0 & -i \\ i & 0 \end{pmatrix} , \tag{2.1}$$

$$\psi_\alpha = \psi^\beta C_{\beta\alpha} \quad , \quad \psi^\alpha = C^{\alpha\beta}\psi_\beta$$

with the same conventions for dotted indices. We also define

$$\psi^2 = \tfrac{1}{2}\psi^\alpha\psi^\beta C_{\beta\alpha} = \tfrac{1}{2}\psi^\alpha\psi_\alpha \tag{2.2}$$

Superspace for N=1 supersymmetry is parametrized by space-time coordinate x^a and anticommuting spinor coordinates θ^α, $\bar\theta^{\dot\alpha}$, that we collectively denote as $z^A \equiv (x^a, \theta^\alpha, \bar\theta^{\dot\alpha})$. The corresponding partial derivatives are $\partial_A \equiv (\partial_a, \partial_\alpha, \partial_{\dot\alpha})$ with $\partial_A z^B = \delta_A{}^B$. Supersymmetry transformations correspond to coordinate transformations in superspace, $z^A \to z'^A$ with $x'^a = x^a - \tfrac{i}{2}(\varepsilon^\alpha\bar\theta^{\dot\alpha} + \bar\varepsilon^{\dot\alpha}\theta^\alpha)$, $\theta'^\alpha = \theta^\alpha + \varepsilon^\alpha$, $\bar\theta'^{\dot\alpha} = \bar\theta^{\dot\alpha} + \bar\varepsilon^{\dot\alpha}$. A superfield $\Psi(x,\theta,\bar\theta)$ is a function on this space that transforms as a scalar under supersymmetry: $\Psi'(z') = \Psi(z)$.

Space-time derivatives are covariant with respect to global supersymmetry, but θ-derivatives are not. We introduce covariant spinor derivatives

$$D_\alpha = \partial_\alpha + \tfrac{1}{2}\bar{\theta}^{\dot{\alpha}} i\partial_{\alpha\dot{\alpha}} \quad , \quad \bar{D}_{\dot{\alpha}} = \partial_{\dot{\alpha}} + \tfrac{1}{2}\theta^{\alpha} i\partial_{\alpha\dot{\alpha}} \tag{2.3}$$

They satisfy the following (anti) commutation relations

$$\{D_\alpha, \bar{D}_{\dot{\alpha}}\} = i\partial_{\alpha\dot{\alpha}}$$
$$\{D_\alpha, D_\beta\} = \{\bar{D}_{\dot{\alpha}}, \bar{D}_{\dot{\beta}}\} = 0 \tag{2.4}$$

Note that $D_\alpha D_\beta D_\gamma = 0$ because of anticommutativity. Other useful relations satisfied by the covariant derivatives are

$$D^\alpha D_\beta = \delta_\beta{}^\alpha D^2 \quad , \quad D^2\theta^2 = -1$$

$$[D^\alpha, \bar{D}^2] = i\partial^{\alpha\dot{\alpha}}\bar{D}_{\dot{\alpha}} \quad , \quad D^2\bar{D}^2 D^2 = D^2 \Box \tag{2.5}$$

$$D^2\bar{D}^2 + \bar{D}^2 D^2 - D^\alpha \bar{D}^2 D_\alpha = \Box$$

where $\Box = 1/2\ \partial^a\partial_a$.

We shall consider real scalar superfields V, which satisfy the condition $V(x,\theta,\bar{\theta}) = V^\dagger(x,\theta,\bar{\theta})$, and chiral and antichiral superfields ϕ, $\bar{\phi}$ subject to the following differential constraints

$$\bar{D}_{\dot{\alpha}}\phi = 0 \quad , \quad D_\alpha \bar{\phi} = 0 \tag{2.6}$$

In order to construct manifestly supersymmetric invariant actions we need to introduce integrals over superspace. We use the definition of Berezin integral over an anticommuting variable

$$\int d\theta\ \theta = 1 \quad , \quad \int d\theta\ 1 = 0 \tag{2.7}$$

or equivalently (note $f(\theta) = a + b\theta$)

$$\int d\theta_\alpha\ f(\theta) = \partial_\alpha f(\theta) \tag{2.8}$$

that is integration is identical to differentiation. Using (2.8) and (2.3), we have the following useful relations

$$\int d^4x\ d\theta_\alpha = \int d^4x\ D_\alpha \tag{2.9}$$

and also

$$\int d^4x\, d^2\theta = \int d^4x\, D^2 \quad , \quad \int d^4x\, d^2\bar{\theta} = \int d^4x\, \bar{D}^2$$

$$\int d^4x\, d^4\theta \equiv \int d^4x\, d^2\theta\, d^2\bar{\theta} \tag{2.10}$$

$$= \int d^4x\, d^2\theta\, \bar{D}^2 = \int d^4x\, d^2\bar{\theta}\, D^2 = \int d^4x\, D^2\bar{D}^2$$

We can define a δ-function in θ space

$$\int d^4\theta\, \delta^4(\theta-\theta')\, f(\theta) = f(\theta') \tag{2.11}$$

which implies

$$\delta^4(\theta-\theta') = (\theta-\theta')^2\, (\bar{\theta}-\bar{\theta}')^2 \tag{2.12}$$

In particular we have

$$D^2\bar{D}^2\, \delta^4(\theta-\theta')_{|\theta=\theta'} = 1 \tag{2.13}$$

whereas if fewer D's act on the δ-function we get zero.

Supersymmetry invariants are obtained as integrals over superspace: For a general superfield Ψ we have

$$S_\Psi = \int d^4x\, d^4\theta\, \Psi \tag{2.14}$$

For chiral and antichiral superfields we define

$$S_\Phi = \int d^4x\, d^2\theta\, \Phi \quad , \quad S_{\bar{\Phi}} = \int d^4x\, d^2\bar{\theta}\, \bar{\Phi} \tag{2.15}$$

The invariance of the integrals in (2.14) and (2.15) is just a consequence of the fact that supersymmetry transformations are coordinate transformations in superspace and we have defined the θ-integration in a translationally invariant way.

From (2.4) and (2.3) we see that dimension of $\partial_\alpha \sim m^{1/2}$ and consequently (see (2.8)) a general integral has dimension $\int d^4x\, d^4\theta = \int d^8z \sim m^{-2}$ and a chiral integral $\int d^4x\, d^2\theta = \int d^6z \sim m^{-3}$.

Since functional methods are very convenient for studying the quantization of superfield theories we will consider in the following section, we give here the basic formulae for functional calculus with superfields. The definition of functional differentiation for unconstrained general superfields is completely analogous to that for ordinary functions

$$\frac{\delta\Psi(z)}{\delta\Psi(z')} = \delta^8(z-z') = \delta^4(x-x')\ \delta^4(\theta-\theta') \tag{2.16}$$

For a chiral or antichiral superfield we define instead

$$\frac{\delta\Phi(z)}{\delta\Phi(z')} = \bar{D}^2\delta^8(z-z') \quad , \quad \frac{\delta\bar{\Phi}(z)}{\delta\bar{\Phi}(z')} = D^2\delta^8(z-z') \tag{2.17}$$

where the $\bar{D}^2$, D^2 factors in (2.17) account for the chirality constraint imposed on Φ and $\bar{\Phi}$.

Functional integration over superfields is defined as a natural extension of the usual definition. The functional measure is given by the product of the functional measures of the independent components contained in the corresponding superfield. The fundamental integrals for real and chiral superfields are (Gates, Grisaru, Rocek and Siegel, 1983)

$$\int \mathcal{D}V \exp\left[\ \tfrac{1}{2}\int d^8z\ VMV\ \right] = (\det M)^{-1/2}$$

$$\int \mathcal{D}\Phi \exp\left[\tfrac{1}{2}\int d^6z\ \Phi^2\right] = 1 \quad , \quad \int \mathcal{D}\bar{\Phi} \exp\left[\tfrac{1}{2}\int d^6\bar{z}\ \bar{\Phi}^2\right] = 1$$

$$\int \mathcal{D}\Phi\mathcal{D}\bar{\Phi} \exp\left[\int d^8z\ \bar{\Phi}\Phi\right] = (\det \Box)^{-1/2} \tag{2.18}$$

This is all we need to perform Gaussian integrations; other functional integrals that one encounters in perturbative calculations are obtained in terms of these by introducing sources and differentiating with respect to them.

3. QUANTIZATION OF N=1 THEORIES: GLOBAL SUPERGRAPHS

In this section we will consider a typical action describing supersymmetric Yang-Mills fields interacting with chiral multiplets. Since the classical action contains gauge superfields, in order to quantize the theory we need to fix the gauge. To maintain manifest supersymmetry we introduce supersymmetric gauge-fixing terms and correspondingly superfield Faddeev-Popov ghosts. We then construct the generating functional associated to the theory which allows us to derive the Feynman rules: Superspace propagators and vertices. At this point the procedure to obtain supergraphs is straightforward. The new feature as compared to ordinary Feynman diagrams is that here the vertices always carry a $d^4x d^4\theta$ integration and that in general spinor derivatives act on the propagators and on the external lines. We can manipulate however these operators and perform the θ-integrations so that at the end the contribution from a supergraph will contain one θ-integration and ordinary loop-momentum integrals.

We consider the following action

$$S = \frac{1}{2}\,\mathrm{tr}\int d^6z\, W^\alpha W_\alpha + \int d^8z\, \bar{\Phi} e^V \Phi$$
$$- \int d^6z\, [\, \frac{m}{2}\Phi^2 + \frac{\lambda}{6}\Phi^3\,] + \text{h.c.} \tag{3.1}$$

where

$$W^\alpha = i\bar{D}^2 e^{-V} D^\alpha e^V = i\bar{D}^2 D^\alpha V + \text{higher order terms} \tag{3.2}$$

is the Yang-Mills field-strength superfield and Φ a chiral superfield multiplet. $V=V^i T_i$, where T's are generators of the gauge group and the chiral superfield carries a representation of the group. The action in (3.1) is invariant under the following set of transformations (provided the mass and chiral self-interaction terms are chosen appropriately)

$$\Phi' = e^{i\Lambda}\Phi \quad , \quad \bar{\Phi}' = \bar{\Phi} e^{-i\bar{\Lambda}} \tag{3.3}$$

and

$$e^{V'} = e^{i\bar{\Lambda}} e^V e^{-i\Lambda} \tag{3.4a}$$

or

$$\delta V = L_{\frac{1}{2}V}\,[-i(\bar{\Lambda}+\Lambda) + \coth L_{\frac{1}{2}V}\; i(\bar{\Lambda}-\Lambda)] \quad , \quad L_X Y = [X,Y] \tag{3.4b}$$

where Λ is the chiral superfield parameter of the gauge group ($\bar{D}_{\dot{\alpha}}\Lambda=0$).

As gauge-fixing function we have to choose a chiral scalar corresponding to the chirality property of the gauge parameter. We use

$$F = \bar{D}^2 V \quad , \quad \bar{F} = D^2 V \tag{3.5}$$

The gauge is then fixed following the standard procedure: We introduce in the functional integral over V and Φ the Faddeev-Popov determinant Δ and factors $\delta(\bar{D}^2V-a)\ \delta(D^2V-\bar{a})$. We then gauge average over a and $\bar{a}$ and replace the Faddeev-Popov determinant by the corresponding integral representation in terms of the ghost action

$$Z = \int \mathcal{D}V\,\mathcal{D}\Phi\,\mathcal{D}\bar{\Phi}\,\mathcal{D}a\,\mathcal{D}\bar{a}\ \ \Delta^{-1}(V)\ \delta(\bar{D}^2V-a)\ \delta(D^2V-\bar{a})$$

$$\exp\left[-\frac{1}{\alpha}\int d^8z\ a\bar{a} + S\right]$$

$$= \int \mathcal{D}V\,\mathcal{D}\Phi\,\mathcal{D}\bar{\Phi}\,\mathcal{D}c\,\mathcal{D}\bar{c}\,\mathcal{D}c'\,\mathcal{D}\bar{c}'\ \exp(S + S_{GF} + S_{FP}) \tag{3.6}$$

where

$$S_{GF} = -\frac{1}{\alpha}\ \mathrm{tr}\int d^8z\ D^2V\bar{D}^2V$$

$$= -\frac{1}{2\alpha}\ \mathrm{tr}\int d^8z\ V(D^2\bar{D}^2 + \bar{D}^2D^2)V \tag{3.7}$$

and

$$S_{FP} = i\ \mathrm{tr}\int d^6z\ c'\bar{D}^2(\delta V) + i\ \mathrm{tr}\int d^6\bar{z}\ \bar{c}'D^2(\delta V) \tag{3.8}$$

In (3.8) the parameters Λ, $\bar{\Lambda}$ in the gauge variation δV are replaced by the chiral ghost superfields c and $\bar{c}$.

The kinetic action of the gauge field is obtained as the sum of (3.7) and the quadratic part $\frac{1}{2}\mathrm{tr}\int d^8z\ VD^\alpha\bar{D}^2D_\alpha V$ from (3.1). In the supersymmetric Fermi-Feynman gauge $\alpha=1$ it is simply given by $-\frac{1}{2}\mathrm{tr}\int d^8z\ V\Box V$. (We used the last identity in (2.5)). Finally, to low orders in V, we obtain the action

$$S + S_{GF} + S_{FP} = \mathrm{tr}\int d^8z\ \Big[-\tfrac{1}{2}V\Box V + \tfrac{1}{2}V\{D^\alpha V,\ \bar{D}^2D_\alpha V\} + \ \ldots$$

$$+ \bar{c}'c - c'\bar{c} + \tfrac{1}{2}(c'+\bar{c}')[V,c+\bar{c}] + \ldots\Big] \tag{3.9}$$

$$+ \int d^8z\ \bar{\Phi}e^V\Phi - \int d^6z\ \left(\tfrac{1}{2}m\Phi^2 + \tfrac{\lambda}{6}\Phi^3\right) + \text{h.c.}$$

The Feynman rules can be derived by constructing the

generating functional (for simplicity we omit the ghost action in (3.9))

$$Z(J,j,\bar{j}) = \int \mathcal{D}V\,\mathcal{D}\Phi\,\mathcal{D}\bar{\Phi}\, \exp\{S(V,\Phi,\bar{\Phi}) + \int d^8z\, JV + \int d^6z\, j\Phi + \text{h.c.}\} \tag{3.10}$$

$$= \exp S_{int}\left(\frac{\delta}{\delta J}, \frac{\delta}{\delta j}, \frac{\delta}{\delta \bar{j}}\right) \int \mathcal{D}V\,\mathcal{D}\bar{\Phi}\,\mathcal{D}\Phi\, \exp\{\int d^8z[-\tfrac{1}{2}V\Box V + JV]$$

$$+ \int d^8z\, \bar{\Phi}\Phi + \int d^6z\, [-\tfrac{1}{2}m\Phi^2 + j\Phi] + \text{h.c.}\}$$

where we have introduced real and chiral sources for the corresponding fields. We perform the integrals by completing squares and using (2.18). We obtain then expressions for the propagators of the gauge and chiral superfields (Gates, Grisaru, Rocek and Siegel, 1983)

$$\langle V(x,\theta)\, V(x',\theta')\rangle = \Box^{-1}\, \delta^4(x-x')\, \delta^4(\theta-\theta') \tag{3.11}$$

$$\langle \bar{\Phi}(x,\theta)\, \Phi(x',\theta')\rangle = -(\Box-m^2)^{-1}\, \delta^4(x-x')\, \delta^4(\theta-\theta')$$

$$\langle \bar{\Phi}(x,\theta)\, \bar{\Phi}(x',\theta')\rangle = -m\bar{D}^2\, [\Box(\Box-m^2)]^{-1}\, \delta^4(x-x')\, \delta^4(\theta-\theta')$$

The vertices are obtained directly from the interaction terms in the action, using (2.16) and (2.17). For example, from the interaction term $\frac{1}{2}\int d^8z\, VD^\alpha V\bar{D}^2 D_\alpha V$ we obtain the vertex in Fig. 1.

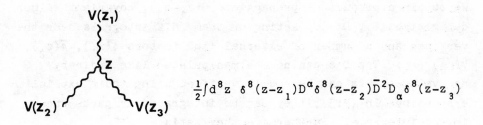

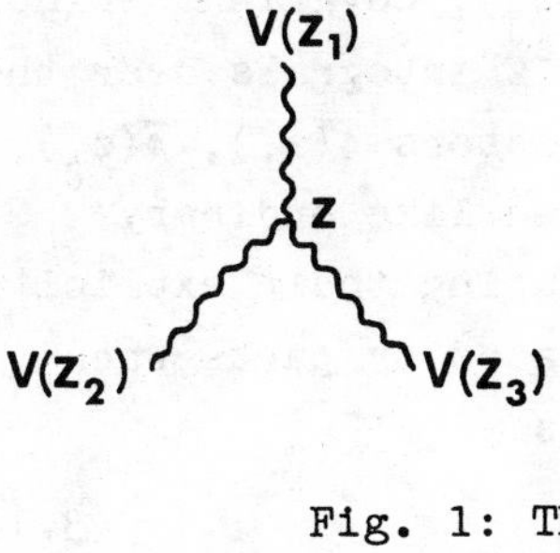

$$\tfrac{1}{2}\int d^8z\, \delta^8(z-z_1) D^\alpha \delta^8(z-z_2) \bar{D}^2 D_\alpha \delta^8(z-z_3)$$

Fig. 1: Three-point gauge vertex.

Purely chiral interaction terms in the action, i.e. $\frac{\lambda}{6}\int d^6z\, \Phi^3$ will correspond to the vertex in Fig. 2.

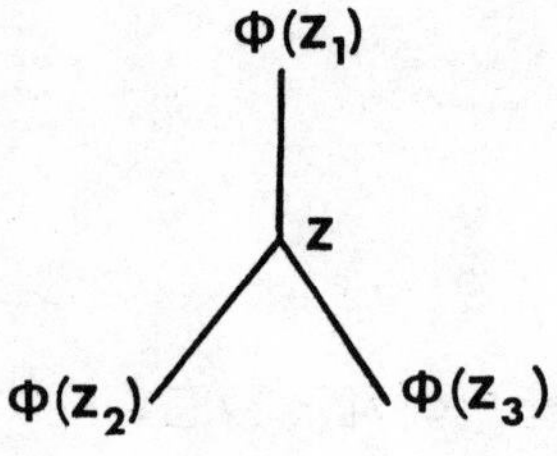

$$\int d^6z\, \bar{D}^2\delta^8(z-z_1)\bar{D}^2\delta^8(z-z_2)\bar{D}^2\delta^8(z-z_3)$$

$$= \int d^8z\, \delta^8(z-z_1)\bar{D}^2\delta^8(z-z_2)\bar{D}^2\delta^8(z-z_3)$$

Fig. 2: Chiral vertex.

To convert the d^6z integral to a full superspace integral d^8z we have used the identity in (2.10). Therefore at every vertex we always integrate over $d^4\theta d^4x$, with the rule that at each (anti) chiral vertex we omit a (D^2) $\bar{D}^2$ factor from one of the lines. These rules give the ingredients necessary to start computing the contribution from a supergraph.

Consider for example the one-loop diagram in Fig. 3.

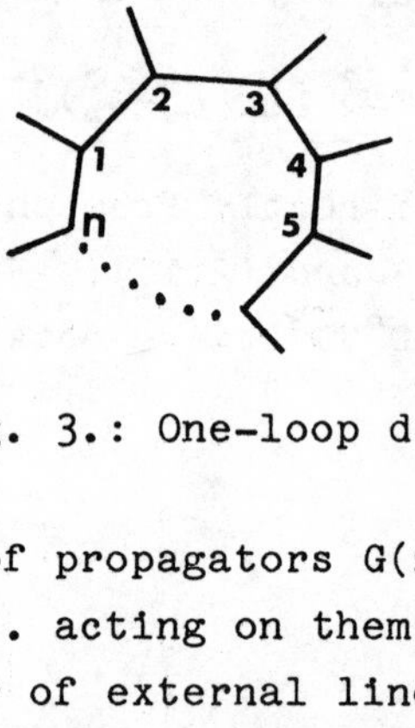

Fig. 3.: One-loop diagram.

We obtain a product of propagators $G(z_i-z_j)$, covariant spinor derivatives D_α, $\bar{D}_{\dot{\alpha}}$ ... acting on them, d^8z integrals over the vertices and a number of external line factors $\Phi(z_i)$, $\bar{\Phi}(z_j)$, $V(z_k)$ etc. The D's can now be manipulated like ordinary derivatives. As can be easily verified using their explicit expressions in (2.3), they can be integrated by parts when inside integrals. Furthermore they satisfy

$$D^\alpha_z\ G(z-z') = -D^\alpha_{z'}G(z-z') \qquad (3.12)$$

Graphically this corresponds to moving a spinor derivative

from one end to the other of a line. Thus for the diagram in Fig. 3, we can start for example freeing the propagator $G(z_1-z_2)$ of any covariant spinor derivatives: We first move the D's from one end to the other on the 1-2 line using (3.12); then we integrate them by parts on the external 2 line and on the next propagator $G(z_2-z_3)$. At this stage, since $G(z_1-z_2) = \square^{-1}\ \delta^4(x_1-x_2)\ \delta^4(\theta_1-\theta_2) = G(x_1-x_2)\ \delta^4(\theta_1-\theta_2)$ we can perform the integration over the θ_1 variable ($\square$ is independent of θ).

This procedure is then successively repeated on the lines of the loop: One generates a sum of terms containing expressions

$$\int d^4x_1 \ldots d^4x_n d^4\theta_{n-1}\ d^4\theta_n\ G(x_1-x_2)G(x_2-x_3)\ldots G(x_n-x_1)$$
$$\delta^4(\theta_{n-1}-\theta_n)\ D\bar{D}\ldots D\ \delta^4(\theta_n-\theta_{n-1}) \tag{3.13}$$

multiplied by the external line factors with, in general, spinor derivatives acting on them, from the integration by parts. Since $(D)^3 = 0$, we can now use the anticommutation relations in (2.4), (2.5) to reduce the number of spinor derivatives in (3.13) to at most 2 D's and 2 $\bar{D}$'s times a number of space-time derivatives. Using the definition of the δ-function in (2.12) and the relation (2.13), it should be clear that we will get a non-vanishing contribution only if we end up with exactly 2 D's and 2 $\bar{D}$'s , in which case

$$\delta^4(\theta_{n-1}-\theta_n)D^2\bar{D}^2\ \delta^4(\theta_n-\theta_{n-1}) = \delta^4(\theta_{n-1}-\theta_n) \tag{3.14}$$

and we can use the δ-function to perform the θ_{n-1} integration. At this point we have reduced the loop to a single point in θ-space.

These techniques apply as well to higher-loop diagrams. The procedure has to be repeated loop by loop, using the $\delta^4(\theta_i-\theta_j)$ to perform all but one of the θ-integrations, ending up with a contribution to the effective action which is local in θ

$$\int d^4\theta d^4x_1 \ldots d^4x_n \; F(x_1,\ldots,x_n)\Phi(x_1,\theta)\ldots \; D^\alpha V(x_i,\theta)\ldots \quad (3.15)$$

F is in general a non-local function of its arguments. At this stage, the contribution in (3.15) can be rewritten in momentum space and evaluated in standard fashion.

The manipulations of the covariant derivatives in the graphs, so called D-algebra, can lead to long intermediate expressions and one has to learn a number of tricks to make the procedure efficient. To get more insights into the D-algebra manipulations one should refer to "Superspace" and to the several papers that describe higher-loop calculations in the Wess-Zumino model and in SSYM (Abbott and Grisaru, 1980; Caswell and Zanon, 1981; Grisaru, Rocek and Siegel, 1981). As a simple exercise, the reader should verify that the supergraph in Fig. 4, describing a chiral superfield contribution to the V self-energy leads to the expression

$$\frac{1}{2}\int \frac{d^4p}{(2\pi)^4} \frac{d^nk}{(2\pi)^n} d^4\theta \; V(-p,\theta) \; \frac{\bar{D}^2D^2 - k^{\alpha\dot{\alpha}}\bar{D}_{\dot{\alpha}}D_\alpha - k^2}{k^2(k+p)^2} \; V(p,\theta) \quad (3.16)$$

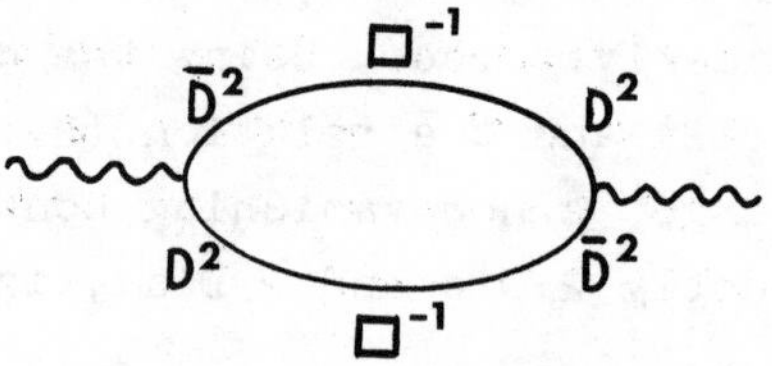

Fig. 4: A self-energy supergraph.

We now discuss some general results that can be derived from the generic contribution in (3.15).

a) The method leads to an effective action that, to any order in perturbation theory, has the form

$$\Gamma = \sum_n \int d^4\theta d^4x_1 \ldots d^4x_n \; F(x_1,\ldots,x_n) \quad (3.17)$$

$$\cdot \; \text{Polynomial in } \Phi(x_i,\theta), \; D^\alpha\Phi(x_j,\theta)\ldots V(x_k,\theta)$$

local in θ and integrated over $d^4\theta$ not $d^2\theta$. This has the consequence that chiral superfield terms $\int d^2\theta(m\Phi^2+\lambda\Phi^3+\zeta\Phi)$ which might be present in the original classical action do not receive any finite or infinite radiative corrections. This is the no-renormalization theorem of supersymmetry. Masses and coupling constants are renormalized only through the wave-function renormalization. This result is valid even for non renormalizable theories, e.g. supergravity.

The fact that we cannot produce terms which are chiral integrals, also implies that supersymmetry cannot be broken by radiative corrections in perturbation theory.

b) Power counting is very simple: $d^4\theta$ has dimension 2, D_α has dimension $\frac{1}{2}$, Φ has dimension 1 while V is dimensionless. Divergent graphs that contain no subdivergences correspond to local terms in the effective action of the form

$$\Gamma_\infty = \int d^4x\, d^4\theta\, P(\Phi,\bar{\Phi},V,D_\alpha\Phi,\ldots) \tag{3.18}$$

where P is a polynomial in the fields and their derivatives. Since the effective action is dimensionless P must have dimension 2. Thus at each order of perturbation theory, graphs with more than two external Φ's are convergent. A contribution containing a factor $\bar{\Phi}\Phi$ will produce a logarithmic divergence and consequently the chiral kinetic term requires in general a wavefunction renormalization. Since dim V=0, dimensional analysis would allow for divergent contribution containing an arbitrary number of V lines. However, V being a gauge field, the divergences are controlled by the Slavnov-Taylor identities so that in fact only the two- and three-point functions require independent renormalization. The analysis of the divergences involving gauge superfields becomes clear in the background field approach that we will describe in the next section. The actual proof of renormalization of the theory is complicated by the presence of infrared divergences. Rigorous results on this subject have been proven recently by Piguet and Sibold (1984).

c) Manifest supersymmetry is maintained at every stage of the calculation: Any regularization scheme that preserves translational invariance in superspace will be supersymmetric and renormalization is simple. For supergraphs the simplest regularization procedure is to use dimensional regularization after the D-algebra has been completed. This procedure is essentially equivalent to the dimensional reduction scheme used for component calculations (Siegel, 1979). However, even though the superfield result is automatically supersymmetric ambiguities may arise at higher-loops (Siegel, 1980).

Supergraph techniques may be extended to higher N supersymmetric theories provided an unconstrained superfield formalism exists. This is necessary in order to quantize the theory. For N=2,3 recent work (Galperin, Ivanov, Kalitzin, Ogievetsky and Sokatchev, 1984) is very promising.

4. BACKGROUND FIELD SUPERGRAPHS

The conventional gauge-fixing procedure for quantization of gauge theories, both in ordinary space and in superspace, spoils the invariance of the classical theory and leads to an effective action which is not gauge invariant in general (of course physical quantities computed from it are invariant). The background field formalism is a method that allows us to fix the gauge in such a manner that quantum corrections maintain explicit gauge invariance. This is achieved by describing the theory in terms of two fields, the background and the quantum fields. No gauge fixing is required for the background which always appears as an external field in the Feynman diagrams. For the quantum field, which is the integration variable in the generating functional, one can choose a background covariant gauge-fixing term and thus obtain an effective action explicitly gauge invariant in the background field variable.

In ordinary Yang-Mills (Abbott, 1981) one starts with the invariant action $S(A_a)$ and splits the field $A_a \rightarrow A_a + B_a$ with B_a the background field. The action is now invariant under the two sets of transformations:

1) quantum transformations

$$\delta A_a = \nabla_a \lambda + i[\lambda, A_a] \quad , \quad \delta B_a = 0 \quad , \quad (\nabla_a = \partial_a - i[B_a, \;]) \tag{4.1}$$

2) background transformations

$$\delta B_a = \nabla_a \lambda \quad , \quad \delta A_a = i[\lambda, A_a] \tag{4.2}$$

The quantization is carried out for the field A_a by choosing a background covariant gauge-fixing function so that the invariance with respect to the background transformations is maintained. This is achieved by background covariantization of the derivatives in the usual gauge-fixing function:

$$\partial^a A_a \rightarrow \nabla^a(B) A_a = \partial^a A_a - i[B^a, A_a] \tag{4.3}$$

The ghost action is also covariantized with respect to the background field. One can show that the conventional effective action is perturbatively obtained by computing one-particle-irreducible graphs with only A_a internal lines and B_a external lines (Abbott, 1981).

The extension of the background field method for ordinary Yang-Mills to the supersymmetric case involves some new features because of the non linearity of the gauge transformations in (3.4). The correct procedure in superspace can be obtained in a way parallel to the ordinary case provided the splitting $A_a \rightarrow A_a + B_a$ is interpreted as a splitting of the covariant derivatives

$$\nabla_a = \partial_a - i[A_a, \;] \rightarrow \partial_a - i[B_a, \;] - i[A_a, \;] = \nabla_a(B) - i[A_a, \;] \tag{4.4}$$

i.e. a replacement of the flat derivative by a background covariant derivative.

In supersymmetric Yang-Mills we proceed in a similar way. We first introduce the covariant derivatives (in "chiral" representation)

$$\nabla_\alpha = e^{-V} D_\alpha e^V = D_\alpha - i\Gamma_\alpha = D_\alpha + D_\alpha V + \ldots$$
$$\bar{\nabla}_{\dot{\alpha}} = \bar{D}_{\dot{\alpha}} \quad , \quad \nabla_{\alpha\dot{\alpha}} = -i\{\nabla_\alpha, \bar{\nabla}_{\dot{\alpha}}\} = \partial_{\alpha\dot{\alpha}} - i\Gamma_{\alpha\dot{\alpha}} \tag{4.5}$$

where we have defined the connection superfields by $\nabla_A = D_A - i\Gamma_A$. The ∇_A's are gauge covariant with respect to the Λ transformations (see (3.4))

$$\nabla'_A = e^{i\Lambda}\nabla_A e^{-i\Lambda} \qquad \text{or} \qquad \delta\nabla_A = i[\Lambda, \nabla_A] \tag{4.6}$$

The background splitting is realized by replacing the usual derivatives in (4.5) by background covariant derivatives $\underset{\sim}{\nabla}_A$

$$\nabla_\alpha \rightarrow e^{-V}\underset{\sim}{\nabla}_\alpha e^V \quad , \quad \bar{\nabla}_{\dot{\alpha}} \rightarrow \underset{\sim}{\bar{\nabla}}_{\dot{\alpha}} \quad , \quad \nabla_{\alpha\dot{\alpha}} \rightarrow -i\{\nabla_\alpha, \bar{\nabla}_{\dot{\alpha}}\} \tag{4.7}$$

Since the gauge field enters nonlinearly in the covariant derivatives, (4.7) corresponds to a nonlinear splitting $V \rightarrow V \oplus V_B$.

The covariant derivatives ∇_A transform covariantly under two kinds of transformations:

1) quantum transformations

$$e^V \rightarrow e^{i\bar{\Lambda}}e^V e^{-i\Lambda} \quad , \quad \underset{\sim}{\nabla}_A \rightarrow \underset{\sim}{\nabla}_A \quad , \quad \nabla_A \rightarrow e^{i\Lambda}\nabla_A\, e^{-i\Lambda} \tag{4.8}$$

with background covariantly chiral parameters $\underset{\sim}{\nabla}_\alpha \bar{\Lambda} = \underset{\sim}{\bar{\nabla}}_{\dot{\alpha}}\Lambda = 0$.

2) background transformations

$$e^V \rightarrow e^{iK}e^V e^{-iK} \quad , \quad \underset{\sim}{\nabla}_A \rightarrow e^{iK}\underset{\sim}{\nabla}_A\, e^{-iK}$$
$$\nabla_A \rightarrow e^{iK}\nabla_A\, e^{-iK} \tag{4.9}$$

with a real parameter $K = \bar{K}$.

Chiral superfields coupled to the gauge fields can also be treated in a background covariant way. We start with the usual action

$$\int d^8z\, \bar{\Phi}_o e^V \Phi_o = \int d^8z\, e^{-V}\bar{\Phi}_o\Phi_o \tag{4.10}$$

where Φ_o and $\bar{\Phi}_o$ are ordinary chiral and antichiral, $D_\alpha\bar{\Phi}_o = 0$ and $\bar{D}_{\dot{\alpha}}\Phi_o = 0$. We then define covariantly chiral and antichiral superfields $\Phi_c = \Phi_o$ and $\bar{\Phi}_c = e^{-V}\bar{\Phi}_o$ which satisfy

$$\bar{\nabla}_{\dot{\alpha}}\Phi_c = 0 \quad , \quad \nabla_\alpha \tilde{\Phi}_c = 0 \tag{4.11}$$

with the covariant derivatives given in (4.5). In terms of these covariantly chiral fields the action becomes $\int d^8z\ \tilde{\Phi}_c\Phi_c$. After the quantum-background splitting in (4.7), it goes into

$$\int d^8z\ \tilde{\Phi}e^V\Phi \tag{4.12}$$

where now V is the quantum field; the Φ's are background covariantly chiral, $\bar{\nabla}_{\dot{\alpha}}\Phi = 0$, and they depend implicitly on the background gauge field.

If we are interested in computing diagrams with external chiral lines, we perform a linear quantum-background splitting

$$\Phi\to\Phi+\Phi_B \quad , \quad \tilde{\Phi}\to\tilde{\Phi}+\tilde{\Phi}_B \tag{4.13}$$

The quantum fields transform under

1) quantum transformations

$$\Phi\to e^{i\Lambda}\Phi \quad , \quad \tilde{\Phi}\to\tilde{\Phi}e^{-i\bar{\Lambda}} \tag{4.14}$$

2) background transformations

$$\Phi\to e^{iK}\Phi \quad , \quad \tilde{\Phi}\to\tilde{\Phi}e^{-iK} \tag{4.15}$$

The chiral field action is invariant under both 1) and 2) transformations.

The Yang-Mills field strength defined in (3.2) is expressible in terms of covariant derivatives as

$$W_\alpha = \tfrac{i}{2}[\bar{\nabla}^{\dot{\alpha}},\{\bar{\nabla}_{\dot{\alpha}},\nabla_\alpha\}] \tag{4.16}$$

Substituting (4.7) into (4.16) we obtain for the gauge action $\frac{1}{2}\mathrm{tr}\int d^6z\ W^\alpha W_\alpha$ the corresponding splitting with the quantum V field explicit and the background field implicit in the covariant derivatives. The quatization is done preserving the background invariance: The gauge is fixed by introducing a background covariantly chiral gauge-fixing function

$$F = \bar{\nabla}^2 V \quad , \quad \bar{F} = \nabla^2 V \tag{4.17}$$

with a gauge-fixing term $\nabla^2 V\bar{\nabla}^2 V$. Since F satisfies a covariant chirality condition, $\underset{\sim}{\bar\nabla}_{\dot\alpha}F=0$, the Faddeev-Popov ghosts c, c' will be covariantly chiral fields. The gauge averaging in (3.6) $\int\mathcal{D}a\mathcal{D}\bar a\ \exp\int a\bar a$ produces a determinant with a non trivial dependence on the background field since $a=\underset{\sim}{\bar\nabla}^2V$ is also covariantly chiral. To normalize the averaging one introduces an additional factor in the generating functional $\int\mathcal{D}b\mathcal{D}\bar b\ \exp\int b\bar b$ where the covariantly chiral superfield b, the Nielsen-Kallosh ghost, has abnormal statistics. The N.K. ghost interacts only with the background and thus it gives contributions only at the one-loop level. After gauge fixing the Yang-Mills action is given by

$$-\tfrac{1}{2}\mathrm{tr}\int d^8z[(e^{-V}\nabla^\alpha e^V)\bar\nabla^2(e^{-V}\nabla_\alpha e^V) + V(\nabla^2\bar\nabla^2+\bar\nabla^2\nabla^2)V] \qquad (4.18)$$

All the derivatives are background covariant and we omit the underline from now on. Using the commutation relations

$$[\nabla_\alpha,\nabla_{\beta\dot\beta}] = C_{\alpha\beta}\bar W_{\dot\beta} \quad , \quad [\bar\nabla_{\dot\alpha},\nabla_{\beta\dot\beta}] = C_{\dot\alpha\dot\beta}W_\beta \qquad (4.19)$$

we can rewrite the action in (4.18) as

$$-\tfrac{1}{2}\mathrm{tr}\int d^8z\ [V(\Box - iW^\alpha\nabla_\alpha - i\bar W^{\dot\alpha}\bar\nabla_{\dot\alpha})V - V\nabla^\alpha V\bar\nabla^2\nabla_\alpha V + \ldots] \qquad (4.20)$$

where $\Box = \frac{1}{2}\nabla^a\nabla_a$ is the background covariant d'Alembertian and W_α is the background covariant field strength.

In order to perform perturbative calculations we write $\nabla_\alpha = D_\alpha - i\Gamma_\alpha$ and also $\nabla_a = \partial_a - i\Gamma_a$ so that the action in (4.20) becomes

$$-\tfrac{1}{2}\mathrm{tr}\int d^8z\ \{V[\Box_o - i\Gamma^a\partial_a - \tfrac{i}{2}(\partial^a\Gamma_a) - \tfrac{1}{2}\Gamma^a\Gamma_a - iW^\alpha D_\alpha - W^\alpha\Gamma_\alpha$$
$$- i\bar W^{\dot\alpha}\bar D_{\dot\alpha} - \bar W^{\dot\alpha}\bar\Gamma_{\dot\alpha}]V - VD^\alpha V\bar D^2 D_\alpha V + iV\Gamma^\alpha V\bar D^2 D_\alpha V + \ldots\} \qquad (4.21)$$

with $\Box_o = \frac{1}{2}\partial^a\partial_a$. We now compute Feynman diagrams with internal quantum lines and keep the background external. The rules are the ones in the previous section: One has to complete the D-algebra in the loops and evaluate momentum integrals.

If chiral superfields are also present it would seem that in order to perform quantum calculations we have to solve the chirality constraint $\bar{\nabla}_{\dot{\alpha}}\Phi=0$ in terms of the background gauge field. We shall see in the next section how this can be avoided.

The background field formalism combined with superspace perturbation theory has been proven quite valuable. As compared to ordinary methods it presents the following advantages:

a) The method leads to improved convergence properties of individual graphs. In supersymmetric Yang-Mills theories, supergraphs computed in conventional fashion give contribubutions that are functions of the prepotential V of dimension 0 and thus can have a high degree of divergence. (The actual degree of divergence of the effective action is controlled by the Slavnov-Taylor identities but this results only as a consequence of cancellations of individual contributions.) In the background field method the background field is always contained in the superfield connections Γ_α, Γ_a of dimensions $\frac{1}{2}$, 1 respectively, and in the field strength W_α of dimension $\frac{3}{2}$. The degree of divergence of individual supergraphs is controlled by the lowest dimension object, the spinor connection Γ_α (dimension $\frac{1}{2}$ as compared to dimension 0 in the conventional approach). Therefore the individual supergraphs are more convergent: Each of them represents sums of ordinary supergraphs.

b) Perturbative calculations are simplified. The improved convergence properties of the diagrams mean that some cancellations have automatically taken place and therefore supergraphs computed using the background field formalism are simpler and fewer in number.

c) Manifest gauge invariance is maintained at every order of perturbation theory and the result is always expressible in terms of gauge invariant quantities. The possibility of preserving gauge invariance gives an additional control in the calculations.

The method is particularly efficient for perturbative calculations in gauge theories. At the one-loop level as one can see from the action in (4.21), the V-loop contributions to the two- and three-point functions are automatically zero since the background-quantum vertices contain only one spinor derivative each and one needs at least two D's and two $\bar{D}$'s to get a non vanishing contribution from a loop. This simplification allowed for example the calculation of the one-loop, four-particle S matrix in N=4 SSYM (Grisaru and Siegel, 1982): The contributions from the three chiral fields Φ_i and the three chiral ghosts c, c' and b, exactly cancel without any calculations because of opposite statistics and the only non zero contribution comes from a box diagram with two external W's and two $\bar{W}$'s.

The method has also been used to calculate the two-loop β function in SSYM (Abbott, Grisaru and Zanon, 1984) and indeed simplifications were found with respect to ordinary methods.

5. COVARIANT SUPERGRAPHS

Further simplifications in higher-loop calculations can be obtained with the introduction of covariant D-algebra for supergraphs (Grisaru and Zanon, 1984a, Grisaru and Zanon, 1984b).

The background field method leads to contributions that are functions of the connections Γ_α, $\Gamma_{\alpha\dot{\alpha}}$ and field strengths W_α. The connections enter explicitly in the calculations since we had to separate the covariant derivatives as $\nabla_\alpha = D_\alpha - i\Gamma_\alpha$ to perform the D-algebra and $\nabla_{\alpha\dot{\alpha}} = \partial_{\alpha\dot{\alpha}} - i\Gamma_{\alpha\dot{\alpha}}$ to perform the momentum integrals. By using covariant D-algebra one never needs separate ∇_α into two parts. Higher loop diagrams computed with the corresponding rules never produce factors of Γ_α of dimension $\frac{1}{2}$, but only the space-time connection $\Gamma_{\alpha\dot{\alpha}}$ and the field strength W_α of dimensions 1 and $\frac{3}{2}$ respectively. They are therefore more convergent.

The derivation of the new rules is obtained starting from the expression of the action in background covariant form as in (4.12) and (4.20). We denote by $\hat{\Box}$ the operator in the kinetic action for V

$$\hat{\Box} = \Box - iW^{\alpha}\nabla_{\alpha} - i\bar{W}^{\dot{\alpha}}\bar{\nabla}_{\dot{\alpha}} \tag{5.1}$$

and define the corresponding covariant propagator

$$\langle VV\rangle = \hat{\Box}^{-1} \tag{5.2}$$

For background covariantly chiral superfields we define first covariant functional derivatives

$$\frac{\delta\Phi(z)}{\delta\Phi(z')} = \bar{\nabla}^2\delta^8(z-z') \quad , \quad \frac{\delta\bar{\Phi}(z)}{\delta\bar{\Phi}(z')} = \nabla^2\delta^8(z-z') \tag{5.3}$$

and the operators $\Box_{\pm}$

$$\bar{\nabla}^2\nabla^2\Phi = \Box_+\Phi \quad , \quad \nabla^2\bar{\nabla}^2\bar{\Phi} = \Box_-\bar{\Phi} \tag{5.4}$$

with the explicit representation

$$\Box_+ = \Box - iW^{\alpha}\nabla_{\alpha} - \tfrac{1}{2}(\nabla^{\alpha}W_{\alpha}) \quad , \quad \Box_- = \Box - i\bar{W}^{\dot{\alpha}}\bar{\nabla}_{\dot{\alpha}} - \tfrac{1}{2}(\bar{\nabla}^{\dot{\alpha}}\bar{W}_{\dot{\alpha}}) \tag{5.5}$$

In this way (with the only exception of some one loop contributions) manifest covariant Feynman rules can be obtained also for covariantly chiral fields (Gates, Grisaru, Rocek and Siegel, 1983). We have the covariant propagators (in the massive case)

$$\langle\bar{\Phi}\Phi\rangle = -(\Box_+ - m^2)^{-1}$$

$$\langle\Phi\Phi\rangle = -m\nabla^2[\Box_+(\Box_+ - m^2)]^{-1} \quad , \quad \langle\bar{\Phi}\bar{\Phi}\rangle = -m\bar{\nabla}^2[\Box_-(\Box_- - m^2)]^{-1} \tag{5.6}$$

The higher-loop background effective action is obtained by computing "vacuum" diagrams with quantum vertices derived from the interaction terms in the action and covariant propagators as in (5.2) and (5.6). For chiral superfields we have ∇^2, $\bar{\nabla}^2$ factors at the vertices from the functional derivative rules in (5.3). As usual, one factor $\bar{\nabla}^2(\nabla^2)$ is omitted at a chiral (antichiral) vertex to convert a d^6z integral into a d^8z integral.

The kinetic operators for the various fields are sums of terms which do not contain any spinor derivative plus terms which contain the operators $W^{\alpha}\nabla_{\alpha}$ and $\bar{W}^{\dot{\alpha}}\bar{\nabla}_{\dot{\alpha}}$. The basic idea of covariant D-algebra is to expand the propagators separating out the $W\nabla$, $\bar{W}\bar{\nabla}$ terms and perform the D-algebra at this stage in terms of <u>covariant</u> <u>spinor</u> <u>derivatives</u>. The ∇'s are manipulated in the loops using essentially the same tricks as for ordinary D's with the additional complication that they do not commute with the covariant $\Box$. Some of the identities that are useful in the calculations are the following

$$\begin{aligned}
\{\nabla_{\alpha},\bar{\nabla}_{\dot{\alpha}}\} &= i\nabla_{\alpha\dot{\alpha}} , \qquad [\nabla_{\alpha}, \nabla_{\beta\dot{\beta}}] = C_{\alpha\beta}\,\bar{W}_{\dot{\beta}} \\
[\nabla_{\alpha},\bar{\nabla}^{2}] &= -i\nabla_{\alpha\dot{\alpha}}\bar{\nabla}^{\dot{\alpha}} + iW_{\alpha} = -i\bar{\nabla}^{\dot{\alpha}}\nabla_{\alpha\dot{\alpha}} - iW_{\alpha} \\
[\nabla_{\alpha},\Box] &= \bar{W}^{\dot{\alpha}}\nabla_{\alpha\dot{\alpha}} + \tfrac{1}{2}(\nabla_{\alpha\dot{\alpha}}\bar{W}^{\dot{\alpha}}) \\
[\nabla_{\alpha},\hat{\Box}] &= \tfrac{1}{2}(\nabla_{\alpha\dot{\alpha}}\bar{W}^{\dot{\alpha}}) - i(\nabla_{\alpha}W^{\beta})\nabla_{\beta} \\
[\nabla_{\alpha},\Box_{+}] &= \bar{W}^{\dot{\alpha}}\nabla_{\alpha\dot{\alpha}} + (\nabla_{\alpha\dot{\alpha}}\bar{W}^{\dot{\alpha}}) - i(\nabla_{\alpha}W^{\beta})\nabla_{\beta} \\
\nabla^{2}\Box_{+}\bar{\nabla}^{2} &= \Box_{-}\nabla^{2}\bar{\nabla}^{2} , \qquad \bar{\nabla}^{2}\Box_{-}\nabla^{2} = \Box_{+}\bar{\nabla}^{2}\nabla^{2}
\end{aligned} \tag{5.7}$$

To understand the general procedure let us consider the graph in Fig. 5, which would contribute to a three-loop background field calculation in SSYM.

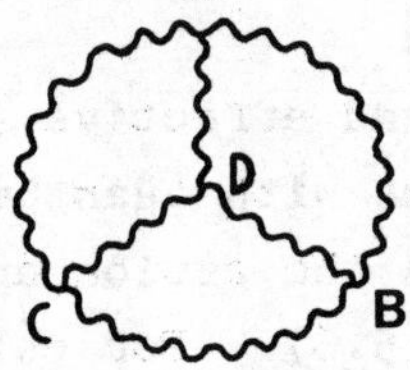

Fig. 5: Three-loop contribution to the background effective action in SSYM

The vertices have the form $V\nabla^\alpha V\bar{\nabla}^2\nabla_\alpha V$ and the propagators are $\hat{\Box}^{-1}$. We start moving the ∇'s from one end to the other of a given line using

$$\begin{aligned}\nabla_\alpha\,\hat{\Box}^{-1} &= \hat{\Box}^{-1}\nabla_\alpha - \hat{\Box}^{-1}\,[\nabla_\alpha,\hat{\Box}]\,\hat{\Box}^{-1} \\ &= \hat{\Box}^{-1}\nabla_\alpha - \tfrac{1}{2}\,\hat{\Box}^{-1}(\nabla_{\alpha\dot{\alpha}}\bar{W}^{\dot{\alpha}})\hat{\Box}^{-1} + \hat{\Box}^{-1}i(\nabla_\alpha W^\beta)\nabla_\beta\hat{\Box}^{-1}\end{aligned} \tag{5.8}$$

In the third term of (5.8), we again commute the ∇_β past the $\hat{\Box}^{-1}$. This procedure can be repeated indefinitely, but since every commutator generates one factor of the background field, it eventually stops if we are interested in an amplitude with a certain number of external lines. Then by integration by parts and successive commutations through the various $\hat{\Box}^{-1}$, we will end up with the ∇'s concentrated for example just at the right of the vertices A,B,C. Using the commutation relations in (5.7) we can reduce their number of $\leqslant 4$ at each vertex. We now have the original diagram plus diagrams containing vertices from the commutators in (5.8). Additional spinor derivatives are still present in the propagators: We expand them at this stage, separating out the $W\nabla$, $\bar{W}\bar{\nabla}$ terms. For each propagator we use the identity

$$\hat{\Box}^{-1} = \Box^{-1} + \Box^{-1}(iW^\alpha\nabla_\alpha + i\bar{W}^{\dot{\alpha}}\bar{\nabla}_{\dot{\alpha}})\hat{\Box}^{-1} \tag{5.9}$$

(note that $\Box$ contains no spinor derivatives) and again commute the ∇'s with the $\hat{\Box}^{-1}$, obtaining new quantum-background vertices from the commutators. Eventually we will generate a number of contributions corresponding to diagrams with $\Box^{-1}$ propagators and background field vertices, with all the spinor derivatives appearing now at the right of the vertices A, B, C in Fig. 5. We complete the covariant D-algebra in every loop, by using the relation

$$\begin{aligned}&\delta^4(\theta_{n-1}-\theta_n)\nabla^2\bar{\nabla}^2\delta^4(\theta_n-\theta_{n-1}) \\ &=\delta^4(\theta_{n-1}-\theta_n)(D-i\Gamma)^2(\bar{D}-i\bar{\Gamma})^2\,\delta^4(\theta_n-\theta_{n-1}) = \delta^4(\theta_n-\theta_{n-1})\end{aligned} \tag{5.10}$$

otherwise the result is zero. After the θ-integrations have been performed, our expressions only contain space-time covariant derivatives ∇_a as well as W_α, $\nabla_\beta W^\alpha$ external field factors. In particular we do not have any terms containing explicit spinor connections Γ_α. We can continue the calculation by separating $\nabla_a = \partial_a - i\Gamma_a$ and $\Box = \Box_o - i\Gamma^a\partial_a - \frac{1}{2}(\partial^a\Gamma_a) - \frac{1}{2}\Gamma^a\Gamma_a$ and expanding the $\Box^{-1}$ propagators in terms of free propagators and additional background vertices.

Supergraphs computed in this fashion are simpler and fewer in number than in ordinary background field calculations. The fact that Γ_α never gets generated implies that the individual contributions are more convergent. Each of them represents sums of ordinary supergraphs; some cancellations have automatically taken place.

As an example of the power of the method we close this section with a two-loop calculation.

We consider the supersymmetric massless QED theory described by the background covariant action (after gauge fixing)

$$S = \int d^8z[\tilde{\eta}_+ e^V \eta_+ + \tilde{\eta}_- e^{-V} \eta_- - \tfrac{1}{2}V\Box_o V] \tag{5.11}$$

where $\eta_\pm$ are a doublet of massless covariantly chiral superfields with opposite charges with respect to the gauge superfield V. We compute the two-loop corrections to the V self-energy. According to our rules the two-loop contribution to the background effective action is given by the "vacuum" diagram in Fig. 6.

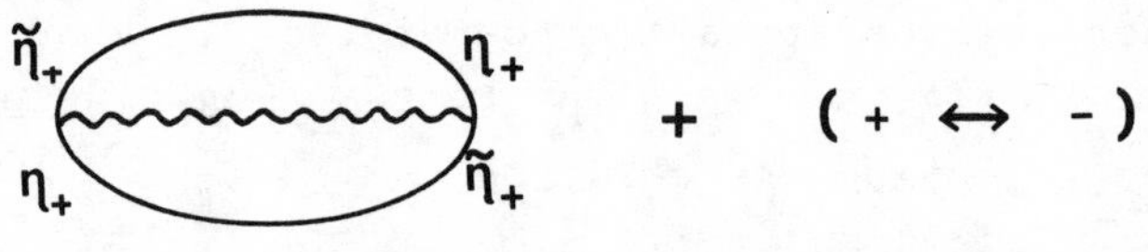

Fig. 6: Two-loop covariant "vacuum" diagram in SQED

Using our covariant Feynman rules for vertices and propagators we obtain the corresponding graph in Fig. 7.

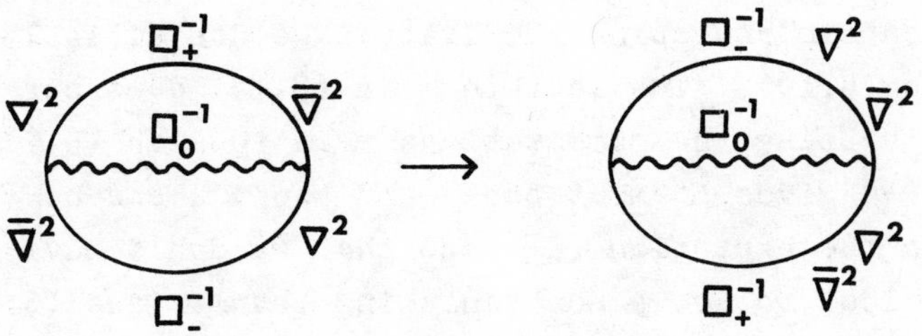

Fig. 7 : Covariant vertices and propagators corresponding to the diagram in Fig. 6

We have used the last identities in (5.7). Note that since the middle line in Fig. 7 carries no spinor derivatives, we can use the δ-function contained in the vector propagator to perform the θ-integration on that line, so that effectively we have now to complete our covariant D-algebra in two-separate loops. We first write

$$\Box_+ = \tilde{\Box}_+ - iW^{\alpha}\nabla_{\alpha} \quad , \quad \Box_- = \tilde{\Box}_- - i\bar{W}^{\dot{\alpha}}\bar{\nabla}_{\dot{\alpha}} \tag{5.12}$$

where (see (4.24))

$$\tilde{\Box}_+ = \Box - \tfrac{1}{2}(\nabla^{\alpha}W_{\alpha}) \quad , \quad \tilde{\Box}_- = \Box - \tfrac{1}{2}(\bar{\nabla}^{\dot{\alpha}}\bar{W}_{\dot{\alpha}}) \tag{5.13}$$

contain no covariant spinor derivative operator. Following the procedure described in the general discussion, we expand the propagators separating the $W\nabla, \bar{W}\bar{\nabla}$ terms and pushing the ∇'s to the right vertex. On the bottom line of the diagram in Fig. 7 we use

$$\Box_+^{-1} = \tilde{\Box}_+^{-1} + \tilde{\Box}_+^{-1} iW^{\alpha}[\nabla_{\alpha}, \Box_+^{-1}] + \tilde{\Box}_+^{-1}\, iW^{\alpha}\, \Box_+^{-1}\nabla_{\alpha} \tag{5.14}$$

$$= \tilde{\Box}_+^{-1} - \tilde{\Box}_+^{-1} iW^{\alpha}\Box_+^{-1}[\bar{W}^{\dot{\alpha}}\nabla_{\alpha\dot{\alpha}} + (\nabla_{\alpha\dot{\alpha}}\bar{W}^{\dot{\alpha}}) - i(\nabla_{\alpha}W^{\beta})\nabla_{\beta}]\Box_+^{-1}$$

$$+ \tilde{\Box}_+^{-1} iW^{\alpha}\Box_+^{-1}\nabla_{\alpha}$$

where we have used the identity in (5.7) for the commutator $[\nabla_\alpha, \square_+^{-1}]$. At this stage the D-algebra in the bottom loop becomes trivial. For the first term in (5.14), since $\tilde{\square}_+^{-1}$ does not contain any spinor derivative we use directly the relation in (5.10). The last term in (5.14) does not contribute: Using the commutations relations in (5.7) the factor $\nabla_\alpha \bar{\nabla}^2 \nabla^2$ gives at most one $\bar{\nabla}$ and two ∇'s and no $\bar{\nabla}$'s are produced in the expansion $\square_+^{-1}$, so that we don't have enough D's in the loop to get a non vanishing result (see (5.10)). The second term is already second order in the background field. Since we are interested in computing a two-point function we can then replace all the propagators with free propagators and the covariant derivatives with ordinary derivatives. Thus the $\nabla_\alpha W^\beta$ term gives a vanishing contribution and for the other terms we again use (5.10) to complete the D-algebra. Exactly the same procedure applies to the top loop of the diagram in Fig. 7. Thus after D-algebra we obtain contributions corresponding to the diagrams in Fig. 8.

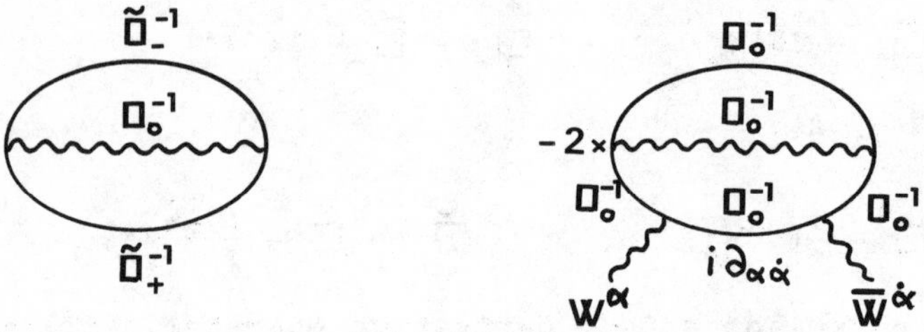

a b

Fig. 8: Diagrams corresponding to the ones in Fig. 7, after D-algebra

The calculation is now almost finished. The diagram in Fig. 8b gives directly a contribution to the effective action of the form

$$2 \int d^4\theta \, \frac{d^4 p}{(2\pi)^4} \frac{d^n q \, d^n k}{(2\pi)^{2n}} W^\alpha(-p) \bar{W}^{\dot{\alpha}}(p) \frac{(p+k)_{\alpha\dot{\alpha}}}{q^2 (q-k)^2 k^4 (k+p)^2} \qquad (5.15)$$

For the diagram in Fig. 8a we expand the $\tilde{\Box}_{\pm}^{-1}$ propagators in terms of free propagators and background vertices up to second order in the external field. The contributions of the second order terms on the same line are

$$\frac{1}{2}\int d^4\theta \frac{d^4p}{(2\pi)^4} \frac{d^nqd^nk}{(2\pi)^{2n}} [\nabla^\alpha W_\alpha(-p)\nabla^\beta W_\beta(p) - \Gamma^a(-p)\Gamma^b(p)(2k+p)_a(2k+p)_b] \frac{1}{q^2 (q-k)^2 k^4 (k+p)^2} \tag{5.16}$$

Finally from the expansion on different lines we obtain

$$\frac{1}{4}\int d^4\theta \frac{d^4p}{(2\pi)^4} \frac{d^nqd^nk}{(2\pi)^{2n}} [\nabla^\alpha W_\alpha(-p)\nabla^\beta W_\beta(p) - \Gamma^a(-p)\Gamma^b(p)(2k-p)_a(2q-p)_b] \frac{1}{(q-k)^2 k^2 (k-p)^2 q^2 (q-p)^2} \tag{5.17}$$

As expected on the basis of our general discussion all the contributions are expressed in terms of the field strengths and space-time connections; terms containing the spinor connections or the prepotential never get generated. Furthermore the covariant D-algebra manipulation was straightforward. Using ordinary Feynman superfield rules, the same calculation would have required the evaluation of the diagrams shown in Fig. 9. All these graphs, with the only exception of the first one, involve a great amount of D-algebra in the loops, many contributions are generated and only at the end the final answer is simple as a result of many cancellations.

Covariant D-algebra and power counting lead to remarkable simpifications in the computation of divergences in SSYM. At any loop order, after removal of subdivergences, the overall UV divergence must be local and by dimensional analysis it can only have the form

$$X \int d^4x \, d^4\theta \, \Gamma^a\Gamma_b(\delta_a{}^b - \hat{\delta}_a{}^b)$$
$$= -\varepsilon X \int d^4x \, d^2\theta \, W^\alpha W_\alpha \tag{5.18}$$

Here $\delta_a{}^b$, $\hat{\delta}_a{}^b$ are 4- and n<4 dimensional Kronecker deltas, X is a numerical (divergent) factor and the above expression is the only gauge invariant structure in n<4 dimensions that has the correct dimension. Whereas, in dimensional reduction $\hat{\delta}_a{}^b$ is generated by momentum integration the $\delta_a{}^b$ term only arises from the $\Gamma^a\Gamma_a$ factor in the expansion of the covariant propagator $\Box^{-1}$ or from a $\Gamma^a\Gamma_a$ term when two contracted vector covariant derivatives act on different lines of a graph. Therefore the complete contribution in (5.18) can be obtained by computing these graphs with an external $\Gamma^a\Gamma_a = \Gamma^a\Gamma_b\delta_a{}^b$ vertex and "covariantizing" $\delta_a{}^b \to \delta_a{}^b - \hat{\delta}_a{}^b$ (Grisaru, Milewski and Zanon, 1984).

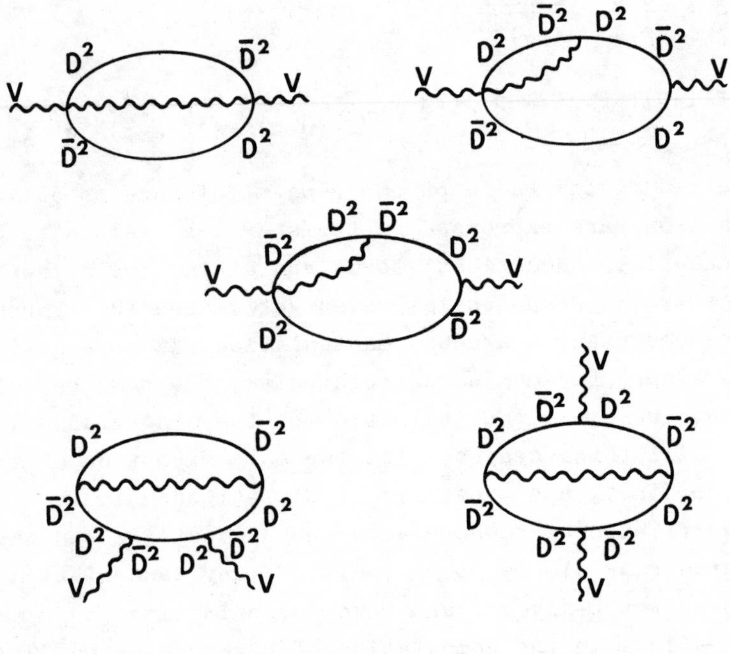

Fig. 9: Diagrams contributing to the two-loop V self-energy using ordinary supergraphs rules

6. SUPERGRAPHS IN LOCALLY SUPERSYMMETRIC THEORIES.

The supergraphs techniques described in the previous Sections can be extended to the case of locally supersymmetric theories (supergravity). The general procedure is no different from that of any gauge theory. The classical action has to be expressed in terms of unconstrained superfields, which are the suitable variables for quantization. The functional integral approach allows to define an effective action in superspace and to derive Feynman rules for perturbative calculations. The background field method can be introduced to work in a manifestly gauge-invariant background field formalism and to derive general results. Finally the use of covariant D-algebra in conjunction with the background field method leads to significant simplifications in the higher-loop calculations of supergraphs. The technical details, rather complicated, for the quantization and the derivation of the supergraph rules are described, for example, in "Superspace", while the methods for covariant supergraphs are given in a recent paper (Grisaru and Zanon, 1984c). In this Section we present a general outline.

In the covariant approach to superspace supergravity one defines the curved superspace covariant derivatives as

$$\nabla_A = E_A{}^M D_M + \phi_A \tag{6.1}$$

The indices A, M denote tangent and world coordinates respectively, $E_A{}^M$ is the vielbein and ϕ_A (=$\phi_A{}^\beta{}_\gamma M^\gamma{}_\beta + \phi_A{}^{\dot\beta}{}_{\dot\gamma} M^{\dot\gamma}{}_{\dot\beta}$ with M's Lorentz-rotation operators) is a Lorentz connection operator. All are functions of x, θ, $\bar\theta$. The torsion and curvature are defined by the graded commutators

$$[\nabla_A, \nabla_B\} = T_{AB}{}^C \nabla_C + R_{AB} \tag{6.2}$$

The representation content with respect to supersymmetry is reduced by requiring that certain components of the torsion and curvature vanish. As a consequence of these constraints one obtains an irreducible representation of supersymmetry that describes the supergravity multiplet, i.e. the gravitino field $\psi_{a,\beta}$, the gravitational field with field strength given by the Riemann tensor, and a set of auxiliary fields, S, P, A_a

in the minimal formulation. The constraints, together with the Bianchi identities also imply that all the nonvanishing torsions and curvatures are expressible in terms of three superfield field strengths R, $G_{\alpha\dot{\alpha}}$, $W_{\alpha\beta\gamma}$ that carry all the physical degree of freedom of the theory.

The superspace action is given by

$$S = \int d^4x\, d^4\theta\, E^{-1} \tag{6.3}$$

where the vielbein superdeterminant is defined by

$$E = \mathrm{sdet}\, E_A{}^M \tag{6.4}$$

Coupling to matter is obtained by direct covariantization. For example the scalar multiplet action becomes

$$\int d^4x\, d^4\theta\, E^{-1}\, \bar{\Phi}\Phi = \int d^4x\, d^4\theta\, E^{-1}\, \Phi_o e^{-iH\cdot\partial}\, \bar{\Phi}_o \tag{6.5}$$

where Φ is covariantly chiral, $\bar{\nabla}_{\dot{\alpha}}\Phi=0$.

The action in (6.3) is not in a form which is suitable for quantization: The constraints imposed on the covariant derivatives imply corresponding constraints on the vielbein $E_A{}^M$ and therefore these are not convenient objects to use as quantum fields in the functional integral. One has to solve the constraints in terms of prepotentials that are unconstrained superfields. The most convenient description makes use of the unconstrained field formalism of Siegel and Gates (1979). The prepotentials are an axial vector superfield H_a and a chiral compensator superfield χ; the constraints on the covariant derivatives are satisfied by expressing $E_A{}^M$ and ϕ_A in terms of them. The expressions for the vielbein allow one to compute the superdeterminant in (6.4) as a power series in H_a and χ and finally to expand the supergravity action in (6.3) to any order in the unconstrained superfields. One finds

$$S = \int d^4x\, d^4\theta\, \{ -3\bar{\chi}\chi - i(\bar{\chi}-\chi)\partial^a H_a - \tfrac{1}{2}H^a \Box H_a + D^2 H^a \bar{D}^2 H_a$$
$$- \tfrac{1}{4}(\partial^a H_a)^2 + \tfrac{1}{12}([\bar{D}^{\dot{\alpha}}, D^\alpha] H_{\alpha\dot{\alpha}})^2 + \text{interaction terms} \} \tag{6.6}$$

The classical action is invariant under the following gauge transformations (at the linearized level)

$$\delta H_{\alpha\dot{\alpha}} = D_\alpha \bar{L}_{\dot{\alpha}} - \bar{D}_{\dot{\alpha}} L_\alpha$$
$$\delta\chi = -\bar{D}^2 L^\alpha D_\alpha \chi - \frac{1}{3}\bar{D}^2 D^\alpha L_\alpha (1+\chi) \tag{6.7}$$

where the gauge parameter L_α is a general spinor superfield. One can choose a gauge-fixing term such that the kinetic action in (6.6) simplifies to give standard propagators for the H_a and χ fields. This can be done, but it is more convenient to discuss the whole quantization in a more general case when background fields are present. The ordinary quantization can be obtained as a special case by setting the background fields to zero.

The quantum-background splitting in supergravity is achieved in a way very similar to that of Yang-Mills. One starts with the expression of the constrained covariant derivatives in terms of ordinary flat-space derivatives and unconstrained prepotentials. The splitting is then obtained by replacing the flat derivatives by background covariant derivatives and reinterpreting the H_a and χ fields as quantum fields. The action is now expressible in terms of these quantum fields and the background which always appears only in gauge invariant forms, i.e. in the field strengths R, $G_{\alpha\dot{\alpha}}$, $W_{\alpha\beta\gamma}$ and implicitly in the background covariant derivatives. If, for simplicity one sets the background field on-shell ($R=G_{\alpha\dot{\alpha}}=0$) the quadratic part of the action becomes

$$S = \int d^4x\, d^4\theta\, E^{-1} \{ -3\bar{\chi}\chi + i(\chi-\bar{\chi})\nabla^a H_a - \frac{1}{2} H^a \Box H_a$$
$$- \frac{1}{4}(\nabla^a H_a)^2 + \bar{\nabla}^2 H^a \nabla^2 H_a + \frac{1}{12}([\bar{\nabla}^{\dot{\alpha}}, \nabla^\alpha] H_{\alpha\dot{\alpha}})^2$$
$$- \frac{1}{2} H^{\alpha\dot{\beta}} (W_\alpha{}^{\gamma\delta} \nabla_\gamma H_{\delta\dot{\beta}} + \bar{W}_{\dot{\beta}}{}^{\dot{\gamma}\dot{\delta}} \bar{\nabla}_{\dot{\gamma}} H_{\alpha\dot{\delta}}) \} \tag{6.8}$$

where $\Box = \frac{1}{2}\nabla^a\nabla_a$, all the derivatives are background covariant and the field strengths $W_{\alpha\beta\gamma}$ and the vielbein determinant are functions of the background fields. Except for the last term this is just a covariantization of (6.6); it is invariant under quantum gauge transformations given by the corresponding covariantized expressions in (6.7).

In principle the quantization procedure is now straightforward, but one has to face a certain number of new features and complications, for example the appearance of a large number of ghosts besides the usual Faddeev-Popov ghosts. Moreover, after gauge-fixing the kinetic terms for the various fields are not in a form which leads to simple propagators, but rather to propagators containing spinorial derivatives in the numerator, and momentum dependence p^{-4}. We also encounter cross-terms between some of the fields. Most of the subsequent work consists in appropriately choosing gauge-fixing terms and making various shifts and field redefinitions so that the final action has a form leading to simple Feynman rules. Without entering the details of the actual computation we briefly outline the various steps: The gauge-fixing procedure introduces Faddeev-Popov ghosts that are spinor superfields. The ghost action has itself a gauge invariance that requires further gauge-fixing and second generation ghosts. The gauge-fixing terms are chosen in a way to achieve the following: a) Eliminate the cross-terms between H and χ in the original action, b) Eliminate all H-H kinetic terms except $H\Box H$. To normalize the gauge-averaging one has to introduce the corresponding Nielsen-Kallosh ghost which couples only to the background and therefore contributes to the effective action only at the one-loop level. One unusual feature is the appearance of a "hidden ghost" due to the occurence of a gauge-fixing function which is a constrained superfield. In the usual gauge-averaging procedure that puts the gauge-fixing function in the exponential of the functional integral it is implicitly assumed that the averaging parameter that one integrates over is unconstrained. When this is not the case the naive 't Hooft averaging may lead to erroneous results. The proper procedure introduces a correction term that can be viewed as a contribution from an additional ghost field. In supergravity the hidden ghost is a chiral spinor σ_α. It has no interaction with the quantum fields and one can show that it actually decouples completely from the background so that it doesn't contribute at all to the effective action.

After quantization the quadratic part of the action has the form

$$S = \int d^4x\, d^4\theta\, E^{-1} [- \tfrac{1}{2}H^a \hat{\Box} H_a - \tfrac{9}{5}\bar{\chi}\chi - \bar{\psi}'^{\dot{\alpha}} i\nabla_{\alpha\dot{\alpha}}\psi^\alpha$$
$$- \psi'^\alpha i\nabla_{\alpha\dot{\alpha}}\bar{\psi}^{\dot{\alpha}} - \bar{\psi}_3^{\dot{\alpha}} i\nabla_{\alpha\dot{\alpha}}\psi_3^\alpha + \sum_1^2{}_i V_i'\Box V_i + \sum_1^4{}_i V_i''\Box V_i''$$
$$+ \sum_1^2{}_i (\tfrac{1}{2}\phi_i^\alpha \nabla^2 \phi_{i\alpha} + \text{h.c.})] \tag{6.9}$$

where

$$\hat{\Box} = \Box + W^\alpha{}_\beta{}^\gamma \nabla_\alpha M_\gamma{}^\beta + \bar{W}^{\dot{\alpha}}{}_{\dot{\beta}}{}^{\dot{\gamma}} \bar{\nabla}_{\dot{\alpha}} M_{\dot{\gamma}}{}^{\dot{\beta}} \tag{6.10}$$

and χ is background covariantly chiral. The fields ψ'_α, ψ_α denote the Faddeev-Popov ghosts, ψ_3^α is the Nielsen-Kallosh ghost and V_i, V_i' are additional real scalar first generation ghosts. They all have abnormal statistics. The real scalars V_i'' and the background covariant chiral spinors $\phi_{i\alpha}$ are second generation ghosts with normal statistics.

Coupling to matter is obtained by expanding the expression in (6.5) in powers of H_a and χ. Up to cubic interactions one obtains

$$S = \int d^4x\, d^4\theta\, \{ \bar{\Phi}\Phi + (\chi+\bar{\chi})\bar{\Phi}\Phi + H^a(\tfrac{1}{2}\bar{\Phi} i\overleftrightarrow{\nabla}_a \Phi - \tfrac{1}{6}[\bar{\nabla}_{\dot{\alpha}}, \nabla_\alpha]\bar{\Phi}\Phi \} \tag{6.11}$$

with Φ background covariantly chiral.

The ordinary quantization result is obtained directly from (6.9) and (6.11) setting the background fields to zero. In this way we set E=1, all the derivatives become flat and chiral superfields are ordinary chiral. The propagators for the various fields are already in standard form. We have

$$\langle H^a H_b \rangle = - \delta^\alpha{}_\beta\, \delta^{\dot{\alpha}}{}_{\dot{\beta}}\, p^{-2}\, \delta^4(\theta-\theta')$$
$$\langle \psi_\alpha \bar{\psi}^{\dot{\alpha}} \rangle = p_\alpha{}^{\dot{\alpha}}\, p^{-2}\, \delta^4(\theta-\theta') \tag{6.12}$$

and the usual propagators for χ, V_i and $\phi_{\alpha i}$. The vertices can be obtained to any order from the expansion of the action; their computation, in principle straightforward, becomes algebraically complicated for higher orders. The cubic supergravity vertices are given in "Superspace", Section 7.5a, while up to the same order, couplings to matter can be read off from (6.11). These expressions allow us to perform some one-loop calculations in supergravity using supergraphs and standard D-algebra.

A simple example is the evaluation of the one-loop supergravity corrections to the chiral self-energy. The relevant vertices are given in (6.11) with covariant derivatives replaced by flat-space derivatives. We have contributions from the diagrams in Fig. 10.

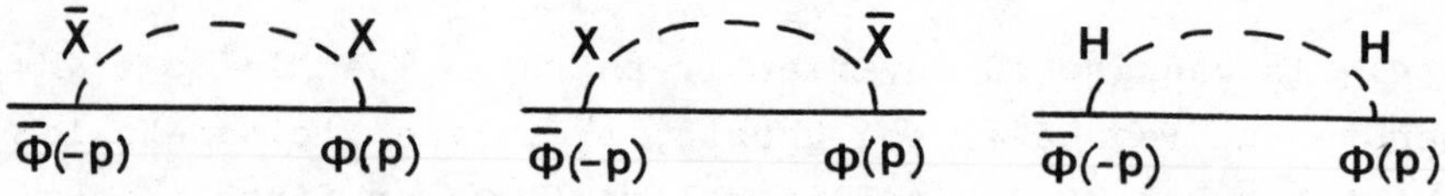

Figure 10: One-loop supergravity corrections to the self-energy of a massless chiral superfield

The first graph does not contribute since after D-algebra it reduces to a vanishing tadpole. From the second diagram we obtain a contribution

$$-\frac{5}{9}\int \frac{d^4p}{(2\pi)^4}\,\frac{d^nk}{(2\pi)^n}\; d^4\theta\; \bar{\Phi}(-p,\theta)\bar{D}^2D^2\Phi(p,\theta)\;\frac{1}{k^2\;(p+k)^2} \tag{6.13}$$

The factor in front of the integral comes from the normalization of the $\bar{\chi}\chi$ propagator in (6.9). Finally the last diagram gives

$$\frac{1}{9}\int \frac{d^4p}{(2\pi)^4}\,\frac{d^nk}{(2\pi)^n}\; d^4\theta\; \bar{\Phi}(-p,\theta)\Phi(p,\theta)\;(-4p^2+p\cdot k)\;\frac{1}{k^2(p+k)^2} \tag{6.14}$$

Performing the k-integral in (6.14) and adding the contribution of (6.13) (note that $\bar{D}^2D^2\Phi(p)=-p^2\Phi(p)$) one finds that for a massless chiral field the supergravity self-energy correction is identically zero.

Other calculations become algebraically complicated,

however they are in general much simpler than the corresponding component calculations.

Since the supergraph rules are the same as for global supersymmetric theories, the no-renormalization theorems still hold in the supergravity case. Since our vertices always carry a $d^4\theta$ integration chiral masses or chiral self-interaction terms do not receive any finite or infinite quantum corrections.

Significant simplifications are achieved by using the background field formalism: Since in the background field method the effective action is gauge invariant, one can derive general results on the degree of divergence of the effective action from the requirement that it be a function of the field strengths and covariant derivatives. A detailed discussion of general properties of the effective action and other results that can be deduced in the background field formalism can be found in "Superspace", Section 7.7.

The actual perturbative calculations are also simplified since the supergraphs one has to compute are fewer in number and less divergent. This last point can be easily understood as follows: In the presence of external supergravity fields ordinary supergraphs lead to contributions to the effective action that are functions of the supergravity prepotentials. Since H_a has dimension -1, individual graph contributions can be highly divergent. In the background field method instead, one derives the supergraph rules expanding the background covariant derivatives in (6.9) and (6.11) in terms of ordinary space-time and spinor derivatives $\partial_{\alpha\dot{\alpha}}, D_\alpha$, and background vielbein $E_A{}^M - \delta_A{}^M$ and connections ϕ_A. From the expansion of the kinetic operators in terms of flat contributions and interactions terms with the background one obtains the usual propagators as in (6.12). At this stage the background fields enter only through quantities that were contained in the covariant derivatives, namely vielbeins and connections, and one never needs to express them in terms of the supergravity prepotentials. Since the lowest dimensional object is the spinor vielbein $E_\alpha{}^m$ of dimension -1/2, supergraphs computed in the background field method will be more convergent as compared to ordinary ones.

The covariant formulation of supergraphs that we introduced in the previous Section for higher-loop calculations in supersymmetric Yang-Mills theories can be extended to the case of supergravity. Using covariant D-algebra a further reduction in the degree of divergence is achieved: the supergraphs give contributions that depend on $E_a{}^m$ and $E_\alpha{}^\mu$ of dimension $\geqslant 0$, but not on $E_\alpha{}^m$ of dimension -1/2. This means that some cancellations have automatically taken place and covariant supergraphs represent sums of ordinary background field supergraphs. The procedure is very similar to that for the Yang-Mills case: One starts with covariant Feynman rules for supergravity or matter systems in the presence of background supergravity as in (6.9) and (6.11). One has covariant propagators such as

$$\langle \bar{\Phi}\Phi \rangle = G_+ \qquad \langle \Phi\bar{\Phi} \rangle = G_- \qquad \langle H^a H_b \rangle = \hat{G}\,\delta^a{}_b \tag{6.13}$$

where for massless fields

$$\begin{aligned} \Box_\pm G_\pm(z,z') &= E\,\delta^8(z-z') \\ \hat{\Box}\,\hat{G}(z,z') &= E\,\delta^8(z-z') \end{aligned} \tag{6.14}$$

The d'Alembertian operator $\hat{\Box}$ is given in (6.10) and $\Box_\pm$ are defined by

$$\Box_+\,\Phi = \bar{\nabla}^2\nabla^2\Phi \qquad \Box_-\,\bar{\Phi} = \nabla^2\bar{\nabla}^2\bar{\Phi} \tag{6.15}$$

The Feynman rules for the effective action are: At one-loop one has the product of the determinants of the kinetic operators. At higher loops one draws vacuum diagrams with vertices from the interaction lagrangians and covariant propagators $G_\pm$ and $\hat{G}$, with additional $\nabla^2, \bar{\nabla}^2$ factors for each chiral or antichiral line leaving a vertex, etc. These diagrams lead to expressions involving products of propagators, covariant spinor and space-time derivatives acting on them from the vertices, and $E^{-1}d^8z$ from the integrals for each vertex. The main idea is to perform at this stage the covariant D-algebra without expressing the spinor covariant derivatives in terms of flat spinor derivatives. A new feature with respect to the Yang-Mills case is due to the fact that spinor derivatives also appear in ∇_a,

$$\nabla_a = E_a{}^m\partial_m + E_a{}^\mu D_\mu + E_a{}^{\dot\mu}\bar D_{\dot\mu} + \phi_a \qquad (6.16)$$

It is possible however to obtain an expression for ∇_a in terms of ∂_m and covariant spinor derivatives $\nabla_\alpha, \bar\nabla_{\dot\alpha}$. From

$$\partial_m = E_m{}^A(\nabla_A - \phi_A) = E_m{}^a(\nabla_a - \phi_a) + E_m{}^\alpha(\nabla_\alpha - \phi_\alpha) + E_m{}^{\dot\alpha}(\bar\nabla_{\dot\alpha} - \phi_{\dot\alpha}) \qquad (6.17)$$

we can write

$$\nabla_a = e_a{}^m\partial_m + \phi_a + e_a{}^m E_m{}^\gamma\phi_\gamma + e_a{}^m E_m{}^{\dot\gamma}\phi_{\dot\gamma} - e_a{}^m E_m{}^\gamma\nabla_\gamma - e_a{}^m E_m{}^{\dot\gamma}\bar\nabla_{\dot\gamma} \qquad (6.18)$$

where $e_a{}^m \neq E_a{}^m$ is the inverse of $E_m{}^a$. It is clear that using the expression in (6.18) for ∇_a the propagators $\hat G$ and $G_\pm$ can be written in the form

$$G = [\, \tilde\Box - \Delta \,]^{-1} \qquad (6.19)$$

where

$$\Delta = A^\alpha\,\nabla_\alpha + \bar A^{\dot\alpha}\,\bar\nabla_{\dot\alpha} + B\,\nabla^2 + \bar B\,\bar\nabla^2 + C^{\alpha\dot\alpha}\,[\nabla_\alpha, \bar\nabla_{\dot\alpha}] \qquad (6.20)$$

and $\tilde\Box$ does not contain any spinor derivatives. For every progagator in the diagram we can use the identity

$$G = \tilde G + \tilde G\Delta G \quad , \quad \tilde G = \tilde\Box^{-1} \qquad (6.21)$$

As in the Yang-Mills case, the perturbative expansion of G with respect to terms involving spinor derivatives only is the starting point for a covariant D-algebra calculation. Details and two-loop examples can be found in a paper by Grisaru and Zanon (1984c).

REFERENCES

Abbott, L. F. 1981, Nucl. Phys. B185 189-203

Abbott, L. F. and Grisaru, M. T. 1980, Nucl. Phys. B169 415-429

Abbott, L. F., Grisaru, M. T. and Zanon, D. 1984, Nucl. Phys. B244 454-468

Caswell, W. E. and Zanon, D. 1981, Nucl. Phys. B182 125-143

Ferrara, S. and Piguet, O. 1975, Nucl. Phys B93 261-302

Galperin, A., Ivanov, E., Kalitzin, S., Ogievetsky, S. and Sokatchev, E. 1984, "Unconstrained N=2 matter, Yang-Mills and supergravity theories in harmonic superspace," Trieste preprint IC/84/43 and "Unconstrained off-shell N=3 supersymmetric Yang-Mills theory," Dubna preprint E2-84-441
Gates, S. J., Grisaru, M. T., Rocek, M. and Siegel, W. 1983, "Superspace," Benjamin-Cummings, Reading, MA
Grisaru, M. T., Milewski, G. and Zanon, D. 1984, "The structure of ultraviolet divergences in Supersymmetric Yang-Mills," Utrecht preprint
Grisaru, M. T., Rocek, M. and Siegel, W. 1979, Nucl. Phys. B159 429-450
Grisaru, M. T., Rocek, M. and Siegel, W. 1981, Nucl. Phys. B183 141-156
Grisaru, M. T. and Siegel, W. 1982, Phys. Lett. 110B 49-53
Grisaru, M. T. and Zanon, D. 1984a, Phys. Lett. 142B 359-364.
Grisaru, M. T. and Zanon, D. 1984b, "Covariant Supergraphs: I - Yang Mills," Brandeis preprint, BRX-TH-163
Grisaru, M. T. and Zanon, D. 1984c, "Covariant Supergraphs: II - Supergravity," Brandeis preprint, BRX-TH-169
Honerkamp, J., Schlindwein, M., Krause, F. and Scheunert, M. 1975, Nucl. Phys. B95 397-426
Piguet, O., and Sibold, K. 1984, "The off-shell infrared problem in N=1 supersymmetric Yang-Mills theories," Université de Genève preprint, UGVA-DPT 04-423
Salam, A. and Strathdee, J. 1975, Phys. Rev. D11 1521-1535
Siegel, W. 1979, Phys. Lett 84B 193-195
Siegel, W. 1980, Phys. Lett. 94B 37-40
Siegel, W. and Gates, S. J. 1979, Nucl. Phys. B147 77-104

THEORIES OF EXTENDED RIGID SUPERSYMMETRY AND THEIR FINITENESS PROPERTIES

P C WEST

1. INTRODUCTION

It was in 1930 that it was first noticed that the probability amplitudes in second quantized field theories were infinite. More than 20 years passed before it was realized how to systematically regulate and renormalize theories with coupling constants of positive dimensions in such a way that all physical amplitudes were finite. Field theories based on point particles with coupling constants having negative dimensions cannot, it seems, be rendered finite by this process. Indeed, the selection of only renormalizable theories to be candidates which describe physics played an important role in the unification of the electromagnetic and weak forces of nature. Furthermore, the most accurately measured number in agreement with prediction is the (g-2) of the electron, and to obtain this number many infinities are regulated and removed.

One must, however, be careful not to place too much emphasis on these successes of renormalization theory. The (g-2) of the electron is a number calculated in a series which is thought to be divergent and in a theory which is possibly inconsistent at higher energies. More generally, if we regulate using a momentum cut-off say, then taking the cut-off to infinity involves probing physics at higher and higher energies. In this sense renormalization theory involves physics at arbitrarily high energies. Of course, as we consider higher energies, we expect to have to take into account all the forces of nature including that of gravity.

Including gravity within the context of point quantum field theories has not succeeded, since this field involves a dimensional coupling with negative dimensions and this leads to theories in which the infinities cannot be removed by renormalization theory. It was hoped for some time that supergravity point quantum field theories would be renormalizable or finite; however, this is generally no longer believed to be true. This belief is founded in the fact that the arguments given to show the finiteness properties of the extended rigid theories do not extend to the supergravity theories precisely because of the presence of the dimensional coupling constant.

From an aesthetic point of view, the existence of infinities which must be regulated and renormalized is a rather unappealing procedure. It involves mutilating the theory by introducing a parameter and then demanding that this parameter is not observable. The operations required to show that this can be achieved in certain theories are rather complicated. These considerations have led a number of people, notably Dirac, to suppose that a fundamental theory of nature should be finite.

Of course, demanding the finiteness of a quantum field theory that contains gauge particles is not a physical criterion, since off-shell Green functions are gauge-dependent. The corresponding meaningful statement is that the quantum theory is conformally or superconformally invariant. Generally speaking, even though a theory is conformally invariant at the classical level, the need to regulate infinities when they occur introduces a dimensional parameter which explicitly breaks the conformal invariance. It is a remarkable fact that there exists a large class of finite i.e. (super) conformally invariant quantum field theories. These theories contain no spins greater than one and possess two supersymmetries (or in the case of one theory, four supersymmetries). It is the purpose of this contribution to establish the superconformal invariance of this special class of theories.

It has long been hoped that the inclusion of gravity would lead to a finite theory of nature. Superstring theories contain gravity as well as Yang-Mills fields, and it is thought likely that these theories may indeed be finite even though they possess coupling constants of negative dimension. The rules of the game are changed, corresponding to the extended nature of the string. These theories live in more than four dimensions, and convincing mechanisms for obtaining effective low-energy four-dimensional theories are unknown. One might speculate, however, that such a low-energy theory should be finite, i.e. a conformally invariant theory. It is certainly true that the world we observe is in effect massless when compared with the large scale of gravity, i.e. the Planck mass, and finite theories are very insensitive to their high-energy effects.

A more unpopular possibility is that gravity is not a fundamental force,

but arises from some kind of dynamical mechanism. In this event one may hope that the theory in which this arises is finite.
In Section 2 we find the theories of extended rigid supersymmetry in x-space and also give their form in superspace. In Section 3 the finiteness properties of these theories are derived using two different arguments. One is based on the superconformal anomaly structure in supersymmetric theories, whilst the other relies on the properties of quantization in superspace. Section 4 shows how to add soft terms that explicitly break supersymmetry but preserve finiteness.

2. THEORIES OF EXTENDED RIGID SUPERSYMMETRY

In this section we wish first to construct the theories of rigid supersymmetry in x-space, their superspace formulations being given later on. The possible on-shell states of supersymmetric massless theories with spins 1 and less are given by the irreducible representations of supersymmetry [1] which are a consequence of the theory of induced representations. These representations are listed in Table 1.

Table 1

Irreducible representations of supersymmetry for particles of spin less than 3/2

Spin \ N	1	1	2	2	4
Spin-1	-	1	-	1	1
Spin-1/2	1	1	2	2	4
Spin-0	2	-	4	2	6

N = number of supercharges.

As noted previously, theories with more than four supercharges must have spins greater than one. Theories with $N \leqslant 4$ are the most symmetric and consistent theories that physicists have constructed so far. In the context of the revised supersymmetric no-go theorem, they are the most symmetric with spins not exceeding 1. Whilst they are consistent in the sense that they are renormalizable and unitary, it may be that some of the supergravity theories that by definition contain a spin-2 state are renormalizable. This, at present, is no more than a hope, and is indeed rather unlikely. It is more possible that some superstring theories are renormalizable and unitary. The inclusion of a spin-3/2 field requires a spin-2 field in order to propagate causally.

The first theory listed under N = 2 has two spin-1/2 Majorana states which are singlets under SU(2), and four spin-0 states which are a complex doublet under SU(2). We call this multiplet the hypermultiplet or N = 2 matter. The second theory listed under N = 2 has one spin-1, two Majorana spin-1/2, and two spin-0 states. These particles are SU(2) singlets, doublets, and singlets, respectively. This multiplet is called the N = 2 Yang-Mills multiplet.

The unique N = 4 multiplet has one spin-1, four Majorana spin-1/2, and six spin-0 in the singlet, vector, and self-dual antisymmetric tensor representations of SU(4).

The above Yang-Mills multiplets must be put in the adjoint representation of the gauge group, whilst that of the N = 2 matter multiplet can belong to any real representation of the gauge group. An important property of these extended theories is that they are vector-like; that is, the fermions of both right- and left-handedness belong to the same representation of the gauge group. This is an inevitable consequence of the fact that the supercharges commute with the gauge group. The hypermultiplet, when considered in the 'rest frame' (m, 0, 0, m), has as its Clifford vacuum the helicity +1/2 state. This state is related to the helicity -1/2 state by the action of two supercharges. Consequently, both the +1/2 and -1/2 states must belong to the same representation of the gauge group. For the super Yang-Mills models all fields are in the same representation as the gauge fields, namely the adjoint representation. This means that if the models contain the particles of the standard model, then for every chiral fermion in the standard model they will contain a fermion of the opposite chirality but in the same representation (the so-called mirror particles). Of course, to find a realistic model one must find some way of splitting the observed fermions from their mirror particles in mass. The inclusion of mirror particles is quite a general feature of all extended theories and can only be avoided by using non-linear realizations -- which is in some sense cheating -- or by including higher spin fields which transform non-trivially under the gauge group.

Before constructing these models we will record the extended supersymmetry algebra that these theories realize:

$$\{Q^{Ai}, Q^{Bj}\} = 2i\varepsilon^{AB}(\Omega^{\ell})^{ij}Z_{e} \tag{2.1a}$$

$$\{Q^{\dot{A}}{}_{i}, Q^{\dot{B}}{}_{j}\} = 2i\varepsilon^{\dot{A}\dot{B}}(\Omega^{\ell})_{ij}Z_{\ell} \tag{2.1b}$$

$$\{Q^{Ai}, Q^{\dot{B}}{}_{j}\} = -2i(\sigma^{\mu})^{A\dot{B}}P_{\mu}\delta^{i}{}_{j} \tag{2.1c}$$

$$[Q^{Ai}, J_{\mu\nu}] = \frac{1}{2}(\sigma_{\mu\nu})^{A}{}_{B}Q^{Bi} , \tag{2.1d}$$

$$[Q^{Ai}, P_{\mu}] = 0 , \quad [Q^{Ai}, T^{r}] = (t^{r})^{i}{}_{j}Q^{Aj} , \tag{2.1e}$$

where $(\Omega^{\ell})^{ij} = -(\Omega^{\ell})^{ji}$ $(i, j = 1 \to N)$, $(\Omega^{\ell})_{ij} = [(\Omega^{\ell})^{ij}]^{*}$, and the central charges Z_{ℓ} are real. For N = 2 we assume that there is only one central charge Z and that $(\Omega)^{ij} = -i\varepsilon^{ij}$.

For the construction of N = 2 models we will be required to write down Majorana conditions on the supercharges, the supersymmetry parameter, etc. The usual Majorana property on a spinor is

$$\bar{\lambda}_{\alpha i} = \lambda^{\beta i}C_{\beta\alpha} . \tag{2.2}$$

The maximal internal symmetry transformation compatible with such a condition is U(N), which is realized as follows:

$$\delta\lambda^{\alpha i} = [(S_{1}')^{i}{}_{j}\delta^{\alpha}{}_{\beta} + (S'')^{i}{}_{j}(\gamma_{5})^{\alpha}{}_{\beta}]\,\lambda^{\beta j} , \tag{2.3}$$

where S' is real and antisymmetric (i.e. $\Lambda_2 i_2 \sigma_2$ for N = 2) whilst S" is complex and symmetric [i.e. $S'' = i\sigma_2(\Lambda_3\sigma_1, \Lambda_0 i\sigma_2, \Lambda_1\sigma_3)$ for N = 2].

For the N = 2 case there exists a numerically invariant second-rank tensor $\varepsilon_{ij} = -\varepsilon_{ji}$, and it is often preferable to use an alternative Majorana condition which is obtained from the one above by a field redefinition. (It should be remembered that any reality condition can be brought to the above form by a field redefinition.) Let us make the redefinition

$$\lambda'^{\alpha j} = e\, \varepsilon^{ji} \left(\frac{1+\gamma_5}{2}\right)^{\alpha}_{\ \beta} \lambda^{\beta i} + d \left(\frac{1-\gamma_5}{2}\right)^{\alpha}_{\ \beta} \lambda^{\beta j} , \tag{2.4}$$

where

$$\varepsilon^{12} = -\varepsilon^{21} = +1 . \tag{2.5}$$

Choosing $+i\, e^* = d$, we find that the Majorana condition becomes

$$\lambda'^{\alpha i} = (i\gamma_5 C)^{\alpha\beta} \varepsilon^{ij} \bar{\lambda}'_{\beta j} , \tag{2.6}$$

where we have adopted the convention that complex conjugation lowers an i or j index.

The symmetry transformation for λ' is deduced from its definition and the transformation of $\lambda^{\alpha i}$, and is given by

$$\delta\lambda'^{\alpha i} = [(c_1)^i{}_j \delta^\alpha{}_\beta + (\gamma_5)^\alpha{}_\beta (c_2)^i{}_j]\, \lambda'^{\beta j} , \tag{2.7}$$

where

$$c_2 = 1\Lambda_0 , \quad c_1 = i(\Lambda_1\sigma_1, \Lambda_2\sigma_2, \Lambda_3\sigma_3) . \tag{2.8}$$

The advantage of this type of Majorana constraint becomes clear. The SU(2) (generated by c_1) is realized without a γ_5, and since ε_{ij} is a numerically invariant tensor of SU(2) it is obviously allowed by the new Majorana constraint. The U(1), however, is realized with a γ_5 transformation. The situation with the original Majorana condition is the opposite; so, although both constraints allow a U(2) symmetry, less of this symmetry is realized in an obvious way with the original Majorana condition. Clearly, in two-component formalism these distinctions are irrelevant. The reader may verify the consistency of the new Majorana condition under complex conjugation.

In the discussion of N = 2 theories which follows, we will adopt the Majorana constraint

$$\varepsilon_{\alpha i} = (i\gamma_5 C)_{\alpha\beta} \varepsilon_{ij} \bar{\varepsilon}^{\beta j} . \tag{2.9}$$

Clearly, having a parameter $\varepsilon_{\alpha i}$ without a Majorana constraint would be equivalent to having four supersymmetries rather than two.

We now construct the rigid theories of N = 2 supersymmetry in turn, starting with N = 2 Yang-Mills theory. We will begin with the on-shell states, find an on-shell supersymmetry linking them, and construct an invariant 'on-shell action'. We then find an off-shell description, i.e. the auxiliary fields and a corresponding invariant action.

2.1 N = 2 Yang-Mills [2]

1) The on-shell states, which belong to representations of (rigid) SU(2), may be represented by the fields A, B, A_μ, and $\lambda_{\alpha i}$. The fields A, B, and A_μ are real and singlets under SU(2), whereas the spin-1/2 is a doublet. To remain with only one doublet of spin-1/2, as required by the on-shell states, we must impose a Majorana condition. This condition is of the form

$$\lambda_{\alpha i} = (i\gamma_5 C)_{\alpha\beta}\varepsilon_{ij}\bar{\lambda}^{\beta j} . \tag{2.10}$$

The symbol $\varepsilon_{ij} = -\varepsilon_{ji} = \varepsilon^{ij}$ can be used to raise and lower SU(2) indices as follows:

$$\lambda_\alpha{}^i = \varepsilon^{ij}\lambda_{\alpha j} \quad \text{and} \quad \lambda_\alpha{}^i\varepsilon_{ij} = \lambda_{\alpha j} . \tag{2.11}$$

We leave it as an exercise for the reader to deduce the U(1) weights of A, B, $\lambda_{\alpha i}$, and A_μ. Since these fields represent the on-shell states, they obey (neglecting interaction terms) their equations of motion,

$$\partial^2 A = \partial^2 B = (\not\partial\lambda_i)_\alpha = \partial^\nu f_{\mu\nu} = 0 , \tag{2.12}$$

where

$$f_{\mu\nu} = \partial_\mu A_\nu - \partial_\nu A_\mu . \tag{2.13}$$

We first examine the linearized, or free, theory as this is a consistent theory which is much simpler in structure than the interacting theory.

2) We now wish to find the supersymmetry transformations between the on-shell states. On grounds of dimension and linearity, these must be of

the form

$$\delta A = i\bar{\varepsilon}^i \lambda_i \ , \quad \delta B = \bar{\varepsilon}^i \gamma_5 \lambda_i \ , \tag{2.14a}$$

$$\delta A_\mu = +\bar{\varepsilon}^i \gamma_\mu \lambda_i \ , \tag{2.14b}$$

$$\delta \lambda_i = -\frac{1}{2} c_1 f^{\mu\nu} \sigma_{\mu\nu} \varepsilon_i - i c_2 \not{\partial}(A - i\gamma_5 B)\varepsilon_i + c_3 \partial_\mu A^\mu \varepsilon_i \ , \tag{2.14c}$$

where c_1, c_2, and c_3 are constants. The linearized theory should be invariant under local Abelian transformations,

$$\delta A^s_\mu = \partial_\mu \Lambda^s \ , \quad \delta A^s = \delta B^s = \delta \lambda^s_{\alpha i} = 0 \ , \tag{2.15}$$

and rigid G_1 transformations with parameter T^r (i.e. T^r are independent of x^μ),

$$\delta A^r = f^{rst} T^s A^t \ , \quad \delta A_\mu{}^r = f^{rst} T^s A_\mu{}^t \ , \quad \text{etc.} \ , \tag{2.16}$$

where f^{rst} are the structure constants of the group G_1 which will become the gauge group of the interacting theory. All fields are in the adjoint representation of G_1, as they must transform the same way as A_μ. This is a consequence of the fact that supersymmetry and the gauge group G_1 commute. In the above supersymmetry transformations and in what follows, the group indices have not been written but are understood to be present.

The supersymmetry transformations, however, must form a closed algebra. In particular, carrying out the commutator of supersymmetry and local transformations on $\lambda_{\alpha i}$ yields

$$[\delta_\Lambda, \delta_\varepsilon]\lambda_{\alpha i} = c_3 (\partial_\mu \partial^\mu \Lambda) \varepsilon_{\alpha i} \ . \tag{2.17}$$

This is not a recognizable symmetry of the theory; thus it implies that $c_3 = 0$. The commutator of two supersymmetries on A yields

$$[\delta_1, \delta_2]A = 2c_2 \bar{\varepsilon}_2{}^i \gamma^\mu \varepsilon_{1i} \partial_\mu A \ , \tag{2.18}$$

and so to agree with the supersymmetry algebra we choose $c_2 = +1$. The terms containing B and $F_{\mu\nu}$ vanish because of subtracting the ($1\leftrightarrow 2$) exchange and the Majorana properties of $\epsilon_{\alpha i}$. On A_μ we find that the commutator of two supersymmetries yields a transformation of the correct magnitude provided $c_1 = +1$, but we also find a gauge transformation. This phenomenon also occurs in the N = 1 Yang-Mills theory when working with only the fields (A_μ, λ_μ, D). It results from working in the Wess-Zumino gauge [3] and is a consequence of the compensating transformations required to maintain this gauge choice under supersymmetry. The situation in N = 2 Yang-Mills is similar. We find that the algebra closes on $\lambda_{\alpha i}$ provided we use the field equations. We thus refer to the above supersymmetry transformations as an 'on-shell algebra'.

3) We must build an invariant action which can only be of the form

$$A = \int d^4x \left[-\frac{1}{4} f_{\mu\nu}^2 - \frac{1}{2} (\partial_\mu A)^2 - \frac{1}{2} (\partial_\mu B)^2 - \frac{1}{2} \bar{\lambda}^i \not\partial \lambda_i \right] . \qquad (2.19)$$

This action is invariant without the use of field equations, and we refer to such actions as 'on-shell actions'.

At this point it is appropriate to comment on the drawbacks of formulating theories using 'on-shell actions'. Although one can calculate with such actions, it is rather difficult to use them to establish general results. For example, one may wish to find all possible interacting theories. However, adding interaction terms modifies the field equations and so requires new on-shell supersymmetry transformations. In effect, one must construct each new interaction term from scratch. This is not such a serious problem for the theories of extended rigid supersymmetry, as the renormalizable couplings are few in number. However, it is a very serious difficulty when trying to find the most general coupling of matter to supergravity.

Another difficulty concerns the quantization of supersymmetric theories. It is much simpler to handle the symmetries of a theory that is being quantized if the symmetries can be manifestly realized. Clearly this is not the case when working with the 'on-shell actions'.

4) In view of the above discussion we now wish to find a supermultiplet that closes without the use of the classical field equations. That is, we must find the auxiliary fields. A useful guide to the number of auxiliary fields is provided by demanding that their addition gives an off-shell multiplet with equal numbers of bosonic and fermionic degrees of freedom.

The spin-1/2, $\lambda_{\alpha i}$, has 8 degrees of freedom, whilst A, B, and A_μ provide only 1 + 1 + 3 = 5, respectively. The vector A_μ provides only 3 degrees of freedom as the Fermi-Bose rule is proved using the algebra relation $\{Q,Q\} \sim P_\mu$. As noted previously, this only holds for gauge-invariant quantities, and hence we must in effect subtract the gauge degree of freedom of A_μ. As such, we require 3 auxiliary bosonic degrees of freedom. One possibility is that we have an SU(2) triplet of auxiliary fields $X^{ij} = X^{ji}$ which are of dimension two. Let us assume that this is the case. We must now construct the supersymmetry transformations among the fields

$$A,\ B,\ A_\mu,\ \lambda_{\alpha i},\ \text{and}\ X^{ij}\ , \tag{2.20}$$

where

$$(X^{ij})^* \equiv X_{ij} = X^{k\ell}\varepsilon_{ki}\varepsilon_{\ell j}\ . \tag{2.21}$$

Demanding linearity, matching dimensions, and using the fact that when $X^{ij} = 0$ the on-shell algebra reduces to that given earlier, we find that

$$\delta A,\ \delta B,\ \text{and}\ \delta A_\mu$$

must be the same as before, whilst

$$\delta\lambda_i = -\frac{1}{2} f^{\mu\nu}\sigma_{\mu\nu}\varepsilon_i - i\partial\!\!\!/(A - i\gamma_5 B)\varepsilon_i - i(\underline{\tau})_i{}^j\varepsilon_j\underline{X}\ , \tag{2.22a}$$

$$\delta\underline{X} = +\, i c_4 \bar{\varepsilon}^i(\underline{\tau})_i{}^j\partial\!\!\!/\lambda_j\ , \tag{2.22b}$$

where $X_i{}^j = (\underline{\tau})_i{}^j\underline{X}$, the $\tau_i{}^j$ are the Pauli matrices, and the constant c_4 is determined to be $c_4 = +1$ by demanding closure.
The corresponding invariant action is

$$A = \int d^4x \left[-\frac{1}{4} f_{\mu\nu}^2 - \frac{1}{2}(\partial_\mu A)^2 - \frac{1}{2}(\partial_\mu B)^2 - \frac{1}{2}\bar{\lambda}^i\partial\!\!\!/\lambda_i + \frac{1}{2}\underline{X}^2\right]\ . \tag{2.23}$$

Having constructed the linearized theory, we now wish to find the corresponding non-linear theory -- that is, the theory invariant under the now local gauge group G_1. Consequently, we must make the parameter T^r space-time dependent, $T^r(x)$. To recover gauge invariance we must introduce covariant derivatives and knit together the linearized local Abelian transformations with the now local G_1 transformations by identifying

$$\Lambda^r(x) = \frac{1}{g} T^r(x) \; . \tag{2.24}$$

The resulting gauge-invariant action is

$$A^1 = \mathrm{Tr} \int d^4x \left[-\frac{1}{4} F^2_{\mu\nu} - \frac{1}{2} \bar{\lambda}^i \not{D} \lambda_i - \frac{1}{2} (D_\mu A)^2 - \frac{1}{2} (D_\mu B)^2 + \frac{1}{2} \underline{X}^2 \right] \; . \tag{2.25}$$

where

$$F_{\mu\nu} = \partial_\mu A_\nu - \partial_\mu A_\nu - g[A_\mu, A_\mu] \; , \tag{2.26a}$$

$$D_\mu A = \partial_\mu A - g[A_\mu, A] \; , \quad \text{etc.} \tag{2.26b}$$

In the above, $A_\mu = A_\mu{}^s T_s$, where T_s are the generators of G in the adjoint representation.
The action A^1 is, however, no longer supersymmetric. We can recover supersymmetry by adding gauge-invariant terms to A^1 and to the supersymmetry transformation laws. Such additions to the action can only be terms that are quartic in the spin-0 fields A and B as well as Yukawa terms.
The final result for the transformation laws is

$$\delta A = i\bar{\varepsilon}^i \lambda_i \; , \quad \delta B = \bar{\varepsilon}^i \gamma_5 \lambda_i \; , \quad \delta A_\mu = \bar{\varepsilon}^i \gamma_\mu \lambda_i \; , \tag{2.27a}$$

$$\begin{aligned} \delta\lambda_i = & - \frac{1}{2} F^{\mu\nu} \sigma_{\mu\nu} \varepsilon_i - i\gamma^\mu D_\mu (A - i\gamma_5 B) \varepsilon_i \\ & + ig[A, B] \gamma_5 \varepsilon_i - i(\underline{I})_i{}^j \varepsilon_j \underline{X} \; , \end{aligned} \tag{2.27b}$$

$$\delta \underline{X} = i\bar{\varepsilon}^i (\underline{I})_i{}^j \{\gamma^\mu D_\mu \lambda_j + g[A - i\gamma_5 B, \lambda_j]\} \; ; \tag{2.27c}$$

and the supersymmetric action is

$$A^{N=2ym} = \mathrm{Tr}\int d^4x\left[-\frac{1}{4}F_{\mu\nu}^2 - \frac{1}{2}(D_\mu A)^2 - \frac{1}{2}(D_\mu B)^2 + \frac{1}{2}X^2\right.$$
$$\left. - \frac{1}{2}\bar{\lambda}^i\gamma^\mu D_\mu\lambda_i - \frac{ig}{2}\bar{\lambda}^i[A-i\gamma_5 B,\ \lambda_i] - \frac{g^2}{2}([A,\ B])^2\right] . \qquad (2.28)$$

The above description of N = 2 Yang-Mills theory does not involve a central charge in the sense that the fields carry an off-shell realization of the supersymmetry algebra of Section 2 in which the central charge is trivially realized. For an alternative description of N = 2 Yang-Mills theory where one of the spin-0 fields is represented by a conserved vector, the reader is referred to ref. [4].

2.2 N = 2 matter [5]

We now repeat in outline the above procedure for N = 2 matter.

1) The on-shell states are represented by the fields A^{ia} and ψ^a as well as their complex conjugates $(A^{ia})^* = A_{ia}$ and $\bar{\psi}_a$. The spin-0 fields are a complex doublet under SU(2), whilst ψ^a is a singlet. The fields A^{ia} and ψ^a can be in any representation R of a group G_2, whilst their complex conjugates are of course in the representation $\bar{R}$, the index a being the group index.

2) The transformation laws are determined by linearity, matching dimensions, and closure to be of the form

$$\delta A^{ia} = +\bar{\varepsilon}^i\psi^a , \qquad (2.29a)$$

$$\delta\psi^a = +2\gamma^\mu\partial_\mu A^{ia}\varepsilon_i . \qquad (2.29b)$$

We note that carrying out the commutator of two supersymmetries on A^{ia} yields the correct result, namely

$$[\delta_1,\ \delta_2]A^{ia} = +2\bar{\varepsilon}_2{}^i\gamma^\mu\partial_\mu A^{ja}\varepsilon_{ij} - (1\leftrightarrow 2)$$
$$= 2\bar{\varepsilon}_2{}^k\not{\partial}\varepsilon_{1k}A^{ia} . \qquad (2.30)$$

Here we have used the identity

$$\delta_i^k \delta_j^\ell = \frac{1}{2} \delta_i^\ell \delta_j^k + \frac{1}{2} (\underline{\tau})_i{}^\ell (\underline{\tau})_j{}^k \tag{2.31}$$

and the fact that $\varepsilon^{ki}(\underline{\tau})_i{}^\ell$ is symmetric in k and ℓ, whilst $\bar{\varepsilon}^k \gamma_\mu \varepsilon^\ell$ is anti-symmetric in k and ℓ owing to the $i\gamma_5$ matrix in the Majorana condition. The reader may verify that the transformation closes on ψ provided $\partial\!\!\!/ \psi = 0$.

3) The on-shell action is

$$A = \int d^4x \, (-|\partial_\mu A^{ia}|^2 - \frac{1}{2} \bar{\psi}_a \partial\!\!\!/ \psi^a) \ . \tag{2.32}$$

4) To find an off-shell formulation we require a contribution of four auxiliary bosonic degrees of freedom. There are, however, several ways of achieving this. We now review the different possible off-shell formulations.

a) We can add a complex doublet of dimension 2 auxiliary fields F^{ia}. The supersymmetry transformations [5] are

$$\delta A^{ia} = +\bar{\varepsilon}^i \psi^a \ , \tag{2.33a}$$

$$\delta\psi^a = +2\gamma^\mu \partial_\mu A^{ia} \varepsilon_i - 2\gamma_5 F^{ia} \varepsilon_i \ , \tag{2.33b}$$

$$\delta F^{ia} = +\bar{\varepsilon}^i \gamma_5 \gamma^\mu \partial_\mu \psi^a \ . \tag{2.33c}$$

In this case the algebra closes without using the equations of motion, but with an off-shell central charge, i.e.

$$\begin{aligned}[\delta_1, \delta_2] A^i &= \bar{\varepsilon}_2{}^i 2\partial\!\!\!/ A^j \varepsilon_{ij} + \bar{\varepsilon}_2{}^i 2i\gamma_5 F^j \varepsilon_{ij} - (1 \leftrightarrow 2) \\ &= 2\bar{\varepsilon}^i \partial\!\!\!/ \varepsilon_{ij} A^i + 2i\bar{\varepsilon}_2{}^j \gamma_5 \varepsilon_{ij} F^i \ .\end{aligned} \tag{2.34}$$

The transformation of the multiplet under the central charge being

$$\delta A^i = \omega F^i \ , \tag{2.35a}$$

$$\delta F^i = \partial^2 A^i \ , \tag{2.35b}$$

$$\delta\psi = i\omega\gamma_5 \not\partial\psi \tag{2.35c}$$

the reader may verify that the commutation of two supersymmetries and a central charge commute, for example

$$[\delta_3, \delta_\omega]A^i = \omega_i \bar{\varepsilon}^i \gamma_5 \not\partial\psi - \bar{\varepsilon}^i \omega_i \gamma_5 \not\partial\psi = 0 \ . \tag{2.36}$$

An off-shell central charge [4] is one that vanishes when the equations of motion are used. In this case the equations of motion are

$$\delta_\omega A^{ia} = \delta_\omega \psi^a = \delta_\omega F^{ia} = 0 \ . \tag{2.37}$$

The corresponding invariant action is

$$A = \int d^4x \, (-|\partial_\mu A^{ia}|^2 - \frac{1}{2} \bar{\psi}_a \gamma^\mu \partial_\mu \psi^a + |F^{ia}|^2) \ . \tag{2.38}$$

Clearly the N = 2 hypermultiplet is composed of two N = 1 Wess-Zumino multiplets. One could consider making the N = 2 matter from one Wess-Zumino multiplet ($A, B; \chi_\alpha; F, G$) and one linear multiplet (C, ζ_α, V_μ) with $\partial^\mu V_\mu = 0$ Although there is no guarantee that this can be made supersymmetric in fact it can and it is our next possible formulation.

b) In this formulation [6] the on-shell states belong to different representations of SU(2). If we start with a doublet, helicity-1/2 Clifford vacuum, |1/2,i⟩ the on-shell states are an SU(2) doublet of spin-1/2 Majorana fermions and the four spin-0 consisting of a triplet and a singlet. We now realize the spin-1/2 by the SU(2) Majorana field $\lambda_{\alpha i}$, and the triplet of spin-0 by $L^{ij} = L^{ji}$, but the singlet spin 0 by V_μ, which is subject to $\partial_\mu V^\mu = 0$. The off-shell theory now requires only two auxiliary fermions which we take to be the fields S and P. The resulting action is of the form

$$A = \int d^4x \left[-\frac{1}{2} (\partial_\mu L^{ij})^2 - \frac{1}{2} \bar{\lambda}^i \not\partial \lambda_i + \frac{1}{2} V_\mu^2 + \frac{1}{2} S^2 + \frac{1}{2} P^2 \right] \ . \tag{2.39}$$

The transformations may be deduced as before (for example, $\delta L^{ij} = \bar{\varepsilon}^i \lambda^j + \bar{\varepsilon}^j \lambda^i$) and close without an off-shell central charge.

c) There exists another off-shell formulation which is an extension of (b). This consists of relaxing the constraint $\partial^\mu V_\mu = 0$ and imposing it by means of a supersymmetric set of Lagrange-multipliers. This formulation is rather complicated [7], but it is important for superspace quantization.

d) There also exists an extension of the type (a) off-shell formulation given above that does not involve an off-shell central charge [8].

There do not appear to be any self-interactions for N = 2 matter [9]. [We note that in formulation (a) the Yukawa term would be $\bar{\psi}A_i\psi$, which is not SU(2) invariant.]

2.3 The general N = 2 rigid theory

We are now in a position to couple the N = 2 matter to N = 2 Yang-Mills. Normally, when one couples two multiplets together, one introduces additional coupling constants. However, in this case the coupling constants are the gauge coupling constants. This follows from the fact that the coupling of A_μ to the N = 2 matter fields is determined by the gauge coupling constant g. By N = 2 supersymmetry, however, the coupling of any component of the N = 2 Yang-Mills multiplet to the N = 2 hypermultiplet is determined by the gauge coupling constant. These theories have the important property of having only one coupling constant; they are truly grand unified theories. The coupling between the multiplets can easily be found by starting from the linearized theory, whose action is the sum of the linearized action of each theory. We now gauge covariantize in accordance with the fact that the N = 2 matter is in the group representation $R(\bar{R})$, and recover supersymmetry by adding terms to the action and transformation laws. The N = 2 super-Yang-Mills fields have the same transformations as before and were given in Eqs. (2.27). The N = 2 matter transformation law is

$$\delta A^{ia} = \bar{\varepsilon}^i\psi^a , \qquad (2.40a)$$

$$\delta\psi^a = +2\gamma^\mu D_\mu A^{ia}\varepsilon_i - 2\gamma_5 F^{ia}\varepsilon_i - 2ig(T_s)^a{}_b A^{ib}(A_s - i\gamma_5 B_s)\varepsilon_i , \qquad (2.40b)$$

$$\delta F^{ia} = \bar{\varepsilon}^i\gamma_5\gamma^\mu D_\mu\psi^a - 2g\bar{\varepsilon}^i\gamma_5\lambda_{js}A^{jb}(T_s)^a{}_b + g\bar{\varepsilon}^i(i\gamma_5 A_s - B_s)(T^s)^a{}_b\psi^b , \qquad (2.40c)$$

where

$$D_\mu A^{ia} = \partial_\mu A^{ia} - g(A_\mu)^a{}_b A^{ib} , \tag{2.41a}$$

$$D_\mu A^*_{ia} = \partial_\mu A^*_{ia} + g(A_\mu)^b{}_a A^*_{ib} , \tag{2.41b}$$

whilst the invariant action is

$$A = A^{ym} + A^{matter} + A^{interaction} + A^{mass} , \tag{2.42}$$

where

$$A^{ym} = \mathrm{Tr} \int d^4x \Big[-\frac{1}{4} F^2_{\mu\nu} - \frac{1}{2} (D_\mu A)^2 - \frac{1}{2} (D_\mu B)^2 - \frac{1}{2} \bar{\lambda}^i \gamma^\mu D_\mu \lambda_i + \frac{1}{2} X^2 - \frac{ig}{2} \bar{\lambda}^i [A - i\gamma_5 B, \lambda_i] - \frac{g^2}{2} ([A, B])^2 \Big] , \tag{2.43a}$$

$$A^{matter} = \int d^4x \, (-D_\mu A^{ia} D^\mu A^*_{ia} - \frac{1}{2} \bar{\psi}_a \gamma^\mu D_\mu \psi^a + |F^{ia}|^2) , \tag{2.43b}$$

$$A^{interaction} = \int d^4x \Big[-g\bar{\lambda}^{is} (T_s)^a{}_b A^*_{ia} \psi^b + g\bar{\psi}_a A^{ib} (T_s)^a{}_b \lambda_{is} + \frac{ig}{2} (T_s)^a{}_b \bar{\psi}_a (B_s - i\gamma_5 A_s) \gamma_5 \psi^b + g^2 A^*_{ia} A^{ic} (T_s T_t)^a{}_c (A_t A_s + B_t B_s) + 4gX^{ij} A^*_{ic} A_j{}^d (T_s)^c{}_d \Big] . \tag{2.43c}$$

$$A^{mass} = m \int d^4x \Big[iF^{ia} A^*_{ia} - iF^*_{ia} A^{ia} + \frac{1}{2} \bar{\psi}_a \gamma_5 \psi^a + 2igB_s (T^s)^a{}_b A^*_{ia} A^{bi} \Big] . \tag{2.43d}$$

The above action for m = 0 is clearly U(2) and dilatation invariant, but, as expected, it is also N = 2 superconformally invariant.
One further term that can be added is a term linear in X^{ij} for those theories where X^{ij} is a gauge singlet. This occurs when X^{ij} comes from a

U(1) N = 2 Yang-Mills multiplet, which can couple to N = 2 matter multiplets that have non-trivial U(1) weights. But this term breaks U(2).

2.4 The N = 4 Yang-Mills theory [10]

The N = 4 Yang-Mills multiplet has as its on-shell states one spin-1, four spin-1/2, and six spin-0. It is clearly composed of an N = 2 Yang-Mills multiplet and one N = 2 matter multiplet in the adjoint representation. Since in the absence of the mass term this coupling is unique, N = 4 Yang-Mills must be given by Eq. (2.43d) for m = 0, with the N = 2 matter being in the adjoint representation. It only remains to cast this action in a manifestly SU(4) or O(4) invariant form and find the N = 4 on-shell supersymmetry transformations. This model does not possess U(4) symmetry owing to it being a CPT self-conjugate multiplet and to the fact that the U(1) factor commutes, for N = 4 with the supercharge.

In the O(4) formulation the fields are A_μ, $\lambda_{\alpha i}$, A_{ij}, and B_{ij}, where $\lambda_{\alpha i}$ is an O(4) Majorana spinor, i.e.

$$\lambda_{\alpha i} = C_{\alpha\beta}\bar{\lambda}^{\beta}{}_{i} \; , \tag{2.44}$$

and A_{ij} and B_{ij} belongs to the triplet representation of O(4); that is, they are real antisymmetric two-rank self-dual tensors,

$$A_{ij} = -\frac{1}{2}\,\varepsilon_{ijk\ell}A_{k\ell} \; , \qquad B_{ij} = +\frac{1}{2}\,\varepsilon_{ijk\ell}B_{k\ell} \; . \tag{2.45}$$

In the SU(4) formulation the fields are A_μ, $\lambda'_{\alpha i}$, and φ_{ij}. Here $\lambda'_{\alpha i}$ is a chiral spinor in the fundamental representation of SU(4), whilst φ_{ij} are in the second-rank antisymmetric self-dual ($\underline{6}$) of SU(4),

$$\varphi_{ij} \quad = -\,\varphi_{ji} \tag{2.46a}$$

$$(\varphi_{ij})^{*} \;\equiv\; \varphi^{ij} = \frac{1}{2}\,\varepsilon^{ijk\ell}\varphi_{k\ell} \; . \tag{2.46b}$$

We leave it as an exercise for the reader to deduce the on-shell N = 4 transformation laws and action. This theory is N = 4 superconformally invariant and has no known off-shell formulation that does not involve constrained fields [11]. An off-shell formulation has been given in Ref. [4].

Having constructed the extended rigid theories in x-space, we will now find their corresponding superspace [12] formulations. We could find these formulations starting in superspace, but it is much easier to deduce them directly from the x-space results found above.

The extended superspace is a manifold parametrized by $z^{\pi} = (x^{\mu}, \theta^{A}{}_{i}, \theta^{\dot{A}i}, z^{\ell})$. The coordinates, x^{μ} and z^{ℓ} are bosonic (commuting) and of dimension -1, whilst $\theta^{A}{}_{i}$ and $\theta^{\dot{A}i}$ are (anticommuting) fermionic coordinates of dimension -1/2. The role of z^{ℓ} will become clear when we formulate particular models in superspace. The reader may readily check that the following supersymmetry transformations lead to a representation of the extended supersymmetry algebra:

$$\delta x^{\mu} = i\varepsilon^{A}{}_{j}(\sigma^{\mu})_{A\dot{B}}\theta^{\dot{B}j} - i\theta^{A}{}_{j}(\sigma^{\mu})_{A\dot{B}}\varepsilon^{\dot{B}j} , \tag{2.47a}$$

$$\delta\theta^{A}{}_{j} = \varepsilon^{A}{}_{j} , \tag{2.47b}$$

$$\delta\theta^{\dot{A}j} = \varepsilon^{\dot{A}j} , \tag{2.47c}$$

$$\delta z^{\ell} = i\varepsilon^{A}{}_{i}\theta_{Aj}(\Omega^{\ell})^{ij} + i\varepsilon^{\dot{A}i}\theta_{\dot{A}}{}^{j}(\Omega^{\ell})_{ij} . \tag{2.47d}$$

Superfields are functions of x^{μ}, θ's, and z^{ℓ}, i.e.

$$\varphi(x^{\mu}, \theta_{Aj}, \theta^{\dot{A}i}, z^{\ell}) \equiv \varphi(z) , \tag{2.48}$$

and scalar superfields are defined by

$$\varphi'(z') = \varphi(z) . \tag{2.49}$$

We may also put internal and Lorentz indices on these superfields. These indices transform by appropriate rigid transformations under the action of the group. The representatives ℓ_{N} of the generators of the supersymmetry group are easily deduced from the above equation:

$$\delta\varphi = \delta g^N X_N \varphi = \delta g^N \ell_N \varphi \tag{2.50}$$

where

$$X_N = f_N{}^\Lambda \partial_\Lambda \tag{2.51}$$

and δg^N is the infinitesimal group element for which

$$z'^\pi = z^\pi + \delta g^N f_N{}^\pi . \tag{2.52}$$

For example, the supersymmetry generator $Q^{\dot{A}j}$ is represented by

$$\ell_{\dot{A}i} = \frac{\partial}{\partial\theta^{\dot{A}i}} + i(\sigma^\mu)_{A\dot{A}}\theta^A{}_i + i\theta_{\dot{A}}{}^j(\Omega^\ell)_{ij}\frac{\partial}{\partial z^\ell} . \tag{2.53}$$

To construct actions it is convenient to introduce covariant derivatives,

$$D_M = (D_m\, D_A{}^i,\, D_{\dot{B}j},\, D_\ell) , \tag{2.54}$$

where

$$D_m = \partial_m , \tag{2.55a}$$

$$D_A{}^i = \frac{\partial}{\partial\theta^A{}_i} - i(\sigma^m)_{A\dot{B}}\theta^{\dot{B}i}\partial_m - i(\Omega^\ell)^{ij}\theta_{Aj}\frac{\partial}{\partial z^\ell} , \tag{2.55b}$$

$$D_{\dot{A}i} = \frac{\partial}{\partial\theta^{\dot{A}i}} - i(\sigma^\mu)_{B\dot{A}}\theta^B{}_i\partial_\mu - i(\Omega^\ell)_{ij}\theta_{\dot{A}}{}^j\frac{\partial}{\partial z^\ell} . \tag{2.55c}$$

$$D_\ell = \frac{\partial}{\partial z^\ell} . \tag{2.55d}$$

These transform covariantly under the above supersymmetry transformations and obey the relations

$$\{D_A{}^i, D_B{}^j\} = +2i(\Omega^\ell)^{ij}\varepsilon_{AB}D_\ell , \tag{2.56a}$$

$$\{D_{\dot{A}i}, D_{\dot{B}j}\} = 2i(\Omega^\ell)_{ij}\varepsilon_{\dot{A}\dot{B}}D_\ell , \tag{2.56b}$$

$$\{D_A{}^i, D_{\dot{B}j}\} = -2i(\sigma^m)_{A\dot{B}}\delta^i{}_j D_m , \tag{2.56c}$$

$$\{D_m, D_A{}^i\} = \{\partial_\mu, D_{\dot{B}j}\} = 0 = \{\partial_m, D_z\} = \{D_z, \text{anything}\} . \tag{2.56d}$$

The component fields contained in a given superfield can be found by Taylor-expanding in the θ's and z's. Although, the θ expansion terminates after 2^{4N} terms, the z expansion will in general contain infinitely many terms. One can, however, avoid having an infinite number of fields of ever increasing dimension either by not having a z dependence or by relating the component fields which occur at a given level to those that were found at a lower level.

An alternative, and often more useful way of extracting the component fields contained in a given superfield φ, is to consider the θ = 0 components of the superfields:

$$\varphi , \quad D_A{}^i\varphi , \quad D^{\dot{A}i}\varphi , \quad D_z\varphi , \; \ldots . \tag{2.57}$$

However, not all these x-space fields are independent or non-zero. The relations between them can be computed using the algebra of the D's and taking into account any constraints which φ may obey. Although the first few components in this approach are the same as those found by using a Taylor expansion, the higher component fields will differ by space-time derivatives of lower-component fields.

We now observe that, given any superfield ψ whose first component is C(x), the supersymmetry variation of C is given by

$$\delta C = \bar{\varepsilon}Q\psi\big|_{\theta=0} = \bar{\varepsilon}D\psi\big|_{\theta=0} . \tag{2.58}$$

This latter result follows from the equality between Q and D at θ = 0. In the latter method of extracting the component fields, all the x-spacefields occur as the first component of superfields; so their supersymmetry transformations can be found by using the above equation and the D algebra alone.

We can now turn to the task in hand, namely to put the theories of extended rigid supersymmetry into superspace. This will only be done for N = 2 theories, and in what follows later, i.e. finiteness, we will regard N = 4 theories as being composed of N = 2 theories. Let us begin with N = 2 Yang-Mills theory. Its x-space content was found to be

$$(u,\ \lambda_{ai},\ F_{\mu\nu},\ C^{ij})\ , \tag{2.59}$$

where u = A + iB. The supersymmetry variation of u is of the form

$$\delta u = \varepsilon^{A}{}_{i}\lambda_{A}{}^{i}\ . \tag{2.60}$$

We must compare this with

$$\delta u = (\varepsilon^{A}{}_{i}D_{A}{}^{i} + \varepsilon^{\dot{A}j}D_{\dot{A}j})W\big|_{\theta=0}\ , \tag{2.61}$$

where $W(x^{\mu},\ \theta^{\dot{A}}{}_{i},\ \theta^{\dot{B}j},\ u)$ is the superfield whose first component is u. We observe that δu possesses no $\varepsilon^{\dot{A}i}$ term and hence

$$D_{\dot{A}j}W\big|_{\theta=0} = 0\ . \tag{2.62}$$

However, a superfield whose first component vanishes must also vanish, namely

$$D_{\dot{A}j}W = 0\ . \tag{2.63}$$

This constraint implies that

$$\{D_{\dot{A}i},\ D_{\dot{B}j}\}W = 2\varepsilon_{\dot{A}\dot{B}}\varepsilon_{ij}D_{z}W = 0\ , \tag{2.64}$$

or that

$$D_{z}W = \frac{\partial}{\partial z}W = 0\ , \tag{2.65}$$

and consequently W has no dependence on the bosonic central charge coordinate z.

We must now see whether W must have any other superspace constraints in order to describe the N = 2 Yang-Mills theory. The x-space content of a complex superfield W which is chiral is found by applying $D_A{}^i$'s and evaluating all possible independent superfields at $\theta = 0$. We then arrive at the fields u, $\lambda_A{}^i$, t^{ij}_{AB}, t^{ijk}_{ABC}, and $t^{ijk\ell}_{ABCD}$, which are respectively the $\theta = 0$ components of the following superfields:

$$W\,,\quad D_A{}^iW\,,\quad D_A{}^iD_B{}^jW\,,\quad D_A{}^iD_B{}^jD_C{}^kW\,,\quad D_A{}^iD_B{}^jD_C{}^kD_D{}^\ell W\,,$$

$$D_A{}^iD_B{}^jD_C{}^kD_D{}^\ell W\,. \tag{2.66}$$

As W has no central charge, we have the symmetries

$$t^{ij}_{AB} = -t^{ji}_{BA}\,,\ \text{etc.} \tag{2.67}$$

Utilizing this symmetry we may express t^{ij}_{AB} as

$$t^{ij}_{AB} = -\frac{1}{2}\,\varepsilon^{iij}F_{(AB)} - \frac{1}{2}\,C^{(ij)}\varepsilon_{AB}\,. \tag{2.68}$$

Carrying out this analysis for all the fields, we find that the x-space field content of the chiral W is

$$u\,,\quad \lambda_A{}^i,\quad C^{ij}\,,\quad F_{AB}\,,\quad \zeta_A{}^i\,,\quad d\,. \tag{2.69}$$

As such, another superspace constraint is required to reduce W to have only the content of the N = 2 Yang-Mills theory.

The lowest-dimensional component that is not of the correct form is C^{ij}, which is complex, whilst in N = 2 Yang-Mills it is real. We therefore impose C^{ij} real, which corresponds to the superspace constraint

$$D^{ij}W = \bar{D}^{ij}\bar{W}\,, \tag{2.70}$$

where $D^{ij} = D^{Ai}D_A{}^j$. The reader may check that this constraint also implies that

$$\zeta_A{}^i = -i(\partial\!\!\!/)_A{}^{\dot{B}}\lambda_{\dot{B}}{}^i \ , \tag{2.71a}$$

$$d = 16 \times 6(-\partial^2 u) \ , \tag{2.71b}$$

and that

$$\partial_{[\varrho}F_{\mu\nu]} = 0 \ . \tag{2.72}$$

To summarize the result [13], the fields of N = 2 Yang-Mills are contained in a complex superfield W which is subject to

$$D_{\dot{A}i}W = 0 \ , \quad D^{ij}W = \bar{D}^{ij}\bar{W} \ . \tag{2.73}$$

The above calculation has been performed for the linearized N = 2 Yang-Mills theory. The results for the full non-Abelian theory must involve the same constraints in the limit that the gauge coupling goes to zero, and so they can only be

$$\mathcal{D}_{\dot{A}i}W = 0 \ , \quad \mathcal{D}^{ij}W = \bar{\mathcal{D}}^{ij}\bar{W} \ , \tag{2.74}$$

where $\mathcal{D}_N$ is the gauge covariant derivative and is given in terms of the gauge potential A_N by

$$\mathcal{D}_N = (D_N + gA_N) = E_N^{\pi}(\partial_\pi + gA_\pi) \ . \tag{2.75}$$

A more geometric discussion of the superspace description of N = 2 Yang-Mills is given later in this section.

Let us now turn our attention to the N = 2 matter sector. In the formulation of the Sohnius hypermultiplet [14], the x-space field content is

$$(A_i,\ \psi_A,\ \psi^*_A,\ f_i) \ . \tag{2.76}$$

The lowest component A^i has the transformation law

$$\delta A^i = \frac{1}{2}\,\varepsilon^{Ai}\psi_A - \frac{1}{2}\,\varepsilon^{\dot{A}i}\psi_{\dot{A}}\ . \tag{2.77}$$

So if we consider A^i to be the first component of the superfield φ^i, then φ^i must possess the constraint

$$D_A{}^i\varphi_j = \frac{1}{2}\,\delta^i_j D_A{}^k\varphi_k\ , \tag{2.78a}$$

$$D_{\dot{A}}{}^i\varphi_j = \frac{1}{2}\,\delta^i_j D_{\dot{A}}{}^k\varphi_k\ . \tag{2.78b}$$

By raising the j index with ε^{ij}, these constraints may be rewritten in the form

$$D_A{}^{(i}\varphi^{j)} = 0 = D_{\dot{A}}{}^{(i}\varphi^{j)}\ . \tag{2.79}$$

The independent x-component fields in φ^i can be shown, after a little thought, to be the lowest components of the superfields

$$\varphi_i\ ,\quad D_A{}^k\varphi_k\ ,\quad D^j{}_j\varphi_i\ , \tag{2.80}$$

and their complex conjugates. This has the same content as the Sohnius hypermultiplet and hence we may conclude that φ_i has only the constraints of Eq. (2.79).

Let us now consider the alternative formulation of the hypermultiplet that is described by the fields $(L^{ij}, \lambda^i{}_A, S, P, V_\mu)$, where $\partial^\mu V_\mu = 0$. The super-transformation of L^{ij}, which is real, must be of the form

$$\delta L^{ij} = \varepsilon^{Ai}\lambda_A{}^j + \varepsilon^{\dot{A}i}\lambda_{\dot{A}}{}^j + (i\leftrightarrow j)\ , \tag{2.81}$$

and as a result, if L^{ij} is the first component of the real superfield also denoted by L^{ij}, it must obey [6]

$$D_A{}^{(i}L^{jk)} = 0\ , \tag{2.82a}$$

$$D_{\dot{A}}{}^{(i}L^{jk)} = 0 \ . \tag{2.82b}$$

The reader may verify that no other superspace constraints are needed. We can, however, relax Eqs.(2.82) -- and thus constraint $\partial_\mu V^\mu = 0$ -- by introducing [7] the superfield $L^{ijk\ell} = L^{(ijk\ell)}$,

$$D_A{}^{(i}L^{jk)} = D_{A1}L^{ijk\ell} \ . \tag{2.83}$$

This multiplet does not have a conserved vector or any other such constraint and can be used to describe N = 2 matter. Of course it involves many more fields; however, the extra fields do not lead to further on-shell states provided one introduces an extra superfield G whose fields act as Lagrange multipliers. The superfield G satisfies the constraints

$$D_A{}^i D_{Bi} G = 0 = [D_A{}^i, D_{\dot{B}i}]G \ . \tag{2.84}$$

For a description of this multiplet when the matter belongs to a complex representation, see Ref. [15]. For further descriptions of N = 2 matter, see Ref. [8].

We now wish to construct actions for the above superspace theories. For a general N = 2 superfield φ which has $\partial_z\varphi = 0$, an invariant quantity is given by

$$\int d^4x \, d^8\theta \, \varphi \ . \tag{2.85}$$

However, if this is to be an action, φ must have dimension zero.

As such, the actions for the above theories must be integrals over only a subspace of superspace or be constructed from superfields with subcanonical dimensions.

In the case of N = 2 Yang-Mills, the action can involve only $D_A{}^i$ and not $D_{\dot{A}}{}^i$ acting on W, and the only candidate of the correct dimension is [13]

$$A^{ym} = \int d^4x \, D^{ij}D_{ij}W^2 + \text{h.c.} \tag{2.86}$$

Clearly this is invariant as

$$\delta A^{ym} = \int d^4x\, (\varepsilon^A{}_k D_A{}^k + \varepsilon^{\dot{A}i} D_{\dot{A}i}) D^{ij} D_{ij} W^2 = 0\ . \tag{2.87}$$

For the Sohnius hypermultiplet, the appropriate superspace action [14] is

$$\int d^4x\, D^{ij} \bar{\varphi}_i D^k{}_k \varphi_j\ , \tag{2.88}$$

whilst for the version with the conserved vector V_μ the action [6] is

$$\int d^4x\, D^{(ij} D^{k\ell)} L_{(ij} L_{k\ell)}\ . \tag{2.89}$$

The invariance of this action is not obvious, but may be verified by using the anticommutations for the D's.

The action for the relaxed formulation of the hypermultiplet is rather complicated.

For an interesting alternative formulation of N = 2 theories based on a different type of superspace to that considered here, see Ref. [16].

In quantizing a theory it is usual to work with unconstrained fields. In the case of superspace this means unconstrained superfields. As such, it is necessary to solve the superspace constraints given above. The important exception to this rule is of course the chiral superfield φ, but here one can also solve the constraint by

$$\varphi = \bar{D}^2 U \tag{2.90}$$

and quantize with the unconstrained prepotential U.

We recall that the constraints of N = 1 Yang-Mills

$$\mathcal{D}_A W_{\dot{B}} = 0\ , \quad \mathcal{D}^A W_A = \mathcal{D}^{\dot{B}} W_{\dot{B}} \tag{2.91}$$

were solved by

$$W_{\dot{B}} = D^2 (e^{-gV} D_{\dot{B}} e^{gV})\ . \tag{2.92}$$

The solution of the superspace constraints of extended superfields is rather complicated and analytic solutions are not known in the non-linear case. For N = 2 Yang-Mills at the linearized level the solution of the constraints on W is given in terms of an unconstrained dimension -2 superfield V^{ij} [17]

$$W = D^{ij}D_{ij}\bar{D}^{k\ell}V_{k\ell} \; . \tag{2.93}$$

The solution of the constraints of the relaxed hypermultiplet formulation involves the prepotential ϱ_{Ai} and $X^{ijk\ell}$ of dimensions -3/2 and -1, respectively [15]. It has been shown that the linearized constraints can be systematically iterated to solve the constraints in the non-Abelian theory [18, 19].

We will now consider the N = 2 Yang-Mills theory from a geometric viewpoint [20]. Let us, as in ordinary Yang-Mills theory, introduce potentials A_N which covariantize the superspace derivatives:

$$\mathcal{D}_N = D_N + gA_N \cdot Y \; , \tag{2.94}$$

where Y is the Yang-Mills generator and $A_N = (A_A{}^i, A_{\dot{B}i}, A_c)$. We can then define super Yang-Mills field strengths by the equation

$$[\mathcal{D}_N, \mathcal{D}_M\} = T_{NM}{}^R\mathcal{D}_R + F_{MN} \cdot Y \; , \tag{2.95}$$

where $T_{MN}{}^R$ is the torsion of rigid superspace and is zero except for the component

$$T_{A\dot{B}} = -2i(\sigma^m)_{A\dot{B}} \; . \tag{2.96}$$

The F_{MN} must then obey the Bianchi identities

$$\sum_{MNR} (\mathcal{D}_M F_{NR} + T_{MN}{}^S F_{SR}) = 0 \; , \tag{2.97}$$

the appropriate symmetrization being understood. Consider, now, N = 2 matter, say in formulation (a). If it transforms under the gauge group the defining condition must be modified, in order to be gauge invariant, to be

$$D_A^{(i}\varphi^{j)} = 0 = D_{\dot{A}}^{(i}\varphi^{j)} = 0 \ . \tag{2.98}$$

Consequently, we find that

$$\sum_{(ijk)} \{D_A^{\ i}, D_{\dot{B}}^{\ j}\}\varphi^k = 0 = \sum_{(ijk)} F_A^{\ (i}{}_{\dot{B}}{}^{j}\varphi^{k)} \ , \tag{2.99}$$

and similarly for the other possible expressions. We must therefore conclude that

$$F_A^{\ (i}{}_{\dot{B}}{}^{j)} = F_A^{\ (i}{}_B{}^{j)} = F_{\dot{A}}^{\ (i}{}_{\dot{B}}{}^{j)} = 0 \ . \tag{2.100}$$

Hence, we have found that in order to have N = 2 matter in the presence of gauge fields we must constrain the super Yang-Mills field strengths [21]. An analysis of formulation (b) of the N = 2 matter requires

$$D_A^{(i}L^{jk)} = D_{\dot{B}}^{(i}L^{jk)} = 0 \ , \tag{2.101}$$

and one finds the same constraints on F_{MN} as given above.
One can also eliminate by a covariant constraint the potential A_m in terms of A_{Bi} and $A_{\dot{C}j}$ [21]. The appropriate constraint is

$$F_A^{\ i}{}_{\dot{B}i} = 0 \ . \tag{2.102}$$

This is similar to the change between the first- and second-order formalism in general relativity. The lowest dimension field strength remaining is of dimension one and is

$$W \equiv F^{Ai}{}_{Ai} \ . \tag{2.103}$$

The Bianchi identities then imply, from the constraints of Eqs. (2.100) and (2.102), that

$$D_{\dot{A}i}W = 0 \quad \text{and} \quad D^{ij}W = \bar{D}^{ij}\bar{W} \ , \qquad (2.104)$$

which was our previous result.

3. FINITENESS PROPERTIES

In the article by Grisaru and Zanon (see contribution in this paper) it was explained how the interacting Wess-Zumino model had even fewer infinite renormalizations than one might naïvely expect by demanding that the renormalization procedure preserve supersymmetry. In fact, it only has one infinite renormalization, as does the N = 1 Yang-Mills theory (in the background field method or in a preferred gauge). As one studies theories with more supersymmetries, one might expect an even more remarkable ultraviolet behaviour.
The most spectacular renormalization properties of supersymmetric theories are the finiteness of a large class of extended rigid supersymmetric theories. At first, attention was focused entirely on the maximally extended N = 4 supersymmetric Yang-Mills theory. The β-function was shown to vanish for this theory at one [22], two [23], and three loops [24]. Soon after the three-loop calculation, an argument for finiteness to all orders was made. This argument [25, 26] was based on the anomaly structure of supersymmetric theories. More recently, two more arguments for the finiteness of N = 4 Yang-Mills theory have been found. One argument relies on a generalization of the N = 1 non-renormalization theorem to extended supersymmetry [27, 19], whilst the other relies on putting N = 4 Yang-Mills theories in a light-cone gauge [28, 29].
It was noticed [30], using the results of Ref. [31], that the β-function of N = 2 Yang-Mills theory vanished at two loops. It was then argued, using the non-renormalization argument, that N = 2 Yang-Mills theory was finite above one loop [27]. While re-examining the anomalies argument it was realized that any N = 2 rigid supersymmetric theory was finite above one loop [15]. In fact, it is possible to arrange the representation content of N = 2 rigid theories so that there are finite N = 2 theories [15]. A modern account of the anomalies argument in a form which applies to N = 2 theories can be found in Ref. [32], whilst the application of the non-renormalization argument [19, 27] to N = 2 theories is given in Ref. [32]. In the following

discussion we will only consider the anomaly and non-renormalization arguments.

These results have been confirmed by an explicit calculation using the N = 1 superfield formalism of N = 2 theories. It has been found that the two-loop β-function of any rigid N = 2 theory vanishes [33].

3.1 The anomalies argument [25, 26]

The strategy which this argument employs utilizes the fact that in any supersymmetric theory the energy-momentum tensor $\theta_{\mu\nu}$, some of the internal currents $j_\mu{}^i{}_j$, and the supercurrent $j_{\mu\alpha i}$ lie in a supermultiplet. Consequently, any superconformal anomalies which these currents possess must also lie in a supermultiplet. Typically this supermultiplet of anomalies will include $\theta_\mu{}^\mu$, $(\gamma^\mu j_{\mu i})_\alpha$ and $\partial^\mu j_\mu{}^i$ for some i, j, corresponding to the breaking of dilation, special supersymmetry, and some of the internal currents, respectively. Clearly, if some of the relevant internal symmetries are preserved (i.e. $\partial_\mu j^{\mu i}{}_j = 0$ for some i, j) then the anomaly multiplet, if it is irreducible, will vanish, and consequently $\theta_\mu{}^\mu = 0$. However, $\theta_\mu{}^\mu$ is proportional to an operator $(F_{\mu\nu}F^{\mu\nu} + \ldots)$ times the β-function, and so the β-function must vanish. From this result we can argue, in specific formalisms such as the background field method, for the finiteness of the theory being considered (see below).

To illustrate how the argument goes, we will first argue for the finiteness of N = 4 Yang-Mills theory using a simplified version of the anomalies argument. To do this we must make the following assumption.

The quantum correction of N = 4 Yang-Mills preserves one supersymmetry and the SU(4) internal symmetry.

Let us first establish that all the chiral currents of the theory are preserved. The maximal symmetry is U(4) = SU(4) x U(1); however, this U(1) factor, whose generator is denoted by B, has for general N the following commutation relation with $Q^j{}_A$:

$$[Q_A{}^j, B] = \frac{i(N-4)}{4} Q^j{}_A \,. \tag{3.1}$$

In the case of N = 4, B and $Q_A{}^j$ commute; and consequently B, which is a chiral rotation, has the same action on all the states of any N = 4 multiplet. However, N = 4 Yang-Mills is a CPT self-conjugate multiplet; so B must have the same action on the +1/2 and -1/2 helicity states, which is possible only if the action of B is zero. Hence, the model has only SU(4) symmetry, which decomposes under O(4) as follows:

$$\underline{15} \text{ of } SU(4) = \underline{6} + \underline{9} \text{ of } O(4) , \tag{3.2}$$

where the $\underline{9}$ of O(4) are chiral currents, whilst the $\underline{6}$ of O(4) are currents that do not involve γ_5. Hence if SU(4) is preserved, the $\underline{9}$ chiral currents are preserved, and so all chiral currents are preserved.

In the O(4) formulation, which has $\underline{4}$ Majorana spinors $\chi_{\alpha i}$, the $\underline{9}$ chiral currents contain the term $\bar{\chi}^{(i}\gamma^\mu\gamma^5\chi^{j)}$, whilst the $\underline{6}$ currents contain the term $\bar{\chi}^{[i}\gamma^\mu\chi^{j]}$.

Consider now the N = 4 Yang-Mills theory when decomposed into N = 1 representations. The R current $j_\mu^{(5)}$ will be one of the $\underline{9}$ and will be preserved, i.e. $\partial_\mu j^{\mu(5)} = 0$.

We must now consider the form of the anomaly equation. It has a right-hand side constructed from the N = 1 Yang-Mills field strengths W_A and the chiral matter fields φ^i. On dimensional grounds it must be of the form

$$D^{\dot{A}}J_{A\dot{A}} = -\frac{1}{3}\frac{\beta(g)}{g}D_A W^2 \quad \text{plus terms involving } \varphi^i . \tag{3.3}$$

However, as previously discussed, the first term is a chiral anomaly and so contains $\theta_\mu{}^\mu$ and $\partial^\mu j_\mu{}^{(5)}$. Since $\partial^\mu j^{\mu(5)} = 0$ we must conclude that $\beta(g) = 0$.

A similar but somewhat stronger argument [25] can be employed starting from the assumption that the quantum N = 4 Yang-Mills theory preserves N = 1 supersymmetry and O(4) symmetry. In this case we must consider N = 2 anomaly multiplets.

We now wish to apply [32] the anomalies argument to N = 2 supersymmetric rigid theories and show the following theorem.

Theorem [15]

The β-function in N = 2 rigid supersymmetric theories vanishes above one loop.

Proof

We shall first establish that the quantum theory preserves N = 2 supersymmetry and U(2) internal symmetry. At the classical level the two supersymmetries are manifestly preserved by the N = 2 matter formulations given earlier. However, in order to quantize the theory in a straightforward manner, let us consider those formulations that do not involve constraints in x-space. That is, we consider the relaxed hypermultiplet formulation consisting of the superfields L, L^{ij}, $L^{ijk\ell}$; or we could also use the new relaxed version of the Sohnius hypermultiplet discussed in Section 2.

We now require a method of regularization that preserves these symmetries. The safest method is that of higher derivatives [34]. That is, we consider the action which in component fields contains

$$-\frac{1}{4}F_{\mu\nu}^2 - \frac{1}{4}\frac{1}{\Lambda^{2r}}F^{\mu\nu}(\partial^2)^r F_{\mu\nu} + \text{supersymmetric extension} . \tag{3.4}$$

In the above, r is an integer which is usually equal to one or two. This method clearly preserves N = 2 supersymmetry and U(2), and regulates all graphs except primitive one-loop diagrams. In general, introducing higher derivatives into a theory alters the infinity structure at one loop, i.e. the β-function of the theory. However, for the N = 2 theories expressed in the formalism given above, the additional massive states introduced by the higher derivatives must belong to the N = 2 supermultiplets with the content one spin-1, four spin-1/2, and five spin-0. This follows from the fact that this is the only available massive multiplet which does not possess a central charge and involves spins less than or equal to one. Consequently, if the β-function at one loop for the theory with higher derivatives is given by [35, 36]

$$\beta(g) = \frac{g^3}{96\pi^2}\sum_{\lambda} (-1)^{2\lambda}C_\lambda(1-12\lambda^2) , \tag{3.5}$$

where the sum is over all helicity states, then the β-function is the same as in the theory without higher derivative. It is argued in Ref. [37] that this formula for the β-function is indeed the correct one.

Hence if the theory is finite at one loop, the theory when higher-derivative regulated will be finite at one loop and thus be rendered finite to all orders by higher derivatives.

As such, an N = 2 one-loop finite theory can preserve N = 2 supersymmetry and U(2) at the quantum level.

It therefore only remains to establish that the multiplet of currents of N = 2 supersymmetric theories has the required form. The N = 2 supercurrent is an object J which has dimension 2; and its anomaly equation has the form

$$D^{ij}J = \text{anomaly} \ , \tag{3.6}$$

where $D^{ij} \equiv D^{A(i}D_A{}^{j)}$. The right-hand side of the above eqution must be of dimension 3 and must be constructed out of the gauge invariant superfields W of N = 2 Yang-Mills and the N = 2 matter superfields and covariant derivatives of these fields. Let us focus on the term involving the N = 2 Yang-Mills fields W. The only candidate is

$$D^{ij}J = -\frac{1}{3}\frac{\beta(g)}{g}\bar{D}^{ij}\bar{W}^2 + \text{terms involving matter fields} \ , \tag{3.7}$$

where $\beta(g)$ is the β-function for the gauge coupling constant g. Now the first term is a set of anomalies that contains a contribution to $\theta_\mu{}^\mu$ as well as to a divergence of one of the chiral U(2) currents. However, these chiral currents are preserved; so $\beta(g) = 0$.

If, on the other hand, the theory has infinities at one loop, then we must regulate these infinities separately. We then expect that the U(2) currents will be preserved above one loop, and, going through the same argument as above, we find that the β-function vanishes above one loop.

There exists a variant of the anomalies argument for finiteness, involving the Adler-Bardeen theorem [38] for the chiral current rather than appealing directly to chiral current conservation. This argument [39] has also been applied to show the finiteness, above one loop, of N = 2 theories.

3.2 The non-renormalization argument

We can now present the extended non-renormalization theorem (see lectures of Grisaru and Zanon).

Theorem [27]

The effective action Γ for any extended supersymmetric theory that possesses an unconstrained superfield formalism can be written as one integral over the whole of extended superspace:

$$\Gamma = \int d^4x_1, \ \ldots, \ d^4x_n \ d^{4N}\theta \ f[\varphi(x_1), \ \ldots, \ \varphi(x_n), \ D\varphi(x_1), \ \ldots]$$
$$\times \ g(x_1, \ \ldots, \ x_n) \ , \tag{3.8}$$

where φ is a generic superfield.

Proof

When a supersymmetric theory admits an unconstrained formalism, we can take the propagators to be δ_{12} and the vertices must have a factor $\int d^{4N}\theta$. We will also have various D factors on the lines leaving the vertices. We can now follow exactly the same argument as was used to prove the N = 1 non-renormalization theorem: we integrate the D's off the δ_{12}'s and shrink the θ-space loops until only one integral remains.

Counter-terms are known to be local, and their contribution to Γ is therefore of the form

$$\Gamma = \int d^4x\, d^{4N}\theta\, f(\varphi, D\varphi, \ldots) \,. \tag{3.9}$$

Of course the effective action must, in the absence of any anomalies, also obey the Ward identities corresponding to the symmetries of the theory. The simplest way of implementing these symmetries is to use the background field method. The background field formalism for the extended theories is very similar to that for N = 1 Yang-Mills, except for the gauge-fixing procedure and corresponding ghosts. This latter point, which is discussed later, leads to the one-loop exception clause in the following theorem.

Theorem [27]

Consider any supersymmetric theory that possesses an unconstrained superfield formulation; then, using the background field formalism, we can express the local quantum contributions, *above one loop*, to the effective action Γ as one integral over the whole of superspace of a gauge-invariant function of the background potentials and matter fields.

Consequently, the counter-terms must be of the form

$$\Gamma = \int d^4x\, d^{4N}\theta\, f(A_N^C, X^C, D_M^C A_N^C, D^C X^C) \,, \tag{3.10}$$

where f is a gauge-invariant function of the background gauge potential A_N^C and represents the matter fields X^C. We will return to the one-loop exception later.

The strategy [27] of the non-renormalization argument is simply to see if dimensional analysis allows any counter-terms of the form of Eq. (3.9). In the absence of dimensional coupling constants this is unlikely, as the measure has a dimension of -4 + 2N. It was argued in Ref. [27] that N = 2 Yang-Mills theory was finite above one loop and that N = 4 Yang-Mills theory would be finite if an N = 4 superfield formalism existed. In Ref. [19] it was argued that N = 4 Yang-Mills theories were finite as a result of the N = 2 extended superfield formalisms discussed previously. The finiteness of N = 2 rigid supersymmetric theories above one loop and the criterion for finiteness at one loop were established in Ref. [15]. We will now apply the non-renormalization argument to N = 2 rigid supersymmetric theories. As discussed in Section 2, the N = 2 Yang-Mills is represented by A_N, which has the same dimension as D_N, namely: one if N = m, and one-half otherwise. The N = 2 matter is represented by the fields L, L^{ij}, $L^{ijk\ell}$, which all have dimension 1. Using the above theorem we find that, above one loop in the background field formalism, the local counter-terms are of the form

$$\Gamma = \int d^8\theta\, d^4x\, f(A_N,\, D_{Bj}A_N,\, L,\, L^{ij},\, L^{ijk\ell},\, \ldots)$$

$$= \int d^8\theta\, d^4x\, f(A_{Bi},\, D_{Aj}\, \ldots,\, D_{Ck}A_{D\ell},\, L,\, D_{Ck}L,\, \ldots)\ . \qquad (3.11)$$

However, on dimensional grounds none of the above terms is allowed, and so we must conclude that the N = 2 rigid supersymmetric theories are finite above one loop and have correspondingly a vanishing β-function above one loop.

The above statement must be interpreted carefully when the theories have one-loop infinities, i.e. a non-zero one-loop β-function. In that case we will find the inevitable higher-loop $1/\varepsilon^N$ poles, where n > 1, which are a consequence of the $1/\varepsilon$ pole at one loop. Also the subtraction procedure must be minimal in order that the β-function does not receive higher-loop contributions due to finite counter-terms inserted in the divergent on-loop graphs.

In order to make the above discussion more concrete, let us do the superspace power counting for N = 2 Yang-Mills theory and verify that it is indeed finite above one loop*). The prepotential of the N = 2 Yang-Mills theory is V^{ij} and is of dimension -2. As such, the action must have the generic form

$$A = \int d^4x\, d^8\theta\, (VD^8V + V^2D^{12}V + \ldots)\ . \tag{3.12}$$

Consequently, the propagators are of the form $D^{-8}\delta_{12}$ and the vertices involve a D^{12} factor as well as a full superspace integration. The external lines in the background formalism above one loop are in terms of $A_{Bi} = D_{Bi}D^4V$. Let there be E external lines, P propagators, and V cubic vertices. The degree of divergence for a graph with only *cubic* vertices is

$$D = 4L - 4P - E\left(2 + \frac{1}{2}\right) + 6V - 4L\ . \tag{3.13}$$

The last factor results from requiring eight D's for every internal loop. Using the relation 3V = 2P + E we find that

$$D = -\frac{E}{2}\ , \tag{3.14}$$

and we conclude that above one loop all graphs involving only cubic vertices and finite. The reader may easily generalize this result to include graphs involving any type of vertices.

Let us now return to the one-loop exception mentioned above. It results from the fact that in the extended theories we must introduce ghosts as usual; however, in these theories the ghost action has a residual invariance. Fixing this invariance and introducing ghosts for ghosts, we find that these new ghosts still possess a gauge invariance. This goes on indefinitely, and we find an infinite number of ghosts for ghosts. Fortunately, for super Yang-Mills fields the second-generation ghosts and all further ghosts only couple to the background fields; and for matter, even the first-generation-matter ghosts do not couple to the quantum fields. Hence

*) P. Howe and P. West: The following is an account of an unpublished discussion.

it is a problem that only affects one-loop graphs. To define the theory, we must truncate the infinite sequence of ghosts for ghosts. This is achieved at the expense of introducing the background gauge prepotential, and so we find that at one loop we can have an explicit occurrence of the background prepotential.

In the N = 2 Yang-Mills the field strength is given in terms of the prepotential by

$$W = \bar{D}^{kl}\bar{D}_{kl}D^{ij}V_{ij} \; . \tag{3.15}$$

The gauge invariance

$$\delta V^{ij} = D_{Ak}\chi^{ijkA} + D_{Ak}\chi^{ijkA} \tag{3.16}$$

leaves W invariant as

$$D_{(i}{}^{A}D_{j}{}^{B}D_{k)}{}^{C} = 0 \; . \tag{3.17}$$

Any gauge-fixing term involving V^{jk} must, however, also be invariant under the gauge transformation

$$\delta\chi^{(ijk)A} = D_{B\ell}\chi^{(ijk\ell)(AB)} \; , \tag{3.18}$$

and so on. The origin of the one-loop exception was given in Ref. [27], and a detailed discussion can be found in Ref. [19].

3.3 Finite N = 2 supersymmetric rigid theories [15]

We have seen in the previous subsection that N = 2 supersymmetric rigid theories are finite if they have vanishing one-loop β-functions. For an N = 1 theory consisting of N = 1 Yang-Mills theory and Wess-Zumino multiplets φ_σ in the representation R_σ of the gauge group, the one-loop β-function is [24]

$$\beta(g) = \frac{1}{(4\pi)^2} g^3 \left[\sum_\sigma T(R_\sigma) - 3C_2(G) \right] . \tag{3.19}$$

An N = 2 rigid theory consists of N = 2 Yang-Mills and N = 2 matter. The N = 2 Yang-Mills theory consists of N = 1 Yang-Mills and one Wess-Zumino multiplet σ in the adjoint representations. The N = 2 matter, on the other hand, consists of Wess-Zumino multiplets X_σ and Y_σ in the representations R_σ and $\bar{R}_\sigma$, respectively. Adjusting the one-loop β-function of Eq. (3.19), we find that the one-loop β-function for N = 2 rigid theories is

$$\beta(g) = \frac{2g^3}{(4\pi)^2} \left[\sum_\sigma T(R_\sigma) - C_2(G) \right] . \tag{3.20}$$

There are many cases for which the β(g) vanishes. For example, in the case of SU(N), $C_2(N) = N$, whilst for the fundamental representation T(R) = 1/2. Hence we can have 2N fundamental representations and the theory will be finite.

It is interesting [40] that most of the groups that have been proposed for grand unification belong to a single sequence of groups. Indeed, of the five sequences of groups into which all Lie groups were classified by Cartan, this is the only finite sequence. This sequence is E_8, E_7, E_6, E_5 = SO(10), E_4 = SU(5), and E_3 = SU(3) × SU(2). The fact that E_3 × U(1) is the group of low-energy physics is intriguing, as the U(1) factor is a typical remnant of some higher symmetry breaking. We will now see [41] which of N = 2 theories that have the above gauge groups are finite. For E_4 = SU(5) we must have p + 3q + 7r = 10, where p, q, and r are the number of hypermultiplets in the $5 + \bar{5}$, $10 + \overline{10}$, and $15 + \overline{15}$ representations, respectively. For SO(10) we can have p and q hypermultiplets in the $10 + \overline{10}$ and $16 + \overline{16}$ representation, respectively, provided p + 2q = 8. The group E_6 will result in a finite theory if we use four hypermultiplets in the $27 + \overline{27}$ representation, whilst the group E_7 requires three hypermultiplets in the 56 + 56 representation. For E_8 the lowest dimensional representation is the adjoint, and so we are only allowed three chiral fields in the adjoint representations. We observe that for the groups SO(10), E_6, E_7, and E_8, the theories that are finite contain the observed fermions. In fact, they contain three or more generations plus their mirrors. Of course it is not clear how many generations would survive massless when the gauge groups are spontaneously broken. Recently, it has been pointed out that there are other possible finite theories when the N = 2 matter is a CPT self-conjugate multiplet [42].

In Ref. [43] it has been shown that an N = 2 mass term introduces no infinities.

It is appropriate, at this point, to discuss the strengths and weaknesses of the above arguments. The anomalies argument in its strongest form relies on the use of the higher derivative method of regularization. However, the implementation of this method to N = 2 superspace theories has not been worked out in every detail. There is a variant [39] of the anomalies argument that uses the Adler-Bardeen theorem, and in this version it has been used to establish the finiteness of a class of N = 1 supersymmetric theories up to two loops; in this sense the anomalies argument has a wider applicability.

The non-renormalization argument, and in its detailed use the anomalies argument, relies on the N = 2 super-Feynman rules. A detailed account of these rules has not been given; but it is clear that they will lead to severe off-shell infrared divergences -- that is, divergences of the form $\int d^4k/k^4$. In order to make sense of this formalism, these divergences must first be regulated and then removed. In doing this we may lose the manifest nature of some of the symmetries of the theory. For example, adding mass terms breaks the gauge invariance, whilst putting the theories in a periodic box [44] breaks Lorentz invariance. It is not clear that infrared divergences may lead to evasion of the non-renormalization theorem [45].

A further problem concerns the fact that when N = 2 matter belongs to a complex representation, the N = 2 matter must be doubled [15]. However, it has been verified [33] by explicit computation that even with odd numbers of complex representations, the theories are finite at two loops. Finally, it has not been examined in detail how regulating one-loop infinities by non-supersymmetric schemes may pollute higher-order diagrams containing these one-loop divergences. These latter objections of course apply to both the anomalies and the non-renormalization arguments.

There has been little discussion of the shortcomings of the light-cone approach. However, it is far from clear how one recovers Lorentz invariance. Presumably, one must verify that the non-linear Ward identities for Lorentz invariance are valid. This may, however, involve adding finite local counter terms to the theory. This argument has yet to be extended to N = 2 theories.

I have stressed these weaknesses, not because I do not believe in the arguments for finiteness, but to show that they are not proofs in a mathematical sense and that there is still room for further work.
Recently [46], the anomalies argument was applied to two-dimensional extended σ-models in order to establish their remarkable finiteness properties [47]. However, since supergravity theories involve dimensional coupling constants, and both the anomalies and the non-renormalization arguments rely on dimensional analysis, it is easily seen that these arguments can be evaded. For a review of the status of infinities in supergravity theories, see Refs. [32] and [48]. An interesting recent calculation in this context is given in Ref. [49]. The above discussion has dealt with the finiteness properties of the extended rigid thories. In fact, it is known that there exist N = 1 supersymmetric theories that are one-loop finite, and it has been further shown that one-loop finiteness implies two-loop finiteness [50, 51]. More recently, it has been proved that n-loop finiteness in N = 1 models implies the vanishing of the gauge β function at n + 1 loops [52, 53]. The Yukawa β-function in these two-loop finite theories has been calculated [54], and which N = 1 theories are three-loop finite is under study. It is important to note that some of these theories contain fermions of a given chirality only.

4. EXPLICIT BREAKING AND FINITENESS

Given the large class of finite theories discussed in the previous section, we now wish to investigate whether adding soft terms can preserve their finiteness. By soft terms we mean terms of dimension 3 or less that are gauge-invariant and parity conserving. Inserting a soft term into any graph lowers its degree of divergence; however, these terms also explicitly break the supersymmetry and internal symmetry that are responsible for the finiteness of these very special theories. We will find that not all soft terms preserve finiteness, but only certain combinations of soft terms.
The first soft terms which were found to maintain finiteness were N = 1 supersymmetric mass terms added to N = 4 Yang-Mills theory [55]. A general analysis giving the necessary and sufficient conditions for a soft term to maintain finiteness in N = 4 Yang-Mills theory was given using the spurion technique in Ref. [56]. However, some authors [57] have independently found finiteness-preserving soft terms for N = 4 Yang-Mills by using the light-

cone formalism, whilst an analysis at the level of component fields, which found some of the soft terms that preserve finiteness, was given in Ref. [58].

One particular soft term of the form (A^2-B^2) which was found in Refs. [56] and [58] was also later found in Ref. [59] by using an extension of the light-cone formalism of Ref. [57].

The analysis of the necessary and sufficient conditions for a soft term to preserve finiteness in the finite class of N = 2 theories was given in Ref. [60] by using the spurion technique. Some of these results were also found by calculating at the level of component fields [61].

We will follow Refs. [56] and [60] and use the spurion technique [62] to investigate the divergences induced by soft terms. Let us consider, to begin with, a general N = 1 supersymmetric theory consisting of N = 1 Yang-Mills V and Wess-Zumino multiplets φ. We will denote the x-space component fields of φ by (A, B, χ_A, F, G) and those of N = 1 Yang-Mills by (A_μ, λ_A, D).

As an example, consider adding the term $\mu^2(A^2-B^2)$ to the supersymmetric action. In order to still work with the superfield formalism, and in particular with the super-Feynman rules, we will rewrite this addition in superspace by introducing the spurion superfield,

$$S = \mu^2\theta^2 \ , \tag{4.1}$$

where $\theta^2 = \theta^A\theta_A$ and μ^2 is a constant. The superfield S is a chiral superfield in the sense that $D_A S = 0$, but it is not a scalar superfield in that it cannot transform correctly under supersymmetry. The $\mu^2(A^2-B^2)$ term is now added by including in the action the term

$$\int d^4x\, d^2\theta\, S\varphi^2 + \text{h.c.} = \int d^4x\, \mu^2(A^2-B^2) \ . \tag{4.2}$$

We can now calculate quantum processes using the super-Feynman rules. We have the same propagators and vertices as before except for the extra vertex given in Eq. (4.2). The S superfield is only an external field, and the additional vertex has only one $-1/4\ \bar{D}^2$ factor associated with one of the two chiral lines as shown in Fig. 1. This assignment of $\bar{D}^2$ factors is made so as to have an integral over the whole of N = 1 superspace at the vertex. Consequently, the non-renormalization

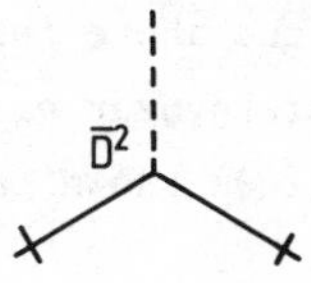

Fig. 1 The $S\varphi^2$ vertex

theorem is still valid. Counting dimensions, we find that the dimension of S is one. The spurions necessary to introduce all other possible soft terms are listed in Table 2. In Table 2, μ_1, m, n, ξ, and e are constants, and the group theory indices are not displayed but are understood to be present in the appropriate places. The factors of V are necessary in order to maintain gauge invariance and are important for obtaining the correct answer when calculating in superspace. Their component expressions, however, have been evaluated for simplicity in the Wess-Zumino gauge.

<u>Table 2</u>

List of spurions

Spurion	Addition	Dimension of spurion
$U_1 = \mu_1^2\theta^2\bar{\theta}^2$	$\int d^4x\, d^4\theta\, U_1\bar{\varphi}(e^{gV})\varphi = \int d^4x\, \mu_1^2(A^2 + B^2)$	0
$S_1 = \theta^2\xi$	$\int d^4x\, d^4\theta\, S_1\varphi^3 + h.c. = \int d^4x\, \xi(A^3 - 3AB^2)$	0
$M = \theta^2\bar{\theta}^2 m$	$\int d^4x\, d^4\theta\, M\mathcal{D}^A\varphi\mathcal{D}_A\varphi + h.c. = \int d^4x\, m\chi^A\chi_A + h.c.$	-1
$N = \theta^2\eta$	$\int d^4x\, d^2\theta\, NW^AW_A + h.c. = \int d^4x\, \eta\lambda^A\lambda_A + h.c.$	0
$E = \theta^2\bar{\theta}^2 e$	$\int d^4x\, d^4\theta\, E(e^{gV}\varphi e^{-gV}\bar{\varphi}^2) + h.c.$ $= \int d^4x\, eA(A^2 + B^2)$	-1

All the above insertions produce new verties which can be used to construct super-Feynman graphs. The $\chi^A\chi_A$ insertion, for example, gives the additional vertices shown in Fig. 2.

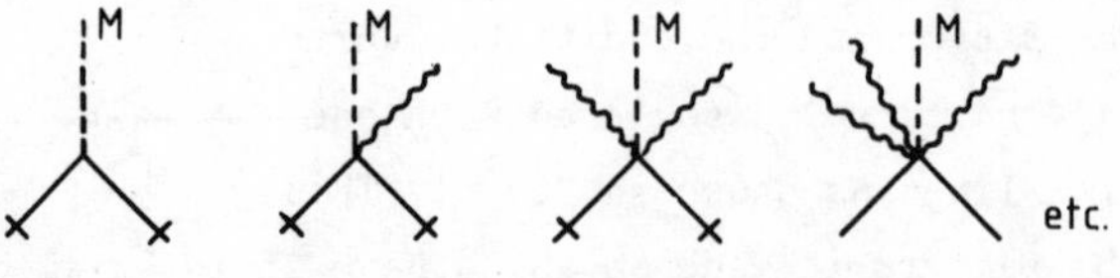

Fig. 2 The $M\mathcal{D}^A\varphi\mathcal{D}_A\varphi$ vertex

Any new infinities induced by the addition of the above insertions will obey the non-renormalization theorem and so be of the generic form

$$\int d^4x\, d^4\theta\; S(D^2S)^r U(D^2\bar{D}^2U)^p f(\varphi,\, V,\, D\varphi,\, \ldots)\ , \tag{4.3}$$

where S is a chiral spurion superfield $S = s\theta^2$, U is a general spurion superfield $U = u\theta^2\bar{\theta}^2$, and f is a gauge-invariant function of φ and V. Let us return to the addition of only the A^2-B^2 term for the purposes of illustration. The induced infinity will be of the form

$$\int d^4x\, d^4\theta\; S(D^2S)^r f(\varphi,\, V,\, D\varphi,\, \ldots)\ . \tag{4.4}$$

As the dimension of S is 1, the only possible infinity is of the form

$$\int d^4x\, d^4\theta\; S\bar{\varphi}\ ; \tag{4.5}$$

however, this term is only gauge-invariant if φ is a gauge singlet. Consequently, in the absence of gauge singlets, any A^2-B^2 insertion induces no new types of infinity. Using an inductive argument of Weinberg [63], let us assume that there are no induced infinities at n loops; then the induced infinities at n + 1 loops arise either as an overall superficial divergence or as subdivergences. The latter infinity is absent by assumption, whilst the former type of infinity is absent since there are no S-dependent infinities. Consequently, we arrive at the following result.

Theorem

In any N = 1 supersymmetric theory an A^2-B^2 term induces no additional infinities if there are no gauge singlets in the theory.

Let us now consider an $A^2 + B^2$ addition. The resulting induced infinity can only be of the form

$$\int d^4x\, d^4\theta\; U\bar{\varphi}\varphi + \left(\int d^4x\, d^4\theta\; u\varphi^2 + \text{h.c.}\right)\ . \tag{4.6}$$

These infinities are of the form $A^2 + B^2$ and $A^2 - B^2$; however, the latter is often forbidden by symmetry arguments.

The $\bar{\chi}\chi$ insertion can induce, in a general theory, the infinities

$$\int d^4x\, d^4\theta\, [U(\varphi\bar{\varphi}^2) + U\varphi D^2\varphi + U(\varphi^3) + U(D^2\bar{D}^2U)\bar{\varphi}\varphi + U(D^2\bar{D}^2U)\bar{\varphi}^2 + \text{h.c.}]\ , \tag{4.7}$$

plus possible linear terms.
The reader is referred to Ref. [56] for the infinities produced by the other additions.
We will now return to the task in hand and consider which soft terms preserve finiteness in those N = 2 rigid supersymmetric theories that are finite. In terms of an N = 1 superfield description, N = 2 Yang-Mills is composed of N = 1 Yang-Mills V and one Wess-Zumino multiplet φ in the adjoint representation, whilst N = 2 matter consists of σ chiral multiplet $X^a_{\ \sigma}$ in the representation R_σ and σ chiral multiplets $Y_{a\sigma}$ in the representation R_σ. The index a labels the elements of the representation R_σ. The action of N = 2 rigid supersymmetric theories written in terms of these N = 1 superfields is

$$A = \text{Tr} \int d^4x\, d^2\theta\, \frac{W^A W_A}{64g^2}$$

$$+ \int d^4x\, d^4\theta\, [\bar{\varphi}^s(e^{gV})_s{}^t\varphi_t + \bar{X}_{a\sigma}(e^{gV_\sigma})^a{}_b X^b{}_\sigma + Y_{a\sigma}(e^{-gV_\sigma})^a{}_b \bar{Y}^b{}_\sigma]$$

$$+ g\int d^4x\, d^2\theta\, \varphi_s (R^\sigma_s)^a{}_b X^b{}_\sigma Y_{a\sigma} + \text{h.c.} + \text{gauge fixing} + \text{ghosts}. \tag{4.8}$$

In the above, $(V^\sigma)^a{}_b = V^s(R^\sigma_s)^a{}_b$ and $(R^\sigma_s)^a{}_b$ are the generators of the group G in the representation R^σ. For the adjoint representation we have, for example,

$$(T_s)_{\ell k} = -f_{s\ell k}\ , \tag{4.9}$$

where $f_{s\ell k}$ are the structure constants of G. We recall that these theories are finite if and only if $C_2(G) = \sum_\sigma T(R^\sigma)$.
Clearly, from our previous discussion the addition of any term of the form $A^2 - B^2$ maintains finiteness. Let us consider adding an $A^2 + B^2$ term, which is achieved by adding the terms

$$\int d^4x\, d^4\theta\, [U_1 \bar{\varphi}^s (e^{gV})_s{}^k \varphi_k + U_{2\sigma} \bar{X}_{a\sigma} (e^{gV}\sigma)^a{}_b X^b{}_\sigma + U_{3\sigma} Y_{a\sigma} (e^{-gV}\sigma)^a{}_b \bar{Y}^b{}_\sigma] , \tag{4.10}$$

here the spurion superfields are of the form

$$U_1 = \mu_1 \theta^2 \bar{\theta}^2 , \quad U_{i\sigma} = \mu_{i\sigma} \theta^2 \bar{\theta}^2 , \quad i = 2, 3 . \tag{4.11}$$

he induced infinities can only be of the generic form

$$\int d^4x\, d^4\theta\, U(\bar{\varphi}\varphi + \bar{X}X + Y\bar{Y}) , \tag{4.12}$$

s a φ^2, X^2, or Y^2 term is ruled out by the symmetry

$$\varphi \to e^{2i\alpha}\varphi , \quad X \to e^{-i\alpha}X , \quad Y \to e^{-i\alpha Y} , \tag{4.13}$$

hich is a symmetry both of the N = 2 action and of the spurion insertion f Eq. (4.10).

e now consider these induced infinities at one loop. The relevant diagrams re given in Fig. 3. These graphs of Fig. 3 can be evaluated most easily sing the following argument. Consider any propagator that contains a U nsertion (Fig. 4) and is part of a larger graph. Using the super-Feynman ules, we may evaluate this part of the graph to contribute the factor

$$\int d^4\theta_2 \left(-\frac{1}{4}\bar{D}_1^2\right)\left(-\frac{1}{4}D_2^2\right)\frac{\delta_{12}}{k^2}\, U(2) \left(-\frac{1}{4}\bar{D}_2^2\right)\frac{\delta_{23}}{k^2} \tag{4.14}$$

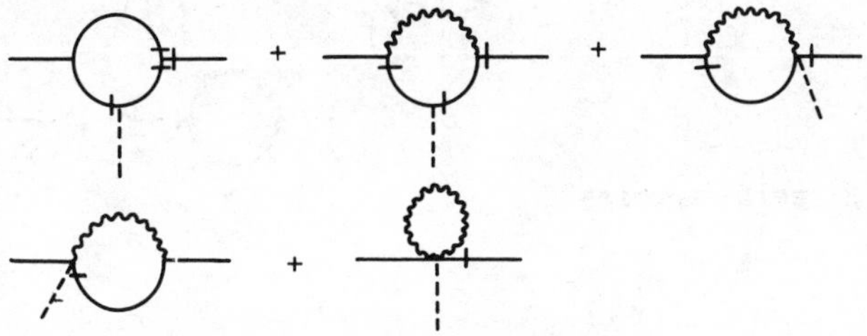

Fig. 3 The $\bar{\varphi}\varphi$ self-energy

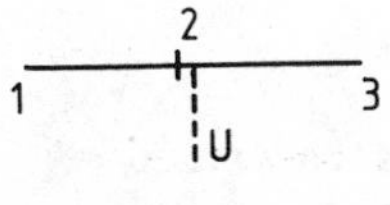

Fig. 4 The U-inserted propagator

to the total expression for the graph. Integrating the D's by parts and using the fact that any infinity cannot contain a D acting on U, we find that the infinite part of the graph contains the factor

$$\int d^4\theta_2 \, \frac{\delta_{12}}{k^2} \, U(2) \, \frac{\bar{D}_2^2 D_2 \bar{D}_2^2}{16 \cdot (-4)} \, \frac{\delta_{23}}{k^2}$$

$$= - \int d^4\theta_2 \, \delta_{12} U(2) \left(- \frac{1}{4} \bar{D}_2^2 \right) \frac{\delta_{23}}{k^2} = - U(1) \left(- \frac{1}{4} \bar{D}_1^2 \right) \frac{\delta_{13}}{k^2} \, . \qquad (4.15)$$

That is, the graph with the U insertion is equal to (-U) times the graph with no U insertion.

Similarly we find that a U insertion on a $\bar{\varphi}(gV)^n \varphi$ vertex yields a graph which is equal to (+U) times the graph with no U insertion.

Taking the case of external $X\bar{X}$ lines, the one-loop graphs are as shown in Fig. 5. Using the above discussion and the one-loop finiteness condition which is diagramatically given in Fig. 6, we find that the induced infinit is of the form

$$C(R) \int d^4x \, d^4\theta \, X_{a\sigma} \bar{X}^{a\sigma} [(-U_1 - U_{3\sigma}) - (-U_{2\sigma}) - (U_{2\sigma}) - (U_{2\sigma}) + (0)]$$

$$= - C(R) \int d^4x \, d^4\theta \, X_{a\sigma} \bar{X}^{a\sigma} (U_1 + U_{2\sigma} + U_{3\sigma}) \, . \qquad (4.16$$

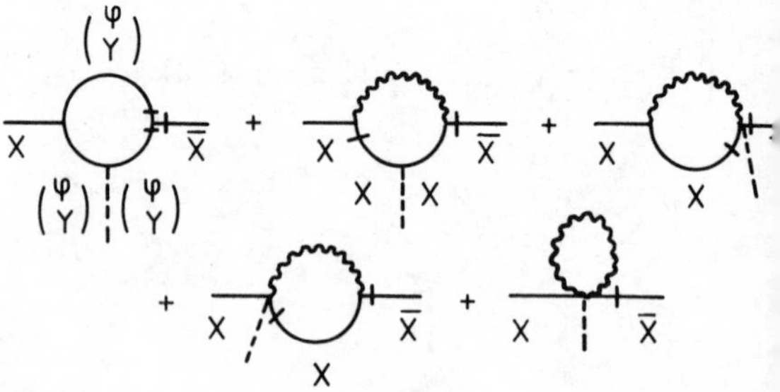

Fig. 5 The $X\bar{X}$ self-energies

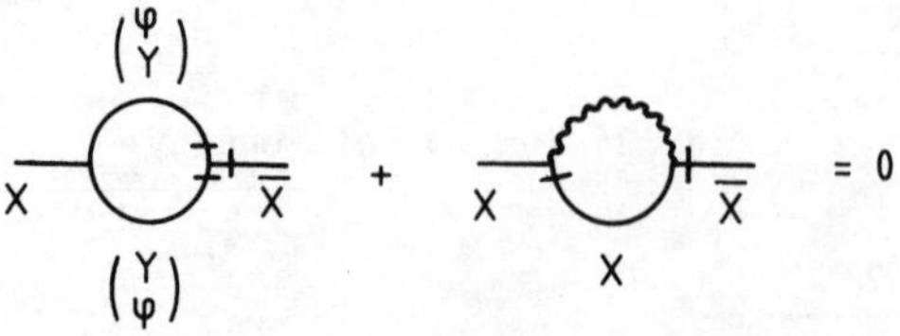

Fig. 6 The one-loop finiteness condition

We note that the last graph vanishes as there are not four D's contained in the vector loop. Hence there are no $\bar{X}X$ infinities at one loop provided

$$U_1 + U_{2\sigma} + U_{3\sigma} = 0 \;\forall\, \sigma \; . \tag{4.17}$$

The $\bar{Y}Y$ infinities also vanish in this case owing to the $X \leftrightarrow Y$, $U_{2\sigma} \leftrightarrow U_{3\sigma}$ interchange symmetry. In fact, the $\bar{\varphi}\varphi$ one-loop infinities also vanish when Eq. (4.17) holds and we use the finiteness condition for N = 2 theories. We can summarize the above discussion by the statement that $A^2 + B^2$ additions preserve finiteness at one loop if and only if Eq. (4.17) holds. By a straightforward generalization of the above discussion it can be shown that an $A^2 + B^2$ term preserves finiteness to all orders provided Eq. (4.17) holds. The reader is referred to Ref. [60] for an account of this argument. For the case of N = 4 Yang-Mills, we have only one species of N = 2 matter fields which are in the adjoint representation, and then the finiteness condition becomes $U_1 + U_2 + U_3 = 0$. This is equivalent to the statement that

$$\mathrm{STr}\; m^2 = \sum_j m_j^{\,2} (-1)^{2j+1} = 0 \; .$$

In N = 2 theories, however, $\mathrm{STr}\; m^2$ is in general non-zero. The addition of all possible soft terms and the resulting induced one-loop infinities, including infinities due to mixed insertions, are given in Ref. [60]. The following is a schematic description of this analysis. Denoting the physical field component content of any of the chiral fields $X^A{}_\sigma$, $Y_{a\sigma}$, φ by A, B, χ, and the spinor in the Yang-Mills multiplet by λ, the one-loop infinities are given in Table 3. A tick indicates the appearance of an infinity. Consider adding a $\chi\chi$ insertion; it gives rise to an infinity of the form $A(A^2 + B^2)$. The only way this infinity can be removed is by adding an appropriate $A(A^2 + B^2)$ soft term and arranging its coefficient such that the $A(A^2 + B^2)$ infinities cancel. Once this has been carried out, it is found that the $A^2 - B^2$ infinities cancel automatically. The remaining $A^2 + B^2$ infinities do not cancel, but they can be removed by adding an appropriate $A^2 + B^2$ soft term. The resulting soft insertions that produce no infinities are of the form

$$m^2(A^2 + B^2) + m\bar{\chi}\chi + mA(A^2 + B^2) \ . \qquad (4.18$$

Examination of the coefficients reveals that this term is none other than an N = 1 supersymmetric mass term, and so can be rewritten in the form

$$m \int d^2\theta \, d^4x \, \mathrm{Tr} \, \varphi^2 + \mathrm{h.c.} \qquad (4.19$$

for the case of φ, and similar terms for X_σ and Y^σ.

Table 3

Infinity structure due to insertions

	Infinity produced					
Insertion	$A^2 - B^2$	$A^2 + B^2$	$\bar{\chi}\chi$	$\bar{\lambda}\lambda$	$A(A^2 + B^2)$	$A^3 - 3AB^2$
$A^2 - B^2$						
$A^2 + B^2$		✓				
$\bar{\chi}\chi$	✓	✓			✓	
$\bar{\lambda}\lambda$	✓	✓				✓
$A(A^2 + B^2)$	✓	✓			✓	
$A^3 - 3AB^2$	✓	✓				✓

An alternative set of soft insertions that also induces no infinities is found by adding a mass term for the gaugino $\bar{\lambda}\lambda$. The resulting $A^3 - 3AB^2$ infinity can only be cancelled by adding a term of the same form, i.e. $A^3 - 3AB^2$ with an exactly chosen coefficient. Again the $A^2 - B^2$ infinity cancels automatically and the remaining $A^2 + B^2$ infinity can be removed by adding an appropriate $A^2 + B^2$ term. The resulting combination of terms is of the generic form

$$m\bar{\lambda}\lambda + m^2(A^2 + B^2) + m(A^3 - 3AB^2) \ . \qquad (4.20$$

Although this is not an N = 1 mass term, it is like an N = 1 mass term in the sense that it is related by O(2) invariance to the mass term of Eq. (4.20); as such, one would expect this term to preserve finiteness. The N = 1 supersymmetric mass terms preserve finiteness to all orders. This results from the fact that a term such as $\int d^4x\, d^2\theta m\varphi^2$, as well as similar terms for X and Y, cannot be generated as a result of the non-renormalization theorem. We can also be confident that the 'N = 1-like mass terms' preserve finiteness to all orders as they are related to N = 1 mass terms by an O(2) rotation.

Consequently, we may summarize this section by listing the necessary and sufficient conditions that preserve finiteness in N = 2 theories: the soft terms must be expressible as a linear combination of

i) N = 1 supersymmetric masses,

ii) N = 1-like mass terms [i.e. ($m\lambda\lambda$ + ...)],

iii) any $A^2 - B^2$ mass,

iv) $A^2 + B^2$ masses provided they satisfy

$$U_1 + U_{2\sigma} + U_{3\sigma} = 0 \ \forall\ \sigma\ . \qquad (4.21)$$

We can also consider whether soft terms preserve the finiteness of the N = 1 theories that have been shown to be two-loop finite [50, 51]. Using arguments similar to those presented above for the N = 2 case, we find that, of the terms listed above, terms of types (i), (iii), and (iv) preserve finiteness to two loops [50], whilst more recently it has been shown that terms of type (ii) preserve one-loop finiteness [64]. Softly broken realistic models of N = 2 supersymmetry have been considered in Ref. [65].

REFERENCES

[1] E.P. Wigner, Ann. of Math. 40, 149 (1939).
A. Salam and J. Strathdee, Nucl. Phys. B80, 499 (1974).
M. Gell-Mann and Y. Ne'eman, unpublished (1974).
W. Nahm, Nucl. Phys. B135, 149 (1978).
For a review, see:
D.Z. Freedman, Proc. Summer Institute on Recent Developments in Gravitation, Cargèse, 1978, eds. M. Lévy and S. Deser (Plenum, New York, 1979), p. 549.
S. Ferrara and C. Savoy, Proc. First School on Supergravity, Trieste, 1981: 'Supergravity '81', eds. S. Ferrara and J. Taylor (University Press, Cambridge, 1982), p. 47.

[2] A. Salam and J. Strathdee, Phys. Lett. 51B, 353 (1974).
P. Fayet, Nucl. Phys. B113, 135 (1976).

[3] J. Wess and B. Zumino, Nucl. Phys. B78, 1 (1974).

[4] M. Sohnius, K. Stelle and P. West, Nucl. Phys. B17, 727 (1980); Phys. Lett. 92B, 123 (1980).

[5] P. Fayet, Nucl. Phys. B113, 135 (1976).

[6] P. Breitenlohner and M. Sohnius, Nucl. Phys. B178, 151 (1981).
M. Sohnius, K. Stelle and P. West, Proc. Nuffield Workshop on Superspace and Supergravity, Cambridge, 1980, eds. S.W. Hawking and M. Roček (University Press, Cambridge, 1981), p. 283.

[7] P. Howe, K. Stelle and P. Townsend, Nucl. Phys. B214, 519 (1983).

[8] P. Howe, K. Stelle and P. West, 'N = 1, d = 6 harmonic superspace', King's College preprint in preparation.

[9] G. Sierra and P.K. Townsend, Nucl. Phys. B233, 289 (1984).
L. Mezincescu and Y.P. Yao, Nucl. Phys. B241, 605 (1984).

[10] F. Gliozzi, J. Scherk and D. Olive, Nucl. Phys. B122, 253 (1977).
L. Brink, J. Schwarz and J. Scherk, Nucl. Phys. B121, 77 (1977).

[11] In this context, see:
M. Roček and W. Siegel, Phys. Lett. 105B, 275 (1981).
V.O. Rivelles and J.G. Taylor, J. Phys. A. Math. Gen. 15, 163 (1982).

[12] A. Salam and J. Strathdee, Nucl. Phys. B80, 499 (1974).

[13] R. Grimm, M. Sohnius and J. Wess, Nucl. Phys. B133, 275 (1978).

[14] M. Sohnius, Nucl. Phys. B165, 483 (1980).

[15] P. Howe, K. Stelle and P. West, Phys. Lett. 124B, 55 (1983).

[16] A. Galperin, E. Ivanov, S. Kalitzin, V. Ogievetsky and E. Sokatchev, Trieste ICTP preprint IC 84-43 (1984).

[17] L. Mezincescu, Dubna JINR report P2-12572 (1979).

[18] J. Koller, Nucl. Phys. B222, 319 (1983); Phys. Lett. 124B, 324 (1983).

[19] P. Howe, K. Stelle and P.K. Townsend, Nucl. Phys. B236, 125 (1984).

[20] J. Wess, Lecture Notes in Physics 77 (Springer, New York, 1978).

[21] J. Gates, K. Stelle and P. West, Nucl. Phys. B169, 347 (1980).

[22] S. Ferrara and B. Zumino, Nucl. Phys. B79, 413 (1974).

[23] D.R.T. Jones, Phys. Lett. 72B, 199 (1977);
E. Poggio and H. Pendleton, Phys. Lett. 72B, 200 (1977).

[24] O. Tarasov, A. Vladimirov and A. Yu, Phys. Lett. 93B, 429 (1980).
M.T. Grisaru, M. Roček and W. Siegel, Phys. Rev. Lett. 45, 1063 (1980).
W.E. Caswell and D. Zanon, Nucl. Phys. B182, 125 (1981).

[25] M. Sohnius and P. West, Phys. Lett. 100B, 45 (1981).

[26] S. Ferrara and B. Zumino, unpublished.

[27] M. Grisaru and W. Siegel, Nucl. Phys. B201, 292 (1982).

[28] S. Mandelstam, Proc. 21st Int. Conf. on High Energy Physics, Paris, 1982, eds. P. Petiau and J. Porneuf (J. Phys. 43, suppl. to No. 12, Paris, 1982), p. C3-331.

[29] L. Brink, O. Lindgren and B. Nilsson, Nucl. Phys. B212, 401 (1983); Phys. Lett. 123B, 328 (1983).
[30] D. Freedman, private communication.
[31] D.R.T. Jones, Nucl. Phys. B87, 127 (1975).
[32] P. West, Proc. II Conf. on Quantum Field Theory and the Fundamental Problems of Physics, Shelter Island, 1983, eds. R. Jackiw, N. Khuri, S. Weinberg and E. Witten (MIT Press, Cambridge, Mass., 1985), p. 127. This work contains a general review of the finiteness properties of supersymmetric theories and in particular an account of the anomalies argument applicable to N = 2 theories. This latter work was performed in collaboration with P. Howe.
[33] P. Howe and P. West, Nucl. Phys. B242, 364 (1984).
[34] A.A. Slavnov, Teor. Mat. Fiz. 13, 1064 (1972).
B.W. Lee and J. Zinn-Justin, Phys. Rev. D5, 3121 (1972).
[35] D.J. Gross and F. Wilczek, Phys. Rev. D8, 3633 (1973).
[36] T. Curtright, Phys. Lett. 102B, 17 (1981);
[37] P. West, CALTEC preprint, CALT-68-1226 (1985).
[38] S.L. Adler and W.A. Bardeen, Phys. Rev. 182, 1517 (1969).
A. Zee, Phys. Rev. Lett. 29, 1198 (1972).
J.H. Lowenstein and B. Schroer, Phys. Rev. D6, 1553 (1972) and D7, 1929 (1973).
[39] M. Grisaru and P. West, Nucl. Phys. B254, 249 (1985).
[40] R. Slansky, Phys. Report 79, 1 (1981).
D. Olive, Proc. Europhysics Study Conf. on the Unification of the Fundamental Particle Interactions (II), Erice, 1981, eds. J. Ellis and S. Ferrara (Plenum, New York, 1983), p. 15.
F. Gursey, Proc. First Workshop on Grand Unification, Durham, 1980, eds. P.H. Frampton, S.H. Glashow and A. Yildiz (Math. Sci. Press, Brookline, Mass., 1980), p. 39.
Mehmet Koca, Phys. Rev. D42, 2636 (1981).
D. Olive and P. West, Nucl. Phys. B217, 248 (1983).
[41] A. Parkes and P. West, Phys. Lett. 127B, 353 (1983).
[42] J.P. Derendinger, S. Ferrara and A. Masiero, Phys. Lett. 143B, 133 (1984).
[43] R. Barbieri, S. Ferrara, L. Maiani, F. Palumbo and A. Savoy, Phys. Lett. 115B, 212 (1982).
[44] L. Susskind, private communication.
[45] S.J. Gates, M.T. Grisaru, M. Roček and W. Siegel, Superspace: or one thousand and one lessons in supersymmetry (Benjamin/Cummings, London, 1983) (vol. 58 of Frontiers in Physics).
[46] L. Alvarez-Gaumé and P. Ginsparg, Harvard preprint HUTP 85A030 (1985).
L. Alvarez-Gaumé, S. Coleman, and P. Ginsparg, Harvard preprint HUTP 85A037 (1985).
[47] L. Alvarez-Gaumé and D.Z. Freedman, Phys. Lett. 94B, 171 (1980); Comm. Math. Phys. 80, 443 (1981).
[48] R. Kallosh, Proc. First School on Supergravity, Trieste, 1981: 'Supergravity '81', eds S. Ferrara and J. Taylor (Univ. Press, Cambridge, 1982), p. 397.
[49] N. Marcus and A. Sagnotti, Phys. Lett. 135B, 85 (1984).
[50] A. Parkes and P. West, Phys. Lett. 138B, 99 (1984).
[51] D.R.T. Jones and L. Mezincescu, Phys. Lett. 138B, 293 (1984).
[52] A. Parkes and P. West, Nucl. Phys. B256, 340 (1985).
[53] M.T. Grisaru, B. Milewski and D. Zanon, The structure of UV divergences in SS YM theories, Utrecht preprint (1985).
[54] A. Parkes, Three loop finiteness conditions in N = 1 super-Yang-Mills theory, Southampton preprint (1985), to be published in Phys. Lett. B.

[55] A. Parkes and P. West, Phys. Lett. 122B, 365 (1983).
[56] A. Parkes and P. West, Nucl. Phys. B222, 269 (1983).
[57] A. Namaize, A. Salam and J. Strathdee, Phys. Rev. D28, 1481 (1983).
[58] J.J. Van der Bij and Y.-P. Yao, Phys. Lett. 125B, 171 (1983).
[59] S. Rajpoot, J.G. Taylor and M. Zaimi, Phys. Lett. 127B, 347 (1983).
[60] A. Parkes and P. West, Phys. Lett. 127B, 353 (1983).
[61] J.-M. Frère, L. Mezincescu and Y.-P. Yao, Phys. Rev. D29, 1196 (1984).
[62] L. Girardello and M. Grisaru, Nucl. Phys. B194, 55 (1982).
O. Piquet, K. Sibold and M. Schweda, Nucl. Phys. B174, 183 (1980).
[63] S. Weinberg, Phys. Rev. D8, 3497 (1973).
[64] D.R.T. Jones, L. Mezincescu and Y.-P. Yao, Phys. Lett. 148B, 317 (1984).
[65] F. del Aguila, M. Dugan, B. Grinstein, L. Hall, G.G. Ross and P. West, Nucl. Phys. B250, 225 (1985).

CONFORMAL SUPERGRAVITIES AS POINCARÉ SUPERGRAVITY IN $d=4$, CHERN–SIMONS TERMS IN $d=3$ AND SPINNING STRINGS IN $d=2$

P VAN NIEUWENHUIZEN

1. INTRODUCTION

In this essay we discuss the conformal supergravities in various dimensions from a unified geometrical point of view. The emphasis is on a clean derivation of the constraints which we must impose on the curvatures of the corresponding superalgebras in order to specify the geometry.

Upon the request of the editor, we also include a self-contained paedagogical discussion of the theory of simple ordinary supergravity in d = 4 dimensional Minkowski spacetime. This part, given in section 6, may serve as a basis for readers who want to be able to read the contributions concerning supergravity in this volume, but who know virtually nothing about supergravity.

In section 2 we discuss the superalgebras which are relevant for supergravity theories. In section 3 we discuss the conformal N = 1 supergravity in d = 4, in section 4 the N = 1 and $N \geq 2$ models in d = 3 while in section 5 we discuss the N = 1 and N = 2 models in d = 2.

Conformal supergravities are the gauge theories of the superconformal algebra. They have no dimensional coupling constant, and hence the purely gravitational part of the action in d dimensions must be a scalar constructed from Riemann curvatures and spin connection $\omega_\mu{}^m{}_n$ which has dimension d. In four dimensions the leading part of the action is proportional to the square of the Weyl tensor and can be written as

$$\int R^{mn}(\omega)\ R^{pq}(\omega)\,\varepsilon_{mnpq} \qquad (1.1)$$

where R^{mn} are the Riemann curvature two-forms. (This expression contains the Weyl tensor since $dx^{\mu}dx^{\nu}dx^{\rho}dx^{\sigma} \sim \varepsilon^{\mu\nu\rho\sigma}\ d^4x$ while the product of two ε-symbols is a sum of products of Kronecker deltas.) In three dimensions, the leading part is proportional to $\int R^{mn} \wedge \omega^{mn}$, or rather, because of Lorentz invariance, it is proportional to the gravitational Chern-Simons form

$$\int (R^{mn}\ \omega^{mn} - \tfrac{1}{3}\omega^{mn}\ \omega^{nk}\ \omega^{km}). \qquad (1.2)$$

In two dimensions, the leading part of any conformal supergravity action can only be proportional to the scalar curvature R, but because this is a total derivative, there are no pure gauge actions of conformal supergravity in d = 2. One can, however, study the couplings of the conformal supergravity gauge fields to matter, and it will be our contention that these couplings reproduce the spinning string. This will explain why the string variables are inert under conformal supersymmetry, and why there are no Brans-Dicke type RA^2 terms in the action of the spinning string and what the origin is of the recent discovery that in the N = 2 gauge action there is one gauge field which gauges two local symmetries simultaneously

$$\delta B_{\mu} = \partial_{\mu}\lambda_B(x) + \det \varepsilon_{\mu\rho}\ \partial^{\rho}\lambda_A(x) + \dots \qquad (1.3)$$

In order to put all these results in the same perspective, we shall first explain which superalgebras correspond to these various conformal supergravities. Various definitions of superalgebras exist. We will introduce them by considering line elements with bosonic and fermionic coordinates, and define the generators of these superalgebras as those linear transformations between the bosonic and fermionic coordinates, which leave the line element invariant. This definition is the natural generalization of the usual definitions of orthogonal, symplectic and unitary Lie algebras, and it is quite easy to understand. It is not always appreciated how natural superalgebras ought to be for a physicist. Let us, therefore, mention the following fact [1]: the canonical transformations which mix bosonic and fermionic creation and absorption operators linearly form a superalgebra, the so-called orthosymplectic algebra Osp(2N|M) with compact SO(2N) and noncompact Sp(M).

2. SUPERALGEBRAS AND CONFORMAL SUPERGRAVITIES

In this section we determine the superalgebras which underlie the conformal supergravities and the dimensions in which they exist.

It is well known that one can define the orthogonal and symplectic Lie groups as those transformations which leave the line elements $x^i \eta_{ij} y^j$ and $\vartheta^\alpha \Omega_{\alpha\beta} \zeta^\beta$ invariant, where η_{ij} is a symmetric real matrix, for example the Minkowski metric, and $\Omega_{\alpha\beta}$ is an antisymmetric real matrix, for example the charge conjugation matrix if d = 4, 3, 2. One can combine the two Lie algebras into one superalgebra, properly called the orthosymplectic algebra, Osp(N|M), (or noncompact versions when they exist).Its definition can again be given in terms of a line element [1], namely the sum of the orthogonal and symplectic line elements, but (to begin with, later we will remove them) it involves Grassmann variables ϑ^α. The line element in question is thus

$$\vartheta^\alpha \Omega_{\alpha\beta} \zeta^\beta + x^i \eta_{ij} y^j \qquad (2.1)$$

The pairs (ζ^α, y^i) and (ϑ^α, x^i) transform in the same way under infinitesimal transformations

$$\delta \begin{pmatrix} \vartheta \\ x \end{pmatrix} = \begin{pmatrix} A & B \\ C & D \end{pmatrix} \begin{pmatrix} \vartheta \\ x \end{pmatrix}, \text{ idem} \begin{pmatrix} \zeta \\ y \end{pmatrix} \qquad (2.2)$$

It follows that A and D have commuting (even) entries, while B, C have anticommuting (odd) entries. To investigate the invariance of the line element, we note that

$$\delta(\vartheta^T, x^T) = (\vartheta^T, x^T) \begin{pmatrix} A^T & -C^T \\ B^T & D^T \end{pmatrix} \equiv (\vartheta, x)^T M^T. \qquad (2.3)$$

The minus sign is crucial, and due to pulling ϑ^T past C^T. Since the line element can be written as $z^T H w$ with $H = \text{diag}(\Omega, \eta)$ and $z = (\vartheta, x)$, we find as condition for the generators of Osp(N|M)

$$M^T H + HM = 0, \; H = \begin{pmatrix} \Omega & 0 \\ 0 & \eta \end{pmatrix}, \; M^T = \begin{pmatrix} A^T & -C^T \\ B^T & D^T \end{pmatrix} \tag{2.4}$$

It is clear from the definition in terms of a line element that if M_1 and M_2 satisfy (2.4) then so does the commutator $[M_1, M_2] = M_1M_2 - M_2M_1$ where supermatrices are multiplied just as ordinary matrices. One may check this also directly, by using that the transposition rule in (2.4) satisfies

$$(M_1M_2)^T = M_2^T M_1^T. \tag{2.5}$$

Hence, the supermatrices M satisfying (2.4) form a closed algebra with the ordinary commutator as the composition rule.

Let us now construct a new definition of the orthosymplectic algebras involving only real numbers, but no Grassmann variables. For this purpose we consider the entries of the supermatrix equation in (2.4) and imagine removing in the B, C sectors an anticommuting variable. We do the same thing for each of the M_i in (2.5). In this way we are led to consider ordinary matrices m with an unusual transposition rule

$$m = \begin{pmatrix} a & b \\ c & d \end{pmatrix}, \; m^T = \begin{pmatrix} a^T & -c^T \\ b^T & d^T \end{pmatrix}, \; a, b, c, d \text{ ordinary matrices.} \tag{2.6}$$

The composition rule is again the commutator except if one deals with two off-diagonal parts, then it is the anticommutator. Formally, the composition rule $\langle m_1, m_2 \rangle$ is defined by $m = m^e + m^o$, $m^e = \begin{pmatrix} a & o \\ o & d \end{pmatrix}$, $m^o = \begin{pmatrix} o & b \\ c & o \end{pmatrix}$ and

$$\begin{aligned} \langle m_1^e, m_2^e \rangle &= [m_1^e, m_2^e], \quad \langle m_1^e, m_2^o \rangle = [m_1^e, m_2^o] \\ \langle m_1^o, m_2^o \rangle &= \{m_1^o, m_2^o\} \end{aligned} \tag{2.7}$$

The generators of Osp(N|M) are now defined by

$$m^T H + Hm = 0, \; m^T \text{ as in (2.6)} \tag{2.8}$$

and one may check that if m_1 and m_2 satisfy (2.8), then so does $\langle m_1, m_2 \rangle$. In other words, the matrices satisfying (2.8) form a closed superalgebra. (Superalgebra, not because there are anticommuting variables involved (there are not), but because the bracket relations can be symmetric as well as antisymmetric.) The bosonic subalgebra consists of a compact SO(N) and a noncompact Sp(M), but versions with compact Sp(M), and other versions, also exist.

The next class of superalgebras are the superunitarity matrices with superdeterminant one. Their bosonic sector consists again of ordinary matrices of SU(N) and SU(M) as well as a U(1). One defines them by considering

$$(\vartheta^{\alpha})^{*}\Omega_{\alpha\beta}\zeta^{\beta} + (z^{i})^{*}\eta_{ij}w^{j} \tag{2.9}$$

as the line element where now ϑ^{α}, ζ^{β} are complex anticommuting variables, and (z^{i},w^{j}) complex commuting variables. The * operation does <u>not</u> interchange two Grassmann variables, per definition. As before we can determine those linear transformations which leave (2.9) invariant. We again need to transpose the transformation law of (ϑ^{*},z^{*}) which again introduces a minus sign in one of the blocks of M^{T}, and one finds

$$M^{\dagger}H + HM = 0, \quad M^{\dagger} = \begin{pmatrix} A^{\dagger} & -C^{\dagger} \\ B^{\dagger} & D^{\dagger} \end{pmatrix} \tag{2.10}$$

At this point B and C are thus anticommuting. The hermitian conjugation rule in (2.10) for supermatrices satisfies

$$(M_1M_2)^{\dagger} = M_2^{\dagger}M_1^{\dagger} \tag{2.11}$$

and it follows from this, that if M_1 and M_2 satisfy (2.10) then so does $[M_1,M_2]$. We can restrict our attention to supermatrices with unit superdeterminant. Their generators have then vanishing supertrace, which is defined as - tr A + tr D. (For these and other properties of Grassmann variables, see the recent book "Supermanifolds" by B·S·DeWitt, Cambridge University Press.) Since always

$$\mathrm{str}\,[M_1,M_2] = 0 \tag{2.12}$$

it follows that those supermatrices M which satisfy (2.10) and str M = 0, form a closed superalgebra under the ordinary commutator. This superalgebra is SU(N|M).

To obtain a definition in terms of ordinary complex numbers we proceed as before. In this way we find as generators of SU(N|M) the ordinary complex matrices m satisfying

$$m^\dagger H + Hm = 0, \qquad m = \begin{pmatrix} a & b \\ c & d \end{pmatrix}, \qquad m^\dagger = \begin{pmatrix} a^\dagger & -c^\dagger \\ b^\dagger & d^\dagger \end{pmatrix}$$

$\langle m_1, m_2\rangle$ as in (2.7).

$$\text{str } m = - \text{tr } a + \text{tr } d = 0 \tag{2.13}$$

The reader may check that when m_1 and m_2 satisfy (2.13), then so does $\langle m_1, m_2\rangle$. In the bosonic sector we find SU(M) x SU(N) x U(1) where U(1) is generated by a diagonal supertraceless matr

Let us now consider the superalgebras which will correspond to supergravity theories. Supergravity theories correspond to superalgebras which contain in their bosonic part a spacetime algebra. There are four spacetime algebras of interest [2]:

(i) the super-Poincaré algebra. It is a contraction of (ii) and will not be considered further;

(ii) the anti-de Sitter algebras SO(d-1,2) in d-dimensional Minkowski spacetime;

(iii) the de Sitter algebras SO(d,1). There exist super-extensions, involving * algebras, but they lead to field theories with ghosts. A thorough discussion can be found in [3]. The basic reason is that the trac of the $\{Q, Q^*\}$ commutator equals minus the energy operator P_0, instead of $+P_0$ as in (ii). We shall therefore no longer consider them;

(iv) the conformal algebra SO(d,2). It is generated by those Killing vectors which leave a maximally symmetric metric invariant up to an overall scale which may be x-dependent. There are only three types of spaces with maximally symmetric metrics: anti-de Sitter spaces; de Sitter spaces and flat space. Conversely, the only metrics which admit a purely conforma Killing vector are the maximally symmetric spaces. (Se the book by Yano.)

The superalgebras we are interested in must thus contain either SO(d-1,2) or SO(d,2) in the A or D blocks. However, we also want the odd generators (the blocks B and C) to transform under the spacetime group as spinors. This is input, based on physical prejudice (the spin-statistics theorem). Thus, the generators of SO(d-1,2) and SO(d,2) must be in the spinor representation.

From the construction given above, it is clear that all generators are in the vector representation of SO(N), Sp(M) and SU(N). However, there are some isomorphisms which relate SO(N) groups to Sp(M) and SU(N) groups. In particular

$$SO(3,2) \simeq Sp(4,R) \qquad SO(2,2) \simeq Sp(2,R)\otimes Sp(2,R)$$
$$SO(4,2) \simeq SU(2,2) \qquad SO(1,2) \simeq Sp(2,R) \tag{2.14}$$

By Sp(2N,R) we mean 2N x 2N real symplectic matrices. Note that they form a noncompact representation of the symplectic groups. This allows us to find superalgebras which contain SO(d-1,2) and SO(d,2) generators in the spinor representation, because the vector representation of the right-hand sides of (2.14) are the spinor representations of the left-hand sides. (We only talk about algebras, local isomorphisms. We say nothing about groups.)

We therefore find the following correspondence between conformal and anti-de Sitter supergravities and superalgebras.

Table 2.1

Minkowski dimension	Conformal supergravity	anti-de Sitter supergravity
2	Osp(N|2) ⊗ Osp(N|2)	Osp(N|2)
3	Osp(N|4)	Osp(N|2) ⊗ Osp(N|2)
4	SU(2,2|N)	Osp(N|4)
5	F_4?	SU(2,2|N)
6	Osp(6,2|N)	F_4?
7	?	Osp(6,2|N)
8	?	?

In d=6 and d=7, the SO(6,2) ≃ SO(8)* has a superextension. Since the vector representation of SO(8) is equivalent to the spinor representation, there is a superalgebra of physical interest, Osp(6,2|N). It involves, however, complex entries (but real structure constants). The superalgebra F_4 was discussed by B S DeWitt and the author [4]. It is conjectured by them and others to be related to a conformal supergravity theory in d=5.

3. d=4 N=1 CONFORMAL SUPERGRAVITY

The first conformal supergravity model constructed was the N=1 d=4 model [5]. From a theoretical point of view, its interest lies in the issue of the constraints one has to impose; these constraints are needed to define the kinematics and can be derived by a general method which we will discuss. From a practical point of view the constraints are useful because many models of ordinary supergravity couplings, including the gauge action of ordinary (= Poincare) supergravity itself, can be obtained by taking the corresponding conformal couplings and fixing the conformal local symmetries by putting certain fields equal to zero [6]. This we shall not discuss further.

The conformal superalgebra contains the following generators and gauge connection $\omega^A X_A = e^m P_m + \frac{1}{2}\omega^{mn} M_{mn} + f^m K_m + bD + \bar{\psi}Q + \bar{\varphi}S + AA$. The corresponding curvatures will be denoted by $R(P)^m$ etc. They are quite generally defined by

$$R^A = d\omega^A - \tfrac{1}{2} f^A{}_{BC}\,\omega^C \omega^B \qquad (3.1)$$

Let us begin by mentioning a remarkable result: the gauge action of N=1 d=4 conformal supergravity has the extremely simple and elegant form [7]

$$I = \int R(M)^{mn}\, R(M)^{pq} \varepsilon_{mnpq} + \bar{R}(Q)\gamma_5 R(S) + R(A)R(D) \qquad (3.2)$$

(omitting wedge symbols).

To verify this, one needs to know how connections transform. It is here that constraints will appear.

The curvatures satisfy the Bianchi identities

$$DR^A = dR^A + f^A{}_{BC}\, R^C \omega^B \equiv 0. \qquad (3.3)$$

The variation of the connections $\omega^A \to \omega^A + \delta\omega^A$ in the curvatures R^A provides one with the definition of a covariant derivative D

$$\delta R^A = (D\delta\omega)^A = d\delta\omega^A - f^A{}_{BC}\delta\omega^C\omega^B. \tag{3.4}$$

The gauge transformations of the connections ω^A are defined as the covariant derivatives of the local gauge parameters $\varepsilon^A(x)$

$$\delta(\text{gauge})\omega^A = (D\varepsilon)^A = d\varepsilon^A + f^A{}_{BC}\varepsilon^C\omega^B. \tag{3.5}$$

This yields the following local gauge algebra

$$\begin{aligned}[\delta(\varepsilon_1^C),\ \delta(\varepsilon_2^B)]\omega^A &= \delta(\varepsilon_1)(D\varepsilon_2)^A - 1\leftrightarrow 2\\ &= \delta(-f^A{}_{BC}\varepsilon_1^C\varepsilon_2^B)\omega^A.\end{aligned} \tag{3.6}$$

It is now clear that it is simply incorrect to define the transformation rules of the connections only as a gauge transformation. The reason is that the basic commutator

$$\{Q^\alpha, Q^\beta\} = -\tfrac{1}{2}(\gamma^m C^{-1})^{\alpha\beta}P_m \tag{3.7}$$

which holds also in the superconformal algebra, would lead to the following commutator of two local supersymmetry transformations

$$[\delta_Q(\varepsilon_1), \delta_Q(\varepsilon_2)]\omega^A = \delta_P(\tfrac{1}{2}\bar{\varepsilon}_2\gamma^m\varepsilon_1)\omega^A. \tag{3.8}$$

However, δ_P with parameters ξ^m is not a symmetry of the theory one wants. For a thorough discussion, see ref [22].

It is easy to find out what δ_P precisely is [5]. To this aim we begin by rewriting a general coordinate transformation as follows

$$\delta_{gc}(\xi^\lambda)\omega^A{}_\mu = \xi^\lambda\partial_\lambda\omega_\mu{}^A + \partial_\mu\xi^\lambda\omega_\lambda{}^A = D_\mu(\xi^\lambda\omega_\lambda{}^A) + \xi^\lambda R_{\lambda\mu}{}^A. \tag{3.9}$$

Recalling that $(D_\mu\varepsilon)^A$ is the complete set of gauge transformations of the connections ω^A under all local symmetries, we can isolate the terms which contain the parameter that corresponds to δ_P(gauge). These terms tell us how ω^A transforms under δ_P(gauge). The remaining terms in $(D_\mu\varepsilon)^A$ are local gauge transformations other than P-gauge transformations. These latter symmetries we want as symmetries in the theory.

Bringing them to the left-hand side of (3.9), we see that we get on the left-hand side a sum of local symmetries of the theory. (We want the theory to be invariant under general coordinate transformations. This is input at this point, based on physical prejudice.) Hence the right-hand side will be a local symmetry. It reads

$$\delta_P^{(\text{gauge})}(\xi^m)\omega^A \quad + \xi^\lambda R^A_{\lambda\mu} \tag{3.10}$$

Thus, $\delta_p^{(\text{gauge})}$ are not local symmetries of the theory but the combinations

$$\delta_P^{(\text{gauge})}(\xi^\lambda e^m_\lambda)\,\omega^A + \xi^\lambda R^A_{\lambda\mu} \tag{3.11}$$

are local symmetries, namely a sum of gauge transformations and general coordinate transformations.

It now becomes clear how one can systematically deduce the constraints needed to specify the theory. Consider the gauge commutator

$$[\delta_Q^{(\text{gauge})}(\varepsilon_1),\, \delta_Q^{(\text{gauge})}(\varepsilon_2)]\omega^A_\mu = \delta_P^{(\text{gauge})}(\tfrac{1}{2}\bar{\varepsilon}_2\gamma^m\varepsilon_1)\omega^A_\mu . \tag{3.12}$$

Since the right-hand side is not a local symmetry, the δ_Q(gauge) are not good local symmetries either. One must find a mechanism to obtain the extra curvatures on the right-hand side (as in (3.10)). This mechanism is to impose constraints on curvatures which one can solve algebraically. If a given constraint can be solved by eliminating a particular connection, say ω_0, then ω_0 no longer transforms as $D\varepsilon$, but rather as follows from the chain rule. Thus

$$\delta(\text{total})\omega^A = \delta(\text{gauge})\omega^A + \delta(\text{extra})\omega^A . \tag{3.13}$$

For the fields which are independent fields, δ(extra) vanishes, but for the fields which are expressed in terms of the independent fields, δ(extra) will be nonvanishing in general.

If there were no constraints, the structure constants of the gauge algebra would be truly constant, given by $-f^A{}_{BC}\varepsilon^C_1\varepsilon^B_2$, and the gauge algebra would coincide with the abstract superalgebra one started with. The imposition of constraints, needed to replace all $\delta_P^{(gauge)}$ on the right-hand side of gauge commutators by "δ_P + curvatures", leads to field-dependent "structure functions". The minimal change required is to change the δ_Q(gauge) into δ_Q(gauge) + δ_Q(extra), but to keep for the other local symmetries δ(extra) vanishing. The reason is that P_m appears on the right-hand side of commutators only in the $\{Q,Q\}=P$ relation. Hence, δ_Q(total) cannot be equal to δ_Q(gauge). It follows that the minimal modifications are:

Theorem: one must modify the transformation laws of only the dependent fields and only for local Q transformations.

It remains at this point to prove that with the minimal change the set of local gauge symmetries (including general coordinate transformation but excluding δ(gauge)P) will form a closed gauge algebra.

Given a constraint, one can solve from it the connection ω_0. If one makes a gauge transformation of the constraint, one can either obtain zero (in general after using the constraints again) or not. If one does not get zero, $\delta(\text{extra})\omega_0$ must cancel this nonzero result, because the total variations of a solved constraint must vanish. If the gauge variation of the constraint is zero, $\delta(\text{extra})\omega_0 = 0$ and the dependent field ω_0 transforms in the same way as if it were independent.

From these arguments we conclude the following. Since $\delta(\text{extra})\omega_0$ must vanish for all local symmetries except Q, the constraints must be invariant under all δ(gauge) except Q. The constraints must be gauge invariant except under Q.

We can now see how to deduce the constraints. One begins with a field which one wants to keep as an independent field in the theory, the vielbein in general. One evaluates the commutator of two local δ_Q(total) on it. The first variation has δ_Q(total) = δ_Q(gauge) because the vielbein is an independent field. If it varies into a set of fields one of which is dependent, one finds in the second variation a $\delta(\text{extra})\omega_0$, otherwise one finds for the $\{Q,Q\}$ commutator only δ(gauge)P.

One subtracts from this result the required result (δ(gauge)P + R^A terms). The difference is then the constraint. One has then to verify whether this constraint is algebraically solvable. If it is, fine. If not, then the program terminates at this point, and the superalgebra considered does not lead to an acceptable field theory.

Let us show in the case of N=1 d=4 what one gets. From $\delta_Q(\text{gauge}) e_\mu^m = \frac{1}{2}\bar{\varepsilon}\gamma^m\psi_\mu$ and the fact that we want ψ_μ as an independent field in the theory we find that both Q variations involve only δ_Q(gauge). Hence, only a δ_P(gauge) is produced, and one must impose the constraint

$$R(P)^m = 0. \tag{3.14}$$

(We will encounter this constraint again in simple ordinary supergravity where it will be found that it should vanish. The procedure outlined in this section explains from a group-theoretical point of view why.)

This constraint is algebraically solvable. One can solve the spin connection

$$\omega_\mu^{mn} = \omega_\mu^{mn}(e, \psi, b). \tag{3.15}$$

It is indeed gauge invariant, except under δ_Q(gauge). From the variation of $R(P)^m$ under δ_Q(gauge) one deduces that

$$\delta(\text{extra})\omega_\mu{}^{mn} = -\tfrac{1}{4}(\bar{\varepsilon}\gamma_m R_{\mu n}(Q) + \bar{\varepsilon}\gamma_\mu R_{mn}(Q) - \bar{\varepsilon}\gamma_n R_{\mu m}(Q)). \tag{3.16}$$

Next one considers

$$[\delta_Q(\text{total}, \varepsilon_1), \delta_Q(\text{total}, \varepsilon_2)]\psi_\mu. \tag{3.17}$$

On ψ_μ, $\delta_Q(\text{total}) = \delta_Q(\text{gauge})$, but since ψ_μ varies into $(D_\mu(\omega) - \frac{1}{2}b_\mu + \frac{3i}{4}\gamma_5 A_\mu)\varepsilon$, we get an $\delta_Q(\text{extra})\omega_\mu{}^{mn}$ term. Thus we get an extra term involving $\delta_Q(\text{extra})\omega_\mu{}^{mn}$, and this term we equate to the required result, namely $\xi^\lambda R_{\lambda\mu}(Q)$. The result is a new constraint [5]

$$\gamma^\mu R_{\mu\nu}(Q) = 0. \tag{3.18}$$

This constraint can again be solved algebraically, and yields

$$\varphi_\mu{}^\alpha = \tfrac{1}{3}\gamma^\nu R'_{\nu\mu}(Q) + \tfrac{1}{12}\gamma_5\gamma^\tau \varepsilon_{\mu\tau}{}^{\rho\sigma} R'_{\rho\sigma}(Q) \tag{3.19}$$

where φ_μ^α is the conformal gravitino given above (3.1) and where $R'_{\mu\nu}(Q)$ equals $R_{\mu\nu}(Q)$ minus the terms with φ. From (3.18) we get an $\delta(\text{extra})\varphi_\mu$. It reads

$$\delta(\text{extra})\varphi_\mu = \tfrac{1}{6}(\gamma^\nu\varepsilon)R_{\nu\mu}(D) + 5 \text{ more terms involving } R(D),\ R(M) \text{ and } R(A). \tag{3.20}$$

Next we consider the chiral field A_μ. There is no curvature with a term of the form e A, hence A_μ cannot be solved algebraically by constraints: it must be an independent field. Similarly we keep b_μ as an independent field. It could have been solved from $R(P)^m=0$, but we already used $R(P)^m=0$ to solve $\omega_{\mu mn}$, hence that opportunity is gone. We find from $\delta\omega_\mu^A = (D_\mu\varepsilon)^A$ that

$$\delta_Q(\text{gauge})b_\mu = \delta_Q(\text{total})b_\mu = \tfrac{1}{2}\bar{\varepsilon}\varphi_\mu \tag{3.21}$$

$$\delta_Q(\text{gauge})A_\mu = \delta_Q(\text{total})A_\mu = -i\bar{\varepsilon}\gamma_5\varphi_\mu. \tag{3.22}$$

Taking $\delta_Q(\text{total})$ of these expressions and forming the commutator, we find expressions involving $\delta_Q(\text{extra})\varphi_\mu$. We also know what the right-hand side should be, namely

$$\delta_P(\tfrac{1}{2}\bar{\varepsilon}_2\gamma^m\varepsilon_1)b_\mu + \tfrac{1}{2}\bar{\varepsilon}_2\gamma^\lambda\varepsilon_1 R_{\lambda\mu}(D) \tag{3.23}$$

and a similar expression for A_μ.

So we know what $\delta_Q(\text{extra})\varphi_\mu$ should have been in order to get the correct result. We also deduced what $\delta_Q(\text{extra})\varphi_\mu$ actually is. The difference between these two expressions for $\delta_Q(\text{extra})\varphi_\mu$ yields the new constraint. It reads

$$(\text{Ricci})_{\mu\nu} + R_{\mu\nu}(D) + \tfrac{1}{2}\bar{\psi}^\lambda\gamma_\mu R_{\nu\lambda}(Q) = 0 \tag{3.24}$$

$$R_{\mu\nu}(D) + \frac{ie}{4}\varepsilon_{\mu\nu\rho\sigma}R(A)^{e\sigma} = 0$$

where $(\text{Ricci})_{\mu\rho} = R(M)_{\mu\nu}{}^{mn}e_m{}^\nu e_{n\rho}$.

This constraint is again gauge-invariant, except under $\delta_Q(\text{gauge})$, and can be solved algebraically, in which case it eliminates $f_\mu{}^m$. This, in turn, leads to a $\delta_Q(\text{extra})f_\mu{}^m$. At this point one has a closed gauge algebra. The independent fields are

$$e_\mu{}^m,\ \psi_\mu^\alpha,\ A_\mu,\ b_\mu \tag{3.25}$$

while $\omega_{\mu mn}$, φ_μ^α and $f_\mu{}^m$ have been eliminated. The constraints read

$$R(P)^m_{\mu\nu} = 0 \ , \ \gamma^\mu R(Q)^\alpha_{\mu\nu} = 0; \ R_{\mu\nu}(D) + \frac{ie}{4}\varepsilon_{\mu\nu\rho\sigma}R^{\rho\sigma}(A) = 0$$
$$(\mathrm{Ricci})_{\mu\nu} + R_{\mu\nu}(D) + \tfrac{1}{2}\bar{\psi}^\lambda\gamma_\mu R_{\nu\lambda}(Q) = 0. \tag{3.26}$$

The Bianchi identities lead to further relations which are not new constraints but follow from these constraints. They read

$$R_{\mu\nu}(Q) = -\tfrac{1}{2}\gamma_5\varepsilon_{\mu\nu}{}^{\rho\sigma}R_{\rho\sigma}(Q); \qquad \gamma_{[\mu}R(Q)_{\nu\rho]} = 0$$
$$R(M)_{[\nu\rho\sigma]}{}^m = -R(D)_{[\nu\rho}e^m{}_{\sigma]} + R(Q) \text{ terms} \tag{3.27}$$

and others.

In general, it is useful to first make a list of all connections which can in principle be solved from constraints on curvatures. These candidate-constraints need not in the actual case be needed, but if there are constraints they must come from this set. This explains a priori that e^m_μ, ψ^α_μ are independent fields, since there are no curvatures with ee or eψ terms, due to $[P_m,P_n] = 0$ and $[P_m,Q^\alpha] = 0$. It also may happen that the basic commutator $[\delta_Q(\text{total})), \delta_Q(\text{total})]$ contains further symmetries on the right-hand side. (In our derivation that simply means that $\delta_P(\text{gauge})$ + curvature + $\delta_A(\text{gauge})$ with some A is a local symmetry.)

4. d = 3 CONFORMAL SUPERGRAVITIES

The N = 1 conformal algebra in d = 3 dimensions is Osp(1|4) and contains the following generators and connections

$$P_m e^m + \tfrac{1}{2}\omega^{mn}M_{mn} + f^m K_m + bD + \bar{\psi}Q + \bar{\varphi}S. \tag{4.1}$$

There is no axial gauge field as in d = 4, which makes sense since in odd dimensions one cannot define chiral transformations. The gauge action should start with a gravitational Chern-Simons form, as we explained. This suggests to consider as complete gauge action for N = 1 conformal supergravity theory the Chern-Simons form for the full superconformal algebra. In this way one is led to consider the following action [8]

$$I = \int\gamma_{AB}R^B\omega^A + \tfrac{1}{6}f_{ABC}\omega^C\omega^B\omega^A. \tag{4.2}$$

The γ_{AB} are the Killing metric for the superconformal algebra

Osp(1|4). It is defined by

$$\gamma_{AB} = f^{P}{}_{AQ} f^{Q}{}_{BP} (-)^{\sigma(P)} \qquad (4.3)$$

where $\sigma(P) = +1$ if X_P is a bosonic generator, and $\sigma(P) = -1$ if X_P is a fermionic generator. Further f_{ABC} are the structure constants of Osp(1|4) with all indices lowered

$$f_{ABC} = \gamma_{AD} f^{D}{}_{BC}. \qquad (4.4)$$

One may verify that f_{ABC} is totally super-antisymmetric. Under a general variation $\omega^A \to \omega^A + \delta\omega^A$, the variation of the action becomes

$$\delta I = \int 2\gamma_{AB} R^B \delta\omega^A \qquad (4.5)$$

In fact, this is one of the definitions of a Chern-Simons form. As far as we know, the only application of the ideas and properties of Chern-Simons terms to superalgebras is this case.

To determine the total transformation laws of the connections, $\delta_B(\text{total})\,\omega^A = \delta_B(\text{gauge})\omega^A + \delta_B(\text{extra})\omega^A$, we must first determine the constraints. Let us ahead of time mention some of the results, and use them to prove that the action is indeed invariant under all 14 local symmetries $\delta_B(\text{total})$ of the conformal superalgebra. Afterwards we will discuss the derivation of these constraints and the closure of the gauge algebra.

The gauge variations $\delta(\text{gauge})\omega^A = (D\varepsilon)^A$ cancel in the action. This result is trivial, since $\int \gamma_{AB} R^B (D\varepsilon)^A$ vanishes, after partial integration, due to the Bianchi identities $DR^B = 0$. There remain the extra variations

$$\delta I = \int 2\gamma_{AB} R^B \, \delta(\text{extra})\omega^A. \qquad (4.6)$$

As we shall show below, the only nonvanishing $\delta(\text{extra})\omega^A$ are for ω^A equal to the spin connection, the conformal gravitino and the conformal boost field

$$\delta(\text{extra})\omega^{mn}, \ \delta(\text{extra})\varphi^{\alpha}, \ \delta(\text{extra})f^{m}. \qquad (4.7)$$

We claim that each of these expressions is multiplied by a curvature which vanishes as a result of the constraints. The constraints which we shall impose, read

$$R(P)^m = 0, \quad R(Q)^\alpha, \quad R(M)^{mn} = 0. \tag{4.8}$$

Now γ_{AB} is only non-zero in the cases that the pair of indices take on the values (M^{mn}, M^{kl}), (P^m, K^n) and (Q^α, S^α). This one can either verify directly, using equation (4.3) and the structure constants of Osp(1|4), or, more easily, by noting that γ_{AB} can only couple curvatures with opposite Weyl weights. Now, P and K, and Q and S have opposite Weyl weights. Further, there is no antisymmetric Lorentz invariant tensor with two indices (mn), so γ_{AB} cannot couple the pair (M^{mn}, D). Hence, in (4.6) the terms in (4.7) are multiplied by the curvatures in (4.8) and thus δI in (4.6) indeed vanishes. The reader may appreciate the simplicity of this proof. All that was needed was the "diagonality" of γ_{AB}, and the constraints in (4.8). In fact, as we shall show, invariance of the action only requires $R(P)^m = R(Q)^\alpha = 0$ but not $R(M)^{mn} = 0$ because from $R(P)^m = 0$ and $R(Q)^\alpha = 0$ one can deduce $\delta_Q(\text{extra})\omega_\mu{}^{mn} = 0$. (In d = 4 it is also true that $\delta_Q(\text{extra})\omega_\mu^{mn}$ is proportional to R(Q), see (3.16).

To deduce the constraints needed to close the gauge algebra (and to specify the kinematics of the theory), we begin again as in the d = 4 case. We deduce from the requirement of closure on the vielbein that $R(P)^m = 0$, and then from the requirement of closure on the gravitino one obtains the constraint $\bar{\varepsilon}t_m R_{n\mu}(Q) = \bar{\varepsilon}t_n R_{m\mu}(Q)$. This is equivalent to $R_{\mu\nu}(Q) = 0$ (since we are in d = 3), which in turn is equivalent to $t^\mu R_{\mu\nu}(Q) = 0$. (By t_m we denote the Pauli matrices).

From $R(P)^m = 0$ and $R(Q)^\alpha = 0$ one can eliminate ω^{mn} and φ^α, respectively. (Note that $R(P)_{\mu\nu}{}^m$ has always as many components as $\omega_\mu{}^{mn}$, while $R(Q)^\alpha_{\mu\nu} = 0$ has as many components as φ^α_μ in d = 3.) Let us count the number of bosonic and fermionic field components.

$$\begin{aligned}
&9e_\mu^m - 3 \text{ gen. coord.} - 3 \text{ local Lorentz} - 1 \text{ local scale} = \\
&\quad 2 \text{ bosonic} \\
&6\psi_\mu^\alpha - 2 \text{ local Q sup} - 2 \text{ local S sup} = 2 \text{ fermionic} \\
&2b_\mu - 2 \text{ local K gauge invariances} = 0.
\end{aligned} \tag{4.9}$$

Thus, the field f_μ^m is too much and must be eliminated. In fact, there are curvatures in which the combination $e_\mu^m f_\nu^n$ appears, namely R(M) and R(D). From these, one can indeed eliminate f_μ^m as an independent field.

Any constraint which eliminates f_μ^m leads to a closed gauge algebra. The reason is that one can always obtain closure on b_μ by adding to the right-hand side of the {Q, Q} gauge commutator a local K-gauge term. Since only b_μ transforms under K^m, and as $\delta_K(\text{gauge})b_\mu = \xi^m_{(K)}(x)e_{m\mu}$, the good results for vielbein and gravitino are unaffected while the last independent field, b_μ, has then by definition a closed gauge algebra. The simplest constraint from which f_μ^m can be eliminated is $R(M)_{\mu\nu}^{mn} = 0$. Note that this constraint has as many components as f_μ^m.

We have thus at this point derived $R(P) = R(Q) = R(M)^{mn} = 0$ and obtained a closed gauge algebra. However, these constraints imply other constraints by means of the Bianchi identities. The Bianchi identity $DR(P)^m = 0$ implies

$$R(D) = 0. \tag{4.10}$$

The Bianchi identity $DR(Q) = 0$ implies

$$t_{[\mu}R_{\nu\rho]}(S) = 0. \tag{4.11}$$

The Bianchi identity $DR(M)^{mn} = 0$ implies

$$\bar{\psi}t^{mn}R(S) = e^m R(K)^n - e^n R(K)^m. \tag{4.12}$$

The Bianchi identity $DR(D) = 0$ implies

$$\bar{\psi}R(S) = e^m R(K)^m. \tag{4.13}$$

In fact, these latter two identities tell us that of the 3

irreducible parts of $R(K)^{m}_{\mu\nu}$, corresponding to the Young tableaux

the first two can be expressed in $R(S)$.

We can even go further and determine the explicit form of the δ(extra) in (4.7) by taking the δ_Q(gauge) of the constraints in (4.8). From the fact that $\delta_Q(\text{extra})\omega_{\mu}{}^{mn}$ depends in general only on $R(Q)$, because $\{Q, Q\} \sim P$, it follows that

$$\delta_Q(\text{extra})\omega_{\mu mn} = 0. \tag{4.14}$$

Similarly, from δ(gauge) $R(Q)$ we find that $\delta(\text{extra})\varphi_{\mu}^{\alpha}$ depends only on $R(D)$ and $R(M)$, but since these curvatures vanish, also

$$\delta_Q(\text{extra})\varphi_{\mu}^{\alpha} = 0. \tag{4.15}$$

In a similar manner, one finds from $\delta_Q(\text{gauge})R(M)^{mn}$ that

$$\delta_Q(\text{extra})f^{m}_{\mu} = \text{constant}[\bar{\varepsilon}_Q\tau^{mk}R_{k\mu}(S) + \tfrac{1}{4}e^{m}_{\mu}\bar{\varepsilon}_Q\tau^{kl}R_{kl}(S)]. \tag{4.16}$$

In fact, since $\delta_Q(\text{gauge})R(D)$ also gives information on $\delta_Q(\text{extra})f^{m}_{\mu}$, namely

$$\delta_Q(\text{extra})f^{m}_{\mu}\, e_{\nu]m} = \text{constant}\ \bar{\varepsilon}_Q R_{\mu\nu}(S) \tag{4.17}$$

we can combine (4.16) and (4.17) to obtain

$$t^{\mu}R_{\mu\nu}(S) = -i\varepsilon_{\nu}{}^{\rho\sigma}R_{\rho\sigma}(S). \tag{4.18}$$

These results concerned the theory of $N = 1$ conformal supergravity in $d = 3$. One can extend these results to the $N = 2$ case without modification, but from $N = 3$ onwards there are no longer equal numbers of bosonic and fermionic field components, and hence for $N \geq 3$ one needs auxiliary fields to close the gauge algebra. We now briefly discuss the $N = 2$ results [9].

One imposes again $R(P)^{m} = R(M)^{mn} = R(Q)^{i} = 0$ with now $i = 1,2$. The algebra is $\text{Osp}(2|4)$, and the action is again invariant because the new gauge field A_{μ} for the $SO(2)$ gauge transformations is physical and hence its $\delta(\text{total})A_{\mu}$ equals $\delta(\text{gauge})A_{\mu}$ which cancels in (4.6) due to the Bianchi identities

The algebra closes on e_μ^m and ψ_μ^i because of $R(P)^m = R(Q)^i = 0$, and on b_μ because from $R(M)^{mn} = 0$ it follows again that $R(D) = 0$. The reason is that the Bianchi identity $DR(P)^m = 0$ does not involve the new curvature $R(A)$ because P_m and the SO(2) generator commute. There is only one field left to check closure on, and that is the SO(2) field A_μ. Counting field components, we see that the bosonic and fermionic components still match; there is one extra gravitino (2 components) and the gauge field A_μ (also 2 components). This very strongly suggests closure on all fields. In fact, from $\delta(\text{total})R(Q)^i = 0$ one finds that $\delta(\text{extra})\varphi$ no longer vanishes, but is proportional to $R(A)$, and the $[\delta_Q(\text{total}, \varepsilon_1), \delta_Q(\text{total}, \varepsilon_2)]A_\mu$ commutator indeed closes. ($\delta_Q(\text{total})A_\mu = \delta_Q(\text{gauge})A_\mu = -2\bar{\varepsilon}^i\psi_\mu^j\varepsilon_{ij}$).

Let us mention that one can use stronger counting criteria than we have used: the bosonic and fermionic field components should not only match, but they should correspond in 1-1 fashion with the states of massive supersymmetry multiplets. For an exposition, see the author's contribution to the proceedings of the 1985 Nuffield Workshop on Supersymmetry and its Applications at Cambridge, England. These criteria are again fulfilled by the above models.

Let us also mention that in the path-integral the measure, when it is defined, according to Fujikawa [10], such that all general coordinate BRST anomalies cancel, should also be invariant under Q-supersymmetry. This is again a test on the field content of the models, and this test is met as well.

Finally we note the curious observation that the action

$$I = \int(\gamma_{AB}R^B\omega^A + \tfrac{1}{6}f_{ABC}\omega^C\omega^B\omega^A) \qquad (4.19)$$

is, in fact, invariant for any of the N - extended Osp(N|4) models, for any N. There is no N = 16 limit [9]. The reason is that all SO(N) gauge fields are physical, hence they only vary into covariant derivatives which cancel in (4.6) due to the Bianchi identities.

5. d = 2 CONFORMAL SUPERGRAVITIES

Continuing our descent we reach the d = 2 case. As we discussed before, the action should contain as its bosonic part the Einstein scalar curvature, but that vanishes in d = 2. Consequently, there does not seem to be a pure gauge action of conformal (and Poincare) supergravity in d = 2. However, we can study the conformal couplings of the gauge fields of conformal supergravity to conformal matter multiplets. We will only discuss the cases N = 1 and N = 2.

The conformal algebra in d = 2 being Osp(1|2) x Osp(1|2), the gauge fields are given by

$$X_A\omega^A = P_m e^m + \tfrac{1}{2}\omega_{mn}M^{mn} + f_m K^m + bD + \bar{\psi}Q + \bar{\psi}S. \qquad (5.1)$$

To specify the kinematics and obtain a closed gauge algebra, we begin by imposing

$$R(P)^m = R(Q) = 0 \qquad (5.2)$$

for this ensures closure on e^m_μ and ψ^i_μ. We can no longer use Bianchi identities to deduce further constraints as a consequence of the "primary" constraints because in d = 2 Bianchi identities collapse to trivialities. The spacetime and Q and S supersymmetries can gauge away the whole vielbein and gravitino ψ_μ, while b_μ is cancelled by local K-gauges. Hence, we count at this point: 4 bosonic components of f^m_μ and 2 fermionic components of $\varphi_\mu - \frac{1}{2}\tau_\mu\tau.\varphi$. The τ-trace of the conformal gravitino is fixed by the R(Q) = 0 constraint

$$\tau.\varphi = -\tau^{\mu\nu}(D_\mu + \tfrac{1}{2}b_\mu)\psi_\nu. \qquad (5.3)$$

We can remove two further components of f^m_μ by imposing the constraint

$$R(M)^{mn} = R(D) = 0. \qquad (5.4)$$

They can be used to eliminate the trace and the antisymmetric parts of f^m_μ, and they also ensure closure on b_μ. At this point one has equal numbers of bosonic and fermionic field components namely the 2 symmetric traceless parts of f^m_μ and the 2 components of the τ-traceless φ_μ. One might therefore expect that the gauge algebra closes. However, this is not the case [22], and further constraints are needed. It is here that we will encounter a new feature, not present in other models. Namely, we will impose constraints directly on fields, rather than on curvatures. These constraints read

$$f_{\mu\nu} + f_{\nu\mu} - g_{\mu\nu} f = 0; \quad \varphi_\mu - \tfrac{1}{2} t_\mu t.\varphi = 0 \tag{5.5}$$

where $f_{\mu\nu} = f_\nu^m e_{m\mu}$ and $f = f_\lambda^\lambda$. With these constraints, the local gauge algebra closes on e_μ^m, ψ_μ^α and b_μ.

Let us consider a scalar multiplet Σ. If it is a rigid conformal scalar multiplet, it consists of the fields (A, χ, F), with A and F real and χ a 2-component Majorana spinor. From superfield methods, or, better, the theory of induced representations, see [2], one deduces the transformation rules of Σ under rigid conformal supersymmetry. The result is the same as in the rigid Poincare supersymmetry case (because the $(Q\ ,\ P_m)$ subalgebra is the same in both cases). One finds

$$\delta_Q A = \tfrac{1}{2}\bar{\varepsilon}_Q \chi, \qquad \delta_Q \chi = (\tfrac{1}{2}(\partial\!\!\!/ A) + \tfrac{1}{2}\, F)\varepsilon_Q \tag{5.6}$$
$$\delta_Q F = \tfrac{1}{2}\bar{\varepsilon}_Q \partial\!\!\!/ \chi.$$

Using next the commutation rules of the generators one deduces how Σ transforms under D, S. Under K it is inert because K lowers the dimension of a field by one unit, so that only $\delta_K F \sim A$ might have been nonvanishing, but the index structure cannot match. ($\delta_K F = [F, \xi^m_{(K)} K_m]$ cannot be equal to $\xi^m_{(K)} \partial_m A$ because $\partial_m A$ has too high a dimension). Under P_m we define $\delta_P A = [A, \xi^m_{(P)} P_m]$, and in this case one finds

$$\delta_S A = 0, \qquad \delta_S \chi = \lambda A \varepsilon_S, \qquad \delta_S F = -\lambda \bar{\varepsilon}_S \chi$$
$$\delta_D A = \lambda A \lambda_D, \qquad \delta_D \chi = (\lambda + \tfrac{1}{2})\chi\lambda_D, \qquad \delta_D F = (\lambda + 1) F \lambda_D. \tag{5.7}$$

The constant parameter λ is arbitrary, and called the scale of Σ. These results, and, in fact, the whole treatment in this section is almost a verbatim repetition of the d = 4 analysis [11] discussed in more detail in ref [1] section 4.

We now construct a <u>local</u> conformal scalar multiplet Σ. It is again given by $\Sigma = (A, \chi, F)$ but in the transformation rules one replaces ordinary derivatives by superconformally covariant derivatives D_μ^C. Thus, for example, since $\delta_S \chi = \lambda A \epsilon_S$ one finds in $D_\mu^C \chi$ a term $-\lambda A \varphi_\mu$, where φ_μ is the gauge field of S-supersymmetry, namely the conformal gravitino.

Next we apply the general rules of the conformal tensor calculus. [11]

Rule 1: If Σ is a (rigid or local) conformal multiplet, then so is

$$\Sigma_1 x\, \Sigma_2 = (A_1 A_2,\ A_1 \chi_2 + A_2 \chi_1,\ A_1 F_2 + A_2 F_1 - \bar{\chi}_1 \chi_2). \quad (5.8)$$

Rule 2: If Σ is a (rigid or local) multiplet, then so is

$$T\Sigma = (F,\ \not{D}^C \chi,\ \Box^C A). \quad (5.9)$$

The $\Box^C$ equals the ordinary Laplacian for a rigid multiplet, but for a local multiplet it is given by

$$\Box^C A = g^{\mu\nu}\ D_\mu^C\ D_\nu^C\ A$$

(For details, see again ref [1], page 305)

Rule 3: If Σ is a local multiplet with scale $\lambda = 1$, then the following is an action invariant under <u>all</u> local symetries (g.c, l.L, Q, S, D and K)

$$I = \int d^2 x \det e\ [F + \tfrac{1}{2}\bar{\psi}.\tau\chi + \tfrac{1}{4}\bar{\psi}_\mu \tau^{\mu\nu} \psi_\nu]. \quad (5.10)$$

We can now very easily construct the coupling of a local Σ to the fields in (5.1). Namely, we evaluate

$$I(\Sigma\ x\ T\Sigma) \quad \text{for } \Sigma \text{ with } \lambda = 0. \quad (5.11)$$

The reason we need Σ with $\lambda = 0$ is that then $\tilde{\Sigma} \equiv \Sigma\ x\ T\Sigma$ has $\lambda = 1$, as required by Weyl or S invariance. At $\lambda = 0$, however all fields A, χ and F are S-inert, see (5.6). Moreover, in $\Box^C(A)$ an RA term, which is present in d = 4, is now absent; the reason can again be traced back to the fact that $\lambda = 0$. (For $\lambda = 0$, there is no b connection in $D_\mu^C A$, so $D_\mu^C A$ is locally K-inert, so that in $D_\nu^C D_\mu^C A$ there is no K-connection. The constraint $R(M)^{mn} = 0$ would have equated this K-connection (the trace of f_μ^m) to the scalar curvature R).

Evaluation of (5.11) now yields the action of the N = 1

spinning string [12]. In particular one now understands

(i) the occurrence of the 'accidental' symmetries (S-supersymmetry, Weyl invariance). As we have shown, the reason is that the spinning string is a genuine conformal theory, and not the coupling of Poincare supergravity fields to matter with accidental symmetries due to the fact that one is in d = 2. Note that Kaluza-Klein reduction of d = 4 Poincare supergravities to d = 2 yields nonconformal d = 2 models. Thus, d = 2 models need not be conformal [13]!

(ii) the reason that the string variables A, χ, F are S-inert.

(iii) the reason that the Brans-Dicke invariant RA^2 is absent. (Note that RA^2 is nonvanishing.)

Let us now extend our considerations to the N = 2 case. The superalgebra is now

$$Osp(2|2) \times Osp(2|2) \approx SU(1|1) \times SU(1|1) \qquad (5.12)$$

with the new SO(2) x SO(2) corresponding to <u>two</u> new local symmetries, chiral invariance A and phase invariance B. Deducing how the gravitino transforms, i.e. working out the covariant derivative for (5.12), one finds

$$\delta_Q\varphi^i_\mu = (D_\mu + \tfrac{1}{2}b_\mu)\varepsilon^i_Q + \varepsilon^{ij}(\tau_3 A_\mu + B_\mu)\varepsilon^j_Q. \qquad (5.13)$$

Further, the combination $B_\mu = A_\mu + e\varepsilon_{\mu\rho} B^\rho$ contains a term $\partial_\mu\lambda_A(x) + e\varepsilon_{\mu\rho}\partial^\rho\lambda_B(x)$ where λ_A and λ_B are the local chiral and phase parameters. Evaluation of the action now shows that, as far as invariance of the action is concerned, the τ_3 in $\delta_Q\psi^i_\mu$ always combines with a τ_λ in the variations to produce a term $\sim\varepsilon_{\lambda\rho}\tau^\rho$. (This is, of course, only possible in d = 2.) Hence, as far as invariance of the action is concerned, one could as well have started with a term $\varepsilon^{ij}\varepsilon_{\mu\nu}A^\nu\varepsilon^j_Q$ in the gravitino law. In this way one understands an observation of Fradkin and Tseytlin [13] who found a new local invariance in the N = 2 spinning string of Brink and Schwarz [14], but also found that one gauge field, B_μ, gaugestwo local symmetries, local chiral and phase symmetry. Our resolution is that there are actually two gauge fields, A_μ and B_μ, corresponding to the two SO(2) groups in (5.12), but that in the action only the linear combination B_μ appears and that the correct transformation law for the gravitino (involving both A_μ and B_μ) may be replaced by a law involving only B_μ [22]. Note that

$$2A_\mu = (g_{\mu\rho} + e\varepsilon_{\mu\rho})A^\rho + (g_{\mu\rho} - e\varepsilon_{\mu\rho})A^\rho \tag{5.14}$$

is an algebraic covariant split of A_μ into two other gauge fields, but that these fields should not be interpreted as the gauge fields for chiral and phase gauge transformations. Rather, we keep the orthodox point of view that <u>any</u> gauge fiel $F_\mu(x)$ always transforms into ∂_μ times a parameter.

One could extend these ideas beyond N = 2. Thus one migh find the conjectured SU(2) string theory of Ademollo et al.

6. SIMPLE POINCARÉ SUPERGRAVITY IN d = 4 [15, 16]

In order to construct the local lagrangian field theory which corresponds according to the table in section 2 to the superalgebra Osp(1|4), we begin by associating to each generator a gauge connection. This was one of the starting points in the construction of supergravity in [15]. The generators o Sp(4) ~ SO(3,2) (the anti-de Sitter Lie algebra in d = 4) are the translation generators P_m and the Lorentz generators M_{mn}. The corresponding connections have thus the index structure $e_\mu^{\ m}$ and $\omega_\mu^{\ mn}$ and are identified with the vielbein and spin connection. In addition there are 4 odd generators Q^α (α = 1,4). They correspond to the 4 x 1 matrix b and the 1 x 4 matrix c i (2.6). Since c^T_η = b according to (2.6), there are only four c diagonal generators, and these we identify with Q^α. The corresponding gauge fields we denote by ψ^α_μ. Thus, we start with the following minimal set of connections

$$\varphi_\mu^{\ m}, \qquad \psi_\mu^{\ \alpha}, \qquad \omega_\mu^{\ mn}. \tag{6.1}$$

In some theories (d = 4, N > 2, for example), there are more <u>physical</u> fields than only the connections which correspond to the gauge algebra, but in this model these will turn out to be all the fields needed.

One would expect equal numbers of bosonic and fermionic states, but one can only define states if one has an action. One could also count the numbers of bosonic and fermionic field components in the theory, but that would require that we know which local symmetries the theory will have. (For example, in electromagnetism, the existence of a local U(1) means that the part $A_\mu \sim \partial_\mu \Lambda$ is simply not present in the theory.) One could try to discover the list of local symmetries while constructing the local gauge algebra, but that would be algebraically not easy. Here we begin with the, in our opinion, simpler approach of starting from an action. In any case that was the way it was first done.

One could choose for the gravitational (bosonic) part of the action any of the actions one has in general relativity. We note that we expect from the $\{Q,Q\} \sim P$ anticommutator that the commutator of local supersymmetry transformations would lead to local translations, i.e. general coordinate transformations, and that therefore the theory would have to be invariant under general coordinate transformations. We could even start from a linearised action and constant linearised transformation rules, as given by the theory of induced representations applied to the rigid supersymmetry algebra, and use the Noether method to construct the whole theory order-by-order in κ. In that way we would reproduce ordinary general relativity, together with supergravity. While that may be beautiful from a purist point of view, it would be needless extra algebra, since we have all known the result since 1917. Thus, in everything we do we will always covariantise with respect to gravity at all stages.

For the gravitational action we will choose Einstein's action. This is a choice, and supergravity theories based on the Weyl action (yielding conformal supergravity) and various R^2-type of action exist, too [23]. The Einstein action we define as follows

$$\mathcal{L}(E) = \frac{-1}{2\kappa^2} eR(e,\omega), \qquad e = \det e_\mu{}^m$$
$$R(e,\omega) = R(\omega)_{\mu\nu}{}^{mn} e_n{}^\mu e_m{}^\nu . \qquad (6.2)$$

At this point ω_μ^{mn} is just some field; it may or may not be a function of other fields.

The curvature is defined as in general gauge theories with connections ω^A and structure constants given by $[X_A, X_B\} = X_C f^C{}_{AB}$, as follows [17]

$$R^A_{\mu\nu} = \partial_\mu \omega_\nu{}^A - \partial_\nu \omega_\mu^A + f^A{}_{BC}\omega_\mu^C \omega_\mu^B$$
$$R(\omega)_{\mu\nu}{}^{mn} = \partial_\mu \omega_\nu{}^{mn} + \omega_\mu{}^m{}_k \omega_\nu{}^{kn} - \mu \leftrightarrow \nu. \tag{6.3}$$

Let us note that one can use forms to rewrite these results in a compact way as follows

$$R^A = d\omega^A - \tfrac{1}{2} f^A{}_{BC}\omega^C\omega^B \tag{6.4}$$

$$\mathcal{L}(E) = \frac{1}{8\kappa^2} R(\omega)^{mn} e^p e^q e_{mnpq}, \qquad e^P \equiv dx^\mu e_\mu{}^P. \tag{6.5}$$

The choice of Einstein action is justified not only because it is the foundation of classical general relativity, but also because its linearised form coincides with the <u>unique</u> field theory for free massless spin 2 fields with two derivatives and with positive energy [19]. Using the order-by-order in κ Noether method, one can reconstruct from this free field theory the full interacting theory of Einstein, as we already mentioned.

To write down the fermionic part of the action we note that, just as for the spin 2 case, also the action for a free massless spin 3/2 field with one derivative and positive energy is unique [20]. It reads, in any dimension,

$$\mathcal{L} = -\tfrac{1}{2}\bar{\psi}_\mu \gamma^{\mu\rho\sigma}\partial_\rho \psi_\sigma; \qquad \{\gamma^\mu, \gamma^\nu\} = 2\eta^{\mu\nu}; \tag{6.6}$$

In four dimensions, the bar on $\bar{\psi}_\mu$ means $\psi_\mu^T C$ where the charge conjugation matrix C satisfies $C\gamma_\mu = -\gamma_\mu^T C$ and $C^T = -C$. This field ψ_μ is real if the Dirac matrices γ_1, γ_2, γ_3, γ_0, are real; in that case $C \sim \gamma_0$. In a general representation ψ_μ is a Majorana field. A Majorana field is a field whose Dirac conjugate $\psi^+\gamma_4$ (and $\gamma_4 = i\gamma_0$) equals its Majorana conjugate $\psi^T C$. Thus

$$\bar{\psi}_\mu = \psi_\mu^T C, \qquad C\gamma_\mu = -\gamma_\mu^T C, \qquad C^T = -C. \tag{6.7}$$

The action in (6.6) has a local gauge invariance, namely

$$\delta\psi_\mu^\alpha \sim \partial_\mu \varepsilon^\alpha(x) \tag{6.8}$$

where the local gauge parameter is a spinor. This identifies the model we are constructing and which is the gauge theory belonging to Osp(1|4), with the gauge theory of local supersymmetry. Recall that the parameter of rigid supersymmetry is a constant 4-component spinor ε^α, and that gauge fields transform always into "∂_μ of the parameter plus more". One may check this statement for the cases of Yang-Mills theory and gravity. We have thus deduced a gauge invariance (local supersymmetry) from the requirement that the free field actions have positive energy. The same holds for the spin 2 case.

It is easy to write down the action for this spin-3/2 field, covariantised with respect to gravity, using the standard rules. The result reads

$$\begin{aligned}\mathcal{L}(RS) &= -\tfrac{1}{2}e\bar{\psi}_\mu\gamma^{\mu\rho\sigma}D_\rho(\omega)\psi_\sigma, \qquad e = \det e_\mu{}^m \\ D_\rho(\omega)\psi_\sigma &= \partial_\rho\varphi_\sigma + \tfrac{1}{4}\omega_\rho{}^{mn}\gamma_{mn}\psi_\sigma.\end{aligned} \tag{6.9}$$

The symbol $\gamma^{\mu\rho\sigma}$ equals $e_m^\mu e_r^\rho e_s^\sigma$ times γ^{mrs}, and γ^{mrs} is the product of γ^m, γ^r and γ^s, antisymmetrical in mrs with strength unity.

One is thus led to consider as action the following expression

$$\mathcal{L} = \mathcal{L}(E) + \mathcal{L}(RS). \tag{6.10}$$

Since the spin-3/2 field is the fermionic companion of the gravitational variables, we will call it the gravitino. Let us now assume that this action has a local nonlinear invariance with parameter $\varepsilon(x)$. This is similar to the Yang-Mills case, where the linearised theory has two invariances, a local invariance $\delta A_\mu{}^a = \partial_\mu\Lambda^a(x)$ and a rigid symmetry $\delta A_\mu{}^a = f^a{}_{bc}A_\mu{}^b\lambda^c$ with λ^c constant. The former corresponds to $\delta\psi_\mu = \partial_\mu\varepsilon(x)$, the latter to the transformation rules of rigid supersymmetry as applied to the linearised spin (2, 3/2) system. An analysis of the linearised spin (2, 3/2) system of rigid supersymmetry was first given in [18].

For the gravitationally covariantised action in (6.10), we use the gravitationally covariantised transformation rule

$$\delta\psi_\mu = \tfrac{1}{\varkappa}D_\mu(\omega)\varepsilon. \tag{6.11}$$

The gravitational coupling constant $\varkappa$ is needed for dimensiona reasons: ψ_μ has dimension-3/2, just like the electron field, ∂_μ has dimension-1, while ε has dimension-1/2. The latter result follows from rigid supersymmetry, where one has δ(bose field) = ε x (fermi field), and by using that the dimensions of a bose and fermi field differ by one half unit of the dimension of a mass. (That result follows from the fact that the canonical action for bosons and fermions have two and one derivative, respectively, while actions, whether bosonic or fermionic, have always the same dimension.)

We can now start the construction of the theory. First we vary the two gravitino fields in $\mathscr{L}$(R S). This requires some algebra, but the structure of the result is clear. Variation of ψ_σ yields a commutator $[D_\rho, D_\sigma]\varepsilon$, hence a curvature, times $\bar\psi$ and ε. Variation of $\bar\psi_\mu$, after partial integration of $D_\rho\psi_\sigma$ yields the same result. The sum of these two terms is thus of the form: curvature times ψ_μ times ε. It is now immediately clear that only a contracted curvature can appear. Indeed, the only possible combinations are

$$(\bar\psi_{\cdot}\gamma_{\cdot}\varepsilon)R^{\cdot\cdot\cdot\cdot} \text{ and } (\bar\psi_{\cdot}\gamma_{\cdot\cdot\cdot}\varepsilon)R^{\cdot\cdot\cdot\cdot} \tag{6.12}$$

and the latter part vanishes (at least, the part with $R_{\mu\nu}{}^{mn}$ depending on the spin connection $\omega(e)$ of general relativity which depends on vielbeins) due to the cyclic identity. But a contracted curvature is proportional to the Ricci tensor $R_{\mu\nu}$, or equivalently, to the Einstein tensor $G_{\mu\nu} = R_{\mu\nu} - \frac{1}{2}g_{\mu\nu}R$. Hence, the result must be of the form $\bar\varepsilon\gamma^m\psi^\mu R_{\mu m}$ or $\bar\varepsilon\gamma^\mu\psi^m R_{\mu m}$ (Note that $R_\mu{}^\eta = R_{\mu\nu}{}^{mn}e_m{}^\nu$, or rather $R_{mn} = e^\mu_m R_{\mu n}$, is not symmetric for arbitrary $\omega_\mu{}^{mn}$.) Parity excludes terms with γ_5 matrices. In fact, even for general $\omega_\mu{}^{mn}$ one finds no terms with $\bar\psi\gamma_{\cdot\cdot\cdot}\varepsilon$. (Technically this is due to the identity $\{\gamma^{mnrs}, \gamma^{kl}\}$ = terms with one or five gamma matrices but not with three gamma matrices.)

The exact result due to varying both gravitinos in $\mathscr{L}$(RS) reads

$$\delta_\psi \mathcal{L}(RS) = \frac{e}{2\kappa}(\bar{\psi}_\mu \gamma_m \varepsilon) G^{m\mu}(\omega, e). \tag{6.13}$$

Let us now vary the vielbeins in $\mathcal{L}(E)$. As is well known the result is proportional to the Einstein tensor. One finds, for an as yet arbitrary transformation law

$$\delta_e \mathcal{L}(E) = + \frac{1}{\kappa^2}\ \delta e_{\mu m} G^{m\mu}(\omega, e). \tag{6.14}$$

Clearly, these two variations cancel if one puts

$$\delta e_{\mu m} = -\frac{\kappa}{2}\ \bar{\psi}_\mu \gamma_m \varepsilon. \tag{6.15}$$

Using that $\bar{\psi}_\mu \gamma_m \varepsilon = \psi_\mu^T C \gamma_m \varepsilon = -\varepsilon^T \gamma_m^T C^T \psi_\mu = -\varepsilon^T C \gamma_m \psi_\mu = -\bar{\varepsilon}\gamma_m \psi_\mu$, one obtains the equivalent result

$$\delta e_{\mu m} = \frac{\kappa}{2} \bar{\varepsilon} \gamma_m \psi_\mu. \tag{6.16}$$

We have thus derived the vielbein transformation rule from that of the gravitino.

There remain precisely four more variations. They come from

(i) the terms of the form $D_\rho e_\mu^m$ due to the partial integration we performed;

(ii) the variation of the vielbeins in $\mathcal{L}(RS)$;

(iii) the variation of $\omega_\mu{}^{mn}$ in $\mathcal{L}(E)$;

(iv) the variation of $\omega_\mu{}^{mn}$ in $\mathcal{L}(RS)$.

In the first derivation of supergravity $\omega_{\mu mn}$ was a sum of a gravitational term $\omega(e)$ and $\bar{\psi}\psi$ terms. In this case variations were explicitly computed, and it was verified that their sum cancelled. However, the algebra was complicated, and a computer was used to achieve this [15]. In the first-order version, $\omega_\mu{}^{mn}$ was considered as an independent field, and its transformation rule was chosen such that the sum of the variation cancelled [16]. Here we will follow two other approaches, closely related. In the first approach, sometimes called 1.5 order formalism, we assume that the field $\omega_{\mu mn}$ is not independent but solves its own nonpropagating field equation, $\delta I/\delta\omega_{\mu mn} = 0$. The result being $\omega = \omega(e,\psi)$, we imagine that we have all along been dealing with this $\omega(e,\psi)$. Nothing changes in the derivation so far. The usefulness of taking this particular $\omega(e,\varphi)$ is that when one varies it in the action, its variations automatically cancel

$$\frac{\delta I}{\delta\omega(e,\varphi)}\left[\frac{\delta\omega(e,\varphi)}{\delta e}\delta e + \frac{\delta\omega(e,\psi)}{\delta\psi}\delta\psi\right] = 0 \tag{6.17}$$

since $\delta I(e,\psi,\omega(e,\psi))/\delta\omega(e,\psi) \equiv 0$. Thus in this approach the terms (iii) and (iv) need not be computed. Note that we are thus dealing with second order formalism: ω is not an independent field; however, it pays not to expand this $\omega(e,\psi)$ but to keep it as a composite field. Some confusion might arise if one compares this treatment with the result one gets from 'gauging the algebra'. If one gauges in general an algebra, with connections ω^A and structure constants $f^A{}_{BC}$, the gauge variations are defined as

$$\delta\omega^A = d\varepsilon^A - f^A{}_{BC}\omega^C\varepsilon^B. \tag{6.18}$$

Applying this recipe to the super-Poincaré algebra, one obtains $\delta e^m_\mu = \frac{1}{2}\bar{\varepsilon}\gamma^m\psi_\mu$ and $\delta\psi_\mu = D_\mu(\omega)\varepsilon$, which is nice, but also $\delta\omega_\mu{}^{mn}=0$ This result is not the correct result in supergravity, where $\delta\omega_\mu{}^{mn} \neq 0$.(It had better be non zero because from $\{Q,Q\} \sim P$ and $P\omega \neq 0$ it follows that $Q\omega \neq 0$). One can give a derivation of the correct result for $\delta\omega$ using group theory but this derivation is more subtle than just using (6.18).

Here we abandon the approach based on the gauging of groups, and treat $\omega(e,\psi)$ as the solution of $\delta I(e,\psi,\omega)/\delta\omega = 0$. The two variations under (i) and (ii) are then proportional to

$$D_\mu(\omega(e,\psi)e^n{}_\nu - D_\nu(\omega(e,\psi)e^m{}_\mu - \frac{\varkappa}{2}\bar{\psi}_\mu\gamma^m\psi_\nu \tag{6.19}$$

and this expression vanishes identically. It is, in fact, just the field equation for $\omega_{\mu mn}$.

In the second approach we now describe, we do not make any statement about $\omega_{\mu mn}$, and we just write '$\delta\omega_{\mu mn}$' for the variation of $\omega_{\mu mn}$. One now evaluates the four remaining variations explicitly, (which is not very hard) and finds then that their sum <u>factorises</u> into two factors

$$\delta(\text{remaining})[\mathcal{L}(E) + \mathcal{L}(RS)] = \varepsilon^{\mu\nu\rho\sigma}\varepsilon_{mnrs}(-2\varkappa^2)^{-1}$$

$$[\delta\omega_\mu{}^{mn}e^r_\nu + \tfrac{\varkappa}{6}\bar{\varepsilon}\gamma^{mnr}D_\mu\psi_\nu][D_\rho e^s_\sigma - \tfrac{\varkappa^2}{4}\bar{\psi}_\rho\gamma^s\psi_\sigma]. \tag{6.20}$$

We can now see in a very simple way that there are just two formulations of supergravity, no more. Either the first factor vanishes, and this leads to the first-order formalism [16] in which $\omega_\mu{}^{mn}$ is an independent field whose transformation law follows from the vanishing of the first factor. Or the second factor vanishes. In this case we have the second-order formalism [15], in which $\omega_{\mu mn}$ is not an independent field, but depends on $e_\mu{}^m$ and ψ_μ^α. In fact, the solution of the requirement that the second factor vanishes coincides with the solution of the ω field equation. Yet another way in which one can characterise this solution is by constructing the curvature associated with the generator P_m, called $R(P)^m$. Then, the following three conditions are equivalent

$$\frac{\delta I(e,\psi,\omega)}{\delta\omega} = 0; \qquad D_\rho(\omega)e^s{}_\sigma - D_\sigma(\omega)e^s{}_\rho = \frac{\kappa^2}{2}\bar{\psi}_\rho\gamma^s\psi_\sigma$$

$$R(P)^m = 0; \qquad R(P)^m = de^m + \omega^m{}_n e^n - \frac{\kappa^2}{4}\bar{\psi}\gamma^m\psi. \tag{6.21}$$

One can now go on and investigate the local gauge algebra [21]. It turns out that in second-order formalism the (Q,Q) commutator on the gravitino contains extra non closure terms, proportional to the field equation of the gravitino. These can be removed, and a closed gauge algebra can be obtained, by introducing so-called auxiliary fields. These and other matters have been discussed at various other places, but the foregoing should give enough background information for the other articles of supergravity in this volume.

Note added in proof: The N = 4 spinning string theory has now been constructed (see Pernici M and van Nieuwenhuizen P 1986 Phys. Lett. B, to be published). The supersymmetrization of Chern-Simons terms in d = 10 has been discussed by Romans L J and Warner N P, Caltech preprint. An analysis of Chern-Simons terms but in d = 5 supergravity has been given by Roček M, van Nieuwenhuizen P and Zhang C S 1986 Phys. Rev. D 33 370.

REFERENCES

[1] van Nieuwenhuizen P 1981, Phys.Rep. 68 276. I thank Dr Hasiewicz for a discussion on superline elements.
[2] For further properties, see P C West and P van Nieuwenhuizen, forthcoming book, (Cambridge: Cambridge University Press)
[3] Hasiewicz P and Lukierski J, Wroclaw preprint
Pilch K, Sohnius M and van Nieuwenhuizen P 1985 Commun. Math.Phys. 98 105
[4] DeWitt B S and van Nieuwenhuizen P 1982 J.Math.Phys. 23 1953
[5] Kaku M, Townsend P K and van Nieuwenhuizen P 1978 Phys. Rev. D 17 3179
[6] Kaku M and Townsend P K 1978 Phys.Lett. 76 B 54
Das A, Kaku M and Townsend P K 1978 Phys.Rev.Lett. 40 1215
[7] van Nieuwenhuizen P 1985 Festschrift for Y Ne'eman
[8] van Nieuwenhuizen P 1985 Phys.Rev.D 32 872
For a nonconformal approach see Deser S and Kay J H 1983 Phys. Lett. 120B 97
[9] Roček M and van Nieuwenhuizen P 1986 Class.Quantum Grav. 3 43
[10] Fujikawa K 1983 Nucl.Phys. B 226 437
[11] Ferrara S and van Nieuwenhuizen P 1978 Phys.Lett. 76 B 304 and Phys.Lett. 78 B 573
[12] Brink L, di Vecchia P and Howe P S 1976 Phys.Lett. 65 B 471
Deser S and Zumino B 1976 Phys.Lett. 65 B 369
[13] Fradkin E S and Tseytlin A A 1981 Phys.Lett. 106 B 63
[14] Brink L and Schwarz J H 1977 Nucl.Phys. B 121 285
[15] Freedman D Z, van Nieuwenhuizen P and Ferrara S 1976 Phys.Rev. D 14 912
[16] Deser S and Zumino B 1976 Phys.Lett. 62 B 335
[17] The treatment which follows and in which $\omega_{\mu mn}$ is primarily considered as being associated with the generator M_{mn}, is described in the following two papers:
Chamseddine A H and West P C 1977 Nucl.Phys. B 129 39
Townsend P K and van Nieuwenhuizen P 1977 Phys.Lett. B 67 439
The factorization property of the remaining variations in (6.20) was found by Townsend P K (Tokyo lectures).
[18] Grisaru M T, Pendleton H and van Nieuwenhuizen P 1977 Phys.Rev. D 15 996
[19] van Nieuwenhuizen P 1973 Nucl.Phys. B 60 478
[20] Berends F A, de Wit B, van Holten J W and van Nieuwenhuizen P 1979 Nucl.Phys. B 154 261
[21] Freedman D Z and van Nieuwenhuizen P 1976 Phys.Rev. D 14 912
[22] van Nieuwenhuizen P, "Connections between supergravity and strings" to be published in J. Modern Phys. A (a new journal)
[23] Ferrara S, Grisaru M T and van Nieuwenhuizen P 1978 Nucl. Phys. B 138 430

SUPERSTRINGS

M B GREEN AND J H SCHWARZ

1. INTRODUCTION

String theories are, by definition, field theories of elementary one-dimensional objects. The study of string theories originated in the late 1960's in an attempt to describe hadrons and their interactions [1]. Strings are characterized by a fundamental length scale L (or string tension $T \sim L^{-2}$), which in the case of hadrons is taken to be of order 10^{-13} cm. We wish to describe the radically different possibility of using strings to describe fundamental quanta including gravity and Yang-Mills fields. In this setting it is natural to suppose that the length scale is of order 10^{-33} cm, the Planck length. This possibility was proposed some ten years ago in a paper [2] that showed, among other things, that string theories contain a massless spin-two mode that interacts appropriately to be identified as a graviton.

There are three basic types of string theories, each of which can be quantized consistently in a particular spacetime dimension. The first to be developed is a purely bosonic theory that requires 26 dimensions. This theory has serious problems beyond its funny dimension and absence of fermions. Its spectrum contains tachyonic states, the loop diagrams are very singular, and normal-ordering effects prevent a local interpretation of the interactions. Despite all these problems, the study of the bosonic string theory is useful pedagogically. The second class of string theories - referred to as "superstring theories" - consists of supersymmetrical ones in 10 dimensions [3]. The third class, also supersymmetrical, requires two-dimensional spacetime and will not be discussed here [4].

The superstring theories do not suffer from the problems that plague the bosonic string theory. The spectra are entirely free from tachyons and ghosts. One-loop diagrams have been investigated with encouraging results. The one-loop diagrams of type I theories are divergent, but the infinite part has just the right form to be absorbed in a renormalization of the string tension. However, in the anomaly-free case of SO(32) the infinities cancel. The one-loop diagrams are completely finite in the case of type II superstring theories. Also, because of cancellations between bosonic and fermionic contributions, there are no normal-ordering problems. As will be explained, the interactions have a straightforward local interpretation, even though the strings themselves are extended objects.

The fact that Einstein's theory of gravity, as well as supergravity extensions in four or more dimensions, has a dimensionful coupling constant and contact interactions appears to prevent a consistent quantum interpretation. The situation is reminiscent of the four-fermion theory of weak interactions, which is also a nonrenormalizable quantum theory. In that case, the solution provided by the Glashow-Salam-Weinberg electroweak theory entails replacing the four-fermion contact terms by massive exchanges. A renormalizable quantum theory results. It is plausible that a similar solution is required in the case of gravity. The exchange of states with mass of the order of the Planck mass should replace the contact terms. This is exactly what happens in superstring theories. A string field is equivalent to an infinite collection of point-particle fields corresponding to the various normal modes. These consist of massless fields plus an infinite tower of massive fields including ones of arbitrarily high spin. In the case of type II superstring theories, there are only cubic interactions, yet N=8 supergravity is (under certain circumstances) the low-energy effective theory. It appears that superstrings provide satisfactory gravitational analogs of the electroweak theory. What is less clear is whether or not they are the unique possibility for reconciling gravity with quantum mechanics.

A necessary mathematical preliminary is a discussion of the properties of spinors in ten dimensions. In general, a spinorial representation of the Lorentz algebra requires $2^{D/2}$ (or $2^{(D-1)/2}$) complex components. However, these representations are sometimes reducible, so that additional restrictions can be imposed. For example, in any even dimension one can define a Dirac matrix γ_{D+1} (generalizing γ_5) that can be used to define chiral projections $(1 \pm \gamma_{D+1})/2$. Eigenstates are called Weyl spinors. In certain dimensions (including four and ten) one can impose a reality condition on spinors. Specifically, there exists a representation of the Dirac algebra called the "Majorana representation," in which the components of the spinor may be real. A spinor satisfying this reality condition is called Majorana. (In an arbitrary representation the rule is $\bar{\psi}^T = C\psi$, where C is the charge conjugation matrix.) The question then arises whether a spinor can be simultaneously Majorana and Weyl. Clearly, the relevant criterion is whether or not γ_{D+1} is real in a Majorana representation. The answer is no for $D = 4$, but yes for $D = 10$. Thus the minimal irreducible spinor in ten dimensions is a Majorana-Weyl spinor with sixteen independent real components [5].

Ten-dimensional supersymmetry theories are characterized by the type of conserved supercharge that they possess. A "type I" theory possesses a supercharge Q that is both Majorana and Weyl. Such a theory is chiral, i.e., left-right asymmetric in the ten-dimensional sense. A four-dimensional interpretation requires a "spontaneous compactification" of six dimensions. It is by no means assured that the low-energy effective four-dimensional theory will itself be chiral just because the ten-dimensional one is. By the way, it is possible for a spinor in ten dimensions to be chiral without carrying any additional internal symmetry indices in contrast to the four-dimensional situation. In ten dimensions a chiral spinor can be self-conjugate, whereas in four dimensions the antiparticle of a chiral spinor has the opposite handedness. (This is a restatement of Majorana and Weyl compatibility.) When a type I theory is compactified on a six-torus in the most straightforward fashion, the effective four-dimensional theory has N=4 supersymmetry. The Majorana-Weyl supercharge decomposes into four 4-component spinors.

The type II theories have twice as much supersymmetry, and therefore are approximated at low energies by N=8 supergravity after a six-torus compactification. There are two cases. If Q is Majorana but not Weyl the theory is called "type IIA." In this case Q can be decomposed into a pair of Majorana-Weyl spinors of opposite chirality

$$Q = \frac{1+\gamma_{11}}{2}\, Q + \frac{1-\gamma_{11}}{2}\, Q. \tag{1}$$

Type IIA theories always have complete left-right symmetry, and so are not chiral. In a type IIB theory Q is Weyl but not Majorana, which implies that Q can be decomposed into a pair of Majorana-Weyl spinors having the same chirality

$$Q = \mathrm{Re}\, Q + i\,\mathrm{Im} Q. \tag{2}$$

A type IIB theory is chiral, of course. This exhausts the possibilities, since it is presumably impossible to construct an interacting field theory (for point particles or strings) with more than 32 fermionic symmetry generators.

2. TEN-DIMENSIONAL POINT-PARTICLE THEORIES

Massless supersymmetric theories in ten dimensions correspond to the massless sectors of corresponding superstring theories. It is therefore helpful to familiarize ourselves with them first. There are two type I point-particle theories. The first, which corresponds to the massless sector of open type I superstrings, is the supersymmetrical Yang-Mills theory [5,6]

$$\int \left(-\frac{1}{4} F^2 + \frac{i}{2}\, \bar{\chi}\, \gamma \bullet D\, \chi\right) d^{10}x. \tag{3}$$

In this expression $F^a_{\mu\nu}$ is the nonabelian field strength formed from a ten-vector potential A^a_μ in the usual fashion. χ is a Majorana-Weyl spinor in the adjoint representation of the gauge group. This theory has global supersymmetry with a Majorana-Weyl supercharge. Trivial dimensional reduction (in which the dependence of the fields on six of the dimensions is dropped) gives rise to N=4 Yang-Mills theory. The latter is known to be ultraviolet finite at every order in perturbation theory, which makes it of special interest. The ten-dimensional theory itself is a singular quantum theory, however, which is not surprising since a loop involves ten momentum integrations. Certain one-loop diagrams for S-matrix elements turn out to have a quadratic divergence that implies nonrenormalizability. Also, hexagon diagrams (analogous to triangles in four dimensions) give rise to anomalous divergences of gauge currents, which also indicates that eq. (3) is not a satisfactory quantum theory.

The second type I point-particle theory, N=1 D=10 supergravity, corresponds to massless type I closed superstrings. We won't give the action (which is known) but simply list the fields. It contains a metric tensor $g_{\mu\nu}$ (or zehnbein e_μ^r) describing 35 polarization states, a Majorana-Weyl gravitino field ψ_μ (the supersymmetry gauge field) with 56 polarization states, an antisymmetric second-rank gauge field potential $B_{\mu\nu}$ with 28 polarization states, a Majorana-Weyl spinor λ (having opposite handedness from ψ_μ) with 8 polarization states and a real scalar field φ (with just one polarization state). This theory also has nonrenormalizable divergences at one loop. Furthermore, hexagon diagrams contribute anomalous divergences to the stress-energy tensor, also ruling it out as a consistent quantum theory.

Type IIA supergravity theory, corresponding to massless type IIA superstrings, is easily deduced from the well-known D=11 supergravity theory [7]. It is given by a trivial dimensional reduction in which the dependence on one of the eleven dimensions is dropped. Doing this, the eleven dimensional graviton gives rise to a graviton $g_{\mu\nu}$, a vector A_μ, and a scalar φ, the third-rank potential gives a third-rank potential $A_{\mu\nu\rho}$ and a second-rank potential $A_{\mu\nu}$, and the eleven-dimensional gravitino gives rise to a pair of gravitinos ψ_μ^i and a pair of spinors λ^i. This theory is nonchiral and hence anomaly-free. It is nonetheless not a satisfactory quantum theory since certain S-matrix elements are quadratically divergent at one loop (almost as bad as in eleven dimensions).

Type IIB supergravity [8], an alternative extended supergravity theory in ten dimensions, cannot be deduced from a higher dimension. In contrast to the IIA theory it has an O(2) = U(1) symmetry that rotates the two supersymmetries into one another. In fact, this is a subalgebra of a noncompact SU(1,1) symmetry analogous to the well-known $E_{7,7}$ of the N=8 D=4 theory. The general rule that maximal supergravity in D dimensions has a noncompact $E_{11-D,11-D}$ symmetry singles out the IIB theory as the "natural" choice in ten dimensions. ($E_{1,1}$ is interpreted to mean SU(1,1).) The fields are conveniently classified by their U(1) charges. The field content of the theory is given in table 1. The fermionic fields λ and ψ_μ are Weyl spinors of opposite handedness. The field $A_{\mu\nu\rho\sigma}$ describes 35 degrees of freedom, rather than 70, because its field strength

$$F_{\mu\nu\rho\lambda\sigma} = 5\, \partial_{[\mu} A_{\nu\rho\lambda\sigma]} + \cdots \qquad (4)$$

is required to be self dual. This possibility is compatible with the Lorentz metric when the dimension is twice an odd number [9]. (Recall that self dual $F_{\mu\nu}$'s in four dimensions require Euclidean metric.) The dots in eq. (4) represent the fact that the self-dual field strength contains additional terms in the interacting theory.

Field	U(1) Charge	Degrees of Freedom
B	2	2 x 1
λ	3/2	2 x 8
$A_{\mu\nu}$	1	2 x 28
ψ_μ	1/2	2 x 56
$g_{\mu\nu}$	0	35
$A_{\mu\nu\rho\lambda}$	0	35

Table 1. Field Content of Type IIB Supergravity.

A satisfactory discussion of the IIB supergravity theory would require a separate article, so we will just just list a few key facts. The covariant field equations are known and raise interesting issues for the spontaneous compactification program. The theory is chiral and thus has potential anomalies. However, it has been shown that the contributions of λ , ψ_μ, and $A_{\mu\nu\rho\lambda}$ to the relevant hexagon diagrams cancel in a most remarkable fashion [10]. Still, this is not a satisfactory quantum theory, as it too has quadratically divergent one-loop diagrams implying nonrenormalizability. The superstring extension definitely improves matters, the massive levels providing even more delicate Bose-Fermi cancellations, as evidenced by the finiteness of the one-loop superstring diagrams.

3. FREE BOSONIC STRINGS

We will describe the field theory of strings in terms of functional field operators $\Phi[x(\sigma)]$. Before we can do so it is necessary to discuss the dynamics of the x's themselves, which already entails some of the complexity of a gauge theory. The motion of a bosonic string sweeps out a two-dimensional surface or "world sheet," whose embedding in spacetime is given by functions $x^\mu(\sigma,\tau)$. The parameter σ is spacelike and has finite range, $0 < \sigma < \pi$, say. τ is a timelike coordinate that goes from $-\infty$ to $+\infty$.

The fundamental principle, emphasized by Nambu, is that string dynamics is essentially determined by the requirement that it not depend on the particular way in which the world-sheet is parametrized. This principle is most conveniently implemented using the mathematics of general relativity. Specifically, the classical mechanics is described by the action

$$S = -\frac{1}{2} T \int d\sigma\, d\tau \sqrt{-g}\, g^{\alpha\beta}\, \partial_\alpha x^\mu\, \partial_\beta x_\mu, \tag{5}$$

where $g_{\alpha\beta}$ is a two-by-two auxiliary metric tensor and T is the string tension. (Indices α and β take the values σ and τ.) Eq. (5) can be viewed as describing D scalar fields in a two-dimensional gravitational background. We regard it instead as a description of the dynamics of strings in D dimensions. The symmetries of eq. (5) are global Poincaré invariance:

$$\delta x^\mu = l^\mu{}_\nu x^\nu + a^\mu \ , \quad \delta g_{\alpha\beta} = 0 \tag{6a,b}$$

and local reparametrization invariance:

$$\delta x^\mu = \xi^\alpha \, \partial_\alpha x^\mu \tag{7a}$$

$$\delta g_{\alpha\beta} = \xi^\gamma \, \partial_\gamma g_{\alpha\beta} + \partial_\alpha \xi^\gamma \, g_{\gamma\beta} + \partial_\beta \xi^\gamma \, g_{\alpha\gamma}. \tag{7b}$$

Classically, one can eliminate $g_{\alpha\beta}$ from eq. (5) by solving its field equations algebraically and substituting in the action. The resulting expression is the Nambu-Goto formula, namely the area of the world sheet. Quantum mechanically, the elimination of g involves doing a path integral. In general, an extra "Liouville" mode is left over [11], except in the special case of D=26, the so-called "critical dimension." Whether or not it is possible to make sense of the bosonic string theory for $D < 26$ is being studied by several groups. No clear-cut conclusions have yet been obtained. In any case, the question is not relevant to these lectures.

Instead of eliminating g, we can use the reparametrization invariance to choose a gauge in which

$$\sqrt{-g} \; g^{\alpha\beta} = \eta^{\alpha\beta}, \tag{8}$$

where $\eta = \begin{pmatrix} -1 & 0 \\ 0 & 1 \end{pmatrix}$ is the two-dimensional Minkowski metric. This simplifies eq. (5) to the trivial action (setting $T = 1/\pi$)

$$S = -\frac{1}{2\pi} \int d\sigma \, d\tau \; \eta^{\alpha\beta} \, \partial_\alpha x^\mu \, \partial_\beta x_\mu \ , \tag{9}$$

supplemented by the Virasoro constraints

$$(\partial_\tau x^\mu \pm \partial_\sigma x^\mu)^2 = 0, \tag{10}$$

which results from using eq. (8) in the $g_{\alpha\beta}$ field equation.

At this point there are two alternative approaches to studying the quantum theory. The first is covariant quantization, which entails using an indefinite-metric Hilbert space containing unphysical states (gauge degrees of freedom) as well as physical ones. The physical subspace is defined by Gupta-Bleuler-type subsidiary conditions, namely that the positive-frequency parts of the operators in eq. (10) annihilate physical states. In this approach loop calculations of the interacting string theory include contributions from ghost strings. The alternative to covariant quantization is to choose an axial (noncovariant) gauge in which the Hilbert space consists entirely of physical states. In the case of bosonic strings the two approaches are of comparable complexity, but in the case of superstrings the axial-gauge approach is much more convenient and maybe even essential.

The field equations corresponding to eq. (9) are

$$\left(\frac{\partial^2}{\partial\sigma^2} - \frac{\partial^2}{\partial\tau^2}\right)x^\mu = 0, \tag{11}$$

which in the case of open strings must be supplemented by boundary conditions

$$\frac{\partial}{\partial\sigma}\, x^\mu = 0 \text{ for } \sigma = 0,\pi, \tag{12}$$

in order that surface terms can be dropped. The general solution of eqs. (11) and (12) is

$$x^\mu(\sigma,\tau) = \hat{x}^\mu(\tau + \sigma) + \hat{x}^\mu(\tau - \sigma), \tag{13}$$

where

$$\frac{\partial}{\partial\tau}\hat{x}^\mu(\tau) = \frac{\partial}{\partial\tau}\hat{x}^\mu(\tau + 2\pi). \tag{14}$$

A remnant of the reparametrization invariance remains. It involves a single periodic function of one variable $\hat{\xi}(\tau)$,

$$\delta\,\hat{x}^\mu(\tau) = \hat{\xi}(\tau)\,\frac{\partial}{\partial\tau}\,\hat{x}^\mu(\tau). \tag{15}$$

Introducing light-cone coordinates

$$x^\pm = \frac{1}{\sqrt{2}}\,(x^0 \pm x^{25}), \tag{16}$$

the $\hat{\xi}$ symmetry is sufficient to make the additional gauge choice

$$\hat{x}^+(\tau) = \frac{1}{2}(x^+ + p^+\tau)\quad , \tag{17}$$

where x^+ and p^+ are constants. Using eq. (13),

$$x^+(\sigma,\tau) = x^+ + p^+\tau, \tag{18}$$

At this point, the constraints (10) can be used to express $x^-(\sigma,\tau)$ in terms of the 24 transverse coordinates $x^I(\sigma,\tau)$, p^+, and a single integration constant x^-.

Canonical quantization gives for the momentum conjugate to x^μ

$$p^\mu(\sigma,\tau) = \frac{1}{\pi}\,\frac{\partial}{\partial\tau}\,x^\mu(\sigma,\tau), \tag{19}$$

which can be identified as the physical momentum density along the string. Thus eq. (18) corresponds to choosing a σ parametrization such that the p^+ momentum density is constant. The general solution of eqs. (11) and (12) for the transverse coordinates is an expansion of the form

$$x^I(\sigma,\tau) = x^I + p^I\tau + i \sum_{n\neq 0} \frac{1}{n}\, \alpha_n^I\, e^{-in\tau} \cos n\sigma. \tag{20}$$

Quantization implies that

$$[\alpha_m^I, \alpha_n^J] = m\, \delta_{m+n,0}\, \delta^{IJ}, \tag{21}$$

and hence that α's with negative indices are raising operators and ones with positive indices are lowering operators.

The formula for $x^-(\sigma,\tau)$, obtained by solving the Virasoro constraints, contains a term linear in τ that gives the mass-shell value of p^-, from which one deduces that

$$\alpha' M^2 = c_0 + \sum_{n=1}^{\infty} \sum_{I=1}^{24} \alpha_{-n}^I\, \alpha_n^I. \tag{22}$$

(The Regge slope α' is related to the string tension T by $2\pi\alpha' T = 1$.) The constant c_0, which reflects an ordering ambiguity in passing from the classical to the quantum theory, is uniquely determined by Lorentz invariance. To see this, let us look at the spectrum in table 2. We see that the first excited level is a vector representation of SO(24). Since SO(D-2) is the little group for a massless particle in D-dimensions, it is necessary to choose $c_0 = -1$. This implies that the ground state is a tachyon. The states with $\alpha' M^2 = c_0 + 2 = 1$ combine to give the single irreducible representation $\square\square$ of SO(25), which has dimension 324. The states with $\alpha' M^2 = 2$ combine to give the SO(25) multiplets $\square\square\square$ (dimension 2900) and $\begin{smallmatrix}\square\\ \square\end{smallmatrix}$ (dimension 300). At $\alpha' M^2 = 3$ one has SO(25) multiplets $\square\square\square\square$ (dimension 20,150), $\begin{smallmatrix}\square & \square\\ \square & \end{smallmatrix}$ (dimension 5175), $\square\square$ (dimension 324), and a singlet. For large M the multiplicity grows exponentially with M, implying the existence of a limiting temperature somewhere around the Planck mass. An obvious, but nonetheless remarkable, fact is that a functional field $\Phi[x^I(\sigma), p^+, p^+]$, expanded in ordinary component fields precisely describes the complete spectrum, without any gauge, off-shell, auxiliary, or ghost degrees of freedom.

$\alpha' M^2$	States	Multiplicity
c_0	$\|0>$	1
c_0+1	$\alpha^I_{-1}\|0>$	24
c_0+2	$\alpha^I_{-2}\|0>$ $\alpha^I_{-1}\alpha^J_{-1}\|0>$	24 300
c_0+3	$\alpha^I_{-3}\|0>$ $\alpha^I_{-2}\alpha^J_{-1}\|0>$ $\alpha^I_{-1}\alpha^J_{-1}\alpha^K_{-1}\|0>$	24 576 2600

Table 2. The first four mass levels of the bosonic open-string spectrum.

4. FREE SUPERSTRINGS

The description of superstrings is based on a superspace generalization of the discussion in the preceding section [12]. Specifically, the world-sheet parametrized by σ and τ, as before, is mapped into superspace by functions $x^\mu(\sigma,\tau)$ and $\theta^A(\sigma,\tau)$ ($A = 1,2$). The Grassmann (anticommuting) coordinates θ^1 and θ^2 are D-dimensional spinors, the Dirac index being implicit. In particular, for reasons to be explained, they must be Majorana-Weyl in the case D=10. The theory is based on an extension of eq. (5) in which the global Poincaré invariance of eq. (6) is extended to a global super-Poincaré invariance

$$\delta\theta^A = \frac{1}{4}\, l_{\mu\nu}\, \gamma^{\mu\nu}\, \theta^A + \varepsilon^A \tag{23a}$$

$$\delta x^\mu = l^\mu{}_\nu\, x^\nu + a^\mu + i\, \bar{\varepsilon}^A \gamma^\mu \theta^A \tag{23b}$$

$$\delta g_{\alpha\beta} = 0. \tag{23c}$$

As usual, the ε's are infinitesimal Grassmann numbers. The formulas are written for the case in which all spinors are Majorana, which implies that

$$\bar{\varepsilon}\gamma^\mu\theta = -\bar{\theta}\gamma^\mu\varepsilon, \tag{24}$$

for example. To deal with other cases, one simply replaces expressions of the form $\bar{\psi}_1\, \gamma^\mu\, \psi_2$ by $(\bar{\psi}_1\, \gamma^\mu\, \psi_2 - \bar{\psi}_2\, \gamma^\mu\, \psi_1)/2$ in all formulas.

A straightforward guess for an extension of eq. (5) is

$$S_1 = -\frac{1}{2\pi} \int d\sigma\, d\tau\, \sqrt{-g}\; g^{\alpha\beta}\, \Pi_\alpha \bullet \Pi_\beta, \tag{25}$$

where

$$\Pi^\mu_\alpha = \partial_\alpha x^\mu - i\, \bar{\theta}^A \gamma^\mu\, \partial_\alpha \theta^A. \tag{26}$$

This obviously has global super-Poincaré invariance and retains the local reparametrization invariance of eq. (7) if we take

$$\delta\theta^A = \xi^\alpha\, \partial_\alpha \theta^A. \tag{27}$$

Nonetheless, this is not the answer we are seeking. Instead we must take $S = S_1 + S_2$, with

$$S_2 = \frac{1}{\pi} \int d\sigma\, d\tau\, \varepsilon^{\alpha\beta} [-\, i\, \partial_\alpha x^\mu (\bar{\theta}^1 \gamma_\mu \partial_\beta \theta^1 - \bar{\theta}^2 \gamma_\mu \partial_\beta \theta^2) + \bar{\theta}^1 \gamma^\mu \partial_\alpha \theta^1\, \bar{\theta}^2 \gamma_\mu \partial_\beta \theta^2], \tag{28}$$

where $\varepsilon^{\alpha\beta}$ is the usual Levi-Civita symbol. The action S_1 by itself describes a complicated interacting two-dimensional theory with too many θ variables to describe the superstring spectrum. The addition of S_2 has the following virtues: 1) it does not affect a bosonic truncation (in which θ's are set to zero) or a point-particle truncation (in which σ derivatives are set to zero); 2) it retains the global super-Poincaré and local reparametrization invariances; 3) it results in additional local fermionic symmetries such that half the θ components become gauge degrees of freedom and S becomes a free theory.

Let us examine the behavior of S_2 under a global ε transformation. All the terms are trivially seen to cancel except for

$$\frac{1}{\pi} \int d\sigma\, d\tau\; \varepsilon^{\alpha\beta}\; \bar{\varepsilon}^1 \gamma^\mu \partial_\alpha \theta^1 \bar{\theta}^1 \gamma_\mu \partial_\beta \theta^1 \tag{29}$$

and a similar term with θ^2's. If we let ψ_1, ψ_2, ψ_3 represent θ^1, $\partial_\sigma\theta^1$, and $\partial_\tau\theta^1$, then it is easy to show that the integrand of eq. (29) is proportional to

$$\bar{\varepsilon}^1 \gamma^\mu \psi_{[1}\; \bar{\psi}_2 \gamma_\mu \psi_{3]} \tag{30}$$

up to a total derivative. The square brackets represent antisymmetrization of indices. This expression must vanish by itself. The same type of antisymmetrized expression occurs in the demonstration of supersymmetry of super Yang-Mills theories, when $\delta A^a_\mu \sim \bar{\varepsilon} \gamma_\mu \lambda^a$ is substituted in $f_{abc}\, A^a_\mu\, \bar{\lambda}^b \gamma^\mu \lambda^c$. From those studies we know that there are four possible solutions [6]:

1) Majorana-Weyl spinors in ten dimensions
2) Weyl spinors in six dimensions (using the rule described after eq. (24))
3) Majorana spinors in four dimensions
4) Majorana spinors in three dimensions.

Classically, each of these four cases provide a consistent superstring theory. (This is the analog of the statement that the bosonic string theory is consistent classically in any dimension.) Quantum mechanically, the $g_{\alpha\beta}$ variables can be integrated out without giving rise to extra Liouville modes for D=10 only. This is the only case that has been studied. Even if the D=4 and 6 cases are mathematically possible, they are unlikely to have a rich enough structure to give a realistic unified theory.

To describe the local fermionic symmetries of $S_1 + S_2$, we introduce projection operators

$$P_{\pm}^{\alpha\beta} = \frac{1}{2}(g^{\alpha\beta} \pm \varepsilon^{\alpha\beta}/\sqrt{-g}). \tag{31}$$

They project out the self-dual and anti-self-dual parts of two-dimensional vectors. For the infinitesimal parameters that describe the local fermionic symmetries we introduce a two-vector of each type

$$\kappa^{1\alpha} = P_{-}^{\alpha\beta}\kappa_{\beta}^{1} \quad \text{and} \quad \kappa^{2\alpha} = P_{+}^{\alpha\beta}\kappa_{\beta}^{2} \tag{32a,b}$$

Taking the κ's to be Majorana-Weyl spinors as well, there are altogether 32 real components. The local fermionic invariance of S is then given by

$$\delta\theta^{A} = 2i\gamma \bullet \Pi_{\alpha}\kappa^{A\alpha}, \quad \delta x^{\mu} = i\bar{\theta}^{A}\gamma^{\mu}\delta\theta^{A} \tag{33a,b}$$

$$\delta(\sqrt{-g}\, g^{\alpha\beta}) = -16\sqrt{-g}\,(P_{-}^{\beta\gamma}\bar{\kappa}^{1\alpha}\partial_{\gamma}\theta^{1} + P_{+}^{\beta\gamma}\bar{\kappa}^{2\alpha}\partial_{\gamma}\theta^{2}). \tag{33c}$$

The demonstration that S_1+S_2 is invariant under these transformations is straightforward and has been fully carried out.

There are a number of remarkable facts about these formulas that ought to be pointed out. The local fermionic symmetries occur even though there are no corresponding gauge fields. Also, none of the variables are spinors in the two-dimensional sense. (The ten-dimensional indices are "internal" from the two-dimensional point of view.) The θ's and x's transform as two-dimensional scalars, and the κ parameters as two-dimensional vectors. The local fermionic symmetries imply that half the θ components are gauge degrees of freedom. This enables us to supplement the gauge choice in eq. (17) with

$$\gamma^{+}\theta^{A} = 0. \tag{34}$$

It is easy to show that in this gauge the equations of motion become

$$\left(\frac{\partial^2}{\partial\tau^2} - \frac{\partial^2}{\partial\sigma^2}\right) x^{I}(\sigma,\tau) = 0 \tag{35}$$

$$\left(\frac{\partial}{\partial\tau} + \frac{\partial}{\partial\sigma}\right)\theta^{1}(\sigma,\tau) = 0\ , \quad \left(\frac{\partial}{\partial\tau} - \frac{\partial}{\partial\sigma}\right)\theta^{2}(\sigma,\tau) = 0, \tag{36a,b}$$

so that we are in fact dealing with a free two-dimensional field theory. Eqs. (36a) and (36b) can be regarded as two components of a two-dimensional Dirac equation. Even though the θ's are two-dimensional scalars in the covariant formalism, in the light-cone gauge they metamorphose into two-dimensional spinors.

The various types of superstring theories are distinguished by boundary conditions. For example, in the case of an open string (in the light-cone gauge), eq. (12) must be supplemented by

$$\theta^1 = \theta^2 \quad \text{for} \quad \sigma = 0, \pi. \tag{37}$$

This implies, in particular, that the two θ variables must be chosen to have the same handedness. It also forces us to equate the parameters ε^1 and ε^2 of the two global supersymmetries, which explains why type I superstrings only have N=1 supersymmetry. The closed-string boundary conditions are just σ periodicity of x^μ and θ^A. In the chiral theories (types I and IIB), θ^1 and θ^2 have the same handedness, whereas in the nonchiral type IIA theory they have opposite handedness.

The gauge conditions in eq. (34) eliminate half the components of the θ's. Let us denote the remaining eight components of θ^1 by θ^a and of θ^2 by $\tilde{\theta}^a$. Here the index a labels an eight-dimensional spinor representation of the transverse SO(8) symmetry algebra. The transverse coordinates x^I belong to the eight-dimensional vector representation. Solving eq. (36) with the open-string boundary conditions of eq. (37) gives expansions of the form

$$\theta^a(\sigma,\tau) = \sum_{-\infty}^{\infty} \theta_n^a \, e^{-in(\tau-\sigma)}, \quad \tilde{\theta}^a(\sigma,\tau) = \sum_{-\infty}^{\infty} \theta_n^a \, e^{-in(\tau+\sigma)}. \tag{38a,b}$$

Canonical quantization of the θ coordinates gives anticommutation relations

$$\{\theta^a(\sigma,\tau), \theta^b(\sigma',\tau)\} = \{\tilde{\theta}^a(\sigma,\tau), \tilde{\theta}^b(\sigma',\tau)\} \sim \delta^{ab}\, \delta(\sigma-\sigma') \tag{39a,b}$$

$$\{\theta^a(\sigma,\tau), \tilde{\theta}^b(\sigma',\tau)\} = 0. \tag{39c}$$

This implies for the coefficients in eq. (38) that

$$\{\theta_m^a, \theta_n^b\} = \frac{1}{p^+}\, \delta_{m+n,0}\, \delta^{ab}, \tag{40}$$

which is a fermionic counterpart of eq. (21).

The θ variables are not yet in suitable form to serve as the arguments of a functional superfield, because they anticommute to give δ functions rather than zero. This means that they may be viewed as describing both "coordinates" and "conjugate momenta." These need to be separated, since only coordinates (or conjugate momenta after Fourier transforming) should be variables of a quantum field. The separation is most conveniently achieved by reducing the manifest SO(8) symmetry to that of an SO(6) × SO(2) ≈ SU(4) × U(1) subalgebra [13]. This means that six of the transverse directions are described differently from the other two. It is natural to regard these six as the ones that we intend to compactify and the other two as the "ordinary" transverse coordinates. With this interpretation, the U(1) quantum number becomes the four-dimensional helicity, and the SU(4) is an "internal" symmetry. The branching rules for the three eight-dimensional SO(8) representations are

$$8_v = 6_0 + 1_1 + 1_{-1}\,,\quad 8_s = 4_{1/2} + \bar{4}_{-1/2}\,,\quad 8_c = 4_{-1/2} + \bar{4}_{1/2}, \tag{41a,b,c}$$

where the subscripts are the helicity quantum numbers. Under this decomposition the eight θ^a become θ_A ($\bar{4}$ representation of SU(4)) and λ^A (4 representation of SU(4)), and eq. (39a) becomes

$$\{\theta_A(\sigma), \theta_B(\sigma')\} = \{\lambda^A(\sigma), \lambda^B(\sigma')\} = 0 \tag{42a}$$

$$\{\theta_A(\sigma), \lambda^B(\sigma')\} = \delta_A^B\, \delta(\sigma-\sigma'). \tag{42b}$$

There are analogous relations for $\tilde{\theta}_A$ and $\tilde{\lambda}^A$. These relations allow us to interpret θ_A and $\tilde{\theta}_A$ as coordinates and λ^A and $\tilde{\lambda}^A$ as conjugate momenta. Thus superstrings can be described by functional superfields [14] of the form $\Phi[x^I(\sigma), \theta_A(\sigma), \tilde{\theta}_A(\sigma), x^+, p^+]$.

Let us examine the spectrum of type I open superstrings. The mass formula analogous to eq. (22) is

$$\alpha' M^2 = c_1 + \sum_{n=1}^{\infty} (\alpha_{-n}^I \alpha_n^I + n p^+ \theta_{-n}^a \theta_n^a). \tag{43}$$

The states are described by vectors in the Fock space built up from all the α and θ oscillators *and* by wave functions of the zero modes $f(x^\mu, \theta_{0A})$. Thus the ground state is a supermultiplet given by $|0\rangle f$. The function f, expanded in powers of θ_0 is 16 functions of x, 8 bosonic and 8 fermionic, corresponding exactly to the 16 physical polarization states of a super Yang-Mills multiplet in D=10. Since such a multiplet must be massless, we must set $c_1 = 0$. Therefore, in contrast to the bosonic case, the superstring spectrum contains no tachyons.

$\alpha' M^2$	States	Multiplicity
0	$\|0\rangle f$	16
1	$\alpha_{-1}^I \|0\rangle f$	8 x 16
	$\theta_{-1}^a \|0\rangle f$	8 x 16
2	$\alpha_{-1}^I \alpha_{-1}^J \|0\rangle f$	36 x 16
	$\theta_{-1}^a \alpha_{-1}^I \|0\rangle f$	64 x 16
	$\theta_{-1}^a \theta_{-1}^b \|0\rangle f$	28 x 16
	$\alpha_{-2}^I \|0\rangle f$	8 x 16
	$\theta_{-2}^a \|0\rangle f$	8 x 16

Table 3. The first three mass levels of the type I open superstring spectrum.

The 128 bosonic states at the first excited level can be assembled into the SO(9) representations 44 + 84, whereas the 128 fermionic states form an irreducible 128 of SO(9). These are exactly the same multiplicities as occur in D=11 supergravity. In that case SO(9) describes massless particles in

11 dimensions, whereas here it describes massive ones in 10 dimensions. Also, 44 + 84 + 128 gives an irreducible representation of the N=1 super-Poincaré group in each case. The 2304 states at the $\alpha' M^2 = 2$ level are described most succinctly by

$$9 \otimes (44 + 84 + 128). \tag{44}$$

The notation means that the 9-vector representation of SO(9) should be added (Kronecker product) to each of the SO(9) multiplets in the fundamental massive supermultiplet. This construction gives an irreducible representation of the super-Poincaré algebra. Similarly, at the $\alpha' M^2 = 3$ level one has

$$(44 + 16) \otimes (44 + 84 + 128), \tag{45}$$

corresponding to two irreducible super-Poincaré multiplets, and so forth.

In addition to the quantum numbers described above, open-string states also carry Yang-Mills group quantum numbers. In particular, the massless modes belong to the adjoint representation. Only certain groups are compatible with type I superstring theory, however. No method is known to incorporate exceptional groups, even in the classical theory. Anomalies rule out all classical groups except for SO(32), as we show in sect. 6. The SO(n) case is especially easy to describe. One associates fundamental representation indices $a,b = 1, 2, ..., n$ with the ends of the string. Then one imposes the generalized antisymmetry condition

$$\Phi_{ab}\,[x^I(\sigma), \theta_A(\sigma), \tilde{\theta}_A(\sigma)] = -\Phi_{ba}\,[x^I(\pi-\sigma), \tilde{\theta}_A(\pi-\sigma), \theta_A(\pi-\sigma)] \tag{46}$$

on the functional superfield. Note that θ and $\tilde{\theta}$ get interchanged when the orientation is reversed. Referring to the mode expansions in eqs. (20) and (38) we see that the variable substitution in eq. (46) corresponds to $\alpha_n^I \rightarrow (-1)^n \alpha_n^I$, and $\theta_{nA} \rightarrow (-1)^n\, \theta_{nA}$. Therefore, in view of eq. (43), the component fields satisfy

$$\varphi_{ab} = -\varphi_{ba} \quad \text{for} \quad \alpha' M^2 = 0, 2, 4, \cdots \tag{47a}$$

$$\varphi_{ab} = \varphi_{ba} \quad \text{for} \quad \alpha' M^2 = 1, 3, 5, \cdots, \tag{47b}$$

Eq. (46) can be interpreted as meaning that a type I open string carries no intrinsic orientation.

Closed strings are described by periodic x's and θ's, which at $\tau = 0$ have expansions of the form

$$x^I(\sigma) = \sum_{-\infty}^{\infty} x_n^I\, e^{2in\sigma} \tag{48a}$$

$$\theta_A(\sigma) = \sum_{-\infty}^{\infty} \theta_{nA}\, e^{2in\sigma}\,, \quad \tilde{\theta}_A(\sigma) = \sum_{-\infty}^{\infty} \tilde{\theta}_{nA}\, e^{2in\sigma}. \tag{48b,c}$$

The functional superfield that creates or destroys a closed string carries no indices. It is subject to the constraint

$$\Psi[x^I(\sigma+\sigma_0),\ \theta_A(\sigma+\sigma_0),\ \tilde{\theta}_A(\sigma+\sigma_0)] = \Psi[x^I(\sigma),\ \theta_A(\sigma),\ \tilde{\theta}_A(\sigma)], \tag{49}$$

reflecting the fact that the particular point on a closed string that is called $\sigma = 0$ has no special role. For a type IIB closed string there are no further constraints. For a type IIA closed string, the coordinates $\tilde{\theta}_A$ must be replaced by $\tilde{\theta}^A$ (a 4 instead of a $\bar{4}$). A type I closed string is a type IIB one subject to the additional constraint

$$\Psi[x^I(-\sigma),\ \theta_A(-\sigma),\ \tilde{\theta}_A(-\sigma)] = \Psi[x^I(\sigma),\ \tilde{\theta}_A(\sigma),\ \theta_A(\sigma)], \tag{50}$$

reflecting the fact that type I superstrings are unoriented. In particular, massless type IIB states are described by the component superfield $\psi(x^\mu, \theta_{0A}, \tilde{\theta}_{0A})$, which when expanded in powers of θ has 2^8 terms that precisely correspond to the 256 physical modes of the type IIB supergravity theory. Imposing symmetry between θ and $\tilde{\theta}$, as required by eq. (50), removes half the modes leaving the type I supergravity multiplet. Similarly, the type IIA supergravity multiplet is described by an unconstrained component superfield $\psi(x^\mu, \theta_{0A}, \tilde{\theta}_0^A)$.

The spectrum of closed-string excited states for each of the three theories can be deduced from the open-string spectrum described in table 3 and the discussion following it. If we denote the set of states occurring at $\alpha' M^2 = n$ in the open-string sector by R_n, then the set of states occurring at $\alpha' M^2 = 4n$ in the type IIB spectrum is given by the tensor product $R_n \otimes R_n$. The type I closed-string spectrum is given by the subset of states given by the symmetric tensor product (in the graded sense). The type IIA closed-string spectrum is given by $R_n \otimes \bar{R}_n$. The bar represents SU(4) conjugation (reversal of D = 10 - handedness), which only makes a difference for the massless sector.

Starting from the covariant string action, it is easy to deduce the super-Poincaré generators by the Noether procedure. Each is expressed as the spatial integral of the time-component of a conserved current. In the present context that means a σ integral of the τ component of a two-vector current. Thus, associated with each generator g, there is a charge density $g(\sigma)$, which describes the distribution of that particular charge along the string. In the light-cone gauge the expressions for the generators have to be suitably modified from the covariant expressions in order to incorporate the effects of compensating local symmetry transformations that are required in order that variations preserve the gauge conditions. It is convenient to choose the σ parametrization in the light-cone gauge such that the density of p^+ along the string is a universal constant. This implies taking the length of the σ interval proportional to p^+. Specifically, we define

$$\alpha = 2p^+ \tag{51}$$

and let the range be $0 \leqslant \sigma \leqslant \pi|\alpha|$. Thus the density $p^+(\sigma)$ is $\varepsilon(\alpha)/2\pi$, where $\varepsilon(\alpha)$ denotes the sign of α. Our convention is that a field operator with $p^+ > 0$ is an annihilation operator and one with $p^+ < 0$ is a creation operator.

In a type II theory there are 32 conserved fermionic charges (two Majorana-Weyl spinors). These can be decomposed into eight four-dimensional SU(4) representations, which in the IIB case are denoted q^{+A}, q^{-A}, $\tilde{q}^{+A}$, $\tilde{q}^{-A}$ (4's of SU(4)) and q_A^+, q_A^-, $\tilde{q}_A^+$, $\tilde{q}_A^-$ ($\bar{4}$'s of SU(4)). The q^+'s are square-roots of p^+ in the sense that

$$\{q^{+A}, q_B^+\} = \{\tilde{q}^{+A}, \tilde{q}_B^+\} = 2p^+\delta_B^A, \tag{52}$$

while all other anticommutators of q^+'s are zero. The densities of the q^+ charges are given by

$$q^{+A}(\sigma) = \frac{1}{\alpha}\lambda^A(\sigma) = \frac{1}{\alpha}\frac{\delta}{\delta\theta_A(\sigma)} \tag{53a}$$

$$q_A^+(\sigma) = \frac{1}{\pi}\,\varepsilon(\alpha)\,\theta_A(\sigma) \tag{53b}$$

and similarly with tildes. The q^-'s are square-roots of p^- in the same sense, except that they don't give p^- but rather its mass-shell value, the light-cone gauge Hamiltonian,

$$h = \frac{1}{\alpha}[(p^I)^2 + M^2], \tag{54}$$

where M^2 is the operator given in eq. (43). Thus

$$\{q^{-A}, q_B^-\} = \{\tilde{q}^{-A}, \tilde{q}_B^-\} = 2h\delta_B^A, \tag{55}$$

with all other anticommutators of q^-'s giving zero. An anticommutator of a q^+ and a q^- gives transverse momentum components.

The q^- densities are given by

$$q^{-A}(\sigma) = \sqrt{2}\,(\rho^i)^{AB}\,[p^i(\sigma) - \frac{1}{\pi}\frac{\partial}{\partial\sigma}x^i(\sigma)]\,\theta_B(\sigma)$$
$$+ 2\pi\varepsilon(\alpha)[p^L(\sigma) - \frac{1}{\pi}\frac{\partial}{\partial\sigma}x^L(\sigma)]\,\lambda^A(\sigma) \tag{56}$$

and three similar expressions for $q_A^-(\sigma)$, $\tilde{q}^{-A}(\sigma)$, and $\tilde{q}_A^-(\sigma)$. The index i refers to the SU(4) six-vector in the decomposition of eq. (41a) and the index L refers to the helicity -1 singlet. The $p^I(\sigma)$ are the transverse momentum densities

$$p^I(\sigma) = -\,i\,\frac{\delta}{\delta x^I(\sigma)}. \tag{57}$$

The matrices $(\rho^i)^{AB}$ are SU(4) Clebsch-Gordon coefficients for coupling a pair of 4's to a 6. Together with conjugate matrices $(\rho^i)_{AB}$, they satisfy a sort of Dirac algebra. The Hamiltonian density that integrates to give the charge in eq. (54) is

$$h(\sigma) = \varepsilon(\sigma)[\pi(p^I(\sigma))^2 + \frac{1}{\pi}(\frac{\partial}{\partial\sigma}x^I(\sigma))^2]$$
$$-\,2i\,[\theta_A(\sigma)\frac{\partial}{\partial\sigma}\lambda^A(\sigma) + \tilde{\theta}_A(\sigma)\frac{\partial}{\partial\sigma}\tilde{\lambda}^A(\sigma)]. \tag{58}$$

Analogous formulas can be given for all 45 of the Lorentz generators as well. Thus one ends up with expressions for all 87 super-Poincaré generators as differential operators involving θ, x, $\frac{\delta}{\delta\theta}$, and $\frac{\delta}{\delta x}$. These formulas allow one to deduce how the various component fields in the functional superfield $\Psi[x^I(\sigma), \theta_A(\sigma), \tilde{\theta}_A(\sigma), x^+, p^+]$ map into one another under super-Poincaré transformations. This explains in a constructive manner how the functional superfield represents the super-Poincaré algebra.

5. SUPERSTRING FIELD THEORY

In the preceding section we have introduced light-cone-gauge functional superfields that create or destroy superstrings (according to the sign of p^+). The construction ensures that they provide representations of the super-Poincaré algebra and that they only contain physical propagating degrees of freedom. The equation of motion of the free theory is

$$h\Phi = i \frac{\partial}{\partial x^+} \Phi. \tag{59}$$

Thus we have a Schrödinger formalism, analogous to nonrelativistic quantum mechanics, in which a Hamiltonian operator generates the x^+ evolution. The energy E of NR quantum mechanics is replaced by the momentum component p^- in this formalism.

The interacting quantum field theory of superstrings is described by an action of the form

$$S = \int \partial_+\Psi\partial_-\Psi dx^+dx^-D^{16}z - \int H\, dx^+ \ . \tag{60}$$

In this formula $D^{16}z$ represents functional integration over the eight transverse x coordinates and the eight θ coordinates

$$D^{16}z \equiv D^8x^I(\sigma)\, D^4\theta_A(\sigma)\, D^4\tilde{\theta}_A(\sigma). \tag{61}$$

H is the light-cone Hamiltonian, built from quantum fields, that gives the x^+ evolution of the second-quantized theory. Eq. (60) is written for closed-string fields, but there is an analogous kinetic term involving Φ in the case of a type I superstring theory. The important point is that H does not involve $\partial_+\Psi$ or $\partial_+\Phi$, so that $\partial_-\Psi$ is canonically conjugate to Ψ and H really is the light-cone Hamiltonian.

For each super-Poincaré generator g, expressed as a differential operator involving the x's and θ's in the preceding section, we associate an operator G built from the quantum fields.

$$g \to G = G_2 + G_3 + \cdots . \tag{62}$$

G_2 represents a term quadratic in fields, G_3 one cubic in fields and so forth. The important principle, which largely determines the theory, is that the G's should satisfy the same algebra as the g's. For the free theory this is easy to achieve. Only the G_2's are nonzero, and they are given by

$$G_2 = \int_0^\infty \alpha\, d\alpha \int D^{16}z\, \Psi_{-\alpha}\, g \Psi_\alpha. \tag{63}$$

The algebra of the g's is transferred isomorphically to the G_2's as a consequence of the canonical commutation rule

$$\left[\Psi[z_1(\sigma), \alpha_1, x^+], \Psi[z_2(\sigma), \alpha_2, x^+]\right]$$

$$= \frac{1}{\alpha_2} \delta(\alpha_1+\alpha_2) \int_0^{\pi|\alpha_2|} \frac{d\sigma_0}{\pi|\alpha_2|} \Delta^{16} [z_1(\sigma) - z_2(\sigma+\sigma_0)]. \tag{64}$$

The symbol Δ represents a functional delta function. The integral in eq. (64) occurs as a consequence of the constraint in eq. (49). The corresponding canonical commutation rule for open-string fields is

$$\left[\Phi_{ab}[1], \Phi_{cd}[2]\right] = \frac{1}{2\alpha_2} \delta(\alpha_1+\alpha_2)\Big(\delta_{ac}\delta_{bd}\, \Delta^{16} [z_1(\sigma)-z_2(\sigma)]$$

$$- \delta_{ad}\delta_{bc}\, \Delta^{16} [z_1(\sigma)-z_2(\pi|\alpha_2| -\sigma)]\Big), \tag{65}$$

which incorporates the constraint in eq. (46).

The isomorphism of the algebra implies, in particular, that corresponding to eq. (55)

$$\{Q^{-A}, Q_{\bar{B}}^{-}\} = 2H \delta^A_B. \tag{66}$$

This is the key formula, since it is H that determines the theory. The solution of this equation for the free theory has already been given. Now we wish to add interaction terms Q_3^{-A}, etc., such that the algebra in eq. (66) is preserved. Complete solutions for each of the superstring theories have been obtained and will be described below. The methodology employed leads to unique solutions. We cannot assert categorically that it is impossible to modify the formulas while preserving the algebra, but this may well turn out to be the case. The question is important, because it could provide the basis for proving finiteness or renormalizability of the theory [37]. The details of the construction cannot be presented in the space available here. They are given in ref. 14. The procedure used there is to expand all functional operators and delta functionals in Fourier modes and to define the functional integrations as integrals over infinite sets of Fourier coefficients. This provides mathematical control over the convergence properties of various expressions that are otherwise very delicate. Because the analysis was done this way, we are confident that no subtle errors slipped in. Here, we will simply describe the results that were obtained expressed in functional form.

Inserting expansions of the form in eq. (62) into eq. (66) and equating terms order by order in the coupling constant, one obtains at the leading order of interaction

$$\{Q_2^{-A}, Q_{3B}^{-}\} + \{Q_3^{-A}, Q_{2B}^{-}\} = 2H_3 \delta^A_B. \tag{67}$$

Let us consider first interaction terms that are cubic in open-string fields. (These would only occur in a type I theory, of course.) Physically, what happens is that two open strings can join ends to give a single open string. Mathematically this is described by an operator of the form

$$G_3 = g\int(\prod_{r=1}^{3} d\alpha_r D^{16}z_r)\,\delta(\sum_1^3 \alpha_r)\,\Delta^{16}[\sum_1^3 \varepsilon_r z_r(\sigma)]K\,tr(\Phi[1]\Phi[2]\Phi[3]). \tag{68}$$

The delta functionals Δ^{16} describe the fact that strings #1 and #2 at the time of interaction are located in 16-dimensional superspace at positions that coincide with that of string #3, which results from their joining, as shown in fig. 1. (ε_r is +1 for an incoming string and -1 for an outgoing string.)

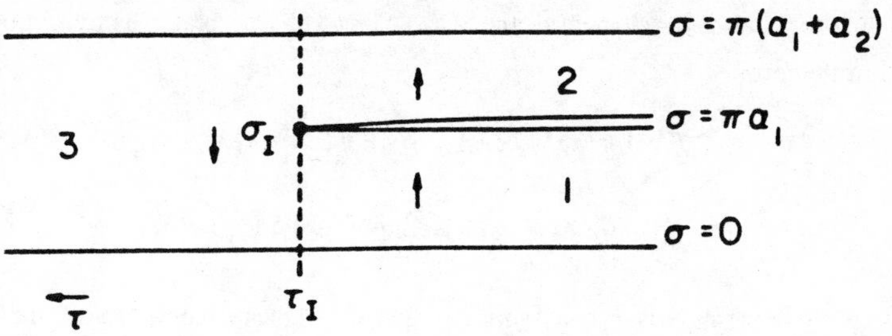

Fig. 1. Light-cone diagram for the three open-string vertex.

Since the massless sector of open superstrings is the super Yang-Mills multiplet, the cubic Yang-Mills vertex should be included as a special case. This means that g is a Yang-Mills coupling constant, whose dimensions are (length)^3 for D = 10. Also, the trace takes account of the group theory - SO(n) indices of lines that join in fig. 1 are identified. K represents additional kinematical details of the interaction. Its form depends on which operator we are describing. In the case of H_3, we know that K must contain a derivative, because the cubic Yang-Mills vertex involves a derivative coupling.

The interaction occurs entirely at σ_I, corresponding to the joining points of the strings - i.e., the interaction is local even though the strings themselves are spatially (superspatially?) extended. Thus the K's should be formed from operators acting at σ_I : $\theta_A(\sigma_I), \dfrac{\delta}{\delta x^I(\sigma_I)}$, etc. From the mode expansions we learn that when these operators act on the delta functionals they are singular. Well-defined finite operators can be defined as follows:

$$Z^I = -\,i\,\sqrt{2}\,\pi \lim_{\varepsilon\to 0} \sqrt{\varepsilon}\frac{\delta}{\delta x^I(\sigma_I-\varepsilon)} \tag{69a}$$

$$Y_A = \frac{1}{\sqrt{2}} \lim_{\varepsilon\to 0} \sqrt{\varepsilon}\,[\,\theta_A(\sigma_I-\varepsilon) + \tilde{\theta}_A(\sigma_I-\varepsilon)]. \tag{69b}$$

In terms of these quantities, the appropriate K operators in eq. (68) that provide a solution of eq. (67) are

$$K(Q_{\bar{3}A}) = Y_A \tag{70a}$$

$$K(Q_{\bar{3}}^{-A}) = \frac{2}{3}\, \varepsilon^{ABCD}\, Y_B Y_C Y_D \tag{70b}$$

$$K(H_3) = \frac{1}{\sqrt{2}}\, Z^L + \frac{1}{2}\, (\rho^i)^{AB} Z^i Y_A Y_B + \frac{1}{6\sqrt{2}}\, Z^R \varepsilon^{ABCD}\, Y_A Y_B Y_C Y_D. \tag{70c}$$

Incidentally, in writing eqs. (69a) and (69b) we did not indicate which of the three string's coordinates should be used. The reason is that in the limit $\varepsilon \to 0$ all three give the same answer.

As a test of our understanding of these formulas, we have used the cubic vertex H_3 to calculate a tree diagram with four on-shell massless external lines. Each external line is characterized by a ten-momentum p^μ (satisfying $p^2 = 0$), four θ coordinates θ_A, and an antisymmetric $n \times n$ matrix λ corresponding to a particular element of the SO(n) Yang-Mills algebra. The tree amplitude can then be written in the form

$$T = [\, tr(\lambda_1 \lambda_2 \lambda_3 \lambda_4)\, A_{1234} + tr(\lambda_1 \lambda_3 \lambda_4 \lambda_2)\, A_{1342}$$

$$+ tr(\lambda_1 \lambda_4 \lambda_2 \lambda_3)\, A_{1423}]\, \delta^{10}(\sum_1^4 p_r)\, \delta^4(\sum_1^4 \alpha_r \theta_r). \tag{71}$$

Each of the three terms has cyclic symmetry for the indicated ordering of the external lines. Although the three terms are related by the obvious interchanges, the calculation in the light-cone formalism for a process in which $1 + 2 \to 3 + 4$ is more complicated for the third term than for the first two. It is distinguished by the fact that the initial lines #1 and #2 are nonconsecutive for the cyclic ordering (1423). The calculation of this term involves a quartic interaction, to be described, whereas the first two are entirely determined by H_3.

The calculation of A_{1234} is given by the sum of the two light-cone perturbation diagrams shown in fig.2. Corresponding to fig. 2a one has

$$A_{1234}^{(a)} = \langle 34 | H_3\, \Delta\, H_3 | 12 \rangle, \tag{72}$$

where

$$\Delta = \frac{1}{H_2 - p^-} = \int_0^\infty e^{-(H_2 - p^-)\tau} d\tau. \tag{73}$$

There is a similar formula for fig. 2b. After appropriate changes of variables the two contributions are given as integrals $\int_0^{x_0} dx$ and $\int_{x_0}^1 dx$ of the same integrand, so that they combine nicely. This is

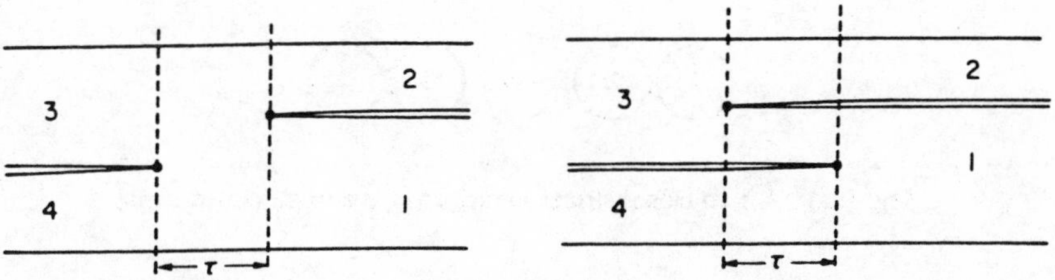

2a 2b

Fig. 2. The two contributions to the A_{1234} tree amplitude.

nontrivial inasmuch as x_0 is a complicated function of the p^+ coordinates of the four lines. The result of the calculation is

$$A_{1234}(string) = \frac{\Gamma(1-\alpha' s)\,\Gamma(1-\alpha' t)}{\Gamma(1-\alpha' s-\alpha' t)}\, A_{1234}(super\ YM), \tag{74}$$

where s and t are the usual Mandelstam invariants. Thus the answer is a simple function times the amplitude of the D = 10 super Yang-Mills field theory. A direct calculation of the Yang-Mills amplitude requires including a quartic contact term. Note that the ratio of Γ functions approaches one in the limit $\alpha' \to 0$, which corresponds to shrinking the strings to points. Also, they contain infinite sequences of poles corresponding to the exchange of massive string modes in the s and t channels.

The analysis of the cubic closed-string vertex can be carried out in a similar manner. When two closed strings touch they can break and rejoin to form a single closed string as depicted in fig. 3a. This is topologically distinct from the cubic open-string interaction. The breaking/joining interaction was seen to be a string generalization of a Yang-Mills interaction. Correspondingly, the exchange interaction of fig. 3a is the string generalization of a gravitational interaction, since massless closed strings include the graviton. Therefore there is a gravitational coupling constant κ associated with this process. Dimensionally, in ten dimensions κ is $(\text{length})^4$. In type I superstring theories, which contain both types of interactions, Lorentz invariance requires that the two couplings are related according to

$$\kappa \sim g^2/\alpha'. \tag{75}$$

The cubic closed-string interaction is given by an expression analogous to eq. (68) with the following important differences: 1) the coupling g is replaced by κ . 2) the $\Phi's$ are replaced by $\Psi's$, which are Yang-Mills singlets so that no trace is required. 3) the mathematical meaning of the delta functionals corresponds to fig. 3b. Wavy lines carrying the same label are identified in superspace. This allows

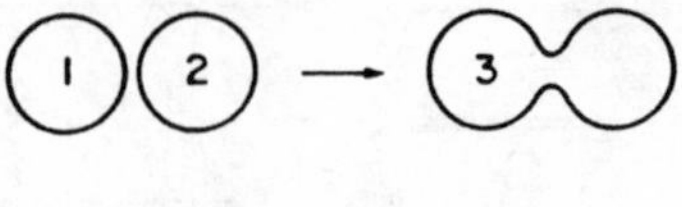

Fig. 3. a) Two closed strings joining to give a single closed string.

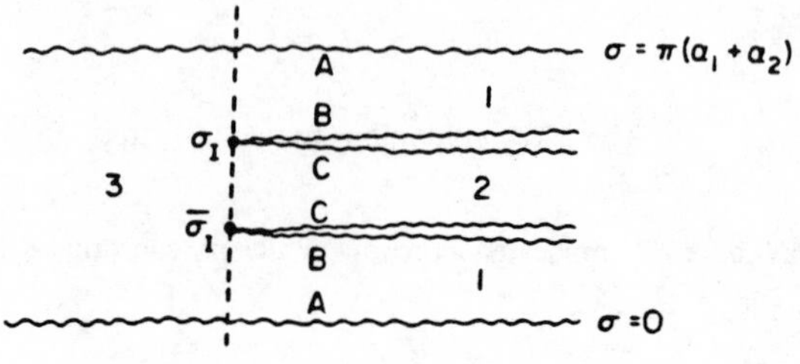

b) Light-cone diagram for the three closed-string vertex. Wavy lines with same labels are identified.

the parametrization of closed strings to be described in a plane. 4) The derivative coupling K_{cl} is one appropriate to the exchange interaction. Since the cubic gravitation coupling involves two derivatives, $K_{cl}(H)$ turns out to be quadratic in derivatives. By solving eq. (67) we are led to the result

$$K_{cl}(H) = K(Y,Z)\,K(\tilde{Y},\tilde{Z}), \tag{76}$$

where $K(Y,Z)$ and $K(\tilde{Y},\tilde{Z})$ refer to the function given in eq. (70c), but for the Y's and Z's one now takes

$$Y_A = \eta \lim_{\varepsilon\to 0} \sqrt{\varepsilon}\,\theta_A(\sigma_I - \varepsilon) \tag{77a}$$

$$\tilde{Y}_A = \eta^* \lim_{\varepsilon\to 0} \sqrt{\varepsilon}\,\tilde{\theta}_A(\sigma_I - \varepsilon) \tag{77b}$$

$$Z^I = \pi\eta \lim_{\varepsilon\to 0} \sqrt{\varepsilon}\,[-i\frac{\delta}{\delta x^I(\sigma_I - \varepsilon)} - \frac{1}{\pi}\frac{\partial}{\partial\sigma}x^I(\sigma_I - \varepsilon)] \tag{78a}$$

$$\tilde{Z}^I = \pi\eta^* \lim_{\varepsilon\to 0} \sqrt{\varepsilon}\,[-i\frac{\delta}{\delta x^I(\sigma_I - \varepsilon)} + \frac{1}{\pi}\frac{\partial}{\partial\sigma}x^I(\sigma_I - \varepsilon)], \tag{78b}$$

where $\eta = e^{i\pi/4}$.

Locality and causality require that the five additional interactions depicted in fig. 4 be included in type I superstring theory. The one in fig. 4a is a $\Phi\Psi$ term of Yang-Mills type with the same local interaction as the cubic open-string coupling. The four interactions in figs. 4b, c, d, e are gravitational in the sense that they involve the same local exchange interaction as the cubic closed-string coupling. As alluded to earlier, the Φ^4 interaction term contributes to the calculation of A_{1423} in eq. (71), but not to A_{1234} and A_{1342}. It vanishes in the $\alpha'\to 0$ limit, whereas the quartic Yang-Mills theory vertex arises as an effective description of the exchange of massive string modes.

Fig 4. a) open string → closed string

b) open string + closed string → open string

c) open string + open string → open string + open string

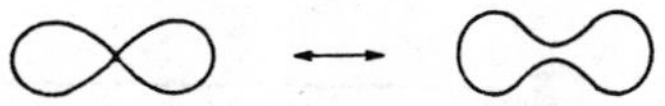

d) closed string → closed string

e) open string → open string.

Of the seven interaction terms that occur in a type I superstring theory (two of Yang-Mills type and five of gravity type) only one, namely the cubic closed-string vertex of fig. 3a can appear in a type II superstring theory. Except for the terms corresponding to figs. 4d and e, this topological classification of possible interactions was identified long ago for the bosonic string theories [15]. For a while, we did not believe it could apply to superstrings because at first sight it appears impossible to reconcile with the supersymmetry algebra of eq. (66). It requires, for example, that

$$\{Q_3^{-A}, Q^{-}_{3B}\} = 0\,, \tag{79}$$

which seems absurd since the anticommutator is a positive-definite operator. (The Φ^4 interaction arises from anticommuting a Q_2 with a Q_4.)

To better understand eq. (79) consider a matrix element between an initial state consisting of strings #1 and #2 and a final state consisting of strings #3 and #4. In this case (assuming $|p_1^+| > |p_4^+|$) the possible intermediate states in the operator product are a one-string state, depicted in fig. 5a, and a three-string state, depicted in fig. 5b. The internal lines in these diagrams represen

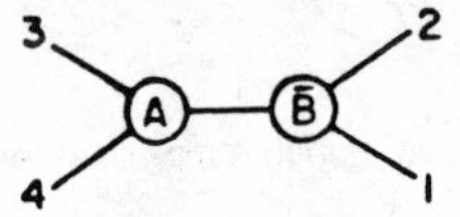

Fig. 5. a) Representation of $Q_{\bar{3}}^{-A} Q_{\bar{3}B}$ with a single-particle intermediate state.

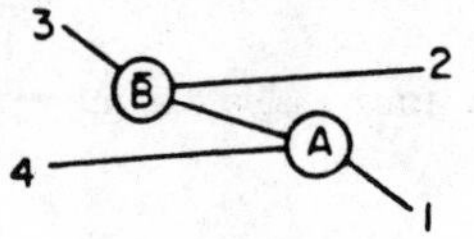

b) Representation of $Q_{\bar{3}B} Q_{\bar{3}}^{-A}$ with a three-particle intermediate state.

operator multiplication. (There is no propagator.) Therefore in a $\sigma - \tau$ diagram the two interaction occur at the same value of τ, so that both diagrams correspond to fig. 6. The cases are distinguishe

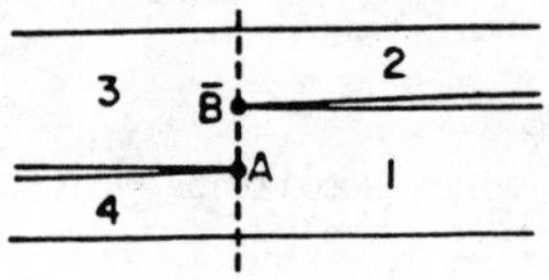

Fig. 6. Light-cone diagram corresponding to both fig. 5a and 5b.

by the order of the factors in the integrand. However,

$$\{K(Q^{-A}), K(Q_{\bar{B}})\} = 0, \qquad (80$$

as is obvious from eqs. (69) and (70). Therefore the matrix element of the anticommutator vanishes.

There is one case in which the reasoning of the previous paragraph breaks down. If the point A and B in fig. 6 coincide, then fig. 5b does not occur and the cancellation cannot be inferred. I this special case one can argue that all components of the momenta and supermomenta go throug unscathed. That is, the result contains the "forward" delta functions.

$$\delta^{10}(p_1 + p_4)\, \delta^4(\alpha_1\theta_1 + \alpha_4\theta_4)\, \delta^{10}(p_2 + p_3)\, \delta^4(\alpha_2\theta_2 + \alpha_3\theta_3). \qquad (81$$

This reconciles the paradox described earlier. The fact that the operator in eq. (79) is positiv definite only implies that a diagonal matrix element is nonzero. In fact, all off-diagonal element vanish. Thus eq. (79) should, more properly, be written as

$$\{Q_{\bar{3}}^{-A}, Q_{\bar{3}B}\} \propto I\, \delta^A_B\,, \qquad (82)$$

where I is a disconnected unit operator. It could be included as an extra term in H_4, but doing so would certainly be rather strange. Presumably, its effects can be taken into account by a suitable rescaling of the wave functions, but the question deserves more study.

The point that we wish to emphasize is that the superstring theory is given entirely by a cubic interaction term (for a type II theory), even though the low-energy effective theory (for a six-torus compactification) is N = 8 supergravity. The latter theory is very complicated with nonpolynomial interactions. The whole collection of contact terms is replaced, in the superstring extension, by the exchange of massive modes. This is completely analogous to the way in which W and Z exchanges replace the four-fermion contact terms in the case of the electroweak theory. Pursuing the analogy further, the superstring field theory appears likely to be well-defined quantum mechanically to all orders in perturbation theory. Finiteness of the type II theories has been proved at one loop. The N = 8 theory is also finite at one loop in four dimensions but is singular in ten dimensions, so this is already an improvement.

The Feynman diagrams that arise in a type II theory can be interpreted as a path integral over all closed oriented two-dimensional world sheets. In particular, at L loops the entire expression corresponds to a sum over surfaces that are topologically equivalent to a sphere with L handles. The potentially dangerous corners of the integration region correspond to singular limits of the topology. The fact that the one-loop diagram has been shown by explicit calculation to be finite means that there is no divergence associated with a corner of the integration region in which a single handle shrinks to a point. What needs to be checked for multiloop diagrams are corners in which several handles simultaneously shrink to the same point. We would be very surprised if any of the integrals turn out to be singular. In any case, the formalism is now developed to the point where such an analysis appears to be feasible.

6. LOOP AMPLITUDES: SUPERSTRING ANOMALY AND DIVERGENCE CANCELLATIONS

In our earlier work we studied certain processes involving four external ground states at one loop in perturbation theory. We found that to this order type II theories are finite [17] unlike ten-dimensional point-like theories, such as type IIB supergravity, which diverge badly at one loop. This illustrated the possibility that superstring theory may be quantum mechanically consistent. The finiteness is due to a conspiracy of divergence cancellations between the *infinite* numbers of supersymmetric fermion and boson states circulating in the loop. The older "bosonic" string theory (which was defined in 26 dimensions) as well as the "spinning" string theory (which only had supersymmetry in the two-dimensional world-sheet of the string) were both badly divergent at one loop. Type I theories with gauge group $G =$ SO(N) or USp(N) have a divergence which can be absorbed into a renormalization of the string tension T [18]. As we will explain later, the systematics of the string perturbation theory makes it plausible that these results will extend to all orders in perturbation theory.

Gauge anomalies can arise in chiral string theories from one-loop hexagon diagrams with external massless open or closed string states. The presence of anomalies in a covariant theory would lead to the coupling of unphysical modes and hence a violation of unitarity. In the light-cone gauge (which is obviously unitary) an anomaly is manifested by a breakdown of Lorentz covariance. (String theories actually have an infinite number of unphysical modes which means that a consistent theory requires an infinite number of anomaly cancellations.).

Although it should be possible to calculate the anomaly in the light-cone gauge formalism it is easier to use a covariant formalism based on the "spinning string" theory [19]. This is not initially ten-dimensionally supersymmetric but it can be made so by projection onto a suitable subspace of the full Hilbert space [5]. The anomalous hexagon diagrams with external open-string (Yang-Mills) states fall into three classes.

The first class contains only the planar diagram in fig. 7 in which the world-sheet is an annulus with the six external states attached to one boundary. There is a group theory factor of

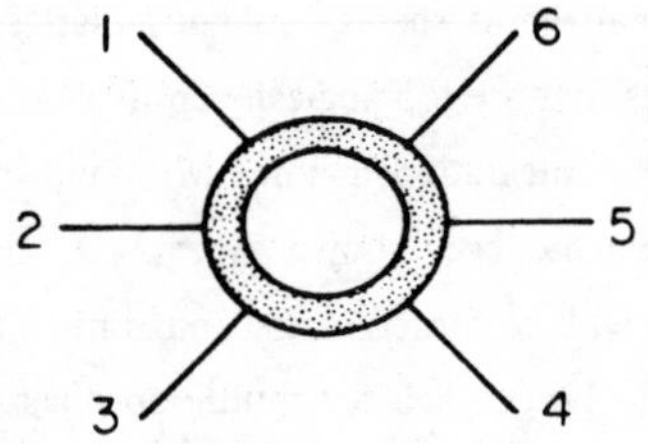

Fig. 7. The anomalous planar open-string hexagon diagram.

$N\,tr(\lambda^1 \cdots \lambda^6)$, associated with this diagram where the gauge group is SO(N) or USp(N) and λ^r is the matrix describing a particle r in the adjoint representation by the pair of indices $\alpha,\beta(= 1, \cdots, N)$ in the fundamental representation. The factor of N comes from $tr(1)$ associated with the inner boundary of fig. 7. The external states are taken to be physical states with momenta k^r and polarizations ζ^r satisfying $k^{r2} = \zeta \bullet k^r = 0$. The anomaly is obtained by taking the divergence on one line, say line #6, which means replacing its polarization vector ζ^6 by its momentum k^6 in its emission vertex. In calculating the anomaly two separate divergence problems arise. The first is the linear divergence of the loop integral that is characteristic of an anomaly. The second is a divergence which arises because the expression for the anomaly has a canceled propagator adjacent to particle #6 (just as in point field theory). The expression therefore contains the ill-defined product of two coincident emission vertices. Both these problems are regularized by a "proper-time" splitting of the vertices which amounts to introducing a gaussian momentum cut-off $\exp(-\tau(p^2 + M^2))$ for a state of mass M where the cut-off $\tau \to 0$ at the end of the calculation. The resulting anomaly is proportional to

$$G = g^6 N \, tr(\lambda_1\lambda_2\lambda_3\lambda_4\lambda_5\lambda_6)\varepsilon^{\mu_1 \cdots \mu_5 \nu_1 \cdots \nu_5}\zeta^1_{\mu_1} \cdots \zeta^5_{\mu_5} k^1_{\nu_1} \cdots k^5_{\nu_5}$$

$$\times \int_0^1 \left[\prod_{i=1}^{5} \theta(\nu_{i+1}-\nu_i)\, d\nu_i\right] \prod_{1 \leqslant i \leqslant j \leqslant 6} (\sin\pi(\nu_j - \nu_i))^{2k^i \bullet k^j/T}$$

$$+ \, permutations, \qquad (83)$$

where $\nu_6 = 1$. This expression reduces to the usual point field theory result in the low-energy limit, $k^i \bullet k^j << T$. It is important, as emphasized earlier, that this expression has poles in the external invariants with residues proportional to anomalies involving massive states.

The second class of diagrams that contributes to the one-loop Yang-Mills anomaly consists of those with non-orientable world-sheets which are Mobius strips. Figure 8 is one example of such a

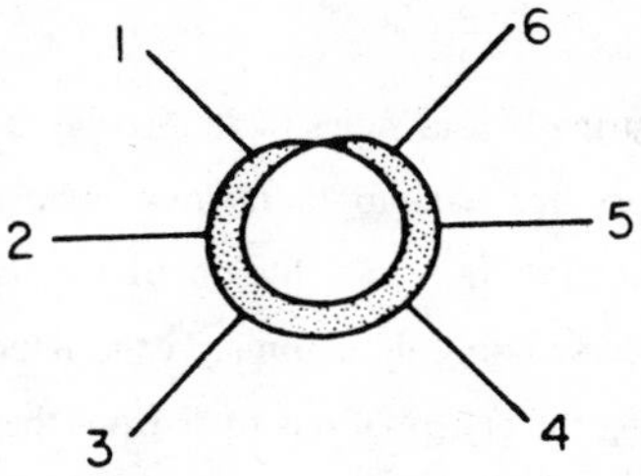

Fig. 8. One of the 32 nonorientable anomalous open-string hexagon diagrams.

diagram which has a single twist in a propagator. In all there are 32 diagrams of this type with an odd number of twists and they all contribute with equal weight to the Bose symmetrized anomaly. In this case the expression for G is identical to that in eq. (3) with the group theory factor replaced by $\pm 32 g^6 tr(\lambda^1 \cdots \lambda^6)$ where the + sign is for USp(N) and the − sign for SO(N). (There is no factor of N because a Möbius strip only has one boundary.)

The last class of diagrams to consider consists of the non-planar orientable diagrams which have an even number of twists such as fig. 9 which has a group theory factor $tr(\lambda^1\lambda^2\lambda^3\lambda^4)tr(\lambda^5\lambda^6)$.

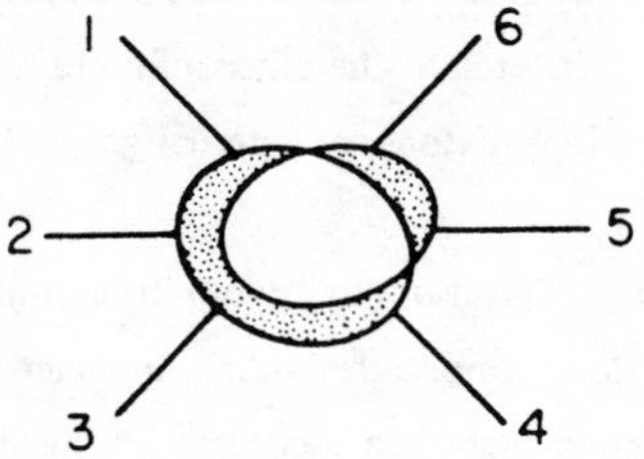

Fig. 9. A nonplanar open-string hexagon diagram which turns out to be free of anomalies for any gauge group.

Remarkably, this turns out to be free of anomalies, a fact which is apparently at odds with th corresponding low-energy point-field theory hexagon diagram limit. The resolution of this apparen contradiction can be seen by distorting the world-sheet into a cylinder as in fig. 10 which shows tha

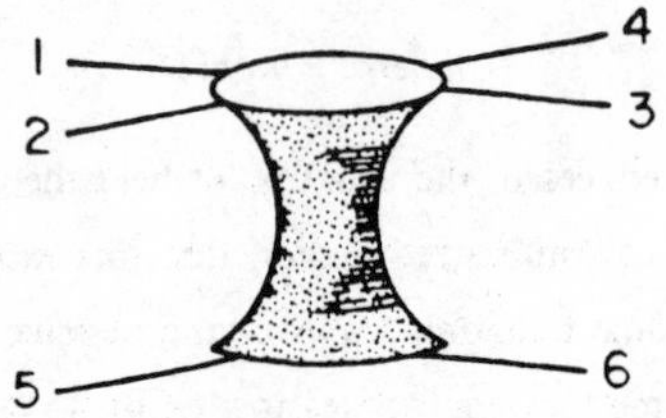

Fig. 10. Another representation of the diagram in fig. 3 which illustrates the closed-string (gravi tational) states in the 5-6 channel.

the 5-6 channel contains closed-string bound states including the massless supergravity multiplet. I the low-energy effective field theory this leads to anomalous *tree* diagrams involving the exchange o the states belonging to the gravity multiplet in addition to the usual hexagon loop. This will b shown later to be the key to understanding the anomaly cancellations. The total Yang-Mills gaug anomaly is obtained by combining the contributions to G from the first two kinds of diagrams (suc as figs. 7 and 8), and we deduce that the anomaly cancels for SO(32).

Anomalies in the divergence of the energy-momentum tensor arise in hexagon diagrams wit two, four or six external gravitons. In superstring theories the graviton occurs as a ground state c type I (unoriented) or type II (oriented) strings. In the type I theory there are four kinds of possibl anomalous string hexagon diagrams with six external gravitons. In these diagrams the gravitons cou ple to interior points on the world sheet. The first kind, described by a world-sheet which is a torus turns out to be anomaly-free. Since this diagram also gives the total anomaly contribution for th type IIb theory, that theory has no anomalies (confirming the expectation based on the anomaly can cellation in the low energy effective field theory [10]). The three other kinds of type I gravitationa hexagon diagrams have world-sheets which are Klein bottles, Mobius strips and annuli. These giv anomalies which cancel for the group SO(32). Details of these calculations are in preparation. Base on the way these anomalies cancel (as well as the other infinities to be discussed next) it is very plau sible that further anomalies involving mixtures of external gauge bosons, gravitini and gravitons wi also cancel for this group.

The SO(32) superstring theory is also singled out from other type I superstring theories b studying the divergences of one-loop amplitudes with four external massless particles. Figure 11 shows a planar loop diagram. The divergence associated with this diagram is contained in

$$g^4 N\, tr(\lambda^1\lambda^2\lambda^3\lambda^4)\int_0^1 \frac{dq}{q} f(q^2)\,, \qquad (84)$$

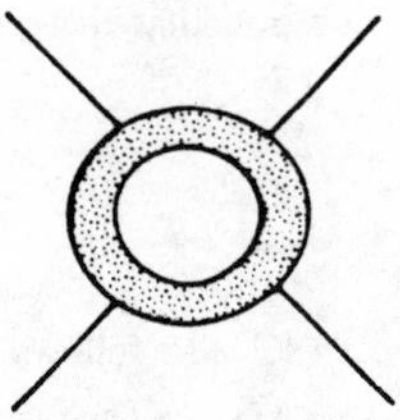

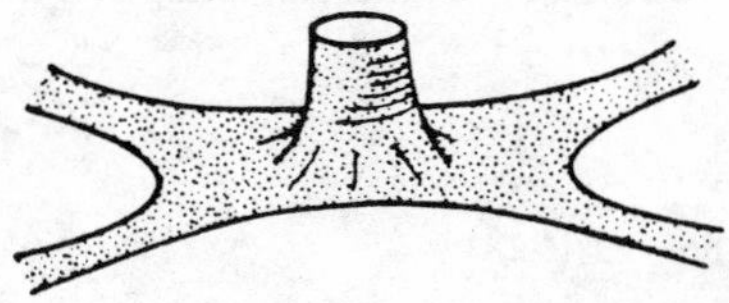

Fig. 11. a) The planar open-string box diagram. b) The world-sheet of fig. 5a distorted to indicate the coupling of a closed string to the vacuum.

where $f(0) = 1$. String theory divergences such as this are not attributed to conventional ultra-violet effects since, no matter how large the loop momentum, there are always states with larger masses. In fact, string divergences can be exhibited in an interesting way as illustrated by fig. 11b, which is a distortion of the world-sheet of the process in fig. 11a. The process is now described in terms of a tree amplitude in which an emitted closed string couples to the vacuum. The loop divergence is replaced by the divergent propagator of the massless closed-string scalar state at zero momentum. The other divergent loops have nonorientable world-sheets (as in fig. 12a) and can be represented by fig. 12b as

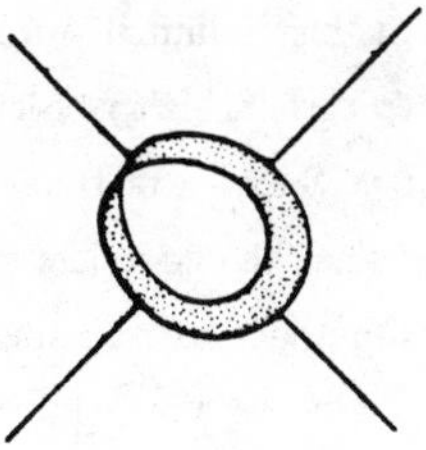

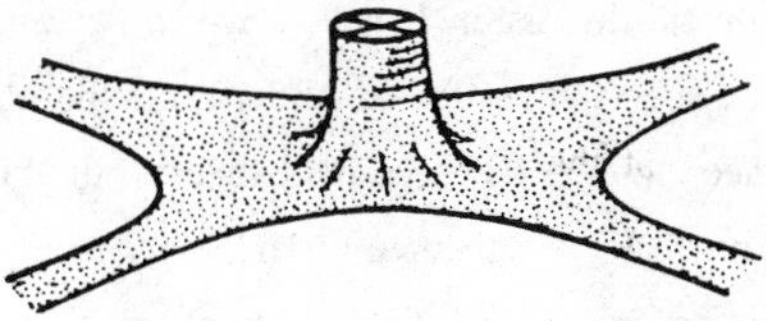

Fig. 12. a) A nonorientable open-string box diagram. b) The world-sheet of fig. 6a distorted to indicate the coupling of a closed string to the vacuum via a cross-cap (indicated by ⊗).

the emission of a closed string which couples to the vacuum via a cross-cap. The total divergence arising from diagrams of this type is contained in

$$\pm 32 g^4 tr(\lambda^1\lambda^2\lambda^3\lambda^4) \int_0^1 \frac{dq}{q} f(-q^2) \tag{85}$$

and hence cancels the divergence of the planar diagram for SO(32) [22].

None of these calculations go beyond one loop. However, general arguments dating back to the early days of string theories [23,24] suggest that all possible string divergences are associated with the zero-momentum emission of the massless closed-string scalar into the vacuum via general tadpoles. This means that if there are divergences the theory is defined with respect to the wrong vacuum which should be redefined to eliminate these tadpoles [24]. The condition that there are no divergences is

 = 0 (86)

where  indicates an on-shell massless scalar state. But eq. (86) also follows if the supersymmetry is unbroken. In the ten-dimensional perturbation theory this has to be true unless the theory has anomalies. Hence there is a profound connection between the possible finiteness of superstring theory at any order in perturbation theory and the absence of anomalies in the theory. This is also borne out by the absence of one-loop anomalies in the chiral N=2 closed-string theory (the type IIB theory) which is also one-loop finite. In conventional point field theories there is no logical connection between the absence of anomalies and finiteness.

7. ANOMALY CANCELLATIONS IN THE LOW-ENERGY EFFECTIVE THEORY

In principle, the superstring field theory can be formulated exactly in terms of its massless states by integrating out the massive modes of Φ and Ψ in the generating functional of the theory. The resulting effective point field action could then be expanded in a power series in *energy*$/\sqrt{T}$. The lowest-order terms in such an expansion correspond to the action for the coupled $N = 1$ super Yang-Mills supergravity system which has been studied previously in this "minimal" form [25,26]. However, in the discussion it will prove to be very important to keep certain higher-order terms in the effective action in order to understand the absence of anomalies. Such extra terms spoil the supersymmetry of the effective field theory but this does not matter since the complete superstring theory is known to be supersymmetric - any truncation probably has to either have anomalies or be non-supersymmetric. The field content of $D = 10, N = 1$ supergravity is: the zehnbein e_μ^m, a second-rank antisymmetric tensor $B_{\mu\nu}$, a scalar φ, the gravitino ψ_μ^+ and a spin-$\frac{1}{2}$ field λ^- (where $\pm$ denote the chiralities of the fields). The Yang-Mills matter content is: the vector potential A^μ and its spin-$\frac{1}{2}$ partner χ^+, both of which are matrices in the adjoint representation of the gauge algebra. The tensor $B_{\mu\nu}$ plays a central role in the cancellation of anomalies.

The action contains the bosonic terms

$$S_0 = -\int d^{10}x\, e\left[\frac{1}{2\kappa^2}R + \frac{1}{\kappa^2}\left(\frac{\partial_\mu\varphi}{\varphi}\right)^2 + \frac{1}{4g^2\varphi}F^a_{\mu\nu}F^{\mu\nu}_a + \frac{3\kappa^2}{2g^4\varphi^2}H_{\mu\nu\rho}H^{\mu\nu\rho}\right] \tag{87}$$

This action has *no free parameters*, since g^2 can be absorbed into a redefinition of φ [27]. This means that the value of the Yang-Mills coupling constant is determined dynamically by the value of the scalar field expectation value, $<\varphi>$. The same is also true for type I string theory, in which case there are also no free parameters. The Yang-Mills field strength $F^a_{\mu\nu}$ is defined by

$$F \equiv \frac{1}{2}\lambda^a F^a_{\mu\nu}\, dx^\mu \wedge dx^\nu = dA + A^2\,, \tag{88}$$

where λ^a_{bc} is the structure constant of the gauge group (i.e. F is a matrix in the adjoint representation) and the language of forms is used to simplify notation. All products of forms are taken to be wedge products. Similarly $R^{mn}_{\mu\nu}$ is defined in terms of the spin connection one-form ω^{mn} by

$$R \equiv \frac{1}{2} R_{\mu\nu} dx^\mu \wedge dx^\nu = d\omega + \omega^2 , \tag{89}$$

where the indices m, n label the ten-dimensional representation of O(1,9). The field strength for the potential $B = B_{\mu\nu} dx^\mu \wedge dx^\nu$ is given by

$$H = dB - X^0_3 , \tag{90}$$

where the Chern-Simons three-form X^0_3 is defined by

$$X^0_3 = \frac{1}{30} \omega^0_{3Y} - \omega^0_{3L} \tag{91}$$

and

$$\omega^0_{3Y} = Tr(AF - \frac{1}{3} A^3) \tag{92}$$

$$\omega^0_{3L} = tr(\omega R - \frac{1}{3} \omega^3) . \tag{93}$$

The notation "Tr" indicates the trace of a matrix in the adjoint representation whereas "tr" indicates a trace in the fundamental representation. The coefficient of the Chern-Simons term, ω^0_{3Y}, in eq. (90) which generates the coupling between B and A in the action is determined by supersymmetry in [26]. The Lorentz group Chern-Simons term, ω^0_{3L}, is an additional feature that emerges from the low-energy behavior of the superstring theory and will be needed later in order to understand the cancellation of gravitational anomalies. It was not included in the minimal formulations of the $N = 1$ point field theory of refs. [25,26] because it is of higher order in derivatives.

The action is invariant under infinitesimal Yang-Mills gauge transformations with parameter Λ and local Lorentz transformations with parameter θ. The field transformations are

$$\delta A = d\Lambda + [A, \Lambda] \tag{94}$$

$$\delta\omega = d\Theta + [\omega, \Theta] \tag{95}$$

$$\delta B = \frac{1}{30} \omega^1_{2Y} - \omega^1_{2L} \equiv -X^1_2 \tag{96}$$

$$\delta X^0_3 = -dX^1_2 \tag{97}$$

where

$$\omega^1_{2Y} = -Tr(A\, d\Lambda) \tag{98}$$

$$\omega^1_{2L} = -tr(\omega d\Theta) . \tag{99}$$

Note that although B has no gauge indices it transforms nontrivially in such a way that

$$\delta H = 0 \tag{100}$$

as required for the invariance of the action.

Nonabelian anomalies in $2n$ dimensions [28] are related, in modern treatments [29], to gauge-invariant Chern-Pontrjagin $(2n + 2)$-forms I_{2n+2} (such as $-\frac{1}{15} Tr\, F^6$ for ten-dimensional Yang-Mills theory). A $(2n + 1)$-form X^0_{2n+1} is then defined by

$$I_{2n+2} = dX^0_{2n+1} \,. \tag{101}$$

X^0_{2n+1} is not invariant under gauge transformations. Its gauge variation can be written as

$$\delta X^0_{2n+1} = -dX^1_{2n} \,, \tag{102}$$

where X^1_{2n} is a function of the gauge parameter. The anomaly G is then given by

$$G = \int X^1_{2n} \,. \tag{103}$$

Any ambiguity in the solution of these equations for G is resolved by imposing Bose symmetry.

I will show below that the anomalies may cancel if the effective point field theory has an I_{12} which *factorizes* into the product of a four-form and an eight-form. For the group SO(32) and also, remarkably, for $E_8 \times E_8$, the Yang-Mills twelve-form is given by

$$I_{12}^{Y.M.} \equiv -\frac{1}{15} Tr\, F^6 = -\frac{1}{720} Tr\, F^2 (Tr\, F^4 - \frac{1}{300} (Tr\, F^2)^2) \,. \tag{104}$$

This relationship between the sixth-order and lower-order symmetric invariants is also true for $E_8 \times E_8$. The twelve-form relevant for the local Lorentz anomalies is given by a combination of terms associated with anomalies generated by ψ^+, λ^- and n species of matter fermions χ^+, where n is the dimension of the adjoint representation of the gauge group. These terms can be obtained from ref. [10] and, in total, give

$$I_{12}^{Lorentz} = \frac{n - 496}{7560} tr\, R^6 + (\frac{1}{8} + \frac{n - 496}{5760}) tr\, R^2 tr\, R^4 + (\frac{1}{32} + \frac{n - 496}{13824})(tr\, R^2)^3 \,. \tag{105}$$

Notice that when n =496 this also factorizes. But this is the dimension of the adjoint representation of both SO(32) and $E_8 \times E_8$! The complete I_{12}, involving mixed Yang-Mills and gravitational terms is given, for both SO(32) and $E_8 \times E_8$,by the factorized expression

$$I_{12}^{TOTAL} = X_4 X_8 \,, \tag{106}$$

where

$$X_4 = (tr\,R^2 - \frac{1}{30} Tr\,F^2) = dX_3^0 \tag{107}$$

and

$$X_8 = \frac{1}{24} Tr\,F^4 - \frac{1}{7200}(Tr\,F^2)^2 - \frac{1}{240} Tr\,F^2 tr\,R^2$$
$$+ \frac{1}{8} tr\,R^4 + \frac{1}{32}(tr\,R^2)^2 \equiv dX_7^0\,. \tag{108}$$

From the earlier discussion (eqs. (23) - (25)) this leads to an anomaly

$$G = c\int (X_2^1 X_8 + 2X_4 X_6^1)\,, \tag{109}$$

where c is a constant and X_2^1 and X_6^1 are defined by eqs. (96), (102), and (108).

This anomaly can be canceled by adding a local polynomial in the fields, S_c, to the original action which is chosen to have an anomalous variation under gauge transformations which cancels the quantum anomaly summarized by G. This term may thus be uniquely determined to be

$$S_c = c\int (-3BX_8 + 2X_3^0 X_7^0)\,. \tag{110}$$

It is easy to see by substituting the variations of the fields (eqs. (94) - (96)) that

$$\delta S_c = -G\,. \tag{111}$$

In terms of scattering amplitudes the one-loop hexagon anomaly is now canceled by an anomaly from a six particle tree diagram in which a $B_{\mu\nu}$ field is exchanged between a vertex with two gauge fields and a vertex with four gauge fields (contained in S_c). This is the tree diagram that was anticipated by the discussion of the string diagram in fig. 10. A remarkable feature of the superstring theory is that it *automatically* contains this diagram with just the right weight to guarantee the cancellation of anomalies without any adjustable parameters. The low-energy effective field theory arising from the superstring theory therefore automatically contains the term S_c with the correct coefficient to ensure anomaly cancellation. In fact, the counterterm S_c and the Chern-Simons terms ω_{3Y}^0 and ω_{3L}^0 originate from similar string theory diagrams as can be seen by factorizing fig. 10 on the massless $B_{\mu\nu}$ state in the 5-6 channel. Each end of the B propagator couples to a disk with Yang-Mills particles coupling to the boundary and gravitational particles coupling to internal points. Another type of disk that contributes to terms with external closed-string states has antipodal points on its boundary identified, which means it is the space RP^2 [22].

8. SPONTANEOUS COMPACTIFICATION

The full superstring field equations should determine the structure of space-time and hopefully lead to a collapse of the extra dimensions. We are far from understanding the structure of these functional equations and only the most trivial toroidal compactifications of string theories have been understood. Even at that level there are interesting effects associated with the fact that a closed string can wind around the compact dimensions in a topologically non-trivial manner [30]. An ambitious approach to finding interesting four-dimensional solutions of the string equations is to "guess" a new first-quantized action which describes an already compactified string theory. This should have the form of a two-dimensional nonlinear σ-model where the string coordinates define the embedding of the world-sheet into space-time and the σ-model metric defines a background geometry with six compact dimensions [31].

It is tempting to try to find hints about string dimensional compactification by studying the low-energy field theory. Unfortunately, the equations of motion of this effective theory (based on the action described earlier) are not entirely meaningful since the terms in the expansion of the string action that are not included could be as important as those that are included. Presumably this effective theory is only a good approximation at distance scales large compared to $(T)^{-1/2}$, which is probably the Planck scale. It is therefore questionable to discuss its compactification while neglecting other effects of the complete string theory for compact dimensions whose size is comparable to the Planck length. Nonetheless, certain general features of this ten-dimensional effective theory may give a useful guide to the lower-dimensional effective theory. For example, there are interesting topological constraints [27] in this point field theory that severely restrict the possible space-time manifolds. One of these can be seen from the definition of H (eq. (90)), which leads to

$$dH = tr\, R^2 - \frac{1}{30} Tr\, F^2 . \tag{112}$$

Thus, in order for H to be globally well defined it is necessary that

$$\int_{M_4} (tr\, R^2 - \frac{1}{30} Tr\, F^2) = 0 , \tag{113}$$

where M_4 is any closed four-dimensional submanifold. This relates topological properties of the space-time manifold to the expectation values of the Yang-Mills fields in the extra dimensions. Remarkably, it turns out that eq. (113) is just the condition needed to ensure that the anomalies do not reappear in the theory when any number of dimensions are compact! The expectation value of F takes values in a subgroup H of $SO(32)$ or $E_8 \times E_8$, breaking the symmetry to a subgroup G. The spectrum of massless fermions in the lower-dimensional effective point field theory can be determined by using index theorems in the compact space. These fermions form representations of G, which can include phenomenologically interesting ones.

One very general feature of this picture that distinguishes it from many other Kaluza-Klein schemes is that the Yang-Mills gauge group can provide all the phenomenologically desirable gauge symmetries. It may then be true that the dimensional compactification produces no further gauge invariances, as would happen if the internal space had no isometries. Ricci-flat spaces have no nonabelian isometries and are natural examples to consider since they also do not give rise to a cosmological constant in the classical compactified theory. Of course, there will almost certainly be quantum sources for a cosmological constant, and so its absence in nature is still a mystery. As a toy example of this scenario we have studied the reduction from ten to six dimensions, using the K_3 manifold as a Ricci-flat internal space [32]. This preserves one six-dimensional supersymmetry. By this means we are able to list many anomaly-free chiral $N = 1$ six-dimensional point field theories with various unbroken gauge groups which are subgroups of SO(32) or $E_8 \times E_8$. When the $E_8 \times E_8$ theory is compactified on a space with no isometries the resulting theory contains two types of matter, originating from the two E_8's, which only interact with each other by forces of gravitational strength. This might have interesting physical implications relating to "dark matter" in the universe [33]. Ricci-flat six-dimensional manifolds with SU(3) holonomy, so-called "Calabi-Yau spaces," have recently been considered with encouraging results [34].

An alternative to the Chan-Paton method of including the internal group theory is to make use of the string vertex representations of Kac-Moody algebras [35]. In this scheme the root lattice of the group is identified with the lattice of the discrete momenta conjugate to internal dimensions which are compactified on a hypertorus. For example, starting with the 26-dimensional bosonic string theory, sixteen dimensions can be chosen to form a torus conjugate to the self-dual lattices $E_8 \times E_8$ or Γ_{16}, which is the SO(32) root lattice together with spinor weights (these are the unique self-dual, rank 16, even, unimodular lattices). In order for the massless states to include vector particles in the adjoint representation of the group it is necessary to use the *unoriented closed-string* theory. The winding numbers conspire with the Kaluza-Klein charges to enlarge the gauge symmetry so that the isometry group $U(1)^{16}$ becomes the maximal torus of the larger, non-abelian group defined by the root lattice. The resulting theory is not very exciting since it is a bosonic theory with tachyonic ground states. However, in an intriguing development [36], a variation of this procedure has been used to construct other $N = 1$ superstring theories with $E_8 \times E_8$ or Γ_{16} root lattices. The latter theory contains SO(32) as the gauge group but it is a very different theory from the type I SO(32) superstring theory since it is made of oriented closed strings only and it has a very different representation content.

The recent developments have confirmed that superstring theories may avoid the inconsistencies associated with quantum point field theories containing gravity. However, our present understanding of these theories is only at the level of perturbation theory about trivial ten-dimensional Minkowski space which is presumably not the most appropriate formulation for obtaining a deep understanding of nonperturbative features (such as compactification) necessary for making testable physical predictions.

The finiteness and anomaly-freedom of the SO(32) (or $E_8 \times E_8$) theory suggest an unusual relationship between the Yang-Mills gauge group and the ten-dimensional space-time group. One tantalizing numerical "coincidence" is the fact that the group of general coordinate transformations in ten dimensions has 992 disconnected components, which is twice the dimension of SO(32) or $E_8 \times E_8$. Is there a more fundamental formulation of the theory in which such relationships between the gauge group and spacetime have an obvious origin?

An important question that has been asked since the earliest days of string theories is why they inevitably contain gauge fields. The only gauge principles used in their construction involve local symmetries of the two-dimensional world sheet swept out by the string. String theories are not based on any ten-dimensional gauge principle, and yet, at distances much larger than $(T)^{-1/2}$, they embody the principle of equivalence as well Yang-Mills gauge invariance. Maybe there is some generalization of the equivalence principle to the space of string configurations that gives a geometrical understanding of the superstring theory, and which reduces to the usual principle at larger distances. An action principle based on such a generalization of the Einstein-Hilbert action would be an important element in understanding the solutions of the theory and hence for understanding how it relates to physics.

REFERENCES

1. For reviews of the early work see: "Dual Theory," Physics Reports reprint vol. 1, ed. M. Jacob (North-Holland, Amsterdam, 1974);
   J. Scherk, Rev. Mod. Phys. 47 (1975) 123.
2. J. Scherk and J.H. Schwarz, Nucl. Phys. B81 (1974) 118.
3. Superstrings, as of mid 1982, are reviewed in: J.H. Schwarz, Phys. Reports 89 (1982) 223;
   M.B. Green, Surveys in H. E. Physics 3 (1983) 127.
4. M. Ademollo et al., Phys.Lett. 62B (1976) 105; Nucl. Phys. B111 (1976) 77.
5. F. Gliozzi, J. Scherk, and D.I. Olive, Nucl. Phys. B122 (1977) 253.
6. L. Brink, J.H. Schwarz, and J. Scherk, Nucl. Phys. B121 (1977) 77.
7. E. Cremmer, B. Julia, and J. Scherk, Phys. Lett. 76B (1978) 409.
8. M.B. Green and J.H. Schwarz, Phys. Lett. 122B (1983) 143;
   J.H. Schwarz and P.C. West, Phys. Lett. 126B (1983) 301;
   J.H. Schwarz, Nucl. Phys. B226 (1983) 269;
   P. Howe and P.C. West, Nucl. Phys. B238 (1984) 181.
9. N. Marcus and J.H. Schwarz, Phys. Lett. 115B (1982) 111.
10. L. Alvarez-Gaumé and E. Witten, Nucl. Phys. B234 (1984) 269.
11. A.M. Polyakov, Phys. Lett. 103B (1981) 207.
12. M.B. Green and J.H. Schwarz, Phys. Lett. 136B (1984) 367.
13. S. Mandelstam, Nucl. Phys. B213 (1983) 149.

14. M.B. Green and J.H. Schwarz, Nucl. Phys. B243 (1984) 475.

15. M. Kaku and K. Kikkawa, Phys. Rev. D10 (1974) 1110; 1823;
E. Cremmer and J.-L. Gervais, Nucl. Phys. B76 (1974) 209, B90 (1975) 410.

16. M.B. Green and J.H. Schwarz, Nucl. Phys. B218 (1983) 43;
M.B. Green, J.H. Schwarz, and L. Brink, Nucl. Phys. B219 (1983) 437.

17. M. B. Green and J. H. Schwarz, Phys. Lett. 109B (1982) 444.

18. M. B. Green and J. H. Schwarz, Nucl. Phys. B198 (1982) 441.

19. P. Ramond, Phys. Rev. D3 (1971) 2415;
A. Neveu and J. H. Schwarz, Nucl. Phys. B31 (1971) 86.

20. M. B. Green and J. H. Schwarz, Phys. Lett. 149B (1984) 117.

21. M. B. Green and J. H. Schwarz, "The Hexagon Gauge Anomaly in Type I Superstring Theory," to be published in Nucl. Phys. B.

22. M. B. Green and J. H. Schwarz, "Infinity Cancellations in SO(32) Superstring Theory," to be published in Phys. Lett. B.

23. J. Shapiro, Phys. Rev. D11 (1975) 2937;
M. Ademollo et al., Nucl. Phys. B124 (1975) 221.

24. M. B. Green, Nucl. Phys. B124 (1977) 461.

25. E. Bergshoeff, M. de Roo, B. de Wit and P. van Nieuwenhuizen, Nucl. Phys. B195 (1982) 97.

26. G. F. Chapline and N. S. Manton, Phys. Lett. 120B (1983) 105.

27. E. Witten, "Some Properties of O(32) Superstrings," Princeton Univ. Preprint.

28. P. H. Frampton and T. W. Kephart, Phys. Rev. Lett. 50 (1983) 1343, 1347;
P. K. Townsend and G. Sierra, Nucl. Phys. B222 (1983) 493.

29. B. Zumino, Les Houches lectures 1983;
R. Stora, Cargèse lectures 1983;
B. Zumino, Y. S. Wu and A. Zee, Nucl. Phys. B239 (1984) 477;
L. Baulieu, Nucl. Phys. B241 (1984) 557;
L. Alvarez-Gaumé and P. Ginsparg, Nucl. Phys. B243 (1984) 449; Harvard Univ. preprint HUTP-84/A016;
L. Faddeev, Phys. Lett. 145B (1984) 81;
M. Atiyah and I. Singer, Proc. Natl. Acad. Sci. USA 81 (1984) 2597.
W. A. Bardeen and B. Zumino, Nucl. Phys. B244 (1984) 421.

30. E. Cremmer and J. Scherk, Nucl. Phys. B103 (1976) 399;
M. B. Green, J. H. Schwarz and L. Brink, Nucl. Phys. B198 (1982) 474.

31. D. Friedan, Z. Qiu and S. Shenker, "Superconformal Invariance in Two Dimensions and the Tricritical Ising Model," Univ. of Chicago preprint EFI-84-35.
D. Friedan and S. Shenker, private communications.

32. M. B. Green, J. H. Schwarz and P. C. West, "Anomaly-free Chiral Theories in Six Dimensions," Caltech preprint to be published in Nucl. Phys. B.

33. E. W. Kolb, D. Seckel, and M. S. Turner, "The Shadow World," Fermilab preprint.

34. P. Candelas, G. T. Horowitz, A. Strominger, and E. Witten, "Vacuum Configurations for Superstrings," U.C. Santa Barbara preprint.

35. I. B. Frenkel and V.G. Kac, Inv. Math. 62 (1980) 23;
P. Goddard and D.I. Olive, DAMTP preprint 83/22, to be published in the Proceedings of the MSRI Workshop on Vertex Operators, ed. J. Lepowski (Springer-Verlag, New York).

36. D. Gross, J. Harvey, E. Martinec and R. Rohm, "The Heterotic String," Princeton Univ. preprint.

37. A.K.H. Bengtsson, L. Brink, M. Cederwall, and M. Ogren, "Uniqueness of Superstring Actions," Göteborg preprint 84-58.

TENSOR CALCULUS AND THE BREAKING OF LOCAL SUPERSYMMETRY

S FERRARA

INTRODUCTION

Supersymmetry, the relativistic symmetry which relates bosons and fermions and their interactions, has received considerable attention in recent years in theoretical elementary particle physics.

From a fundamental point of view the quest for supersymmetry at the level of fundamental interactions is at least twofold.

i) It is the only symmetry principle which enlarges, in a non-trivial way, the space-time Poincaré symmetry through the mathematical concept of graded Lie algebra. Gauged supersymmetry may therefore provide a non-trivial link between particle physics and gravitation[1]. This link emerges both in N-extended supergravities in which all elementary fields are in the same N-extended supermultiplet, as well as in 'low-energy supergravity'[2] in which the supergravity couplings induced by the super-Higgs effect, i.e. the gravitino mass, control the particle mass splitting as well as the pattern of symmetry breaking of the non-gravitational gauge interactions.

ii) Supersymmetry provides unique examples of field theories in which ultraviolet divergences are milder than in ordinary field theories owing to mutual cancellations of boson and fermion loops. One aspect of these cancellations is that supersymmetry avoids quadratic divergences for elementary scalar masses, thus avoiding unnatural fine-tuning for the scalar Higgs v.e.v.'s. These fine-tunings are essential, in the Standard Model of electroweak and strong interactions, for the theory to be meaningful beyond the 1 TeV scale. Another hope, related to this better ultraviolet behaviour, is that some theory based on local supersymmetry may achieve the final unification of all forces including gravity, and therefore provide, as a by-product, a meaningful quantum theory for the gravitational force[1].

The present contribution will be confined to the theoretical setting for the construction of general couplings of N = 1 supergravity to Yang-Mills and matter multiplets describing ordinary non-gravitational interactions.

This framework has been extensively used in recent times since it provides natural extensions of the Standard Model in which the Fermi scale and the electroweak symmetry breaking are controlled by the supersymmetry-breaking scale. The theoretical motivations for the scheme to be suitable for describing physics at the 1 TeV scale will be given in the second part of this section. This contribution is organized as follows: In Sections 2, 3, and 4 the general form of locally supersymmetric N = 1 matter couplings are derived from the tensor calculus of N = 1 local supersymmetry. In Section 5 the super-Higgs effect is described, and in Section 6 the low-energy limit of locally supersymmetric theories is obtained. In Section 7, examples of the super-Higgs sector of the theories are given.

Why supersymmetry?

Non-gravitational interactions are nowadays described, in the framework of relativistic quantum field theory, by gauge forces.

In this context one of the fundamental problems is the origin of the several physical energy scales. In the case of the strong forces, the typical scale, i.e. the inverse hadronic size or interaction range, is satisfactorily accounted for by the introduction of the Λ parameter in Quantum Chromodynamics (QCD). This parameter arises from a scale-invariant Lagrangian through the phenomenon of dimensional transmutation, which relates it to a basic dimensionless coupling constant with the help of the renormalization group.

In the case of the electroweak phenomena, the relevant energy scale is given by

$$G_F^{-\frac{1}{2}} \simeq 293 \text{ GeV} , \tag{1.1}$$

where G_F is the Fermi coupling constant.

A possible way of explaining this new physical scale, in analogy to QCD, would be to invoke a new strong force, based on a non-Abelian gauge theory, with a related Λ parameter of order $G_F^{-\frac{1}{2}}$. This force might give rise to the binding of a composite Higgs field, responsible for a dynamical breaking of the electroweak symmetry, and/or even perhaps of composite quarks, leptons and weak gauge vector bosons. The problem of this viewpoint is that, in order to make it more concrete, one has to rely on strong assumptions on the essentially unknown dynamics of the hypothetical new force. These assumptions can at best be justified on the basis of naïve analogies with QCD, and this only in some cases.

The alternative possibility in describing the physical origin of the Fermi scale relates it to the vacuum expectation value of a fundamental scalar field ϕ, such that

$$\langle \phi \rangle = (\sqrt{2}\, G_F)^{-\frac{1}{2}} = 246 \text{ GeV}. \tag{1.2}$$

This is the case of the standard $SU(2)_L \times U(1)$ model with the Higgs doublet -- whose electrically neutral component is ϕ -- taken as a basic degree of freedom rather than an effective field.

The main advantage of this viewpoint is that it can be realized in the context of a weak-coupling field theory. This in turn makes possible the application of the usual perturbative techniques to give physical predictions.

However, in the framework of perturbative field theory, scalar masses usually receive huge (quadratically divergent) radiative corrections. This is in sharp contrast to the case of fermion masses, which control suitable chiral symmetries and are for this reason protected from acquiring big radiative corrections. More specifically, the typical expressions for the one-loop contributions to these masses are

$$\delta m_S^2 = A\, g^2\, \Lambda^2 \tag{1.3}$$

$$\delta m_F = B\, g^2\, m_F \log \frac{\Lambda}{m_F} \tag{1.4}$$

where A,B are numerical constants of order unity, g is a dimensionless gauge or Yukawa coupling, m_F is a tree-level fermion mass, and Λ is the momentum cut-off.

As a consequence of Eq. (1.3), barring accurate cancellations in A or even between the correction δm_S^2 and a bare mass term, the scalar mass itself will be at least of order $g\Lambda$ and therefore

$$\langle \phi \rangle = \frac{m_S}{\sqrt{2\lambda}} = O\left(\frac{g\Lambda}{\sqrt{2\lambda}}\right), \tag{1.5}$$

where λ is a dimensionless scalar self-coupling constant. This equation, together with Eq. (1.2) implies that in a weak coupling theory ($\lambda < 1$), the cut-off Λ cannot exceed by orders of magnitude $G_F^{-\frac{1}{2}}$. In other words, the validity of the weakly coupled fundamental Higgs scalar model cannot be pushed much further up than the Fermi scale itself*).

An important factor to notice at this point is that the numerical parameter A defined in Eq. (1.3) is not positive definite but it receives contributions opposite in sign from virtual boson and fermion exchanges, as is the case for the total vacuum energy of a given matter system of bosons and fermions. This observation makes the vanishing of A possible -- thus avoiding the scale problem -- which would, however, be accidental unless related to a symmetry principle. This symmetry, which has also to keep under control the

---

*) In theories with a scale M much bigger than $G_F^{-\frac{1}{2}}$, the scalar mass requires additional contributions O(gM). In the context of Grand Unified Theories (GUTs) this is called the hierarchy problem[7].

potential quadratic divergences introduced in higher orders of perturbation theory, must organize bosons and fermions in suitable multiplets with related masses and coupling constant.

The only candidate for such an invariance is an extension of the Fermi-Bose symmetry displayed by the free kinetic Lagrangian of an equal number of bosons and fermions, i.e.

$$\mathcal{L} = -(\tfrac{1}{2}|\partial_\mu z_i|^2 + \bar{\chi}_{Li} \not\partial \chi_R^i) \tag{1.6}$$

$$\delta z_i = \bar{\epsilon}_R \chi_{Li} ,$$
$$\delta \chi_{Li} = + \frac{1}{2}(\not\partial z_i) \epsilon_R , \tag{1.7}$$

where ϵ_R is a constant anticommuting Weyl spinor. Under Eqs. (1.7) $\mathcal{L}$ changes by a total derivative. The symmetry given by Eqs. (1.7) has an extension to interacting Lagrangians and it gives rise to an algebraic structure called supersymmetry.

Unbroken supersymmetry implies relations amongst coupling constants and, most importantly, degeneracy between bosons and fermions belonging to the same supermultiplet. A realistic particle spectrum therefore demands the breaking of supersymmetry characterized by intramultiplet mass splittings Δm_i.

As we have said, supersymmetry may indeed require that the coefficient A in Eq. (1.3) vanishes identically with at most a logarithmic divergence analogous to the fermionic case [Eq. (1.4)]. In the presence of supersymmetry breaking, however, the finite corrections to the scalar mass may be sizeable, being of order $\delta m_S^2 = O(g^2 \Delta m_i^2)$. One can say that the quadratic divergence is replaced by the typical square mass splitting, which is how the supersymmetry breaking enters the radiative correction diagrams.

The same line of the previous arguments now leads to the conclusion that the individual Δm_i cannot exceed, as an order of magnitude, the Fermi scale itself. With broken supersymmetry the absence of fine tuning in δm_S^2 requires that all the known particles find their superpartners in the energy range of the Fermi scale.

The detailed prediction of the mass spectrum of all these new particles requires an understanding of the mechanism of supersymmetry breaking. Most predictive, in particular, are the theories where the breaking is spontaneous. In such a case, and restricting ourselves to a weak coupling model, the following 'low-energy theorem' holds:

$$\Delta m_i^2 = g_i M_S^2 \ , \tag{1.8}$$

where g_i is the coupling to the ith multiplet of the Goldstond fermion (goldstino) ψ_g, and M_S is the supersymmetry-breaking scale defined by the goldstino term in the spinorial supersymmetry current,

$$J_{\mu A} = M_S^2 \ (\gamma_\mu \psi_g)_A \ . \tag{1.9}$$

Two schemes of spontaneous supersymmetry breaking can be considered, depending on the size of the primordial supersymmetry-breaking scale M_S. In the first case

$$M_S^2 = O(\Delta m_i^2) = O(G_F^{-1}) \ , \tag{1.10}$$

and g_i is a coupling of the same order of the electroweak gauge or Yukawa couplings. In the second case[3]

$$M_S^2 = O(M \ G_F^{-\frac{1}{2}}) \ , \tag{1.11}$$

where M is a mass scale much bigger than $G_F^{-\frac{1}{2}}$. Here g_i must consequently be very small, $g_i = O(G_F^{-\frac{1}{2}}/M)$. In suitable supersymmetric theories the coupling g_i may indeed be given as a ratio of different mass scales, to all orders of perturbation theory (decoupling theorem).

A close inspection of models for particle interactions based on a 'low' supersymmetry-breaking scale, $M_S = O(G_F^{-\frac{1}{2}})$, leads to a series of difficulties, both theoretical and phenomenological.

In many respects, the main problem is that in models based on the minimal gauge group SU(3) × SU(2) × U(1) and with $M_S = O(G_F^{-\frac{1}{2}})$, the tree-level mass spectrum is unrealistic because of at least one unacceptably light scalar partner of the quarks. Possible ways of avoiding this difficulty seem to introduce a host of new problems.

In the alternative point of view, $M_S^2 = O(M\, G_F^{-\frac{1}{2}})$, $M \gg G_F^{-\frac{1}{2}}$, the most promising situation is met by identifying M with the Planck mass M_P. This may happen in theories where there are scalar fields acquiring vacuum expectation values of order M_P. The real interest of this situation lies in the fact that supergravity is brought in, since the effects of couplings vanishing as inverse powers of M_P are in such a case no longer negligible.

In order to explore the physical consequences of a theory with a supersymmetry-breaking intermediate bewteen the Fermi and the Planck scale[4], the proper framework is set by a supersymmetric Yang-Mills Lagrangian coupled to supergravity[5,6].

TENSOR CALCULUS

The action and transformation laws of locally supersymmetric systems are constructed by the use of the N = 1 tensor calculus of supergravity[8]. A major simplification occurs by employing the so-called conformal tensor calculus which uses the Weyl multiplet for the gravitational sector, i.e. the multiplet which contains only the gauge fields of conformal supergravity[9].

The N = 1 Weyl multiplet contains the following set of fields,

$$e_{m\mu},\ \psi_{\mu a},\ A_\mu,\ b_\mu) \ , \qquad (2.1)$$

which are respectively the gauge fields for general coordinate transformations (vierbien), for Q supersymmetry (gravitino), for chiral U(1) (R-symmetry), and dilatations.

The additional gauge fields $\omega_{\mu mn}$, $\Phi_{\mu a}$, and $f_{m\mu}$ for local Lorentz rotations, special supersymmetry, and special conformal transformations M_{mn}, S_a, and K_m, respectively, can be solved in terms of the gauge fields in Eq. (2.1) by means of the conventional constraints*)

$$R^{m}_{\mu\nu}(P) = 0 \,, \tag{2.2a}$$

$$\gamma^{\mu} R_{\mu\nu}(Q) = 0 \,, \tag{2.2b}$$

$$\hat{R}_{\mu\nu mn}(M)\, e^{\nu n} + \frac{i}{3}\hat{R}_{\mu m}(A) = 0 \,, \tag{2.2c}$$

$$\left(\tilde{T}_{mn} = \frac{1}{2}\varepsilon_{mnpq} T_{pq}\right) , \tag{2.2d}$$

where $\hat{R}_{\mu\nu mn}$ denotes the fully convariant Lorentz connection (Riemann tensor). The curvatures are covariant, but since the constraints in Eqs. (2.2) are not Q-invariant, the variations $\delta_Q \omega_{\mu nn}$, $\delta_Q \Phi_{\mu a}$ and $\delta_Q f_{m\mu}$ are not simply given by the group rule. In the conformal tensor calculus we use two basic multiplets other than the one of Eq. (2.1). The scalar (or chiral) multiplet S = (z, χ_L, h), with Weyl weight w (w is the Weyl weight of the first component z of S) transforming as

$$\delta z = \bar{\varepsilon}_L \chi_L - \frac{1}{3} i w \alpha z \,, \tag{2.3a}$$

$$\delta\chi_L = \frac{1}{2}\hat{\not{\partial}} z \varepsilon_R + \frac{1}{2} h \varepsilon_L + w z \eta_L + i\alpha\left(\frac{1}{2} - \frac{1}{3} w\right)\chi_L \,, \tag{2.3b}$$

$$\delta h = \bar{\varepsilon}_R \hat{\not{\partial}} \chi_L + 2\bar{\eta}_L \chi_L (1 - w) + i\alpha\left(1 - \frac{1}{3} w\right) h \,, \tag{2.3c}$$

where ε is a space-time dependent anticommuting Weyl (or Majorana) spinor corresponding to Q transformations, and α is the space-time dependent parameter of the U(1) chiral group of the superconformal algebra, normalized such that for the U(1) gauge field A_μ

$$\delta_\alpha A_\mu = \partial_\mu \alpha \,. \tag{2.4}$$

---

*) We use the conventions of Ref. 2. We also use the following conventions: a = 1,2, two-component spinor indices; m = 1, ..., 4 flat Lorentz indices; μ = 1, ..., 4, world indices; α = 1 ... dim gauge group indices; i,j = 1 ... dim R, G-group representation indices, R being a finite unitary representation of G. We also set, in most of our formulae, the gravitational constant k = 1; k is related to the Planck mass m_P as follows: $k = \sqrt{8\pi}/m_P$.

In Eq. (2.3) η is a space-time dependent anticommuting Weyl (or Majorana) spinor corresponding to special S_a supersymmetry transformations, and $\hat{D}_\mu$ denotes the superconformal covariant derivative.

The second multiplet is a real vector multiplet V: V = (C, S_L, H, N_m, λ_R, D). Its transformation is

$$\delta G = \frac{1}{2} i\bar{\varepsilon}_L S_L - \frac{1}{2} i\bar{\varepsilon}_R S_R , \tag{2.5a}$$

$$\delta S_L = \frac{1}{2} iH\varepsilon_L - \frac{1}{2} \not{B}\varepsilon_R - \frac{1}{2} i\hat{\not{D}}C\varepsilon_R - iwC\eta_L + \frac{1}{2} i\alpha S_L , \tag{2.5b}$$

$$\delta H = -i\bar{\varepsilon}_R\hat{\not{D}}S_L - i\varepsilon_R\lambda_R + i(-2 + w)\bar{\eta}_L S_L + i\alpha H , \tag{2.5c}$$

$$\delta B_m = \left[- \frac{1}{2} \bar{\varepsilon}_L\hat{D}_m S_L - \frac{1}{2} \bar{\varepsilon}_L\gamma_m\lambda_R + \frac{1}{2}(1 + w)\bar{\eta}_L\gamma_m S_R\right] + \text{h.c.} , \tag{2.5.d}$$

$$\delta\lambda_R = \frac{1}{2}(\sigma\cdot\hat{F} - iD)\varepsilon_R + \frac{1}{2} w(iH\eta_R - \not{B}\eta_L + i\hat{\not{D}}C\eta_L) +$$

$$+ \frac{1}{2} i\alpha\lambda_R - w\not{\Lambda}_K S_L , \tag{2.5e}$$

$$\delta D = \left[\frac{1}{2} i\bar{\varepsilon}_L\hat{\not{D}}\lambda_R + iw\bar{\eta}_L\left(\lambda_L + \frac{1}{2} \hat{\not{D}}S_R\right) - w\Lambda_{Km}\hat{D}_m C\right] + \text{h.c.} , \tag{2.5f}$$

with

$$\hat{F}_{mn} = 2\hat{D}_{[m}B_{n]} + i\varepsilon_{mnpq}\hat{D}_p\hat{D}_q C , \quad [mn] = \frac{1}{2}(mn - nm) , \tag{2.5g}$$

where the symbol [ ] denotes antisymmetrization, and Λ_{Km} is the space-time dependent parameter for special conformal transformations.

The scalar multiplet S, for Weyl weight w = 3, has its last component h with Weyl weight w = 4, and it can be extended to a superconformal density

$$e^{-1}L_{S(w=3)} = h + \bar{\psi}_R\cdot\gamma\chi_L + \bar{\psi}_{\mu R}\sigma_{\mu\nu}\psi_{\nu R}z . \tag{2.6}$$

Analogously we can embed a vector multiplet V with w = 2 into a scalar multiplet S_V with w = 3, whose components are

$$(-H^*, \quad -i\lambda_L - i\hat{\not{D}}S_R, \quad D + \hat{\Box}C + I\hat{D}_m B^m) . \tag{2.7}$$

Then we can apply Eq. (2.6) to the multiplet S_V to obtain a density formula for the vector multiplet with w = 2:

$$e^{-1}L_{V(w=2)} = D + \hat{\Box}C + \left[\frac{1}{2}\, i\bar{\psi}\cdot\gamma(\lambda_R + \hat{\not{D}}S_L) - \frac{1}{2}\,\psi_{\mu L}\sigma_{\mu\nu}\psi_{\nu L}H + \text{h.c.}\right]. \tag{2.8}$$

We can also get a vector embedding of a scalar multiplet S with w = 0 into a vector multiplet V_S with w = 0, whose components are

$$V_S = \left[i(z - z^*),\ \ 2\chi_L,\ \ -2ih,\ \ -\hat{D}_m(z + z^*),\ 0,\ 0\right]. \tag{2.9}$$

The multiplication rules of scalar and vector multiplets are given respectively by

$$S_1S_2 = (z_1z_2,\ \ z_1\chi_{2L} + z_2\chi_{1L},\ \ z_1h_2 + z_2h_1 - 2\chi_{1L}\chi_{2L})\ ; \tag{2.10a}$$

$$\begin{aligned} V_1V_2 = \Big[&C_1C_2,\ \ C_1S_{2L} + C_2S_{1L},\ \ C_1H_2 + C_2H_1 - \bar{S}_{1L}S_{2L},\\ &C_1B_{m2} + \frac{1}{2}\, i\bar{S}_{1L}\gamma_m S_{2R} + (1 \leftrightarrow 2),\\ &C_1\lambda_{2R} + \frac{1}{2}(-\hat{\not{D}}C_1S_{2L} + H_1S_{2R} - i\not{B}_1S_{2L}) + (1 \leftrightarrow 2)\ ,\\ &C_1D_2 - \bar{S}_{1L}\lambda_{2L} - \bar{S}_{1R}\lambda_{2R} + \frac{1}{2}\, H_1H_2^* - \frac{1}{2}\, B_1\cdot B_2 - \frac{1}{2}\,\hat{D}_mC_1\hat{D}^mC_2\\ &- \frac{1}{2}\,\bar{S}_{1L}\hat{\not{D}}S_{2R} - \frac{1}{2}\,\bar{S}_{1R}\hat{\not{D}}S_{2L} + (1 \leftrightarrow 2)\Big]\ ; \end{aligned} \tag{2.10b}$$

$$\begin{aligned} \frac{1}{2}(S_1\bar{S}_2 + \bar{S}_1S_2) = \Big[&\frac{1}{2}\, z_1z_2^*,\ \ -iz_2^*\chi_{1L},\ \ -z_2^*h_1\ ,\\ &\frac{1}{2}\, i\left(z_2^*\hat{D}_mz_1 - z_1\hat{D}_mz_2^*\right) - i\bar{\chi}_{2R}\gamma_m\chi_{1L}\ ,\\ &-ih_1\chi_{2R} + i\hat{\not{D}}z_2^*\chi_{1L}\ ,\\ &h_1h_2^* - \hat{D}_mz_1\hat{D}^mz_2^* - \chi_{1L}\overleftrightarrow{\hat{\not{D}}}\chi_{1R}\Big] + (1 \leftrightarrow 2)\ . \end{aligned} \tag{2.11}$$

The resulting multiplets have Weyl weight $w_1 + w_2$.

The rules we have given so far are sufficient for constructing all superconformal invariant actions. However, if we want to restri our tensor calculus to Poincaré supersymmetry only, we have to intro duce Poincaré rules which will enable us to construct Poincaré super symmetric local densities for arbitrary Weyl weight of the fields involved. This is most easily achieved by introducing a new compensating chiral multiplet S_0 with Weyl weight w = 1, and then using it to fix a special superconformal gauge. This choice of compensating multiplet corresponds to the so-called old minimal formulation of N = 1 supergravity originally proposed in Refs. 7. The connection of this formulation with the new minimal formulation of Sohnius and West[10], which uses a different compensating multiplet to fix the superconformal gauge, has been given in general elswehere[11].

The chiral compensating multiplet has components

$$S_0 = \left(a,\ \xi,\ \frac{1}{3}\,u\right), \tag{2.12}$$

where u = S - iP is the complex auxiliary field of Poincaré supergravity defined by Cremmer et al.[5].

To get Poincaré supergravity from the superconformal tensor calculus we fix a superconformal gauge by fixing the gauges of dilatation, U(1) chiral rotation, special S-supersymmetry transformations, and conformal boosts, by the conditions

$$a = 1,\quad \xi = 0\ ,\quad b_\mu = 0\ . \tag{2.13}$$

These conditions are preserved under a newly defined Poincaré Q local supersymmetry with the parameter

$$\begin{aligned}\delta^P(\varepsilon) &= \delta_Q(\varepsilon) + \delta_S\!\left(\eta_L = -\frac{1}{6}\,u\varepsilon_L - \frac{1}{6}\,i\not{A}\varepsilon_R\right)\\ &\quad + \delta_K\!\left(\Lambda_{K\mu} = \frac{1}{4}\,\bar{\varepsilon}\phi_\mu - \frac{1}{4}\,\bar{\eta}\psi_\mu\right)\ .\end{aligned} \tag{2.14}$$

Equation (2.14) provides the rule for computing a Poincaré local supersymmetry transformation in terms of the local superconformal rules.

To compute the density formula for a chiral multiplet S of Weyl weight w = 0 we first construct the new chiral multiplet SS_0^3 with Weyl weight w = 3. Since in the Poincaré gauge defined by Eq. (2.13), using Eq. (2.10), we have

$$S_0^n = \left(1,\ 0,\ \frac{1}{3}\,nu\right)\ , \tag{2.15}$$

we obtain

$$SS_0^n = \left(z,\ \chi_L,\ h + \frac{1}{3}\,nuz\right)\ . \tag{2.16}$$

Then the Poincaré density formula for S is given by

$$\begin{aligned}[SS_0^3]_{F,SC} &= [S]_{F,\text{Poincaré}} \\ &= L_{S(w=0)} = eh + euz + e\bar{\psi}_R\cdot\gamma\chi_L + e\bar{\psi}_{\mu R}\sigma_{\mu\nu}\psi_{\nu R}z\ .\end{aligned} \tag{2.17}$$

For the vector multiplet V, we take a Weyl weight $w = 0$ and then introduce the compensating vector multiplet $S_0\bar{S}_0$ with $w = 2$ having components

$$S_0\bar{S}_0 = \left[1,\ 0,\ -\frac{2}{3}\,u,\ -\frac{2}{3}\,A_m,\ 0,\ \frac{2}{9}\left(uu^* - A_mA^m\right)\right]\ . \tag{2.18}$$

We then get a Poincaré density formula for V by using a superconformal density formula for $VS_0\bar{S}_0$, with $w = 2$, and using the multiplication rule given by Eq. (2.11) in the Poincaré gauge [Eq. (2.13)]:

$$\begin{aligned}[VS_0\bar{S}_0]_{D,SC} &= [V]_{D,\text{Poincaré}} = e^{-1}\ L_{V(w=0)} \\ &= D - \frac{1}{2}\,i\bar{\psi}\cdot\gamma\gamma_5\lambda - \frac{1}{3}(u^*H + uH^*) + \frac{2}{3}\,B_mA^m \\ &\quad - \frac{1}{3}\,i\bar{S}\gamma_5\gamma R^P + \frac{1}{4}\,i\varepsilon^{mnrs}\bar{\psi}_m\gamma_n\psi_r\left(B_s - \frac{1}{2}\,\bar{\psi}_s S\right) - \frac{2}{3}\,C\ e^{-1}L_{SC}\end{aligned} \tag{2.19}$$

where

$$\begin{aligned}R_\mu^P &= e^{-1}\ \varepsilon_{\mu\nu}{}^{\rho\sigma}\gamma_5\gamma^\nu\Big[\partial_\rho + \frac{1}{2}\,\omega_{\rho mn}(e_1\psi)\sigma^{mn} \\ &\quad + \frac{1}{2}\,i\gamma_5A_\rho + \frac{1}{6}\,\gamma_\rho(\omega - i\not{A}\gamma_5)\Big]\psi_0\ ,\end{aligned} \tag{2.20}$$

$$\begin{aligned}L_{SC} &= -\frac{1}{2}\,eR[e,\omega(e,\psi)] \\ &\quad - \frac{1}{2}\,\varepsilon^{\mu\nu\rho\sigma}\bar{\psi}_\mu\gamma_5\gamma_\nu D_\rho[\omega(e,\psi)]\psi_\sigma - \frac{1}{3}\,euu^* + \frac{1}{3}\,eA_mA^m\ .\end{aligned} \tag{2.20}$$

We remark that the Poincaré density formula as well as the Poincaré local supersymmetry transformations are w-independent. If S has weight w, then $S' = SS_0^{-w}$ has w = 0. Equation (2.15) defines the components of S′ with $h' = h - \frac{1}{3}wuz$. This defines the w-independent transformation rules of Ref. 6. Analogously for a vector multiplet V with Weyl weight w, we define the new multiplet $V' = V(S_0\bar{S}_0)^{-w/2}$ with w = 0.

The components of V′ are easily computed to be

$$V' = (C,\ S_L,\ H',\ B'_m,\ \lambda'_R,\ D')\ , \tag{2.21}$$

with

$$H' = H + \frac{1}{3}\,wCu,\quad B'_m = B_m + \frac{1}{3}\,wCA_m\ , \tag{2.22a}$$

$$\lambda'_R = \lambda_R + \frac{1}{6}\,w(uS_R - i\not{A}S_L)\ , \tag{2.22b}$$

$$D' = D + \frac{1}{6}\,w\left(Hu^* + H^*u - 2B_mA^m + \frac{1}{2}\,i\bar{S}\gamma_5\gamma\cdot R^P\right) + \frac{1}{18}\,w^2C(uu^* - A_mA^m)\ . \tag{2.22c}$$

The density formulae for S′ and V′ as given by Eq. (2.17) (with h = h′) and Eq. (2.19) (with $H = H'$, $B_m = B'_m$, $\lambda_R = \lambda'_R$, $D = D'$) are local supersymmetric densities with respect to the following Poincaré local supersymmetry transformations:

$$\delta^P z = \bar{\varepsilon}_L\chi_L\ , \tag{2.23a}$$

$$\delta^P\chi_L = \frac{1}{2}\,\hat{\not{D}}^P z\varepsilon_R + \frac{1}{2}\,h'\varepsilon_L\ , \tag{2.23b}$$

$$\delta^P h' = \bar{\varepsilon}_R\hat{\not{D}}^P\chi_L - \frac{1}{3}\,\bar{\varepsilon}_L\chi_L u - \frac{1}{6}\,i\bar{\varepsilon}_R\not{A}\chi_L\ , \tag{2.23c}$$

$$\delta^P C = \frac{1}{2}\,i\bar{\varepsilon}_L S_L + \text{h.c.}\ , \tag{2.23d}$$

$$\delta^P S_L = \frac{1}{2}\,iH'\varepsilon_L - \frac{1}{2}\,\not{B}'\varepsilon_R - \frac{1}{2}\,i\hat{\not{D}}^P C\varepsilon_R\ , \tag{2.23e}$$

$$\delta^P H' = -i\bar{\varepsilon}_R\hat{\not{D}}^P S_L - i\bar{\varepsilon}_R\lambda'_R + \frac{1}{3}\,i\bar{S}_L\varepsilon_L u + \frac{1}{6}\,\bar{S}_L\not{A}\varepsilon_R\ , \tag{2.23f}$$

$$\delta^P B'_m = \left[-\frac{1}{2}\bar{\varepsilon}_L\left(\hat{D}^P_m - \frac{1}{2}iA_m\right)S_L - \frac{1}{2}\bar{\varepsilon}_L\gamma_m\lambda'_R + \frac{1}{12}\bar{S}_R\gamma_m(u\varepsilon_L + i\not{A}\varepsilon_R)\right] + \text{h.c.}, \quad (2.23g)$$

$$\delta^P\lambda'_R = \frac{1}{2}(\sigma\cdot\hat{F}^P - iD')\varepsilon_R \quad (2.23h)$$

$$\delta^P D' = \left[\frac{1}{2}i\bar{\varepsilon}_L\left(\hat{\not{D}}^P - \frac{1}{2}i\not{A}\right)\lambda'_R\right] + \text{h.c.}\ , \quad (2.23i)$$

with

$$\hat{F}^P_{mn} = 2D_{[m}B'_{n]} + \bar{\psi}_{[m}\hat{D}^P_{n]}S + \bar{\psi}_{[m}\gamma_{n]}\lambda' + \bar{S}e^\mu_{[m}D_{n]}\psi_\mu\ , \quad (2.24)$$

We now extend our rules of tensor calculus by introducing a Yang-Mills group[6]. To maintain both supersymmetry an d gauge invariance, the space-time dependent parameters of Yang-Mills transformations must be assigned to an entire chiral multiplet.

The Lie algebra-valued transformation parameters are

$$\Lambda = \tilde{g}^\alpha\Lambda^\alpha T^{\alpha j}_i\ , \quad (2.25)$$

where $\tilde{g}^\alpha$ are gauge coupling constants, T^α are representation matrices for the generators of the group G, and Λ^α are scalar multiplets with Weyl weight w = 0. For a simple group, $\tilde{g}^\alpha = \tilde{g}$. We will often use the simpler notation $\tilde{g}$ for the set of gauge coupling constants $\tilde{g}^\alpha$ in the case of semi-simple or not semi-simple gauge groups G.

The components of the parameter multiplet Λ^α are denoted by

$$\Lambda^\alpha = (y^\alpha, \rho^\alpha, v^\alpha)\ . \quad (2.26)$$

For the Yang-Mills connection (gauge potential) we introduce a Lie algebra-valued vector multiplet

$$V = \tilde{g}V^\alpha T^{\alpha j}_i$$

which, under a finite Yang-Mills variation, transforms as

$$e^{2V} \to e^{-i\bar{\Lambda}}\, e^{2V}\, e^{i\Lambda}\ . \quad (2.28)$$

In the infinitesimal, Eq. (2.28) reads

$$\delta V = \frac{i}{2}\, i(\Lambda - \bar{\Lambda}) + \frac{1}{2}\, i[V, \Lambda + \bar{\Lambda}] + \frac{1}{6}\, i[V,[V, \Lambda - \bar{\Lambda}]] + O(V^3) \; . \tag{2.29}$$

If we are interested in discussing a supersymmetric Yang-Mills theory, we can take gauge choices for all Λ transformations except the real scalar Re y, which actually corresponds to the standard Yang-Mills parameter in ordinary space-time.

The Wess-Zumino gauge is defined by the following gauge choice:

$$C^\alpha = S^\alpha = \mathcal{H}^\alpha = 0 \tag{2.30}$$

for the components of V^α.

The constraints in Eq. (2.30) are not invariant under supersymmetry and gauge transformations, since under these combined transformations we get

$$\delta C^\alpha = \frac{1}{2}\, i(y - y^*) \; , \tag{2.31a}$$

$$\delta S^\alpha_L = \rho^\alpha_L - \frac{1}{2}\, \not{B}\varepsilon_R \; , \tag{2.31b}$$

$$\delta \mathcal{H}^\alpha = -iv^\alpha - \frac{1}{2}\, i\bar{\varepsilon}_R \gamma^\mu \not{B}^\alpha \psi_{\mu R} - i\bar{\varepsilon}_R \lambda^\alpha_R \; . \tag{2.31c}$$

Therefore the Wess-Zumino gauge is only invariant for $y = y^*$, and under a redefined supersymmetry transformation

$$\delta(\varepsilon) = \delta_\Omega(\varepsilon) + \delta\left(\rho^\alpha_L = \frac{1}{2}\, \not{B}^\alpha \varepsilon_R\right) + \delta\left(v^\alpha = -\frac{1}{2}\, \bar{\varepsilon}_R \gamma^\mu \not{B}^\alpha \psi_{\mu R} - \bar{\varepsilon}_R \lambda^\alpha_R\right) \; . \tag{2.32}$$

By applying this rule to the components of V^α in the Wess-Zumino gauge

$$V^\alpha_{wz} = (0,\, 0,\, 0,\, \lambda^\alpha_R,\, B^\alpha_\mu,\, D^\alpha) \; , \tag{2.33}$$

we get

$$\delta B_\mu = -\frac{1}{2}\, \bar{\varepsilon}_L \gamma_\mu \lambda_R - \frac{1}{2}\, \bar{\varepsilon}_R \gamma_\mu \lambda_L - D_\mu y \; , \tag{2.34a}$$

$$\delta\lambda_R = \frac{1}{2}(\sigma\cdot\hat{F} - ID)\varepsilon_R + i[\lambda_R, y] , \qquad (2.34b)$$

$$\delta D = \frac{1}{2} i\bar{\varepsilon}_L \hat{\not{D}}\lambda_R - \frac{1}{2} i\bar{\varepsilon}_R \hat{\not{D}}\lambda_L + i[D, y] , \qquad (2.34c)$$

where $\hat{F}_{\mu\nu}$ and $\hat{D}_\mu$ are now covariant derivatives with respect to local supersymmetry and Yang-Mills transformations.

For chiral multiplets S^i transforming according to some (generally complex) representation of G, we have the finite transformation

$$S \to e^{-i\Lambda} S \qquad (2.35)$$

or in the infinitesimal,

$$\delta S = -i\Lambda S . \qquad (2.36)$$

Equation (2.36), when written in components, reads

$$\delta S' = -i(yz,\ y\chi_L + \rho_L z,\ yh + vz - \bar{\rho}_L\chi_L) . \qquad (2.37)$$

Therefore in the transformation law given by Eq. (2.3) all derivatives are now replaced by Yang-Mills covariant derivatives, and in δh there is an extra term due to the last term in Eq. (2.32):

$$\delta' h = i\bar{\varepsilon}_R\lambda_R z = i\tilde{g}\bar{\varepsilon}_R\lambda_R^\alpha T_i^{\alpha j} z_j , \qquad (2.38)$$

We end this section by giving the vector multiplication of two scalar multiplets which can actually be converted in a chiral multiplication of a chiral multiplet S_1 with the "kinetic multiplet" of S_2, $T(S_2)$, which is the curved generalization of the flat space superfield $\overline{DDS}$.

The components of the multiplet T(S) are

$$T(S_i) = \left(h^{*i} + \frac{1}{3} u^* z^{*i},\ \tilde{\psi}_L^i,\ \tilde{H}^i\right) , \qquad (2.39)$$

where

$$\tilde{\psi}_L^i = \hat{\not{D}}^P \chi_R^i + \frac{1}{6} i\not{A}\chi_R^i + \frac{1}{6}\gamma\cdot R_R^P z^{*i} - i\tilde{g} z^{*j} T_j^{\alpha i}\lambda_L^\alpha , \qquad (2.40)$$

$$\tilde{H}^i = \hat{\Box}^c z^{*i} - \frac{2}{3} u\left(h^{*i} + \frac{1}{3} u^* z^{*i}\right) + \tilde{g} z^{*j} T_j^{\alpha i} D^\alpha - 2i\tilde{g}\bar{\chi}_R^j T_j^{\alpha i} \lambda_R^\alpha , \quad (2.41)$$

and $\hat{\Box}^c$ denotes the conformal d'Alembertian with Yang-Mills covariant derivatives.

Using the chiral density formula for the new chiral multiplet

$$C_j^i S_{1i} T(S_{2j}) + C_j^{i*} S_{2j} T(S_{1i}) \quad (2.42)$$

we get a new local density corresponding to the vector multiplication of chiral multiplets instead of their chiral multiplication as given by Eq. (2.10). The final result for the local density associated with the chiral multiplet Eq. (2.42) is

$$e^{-1} L\; C_j^i S_{1i} T(S_{ij}) + C_j^{i*} S_{2j} T(S_{1i}) =$$

$$\begin{aligned}
& C_j^i\Big[\frac{1}{6}(R + e^{-1}\, \varepsilon^{\mu\nu\rho\sigma}\bar{\psi}_\mu \gamma_5 \gamma_\nu D_\rho \psi_\sigma) z_{1i} z_2^{*j} - \hat{D}_\mu z_{1i} \hat{D}^\mu z_2^{*j} \\
& - \bar{\chi}_{1Li} \not{D} \chi_{2R}^j - \bar{\chi}_{2R}^j \not{D} \chi_{1Li} + \left(h_{1i} + \frac{1}{3} u z_{1i}\right)\left(h_2^{*j} + \frac{1}{3} u^* z_2^{*j}\right) \\
& - \frac{1}{9} A_m^2 z_{1i} z_2^{*j} + \frac{1}{3} i A^m\left(-\bar{\chi}_{1Li} \gamma_m \chi_{2R}^j + z_2^{*j} \hat{D}_m z_{1i} - z_{1i} \hat{D}_m z_2^{*j}\right) \\
& + \frac{1}{8} e^{-1}\, \varepsilon^{\mu\nu\rho\sigma} \bar{\psi}_\mu \gamma_\nu \psi_\rho \left(z_{1i} \hat{D}_\sigma z_2^{*j} - z_2^{*j} \hat{D}_\sigma z_{1i}\right) \\
& + \frac{1}{3} \bar{\chi}_{1Li} \sigma^{\nu\mu} D_\mu \psi_\nu z_2^{*j} + \frac{1}{3} \bar{\chi}_{2R}^j \sigma^{\nu\mu} D_\mu \psi_\nu z_{1i} \\
& + \bar{\chi}_{1Li} \sigma^{\nu\mu} \psi_\nu \hat{D}_\mu z_2^{*j} + \bar{\chi}_{2R}^j \sigma^{\nu\mu} \psi_\nu \hat{D}_\mu z_{1i} - D_\mu \bar{\chi}_{1Li} \sigma^{\nu\mu} \psi_\nu z_2^{*j} \\
& - D_\mu \bar{\chi}_{2R}^j \sigma^{\nu\mu} \psi_\nu z_{1i} + 2i\tilde{g} T_i^{\alpha k} \bar{\lambda}_L^\alpha \chi_{1Lk} z_2^{*j} - 2i\tilde{g}\bar{\chi}_{2R}^k T_k^{\alpha j} \lambda_R^\alpha z_{1i} \\
& + \tilde{g} D^\alpha T_i^{\alpha k} z_{1k} z_2^{*j} - \frac{1}{2} i\tilde{g}\bar{\psi}\cdot\gamma\gamma_5 \lambda^\alpha T_i^{\alpha k} z_{1k} z_2^{*j}\Big] + \text{h.c.} , \quad (2.43)
\end{aligned}$$

where $C_j^i z_{1i} z_2^{*j}$ is invariant under G, i.e.

$$C_j^i T_i^{\alpha k} z_{1k} z_2^{*j} - C_j^i T_k^{\alpha j} z_{1j} z_2^{*k} = 0 . \quad (2.44)$$

The other (antisymmetric) combination of the chiral multiplets,

$$C_j^i S_{1i} T(S_{2j}) - C_j^{i*} S_{2j} T(S_{1i}) , \quad (2.45)$$

gives rise to a density formula which is a total derivative.

Equation (2.43) produces the kinetic terms of the chiral field and their non-polynomial generalizations.

3. MANIFESTLY INVARIANT ACTIONS AND TRANSFORMATION LAWS

In this section we give the matter-coupled supergravity Lagrang in its superspace form and its component form together with the tra formation laws of the component fields under local Poincaré super-symmetry transformation[6].

Let us start with the most general supersymmetric superspace action for a set of n chiral multiplets. This is given in flat supe space by

$$\int d^4x\, d^4\theta\, \Phi(S,\bar{S}) + \int d^4x\, \mathrm{Re} \int d^2\theta\, g(S) \;, \qquad (3.1)$$

where $\Phi(S,\bar{S})$ is an arbitrary real function of the chiral multiplets S_i and their complex conjugate $\bar{S}^i$, and g(S) is a chiral function constructed out of the chiral multiplets S^i. The first component of g(S), $g(z_i)$, is often called the superpotential, The first term in Eq. (3.1) gives rise to a D-type density, whilst the second term gives rise to an F-type density. They are, respectively, the non-polynomial generalizations of the scalar kinetic term $S_i\bar{S}^i$ of globa supersymmetry and the self-interaction term (including the mass terr of the chiral multiplets S^i. If the chiral multiplets S_i transform according to some representation of a compact Lie group G, the acti is assumed to be G-invariant. This requires the following group pr perties on the scalar functions $\Phi(z,z^*)$ and g(z):

$$z^{*j} T^{\alpha i}_{j} \Phi_i = \Phi^j T^{\alpha i}_{j} z_i \;, \qquad (3.2a)$$

$$g^j T^{\alpha i}_{j} z_i = 0 \;, \qquad (3.2b)$$

$$\left(\Phi^i = \frac{\partial \Phi}{\partial z_i} \;, \quad \Phi_i = \frac{\partial \Phi}{\partial z^{*i}} \;, \quad g^i = \frac{\partial g}{\partial z_i} \right) \qquad (3.2c)$$

The extension of the action [Eq. (3.1)] to local supersymmetry and to a local Yang-Mills group G needs the following modification to Eq. (3.1) [12]:

$$\int d^4x\, d^4\theta\, E\left\{\Phi(S,\bar{S}\, \bar{e}^{2V}) + \mathrm{Re}\left[\frac{1}{R}\, g(S)\right]\right\}, \tag{3.3}$$

where V is the gauge vector multiplet of G, E is the superspace determinant, and R is the chiral scalar curvature superfield[12]. The component form of Eq. (3.3) can be computed with the rules of the tensor calculus developed in the previous section.

The remaining part of the Yang-Mills supergravity coupling corresponds to the supergravity extension of the flat Yang-Mills action[12]:

$$\int d^4x\, \mathrm{Re} \int d^2\theta\, W_a^\alpha \varepsilon^{ab} W_b^\alpha\,, \tag{3.4}$$

where W_a^α is the field-strength chiral multiplet.

We can extend Eq. (3.4), including non-polynomial interactions with the chiral multiplets S_i, as follows[6]:

$$\int d^4x\, \mathrm{Re} \int d^2\theta \left[f_{\alpha\beta}(S) W_a^\alpha \varepsilon^{ab} W_b^\beta\right], \tag{3.5}$$

where $f_{\alpha\beta}(S)$ is a chiral superfield transforming as the symmetric product of the adjoint representation of G.

The curved space generalization of Eq. (3.5) is

$$\int d^4x\, d^4\theta\, E\, \mathrm{Re}\left[\frac{1}{R}\, f_{\alpha\beta}(S) W_a^\alpha \varepsilon^{ab} W_b^\beta\right] \tag{3.6}$$

and we can compute Eq. (3.6) using the local density formula for chiral multiplets obtained in Section 2. In principle, under the requirement that the field-Lagrangian contains only first derivatives in fermion fields and second derivatives in the scalar fields, we could also have new invariants whith higher power of the field strength multiplet W_a^α, such as $f_{\alpha_1 \ldots \alpha_n}(S) W^{\alpha_1} \cdots W^{\alpha_n}$. We will not consider these terms here. All other possible modifications of Eqs. (3.3) and (3.6) would necessarily contain higher derivatives of boson and fermion fields. These terms cannot appear in the tree-level Lagrangian if we require normal propagation of the physical fields, but they may well appear in the effective action as a result of radiative corrections, in exact analogy to the high curvature terms which appear in the quantum loop expansion of the Einstein theory.

In global supersymmetry, if the gauge group G contains an Abelian factor U(1), there is an extra possible invariant -- the so-called Fayet-Iliopoulos term[13]

$$\int d^4x \ d^4\theta \ V . \tag{3.7}$$

The naive extension of Eq. (3.7) to curved space,

$$\int d^4x \, d^4\theta \, EV , \tag{3.8}$$

is not gauge invariant, since under a U(1) transformation $E \to E$ and $V \to V + (\Lambda - \bar{\Lambda})$. Then we have to modify both Eq. (3.8) and the U(1) gauge transformation in order to preserve a U(1) gauge invariance[14]

The Fayet-Iliopoulos term can be introduced in curved superspace as the result of the gauging of an R-symmetry[15], which is defined as follows

$$V_R \to V_R + i/g_R(\Lambda - \bar{\Lambda}) , \tag{3.9a}$$

$$S_0 \to e^{i\Lambda} S_0, \quad \bar{S}_0 \to e^{-i\bar{\Lambda}} \bar{S}_0 , \tag{3.9b}$$

$$S_i \to e^{-in_i\Lambda} S_i, \quad \bar{S}^i \to e^{in_i\Lambda} \bar{S}_i , \tag{3.9c}$$

and V_R is the gauge multiplet for this U(1) gauge group. The extension of Eqs. (3.3) and (3.5) to local R-symmetric interactions requires the following condition on the functions $\Phi(S,\bar{S})$, $g(S)$, and $f_{\alpha\beta}(S)$[11]:

$$\Phi(S_i,\bar{S}^i) = \Phi \ e^{-in_i q} S_i, \ e^{in_i q} \bar{S}^i , \tag{3.10}$$

$$g(S_i) = e^{3iq} \ e^{-in_i q} S_i , \tag{3.10}$$

$$f_{\alpha\beta}(S_i) = f_{\alpha\beta} \ e^{-in_i q} S_i , \tag{3.10}$$

where q is a real constant parameter.

Equations (3.10) imply that Φ and $f_{\alpha\beta}$ are R-symmetric, but the superpotential g(S) must transform with a given phase under R-symmetry. This is due to the non-invariance of the chiral measure E/R in Eq. (3.3) or, equivalently, to the transformation property of the compensating multiplet S_0 in the density formula given by Eq. (2.17).

Under the restriction (3.10) the most general Lagrangian density invariant under the Yang-Mills group $G \times U_R(1)$ is[11]

$$\int d^4x\, d^4\theta\, E\left[e^{-g_R V_R}\Phi\left(S_i,\, \bar{S}^i\, e^{n_i g_R V_R}\right)e^{2V}\right] + \mathrm{Re}\left[\frac{1}{2}\, g(S_i) + f_{\alpha\beta}(S_i)\, W^\alpha W^\beta + f_R(S_i) W_R^2\right] \tag{3.11}$$

or equivalently, using the density formulae of Section 2,

$$L = -\frac{1}{2}\left[S_0\, e^{-g_R V_R}\, \bar{S}_0\, \Phi\left(S_i,\, \bar{S}^i\, e^{n_i g_R V_R}\, e^{2V}\right)\right]_D + \left[g(S_i) S_0^3\right]_F - \left[f_{\alpha\beta}(S_i) W^\alpha W^\beta\right]_F - \left[f_R(S_i) W_R^2\right]_F , \tag{3.12}$$

where W_R is the field-strength multiplet for V_R, $g_R = k^2\xi$, where ξ is a dimensional parameter (dim ξ = 2), and g_R is the R-gauge coupling constant. We will now derive the component expression of Eqs. (3.3) and (3.6), and comment at the end on their generalization to the case of a gauged R-symmetry [Eq. (3.12)].

In order to compute the following density formulae,

$$\left[S_0 \bar{S}_0 \Phi(S, \bar{S}\, e^{2V})\right]_{\hat{D}} , \tag{3.13}$$

$$\left[g(S) S_0^3\right]_F , \tag{3.14}$$

$$\left[f_{\alpha\beta} W^\alpha W^\beta\right]_F = \left[f W^2\right]_F , \tag{3.15}$$

we have to work out the components of the vector multiplet $\Phi(S, \bar{S}\, e^{2V})$ and of the chiral multiplets g(S) and fW^2, respectively.

This can be done using the multiplication rules (2.10), (2.11), and (2.42), and the Wess-Zumino gauge choice for the vector multiplet V^α. In this gauge $V^\alpha V^\beta V^\gamma = 0$, and we get

$$\Phi(S,\bar{S}e^2) = \Phi(S,\bar{S}) + 2\tilde{g}\Phi_i \bar{S}^j T^{\alpha i}_{j} V^\alpha$$
$$+ 2\tilde{g}^2(\Phi_{ij}\bar{S}^k T^{\alpha i}_{k} \bar{S}^\ell T^{\beta j}_{\ell} + \Phi_i \bar{S}^k T^{\alpha j}_{k} T^{\beta i}_{j})V^\alpha V^\beta \; . \qquad (3.16)$$

The components $C(\Phi)$, $S_L(\Phi)$, $H(\Phi)$, $B_m(\Phi)$, $\lambda_R(\Phi)$, and $D(\Phi)$ of the local vector multiplet given in Eq. (3.16) are given by[6]

$$C(\Phi) = \Phi(z,z^*) \; , \qquad (3.17a)$$

$$S_L(\Phi) = -2i\Phi^i \chi_{Li} \; , \qquad (3.17b)$$

$$H(\Phi) = -2\Phi^i h_i + 2\Phi^{ij}\bar{\chi}_{Li}\chi_{Lj} \; , \qquad (3.17c)$$

$$B_m(\Phi) = i\Phi^i \hat{D}_m z_i - i\Phi_i \hat{D}_m z^{*i} - 2i\Phi^j_i \bar{\chi}^i_R \gamma_m \chi_{Lj} \; , \qquad (3.17d)$$

$$\lambda_R(\Phi) = -2i\Phi^j_i h_j \chi^i_R + 2i\Phi^{ij}_k \chi^k_R \bar{\chi}_{Li}\chi_{Lj}$$
$$+2i\Phi^j_i \hat{\not{D}} z^{*i}\chi_{Lj} + 2\tilde{g}\lambda^\alpha_R \tilde{D}^\alpha \; , \qquad (3.17e)$$

$$D(\Phi) = 2\Phi^i_j h_i h^{*j} - 2\Phi^{ij}_k \bar{\chi}_{Li}\chi_{Lj} h^{*k} - 2\Phi^k_{ij}\bar{\chi}^i_R \chi^j_R h_k$$
$$+ 2\Phi^{ij}_{k\ell}\bar{\chi}_{Li}\chi_{Lj}\bar{\chi}^k_R \chi^\ell_R - 2\Phi^j_i \hat{D}_\mu z^{*i}\hat{D}^\mu z_j$$
$$- 2\Phi^j_i \bar{\chi}_{Lj} \overleftrightarrow{\hat{\not{D}}} \chi^i_R - 2\Phi^i_{jk}\hat{D}_\mu z^{*k} - \Phi^{ik}_j \hat{D}_\mu z_k \bar{\chi}_{Li}\gamma^\mu \chi^j_R$$
$$+ 2\tilde{g}D^\alpha \tilde{D}^\alpha + 4i\tilde{g}\bar{\lambda}^\alpha_L \tilde{D}^{\alpha i}\chi_{Li} - 4i\tilde{g}\bar{\lambda}^\alpha_R D^\alpha_i \chi^i_R \; ; \qquad (3.17f)$$

where

$$\tilde{D}^\alpha = \Phi_i T^{\alpha i}_{j} z^{*j} = \Phi^i T^{\alpha j}_{i} z_j \; , \qquad \tilde{D}^{\alpha i} = \frac{\partial \tilde{D}^\alpha}{\partial z} \; \tilde{D}^\alpha_i = \frac{\partial \tilde{D}^\alpha}{\partial z^{*i}}$$

(the derivatives $\hat{D}_\mu$ are supersymmetric and Yang-Mills covariant).

Analogously, the components of the two chiral multiplets $g(S)$ amd fW^2 are given respectively by[6]

$$g(S) = \left[g(z),\ \ \chi_L(g) = g^i\chi_{Li},\ \ h(g) = g^i h_i - g^{ij}\bar{\chi}_{Li}\chi_{Lj}\right], \tag{3.18}$$

$$fW^2 = \left[z(fW^2),\ \chi_L(fW^2),\ h(fW^2)\right] \tag{3.19}$$

with

$$z(fW^2) = -\frac{1}{2} f_{\alpha\beta}\bar{\lambda}^\alpha_L\lambda^\beta_L\ , \tag{3.20a}$$

$$\chi_L(fW^2) = \frac{1}{2} f_{\alpha\beta}(\sigma\cdot\hat{F}^{-\alpha} - iD^\alpha)\lambda^\beta_L - \frac{1}{2} f^i_{\alpha\beta}\chi_{Li}\bar{\lambda}^\alpha_L\lambda^\beta_L\ , \tag{3.20b}$$

$$\begin{aligned} h(fW^2) &= f_{\alpha\beta}\left(-\bar{\lambda}^\alpha_L\hat{\not{D}}\lambda^\beta_R - \frac{1}{2}\hat{F}^{-\alpha}_{\mu\nu}F^{-\beta}_{\mu\nu} + \frac{1}{2}D^\alpha D^\beta\right) \\ &\quad + f^i_{\alpha\beta}\bar{\chi}_{Li}(-\sigma\cdot\hat{F}^{-\alpha} + iD^\alpha)\lambda^\beta_L \\ &\quad - \frac{1}{2} f^i_{\alpha\beta}h_i\bar{\lambda}^\alpha_L\lambda^\beta_L + \frac{1}{2} f^{ij}_{\alpha\beta}\bar{\chi}_{Li}\chi_{Lj}\bar{\lambda}^\alpha_L\lambda^\beta_L\ . \end{aligned} \tag{3.20c}$$

Using the density formulae (2.17) and (2.19), we finally get

$$\begin{aligned} e^{-1}L(\Phi) = &-\frac{1}{6}\Phi\, e^{-1}L_{SC} + \Phi^i_j\left(-\frac{1}{2}D_\mu z_i D^\mu z^{*j} - \bar{\chi}_{Li}\not{D}\chi^j_R + \frac{1}{2}h_i h^{*j}\right) \\ &- \Phi^{ij}_k\bar{\chi}_{Li}\chi_{Lj}h^{*k} + \Phi^{ij}_k\bar{\chi}_{Li}\not{D}z_j\chi^k_R \\ &+ \frac{1}{2}\Phi^{ij}_{k\ell}\bar{\chi}_{Li}\chi_{Lj}\bar{\chi}^k_R\chi^\ell_R + \frac{1}{3}u^*(\Phi^i h_i - \Phi^{ij}\chi_{Li}\chi_{Lj}) \\ &+ \frac{1}{3}iA^\mu\left[\frac{1}{2}\Phi^{i}_{j}\bar{\chi}^j_R\gamma_\mu\chi_{Li} + \Phi^i(D_\mu z_i - \bar{\psi}_{\mu L}\chi_{Li})\right] \\ &+ \Phi^i_j\bar{\psi}_{\mu L}\not{D}z^{*j}\gamma_\mu\chi_{Li} - \frac{4}{3}\Phi^i\bar{\chi}_{Li}\sigma^{\mu\nu}D_\mu\psi_{\nu L} \\ &- \frac{1}{8}e^{-1}\varepsilon^{\mu\nu\rho\sigma}\bar{\psi}_\mu\gamma_\nu\psi_\rho\left(\Phi^i D_\sigma z_i + \frac{1}{2}\Phi^j_i\bar{\chi}^i_R\gamma_\sigma\chi_{Lj}\right) \\ &- \frac{1}{2}\Phi^i_j\bar{\psi}_{\mu R}\chi^j_R\bar{\psi}^\mu_L\chi_{Li} + \frac{1}{6}\Phi^i\bar{\chi}_{Li} \\ &\times\left(\bar{\psi}_{\mu L}\bar{\psi}\cdot\gamma\psi^\mu + \frac{1}{2}\sigma^{\mu\nu}\psi^\rho_L\bar{\psi}_\nu\gamma_\rho\psi_\mu + \sigma^{\mu\nu}\psi_{\nu L}\bar{\psi}_\mu\gamma\cdot\psi\right) \\ &+ \frac{1}{2}\tilde{g}\Phi^i T^{\alpha j}_i z_j(D^\alpha + i\bar{\psi}_L\cdot\gamma\lambda^\alpha_R) - 2i\tilde{g}\Phi^j_i T^{\alpha k}_j z_k\bar{\lambda}^\alpha_R\chi^i_R + \text{h.c.}\ , \end{aligned} \tag{3.21}$$

$$\begin{aligned} e^{-1}L(g) = &-\frac{1}{2}g^{ij}\bar{\chi}_{Li}\chi_{Lj} + \frac{1}{2}g^i h_i + \frac{1}{2}gu + \frac{1}{2}\bar{\psi}_R\cdot\gamma\chi_{Li} \\ &+ \frac{1}{2}g\bar{\psi}_{\mu R}\sigma^{\mu\nu}\psi_{\nu R} + \text{h.c.}\ , \end{aligned} \tag{3.22}$$

$$e^{-1} L(fW^2) = \frac{1}{2} f_{\alpha\beta}\Big[-\frac{1}{4} F^{\alpha}_{\mu\nu}F^{\beta}_{\mu\nu} - \frac{1}{2}\bar{\lambda}^{\alpha}\tilde{\not{D}}\lambda^{\beta} + \frac{1}{2} D^{\alpha}D^{\beta}$$
$$+ \frac{1}{4} F^{\alpha}_{\mu\nu}\bar{\psi}_{\rho}\sigma^{\mu\nu}\gamma^{\rho}\lambda^{\beta} + \frac{1}{4} F^{\alpha}_{\mu\nu}\tilde{F}^{\beta}_{\mu\nu} - \frac{1}{2} D_{\mu}(\bar{\lambda}^{\alpha}_{L}\gamma^{\mu}\lambda^{\beta}_{R})\Big]$$
$$+ \frac{1}{2} f^{i}_{\alpha\beta}\Big[\bar{\chi}_{Li}(-\sigma\cdot\hat{F}^{-\alpha} + iD^{\alpha})\lambda^{\beta}_{L} - \frac{1}{2} h_{i}\bar{\lambda}^{\alpha}_{L}\lambda^{\beta}_{L}$$
$$- \frac{1}{2}\bar{\psi}_{R}\cdot\gamma\chi_{Li}\bar{\lambda}^{\alpha}_{L}\lambda^{\beta}_{L}\Big] + \frac{1}{4} f^{ij}_{\alpha\beta}\bar{\chi}_{Li}\chi_{Lj}\bar{\lambda}^{\alpha}_{L}\lambda^{\beta}_{L} + \text{h.c.} \qquad (3.23)$$

Here $\tilde{D}_{\mu}\lambda^{\alpha}$ is a supersymmetric and Yang-Mills covariant derivative but without the $\psi_{\mu}D^{\alpha}$ term. The other derivatives are Yang-Mills and general coordinate covariants but they have no ψ_{μ} torsion.

The over-all Lagrangian corresponding to the superspace expressions of Eqs. (3.3) and (3.6) is therefore given by[6]

$$L = L(\Phi) + L(g) + L(fW^2) \ . \qquad (3.24)$$

The three terms in Eq. (3.24) are separately invariant under local supersymmetry transformations.

These transformations have the following form, for the several fields involved in Eq. (3.24):

$$\delta B^{\alpha}_{\mu} = -\frac{1}{2}\bar{\varepsilon}_{L}\gamma_{\mu}\lambda^{\alpha}_{R} - \frac{1}{2}\bar{\varepsilon}_{R}\gamma_{\mu}\lambda^{\alpha}_{L} \ , \qquad (3.2$$

$$\delta\lambda^{\alpha}_{R} = \frac{1}{2}\sigma^{\mu\nu}\hat{F}^{\alpha}_{\mu\nu}\varepsilon_{R} - \frac{1}{2} iD^{\alpha}\varepsilon_{R} \ , \qquad (3.2$$

$$\delta D^{\alpha} = \frac{1}{2} i\varepsilon_{L}\Big(\hat{\not{D}}^{P} - \frac{1}{2} i\not{A}\Big)\lambda_{R} - \frac{1}{2} i\bar{\varepsilon}_{R}\Big(\hat{\not{D}}^{P} + \frac{1}{2} i\not{A}\Big)\lambda^{\alpha}_{L} \ , \qquad (3.2$$

$$\delta z_{i} = \bar{\varepsilon}_{L}\chi_{Li} \ , \qquad (3.2$$

$$\delta\chi_{Li} = \frac{1}{2}\hat{\not{D}}z_{i}\varepsilon_{R} + \frac{1}{2} h_{i}\varepsilon_{L} \ , \qquad (3.2$$

$$\delta h_{i} = \bar{\varepsilon}_{R}\Big(\hat{\not{D}}^{P} - \frac{1}{2} i\not{A}\Big)\chi_{Li} - \frac{1}{3}\bar{\chi}_{Li}(u\varepsilon_{L} + i\not{A}\varepsilon_{R} + i\tilde{g}\bar{\varepsilon}_{R}\lambda^{\alpha}_{R}T^{\alpha j}_{i}z_{j}) \ , \qquad (3.2$$

$$\delta e^{m}_{\mu} = \frac{1}{2}\bar{\varepsilon}_{L}\gamma^{m}\psi_{\mu R} + \frac{1}{2}\bar{\varepsilon}_{R}\gamma^{m}\psi_{\mu L} \ , \qquad (3.2$$

$$\delta\psi_{\mu L} = \left[\partial_\mu + \frac{1}{2}\omega_{\mu mn}(e,\psi)\sigma^{mn} + \frac{1}{2}iA_\mu\right]\varepsilon_L$$
$$+ \frac{1}{6}\gamma_\mu(u^*\varepsilon_R - i\not{A}\varepsilon_L) , \tag{3.25h}$$

$$\delta A_\mu = \frac{3}{4}i\bar{\varepsilon}_L\left(R^P_\mu - \frac{1}{3}\gamma_\mu\gamma\cdot R^P\right)_L + \text{h.c.} , \tag{3.25i}$$

$$\delta u = \frac{1}{2}\bar{\varepsilon}_R\gamma\cdot R^P_L . \tag{3.25j}$$

The over-all Lagrangian given by Eq. (3.24) contains auxiliary field components of the multiplets S^i, V^α, and the supergravity multiplet

$$(e_{\mu m}, \psi_{\mu a}, u, A_m) . \tag{3.26}$$

In addition, an appropriate Weyl rescaling of the vierbein field, as well as redefinitions of the fermion fields, have to be performed in the first term of Eq. (3.24), in its standard form[5]. These redefinitions, as well as the elimination of the auxiliary fields h_i, D^α, u, and A_m, will be done in the next section.

These manipulations will allow us to recast the final Lagrangian and transformation rules in a simple form in terms of the particle fields

$$(B^\alpha_m, \lambda^\alpha_R), \quad (z_i, \chi_{Li}), \quad (e_{\mu m}, \psi_{\mu a}) . \tag{3.27}$$

We end this section by showing (using simple superspace arguments) that the action given by Eq. (3.24) depends only on a particular combination of the function $\Phi(S,\bar{S})$ and $g(S)$ [5].

To prove this, we make a super-Weyl rescaling on the supervierbein so that its superdeterminant transforms as

$$E \to E' \exp(\Sigma + \bar{\Sigma}) , \tag{3.28}$$

where Σ is a chiral parameter superfield. Under the same rescaling the second term in Eq. (3.3) transforms as

$$\int d^4x\, d^4\theta\, \text{Re}\left[\frac{E'}{R'}(\exp 3\Sigma)g(S)\right] , \tag{3.29}$$

whilst the Yang-Mills part [Eq. (3.6)] keeps the same form:

$$\int d^4x\, d^4\theta\, \mathrm{Re}\, \frac{\bar{E}'}{\bar{R}'} f_{\alpha\beta}(S) W^{\alpha} W^{\beta} \,. \tag{3.30}$$

If we now choose Σ such that

$$\exp 3\Sigma g(S) = 1, \quad \text{i.e. } \exp \Sigma = g(S)^{-1/3} \,, \tag{3.31}$$

we finally obtain that Eq. (3.3) becomes

$$\int d^4x\, d^4\theta\, E' \exp \frac{1}{3} G(S, \bar{S}\, e^{2V}) + \int d^4x\, d^4\theta\, \mathrm{Re}\left(\frac{E'}{R'}\right) , \tag{3.32}$$

where the function G is given by

$$G(z,z^*) = 3 \log \Phi(z,z^*) - \log |g(z)|^2 \,, \tag{3.33}$$

so that

$$\exp \frac{1}{3} G(z,z^*) = \frac{\Phi(z,z^*)}{|g(z)|^{2/3}} \,. \tag{3.34}$$

The use of the function $G(S,\bar{S})$ makes the introduction of the Fayet-Iliopoulos term in formula (3.11) trivial[11,16]. In fact, in this case we simply get

$$\exp\left[\frac{1}{3} G(S,\bar{S} \exp n_i g_R V_R \exp 2V)\right] = \frac{(\Phi(S_i,\bar{S} \exp n_i g_R V_R \exp 2V)}{[g^*(\bar{S} \exp n_i g_R V_R \exp 2V) g(S)]^{1/3}} \,. \tag{3.35}$$

This means that in the G superfield we simply have to replace $\bar{S}\, e^{2V}$ with $\bar{S}\, e^{n_i g_R V_R} e^{2V}$, i.e. we have to covariantize with respect to th full gauge group $G \times U_R(1)$.

We note that this simple rule does not hold in the case of a vanishing superpotential, $g(S_i) = 0$, since in this case the rescaling given by formula (3.31) can no longer be performed[11]. At the end of the next section we will give simple substitution rules for the Lagrangian in terms of physical fields which will enable us to go from the general form of $G(S,\bar{S})$ to the limiting case correspondi to $g(S) = 0$.

FINAL FORM OF THE LAGRANGIAN, TRANSFORMATION LAWS, AND GAUGED R-SYMMETRY

The over-all action of the coupled-matter Yang-Mills supergravity multiplets is given in Eq. (3.24).

In order to obtain the final form of the Lagrangian, we have to eliminate, by means of their field equations, the auxiliary fields h_i, D^α, u, and A_m of the multiplets involved.

It is important to notice that, after their elimination, only the total Lagrangian (3.24) -- but not the three separate pieces given by Eqs. (3.21) to (3.23) -- will be invariant under local supersymmetry transformations of the physical fields.

The part of the Lagrangian (3.24) which contains the auxiliary fields is

$$\begin{aligned} e^{-1} L_{aux} = & \frac{1}{18} \Phi(uu^* - A_m A^m) + \frac{1}{2} \Phi^i_j h_i h^{*j} \\ & - \Phi^{ij}_k h^{*k} \bar{\chi}_{Li} \chi_{Lj} + \frac{1}{2} g^i h_i - \frac{1}{4} f^i_{\alpha\beta} h_i \bar{\lambda}^\alpha_L \lambda^\beta_L \\ & + \frac{1}{3}\left(\Phi^i h_i - \Phi^{ij} \bar{\chi}_{Li} \chi_{Lj} + \frac{3}{2} g^*\right) \\ & + \frac{1}{3} iA^\mu \left[\frac{1}{2} \Phi^i_j \bar{\chi}^j_R \gamma_\mu \chi_{Li} + \Phi^i (D_\mu z_i - \psi_{\mu L} \chi_{Li}) + \frac{3}{4} \bar{\lambda}^\alpha_L \gamma_\mu \lambda^\beta_R f_{\alpha\beta}\right] \\ & + \frac{1}{2} \tilde{g} \Phi^i T^{\alpha j}_i z_j D^\alpha + \frac{1}{4} f_{\alpha\beta} D^\alpha D^\beta + \frac{i}{2} f^i_{\alpha\beta} \bar{\chi}_{Li} \lambda^\alpha_L D^\alpha + \text{h.c.} \end{aligned} \tag{4.1}$$

If we define the new expressions

$$J = 3 \log - \frac{\Phi}{3} \tag{4.2}$$

and

$$\tilde{u} = u + J^i h_i , \tag{4.3}$$

then Eq. (4.1) reads

$$\begin{aligned} e^{-1} L_{aux} = & \frac{1}{18} \Phi(\tilde{u}\tilde{u}^* - A_m A^m) + \frac{1}{6} \Phi J^i_j h_i h^{*j} \\ & + \frac{1}{2} g\left[\tilde{u} - h_i\left(J^i - \frac{g^i}{g}\right)\right] - \frac{1}{g} u^*\left(J^{ij} + \frac{1}{3} J^i J^j\right) \Phi \bar{\chi}_{Li} \chi_{Lj} \end{aligned}$$

(cont.)

$$- \frac{1}{3}\,\Phi h^{*k}\bar{\chi}_{Li}\chi_{Lj}\left(J^{ij}_{k} + \frac{2}{3}\,J^{i}_{k}J^{j}\right) - \frac{1}{4}\,f^{i}_{\alpha\beta}h_{i}\bar{\lambda}^{\alpha}_{L}\lambda^{\beta}_{L}$$

$$+ \frac{1}{3}\,iA^{\mu}\left[\frac{1}{2}\,\Phi^{i}_{j}\bar{\chi}^{j}_{R}\gamma_{\mu}\chi_{Li} + \Phi^{i}(D_{\mu}z_{i} - \bar{\psi}_{\mu L}\chi_{Li}) + \frac{3}{4}\,\bar{\lambda}^{\alpha}\gamma_{\mu}\lambda^{\beta}_{R}f_{\alpha\beta}\right]$$

$$+ \frac{1}{2}\,\tilde{g}\Phi^{i}T^{\alpha j}_{i}z_{j}D^{\alpha} + \frac{1}{4}\,f_{\alpha\beta}D^{\alpha}D^{\beta} + \frac{i}{2}\,f^{i}_{\alpha\beta}\bar{\chi}_{Li}\lambda^{\alpha}_{L}D^{\beta} + \text{h.c.} \qquad (4.4)$$

The field equations for the auxiliary fields are

$$\tilde{u} = \frac{2}{\Phi}\left[-\frac{1}{2}\,g^{*} + \frac{1}{g}\left(J^{ij} + \frac{1}{3}\,J^{i}J^{j}\right)\Phi\bar{\chi}_{Li}\chi_{Lj}\right], \qquad (4.5a)$$

$$\frac{1}{3}\,\Phi h_{i}J^{i}_{k} = -\frac{1}{2}\,g^{*}\left(\frac{g^{*}_{k}}{g^{*}} - J_{k}\right) + \frac{1}{3}\,\Phi\left(J^{ij}_{k} + \frac{2}{3}\,J^{i}_{k}J^{j}\right)\bar{\chi}_{Li}\chi_{Lj}$$

$$+ \frac{1}{4}\,f^{*}_{\alpha\beta k}\bar{\lambda}^{\alpha}_{R}\lambda^{\beta}_{R}, \qquad (4.5b)$$

$$\frac{2}{3}\,\Phi A_{\mu} = \frac{1}{2}\,i\Phi^{i}_{j}\bar{\chi}^{j}_{R}\gamma_{\mu}\chi_{Li} + i\Phi^{i}(D_{\mu}z_{i} - \bar{\psi}_{\mu L}\chi_{Li})$$

$$+ \frac{3}{4}\,if_{\alpha\beta}\bar{\lambda}^{\alpha}_{L}\gamma_{\mu}\lambda^{\beta}_{R} + \text{h.c.}, \qquad (4.5c)$$

$$-\text{Re}\,f_{\alpha\beta}D^{\beta} = \tilde{g}\Phi^{i}T^{\alpha j}_{i}z_{j} + \frac{1}{2}\,if^{i}_{\alpha\beta}\bar{\chi}_{Li}\lambda^{\beta}_{L} - \frac{1}{2}\,if^{*}_{\alpha\beta i}\bar{\chi}^{i}_{R}\lambda^{\beta}_{R}. \qquad (4.5d)$$

The insertion of Eqs. (4.5) into Eq. (4.4) finally gives

$$e^{-1}L = e^{-1}L_{\text{aux}} + e^{-1}\hat{L}, \qquad (4.6)$$

where

$$e^{-1}L_{\text{aux}} = -\frac{g}{\Phi}\left|\frac{1}{2}\,g^{*} - \frac{1}{g}\left(J^{ij} + \frac{1}{3}\,J^{i}J^{j}\right)\Phi\bar{\chi}_{Li}\chi_{Lj}\right|^{2} - \frac{3}{\Phi}\,(J^{-1})^{k}_{\ell}$$

$$\times\left[\frac{1}{2}\,g^{*}\left(\frac{g^{*}_{k}}{g^{*}} - J_{k}\right) - \frac{1}{3}\,\Phi\left(J^{ij}_{k} + \frac{2}{3}\,J^{i}_{k}J^{j}\right)\bar{\chi}_{Li}\chi_{Lj} - \frac{1}{4}\,f^{*}_{\alpha\beta}\,\bar{\lambda}^{\alpha}_{R}\lambda^{\beta}_{R}\right]$$

$$\times\left[\frac{1}{2}\,g\left(\frac{g^{\ell}}{g} - J^{\ell}\right) - \frac{1}{3}\,\Phi\left(J^{\ell}_{mn} + \frac{2}{3}\,J^{\ell}_{m}J_{n}\right)\bar{\chi}^{m}_{R}\chi^{n}_{R} - \frac{1}{4}\,f^{\ell}_{\gamma\delta}\bar{\lambda}^{\gamma}_{L}\lambda^{\delta}_{L}\right]$$

$$-\frac{1}{4\Phi}\left[\Phi^{i}_{j}\bar{\chi}^{j}_{R}\gamma_{\mu}\chi_{Li} + \Phi^{i}(D_{\mu}z_{i} - \bar{\psi}_{\mu L}\chi_{Li}) - \Phi_{i}(D_{\mu}z^{*i} - \bar{\psi}_{\mu R}\chi^{i}_{R})\right.$$

$$\left.+ \frac{3}{2}\,\text{Re}\,f_{\alpha\beta}\bar{\lambda}^{\alpha}_{L}\gamma_{\mu}\lambda^{\beta}_{R}\right]^{2} - \frac{1}{2}\,(\text{Re}\,f)^{-1}_{\alpha\beta}\left(\tilde{g}\Phi^{i}T^{\alpha j}_{i}z_{j} + \frac{i}{2}\,f^{i}_{\alpha\gamma}\bar{\chi}_{Li}\lambda^{\gamma}_{L}\right.$$

$$\left.-\frac{1}{2}\,if^{*}_{\alpha\gamma i}\bar{\chi}^{i}_{R}\lambda^{\gamma}_{R}\right)\times\left(\tilde{g}\Phi^{k}T^{\beta\ell}_{k}z_{\ell} + \frac{1}{2}\,if^{k}_{\beta\delta}\bar{\chi}_{Lk}\lambda^{\delta}_{L} - \frac{1}{2}\,if^{*}_{\beta\delta k}\bar{\chi}^{k}_{R}\lambda^{\delta}_{R}\right)$$

$$(4.7a)$$

$$e^{-1}\hat{L} = \frac{1}{12}\Phi\Big\{R[\omega(e)] + \bar{\psi}_\mu\gamma_5\gamma_\nu D_\rho\psi_\sigma\varepsilon^{\mu\nu\rho\sigma} + \frac{1}{16}(\bar{\psi}_\mu\gamma_\lambda\psi_\rho$$

$$+ 2\bar{\psi}_\lambda\gamma_\mu\psi_\rho)\bar{\psi}^\mu\gamma^\lambda\psi^\rho - \frac{1}{4}(\bar{\psi}_\mu\gamma\cdot\psi)^2 + e^{-1}\partial_\mu(e\bar{\psi}\cdot\gamma\psi^\mu)\Big\}$$

$$+ \Phi^i_j\left(-\frac{1}{2}D_\mu z_i D^\mu z^{*j} - \bar{\chi}_{Li}\not{D}\chi^j_R\right) + \Phi^{ij}_k\bar{\chi}_{Li}\not{D}z_j\chi^k_R$$

$$+ \Phi^i_j\bar{\psi}_{\mu L}\not{D}z^{*j}\gamma^\mu\chi_{Li} - \frac{4}{3}\Phi^i\bar{\chi}_{Li}\sigma^{\mu\nu}D_\mu\psi_{\nu L}$$

$$- \frac{1}{8}e^{-1}\varepsilon^{\mu\nu\rho\sigma}\bar{\psi}_\mu\gamma_\nu\psi_\rho\left(\Phi^i D_\sigma z_i + \frac{1}{2}\Phi^j_i\bar{\chi}^i_R\gamma_\sigma\chi_{Lj}\right)$$

$$- \frac{1}{2}\Phi^i_j\bar{\psi}_{\mu L}\chi_{Li}\bar{\psi}^\mu_R\chi^j_R + \frac{1}{2}\Phi^{ij}_{k\ell}\bar{\chi}_{Li}\chi_{Lj}\bar{\chi}^k_R\chi^\ell_R$$

$$+ \frac{1}{6}\Phi^i\chi_{Li}\left(\psi_{\mu L}\bar{\psi}\cdot\gamma\psi^\mu + \frac{1}{2}\sigma^{\mu\nu}\psi^\ell_L\bar{\psi}_\nu\gamma_\rho\psi_\mu + \sigma^{\mu\nu}\psi_{\nu L}\bar{\psi}_\mu\gamma\cdot\psi\right)$$

$$+ \frac{1}{2}i\tilde{g}\Phi^i T^{\alpha j}_i z_j\bar{\psi}_L\gamma\lambda^\alpha_k - 2i\tilde{g}\Phi^i_k T^{\alpha j}_i z_j\bar{\lambda}^\alpha_R\chi^k_R$$

$$- \frac{1}{2}g^{ij}\bar{\chi}_{Li}\chi_{Lj} + \frac{1}{2}g^i\bar{\psi}_R\cdot\gamma\chi_{Li} + \frac{1}{2}g\bar{\psi}_{\mu R}\sigma^{\mu\nu}\psi_{\nu R}$$

$$+ \frac{1}{2}f_{\alpha\beta}\Big[-\frac{1}{4}F^\alpha_{\mu\nu}F^\beta_{\mu\nu} - \frac{1}{2}\bar{\lambda}^\alpha\not{D}\lambda^\beta + \frac{1}{2}\bar{\lambda}^\alpha\gamma^\mu\sigma^{\rho\sigma}\psi_\mu F^\beta_{\rho\sigma}$$

$$+ \frac{1}{4}\bar{\lambda}^\alpha\gamma^\mu\sigma^{\rho\sigma}\psi_\mu\bar{\psi}_\rho\gamma_\sigma\lambda^\alpha - \frac{1}{16}\bar{\lambda}^\alpha\gamma^\mu\sigma^{mn}\lambda^\beta(2\bar{\psi}_\mu\gamma_m\psi_n$$

$$+ \bar{\psi}_m\gamma_\mu\psi_n) + \frac{1}{4}F^\alpha_{\mu\nu}\tilde{F}^\beta_{\mu\nu} - \frac{1}{2}D_\mu(\bar{\lambda}^\alpha_L\gamma^\mu\lambda^\beta_R)\Big]$$

$$+ \frac{1}{2}f^i_{\alpha\beta}\Big[-\bar{\chi}_{Li}\sigma_{\mu\nu}\lambda^\beta_L\left(F^{-\alpha}_{\mu\nu} - \frac{1}{2}\bar{\psi}_{\rho L}\sigma_{\mu\nu}\gamma^\rho\lambda^\alpha_R + \frac{1}{2}\bar{\psi}_R\cdot\gamma\sigma_{\mu\nu}\lambda^\alpha_L\right)$$

$$- \frac{1}{2}\bar{\psi}_R\cdot\gamma\chi_{Li}\bar{\lambda}^\alpha_L\lambda^\beta_L\Big] + \frac{1}{4}f^{ij}_{\alpha\beta}\bar{\chi}_{Li}\chi_{Lj}\bar{\lambda}^\alpha_L\lambda^\beta_L + \text{h.c.} \tag{4.7b}$$

In order to recast the Einstein term and the Rarita-Schwinger action in canonical form, as well as to keep the gaugino and chiral fermion terms in a quasi-canonical form, the following Weyl rescaling on the vierbein and the fermion fields must be performed:

$$e_{m\mu} \to e_{m\mu}e^\sigma\,, \quad e \to e^{4\sigma}e\,, \tag{4.8a}$$

$$\lambda \to e^{-3\sigma/2} \lambda\ , \tag{4.8b}$$

$$\chi \to e^{-\sigma/2} \chi\ , \tag{4.8c}$$

$$\psi_\mu \to e^{\sigma/2} \psi_\mu\ , \tag{4.8d}$$

with

$$e^{2\sigma} = -\frac{3}{\Phi}\ , \qquad \sigma = -\frac{1}{6} J\ . \tag{4.9}$$

Under the Weyl rescaling of the vierbein we get the standard change for the Einstein-curvature term and of the Lorentz connection:

$$\frac{1}{6}\, e\Phi R \to -\frac{1}{2}\, eR - \frac{3}{4}\, e(\partial_\mu \log \Phi)^2 + \text{4-div.}\ , \tag{4.10}$$

$$\omega_{\mu mn} \to \omega_{\mu mn} - 2e^\nu{}_{[m} e_{n]\mu} \partial_\nu \sigma\ . \tag{4.10}$$

After the substitutions (4.8), the total Lagrangian becomes the sum of a pure bosonic part and a fermionic part:

$$e^{-1}\, L = e^{-1}\, L_B + e^{-1}\, L_F\ , \tag{4.11}$$

where

$$\begin{aligned} e^{-1}\, L_B = {} & -\frac{1}{2} R + J^i_j D_\mu z_i D_\mu z^{*j} - \frac{1}{4} \operatorname{Re} f_{\alpha\beta} F^\alpha_{\mu\nu} F^\beta_{\mu\nu} + \frac{1}{4}\, i \operatorname{Im} f_{\alpha\beta} F^\alpha_{\mu\nu} \tilde{F}^\beta_{\mu\nu} \\ & + \frac{1}{4} |g|^2\, e^{-J} \left[3 + (J^{-1})^k_\ell \left(\frac{g^*_k}{g^*} - J_k \right) \left(\frac{g^\ell}{g} - J^\ell \right) \right] \\ & - \frac{1}{2} \tilde{g}^2 \operatorname{Re} f^{-1}_{\alpha\beta} (J^i T^{\alpha j}_i z_j)\ (J^k T^{\beta\ell}_k z_\ell) \end{aligned} \tag{4.12}$$

and

$$\begin{aligned} e^{-1}\, L_F = {} & \frac{1}{2}\, e^{3\sigma} \left(J^{ij} + \frac{1}{3} J^i J^j \right) g \bar{\chi}_{Li} \chi_{Lj} \\ & + \frac{1}{2}\, g \left(\frac{g^\ell}{g} - J^\ell \right) e^{3\sigma}\, \bar{J}^{-1k}_\ell \left(J^{ij}_k + \frac{2}{3} J^i_k J^j \right) \bar{\chi}_{Li} \chi_{Lj} \\ & + \frac{1}{2}\, g \left(\frac{g^\ell}{g} - J^\ell \right) \frac{3}{4\Phi}\, e^\sigma\, J^{-1k}_\ell f^*_{\alpha\beta k} \bar{\lambda}^\alpha_R \lambda^\beta_R - \frac{1}{18}\, \Phi\, e^{2\sigma} \\ & \times \left(J^{ij} + \frac{1}{3} J^i J^j \right) \bar{\chi}_{Li} \chi_{Lj} \left(J_{k\ell} + \frac{1}{3} J_k J_\ell \right) \bar{\chi}^k_R \chi^\ell_R - \frac{1}{6}\, \Phi\, e^{2\sigma}\, J^{-1k}_\ell \end{aligned}$$

$$\times \left(J^{ij}_{k} + \frac{2}{3} J^{i}_{k}J^{j}\right)\bar{\chi}_{Li}\chi_{Lj}\left(J^{\ell}_{mn} + \frac{2}{3} J^{\ell}_{m}J_{n}\right)\bar{\chi}^{m}_{R}\chi^{n}_{R}$$

$$- \frac{1}{4} f^{\ell}_{\alpha\beta}\bar{\lambda}^{\alpha}_{L}\lambda^{\beta}_{L}J^{-1k}_{\ell}\left(J^{ij}_{k} + \frac{2}{3} J^{i}_{k}J^{j}\right)\bar{\chi}_{Li}\chi_{Lj}$$

$$- \frac{3}{32}\frac{1}{\Phi} e^{-2\sigma} J^{-1k}_{\ell} f^{\ell}_{\alpha\beta}\bar{\lambda}^{\alpha}_{L}\lambda^{\beta}_{L} f^{*}_{\gamma\delta k}\bar{\lambda}^{\gamma}_{R}\lambda^{\delta}_{R}$$

$$- \frac{1}{2\Phi} e^{2\sigma} \Phi^{i}_{j}\bar{\chi}^{j}_{R}\gamma^{\mu}\chi_{Li}\Phi^{k}D_{\mu}z_{k} - \frac{1}{2\Phi} e^{2\sigma} \Phi^{i}D_{\mu}z_{i}$$

$$\times(\Phi_{J}\bar{\psi}_{\mu R}\chi^{j}_{R} - \Phi^{\bar{j}}\psi_{\mu L}\chi_{Lj}) - \frac{3}{4\Phi} \mathrm{Re}\ f_{\alpha\beta}\Phi^{i}D_{\mu}z_{i}\bar{\lambda}^{\alpha}\gamma^{\mu}\lambda^{\beta}_{R}$$

$$- \frac{1}{8\Phi}(\Phi^{i}_{j}\bar{\chi}^{j}_{R}\gamma_{m}\chi_{Li})^{2} + \frac{1}{2\Phi} \Phi^{i}_{j}\bar{\chi}^{j}_{R}\gamma^{\mu}\chi_{Li}\Phi^{k}\bar{\psi}_{\mu L}\chi_{Lk}\ e^{2\sigma}$$

$$- \frac{3}{8\Phi} \Phi^{i}_{j}\bar{\chi}^{j}_{R}\gamma^{m}\chi_{Li}\ \mathrm{Re}\ f_{\alpha\beta}\bar{\lambda}^{\alpha}_{L}\gamma_{m}\lambda^{\beta}_{R}$$

$$+ \frac{3}{4\Phi} \Phi^{i}\bar{\psi}_{\mu L}\chi_{Li}\ \mathrm{Re}\ f_{\alpha\beta}\bar{\lambda}^{\alpha}_{L}\gamma^{\mu}\lambda^{\beta}_{R}$$

$$+ \frac{1}{4\Phi} e^{2\sigma} \Phi^{j}\bar{\psi}_{\mu L}\chi_{Lj}(\Phi_{i}\bar{\psi}_{\mu R}\chi^{i}_{R} - \Phi^{i}\bar{\psi}_{\mu L}\chi_{Li})$$

$$- \frac{9}{32\Phi} \mathrm{Re}\ f_{\alpha\beta}\ \mathrm{Re}\ f_{\gamma\delta}\bar{\lambda}^{\alpha}_{L}\gamma^{m}\lambda^{\beta}_{R}\bar{\lambda}^{\gamma}_{L}\gamma_{m}\lambda^{\delta}_{R}\ e^{-2\sigma}$$

$$- \frac{1}{2} i\ \mathrm{Re}\ f^{-1}_{\alpha\beta}\tilde{g}\Phi^{i}T^{\alpha j}_{i}z_{j}f^{k}_{\beta\gamma}\bar{\chi}_{Lk}\lambda^{\gamma}_{L}\ e^{2\sigma}$$

$$+ \frac{1}{8} \mathrm{Re}\ f^{-1}_{\alpha\beta}f^{i}_{\alpha\gamma}\chi_{Li}\lambda^{\gamma}_{L}(f^{j}_{\beta\delta}\bar{\chi}_{Lj}\lambda^{\delta}_{L} - f^{*}_{\beta\delta j}\bar{\chi}^{j}_{R}\lambda^{\delta}_{R})$$

$$- \frac{1}{4} \bar{\psi}_{\mu}\gamma_{5}\gamma_{\nu}D_{\rho}\psi_{\sigma}\sigma^{\mu\nu\rho\delta}\ e^{-1} - \frac{1}{64} (\bar{\psi}_{\mu}\gamma_{\lambda}\psi_{\rho} + 2\bar{\psi}_{\lambda}\gamma_{\mu}\psi_{\rho})\bar{\psi}^{\mu}\gamma^{\lambda}\psi^{\rho}$$

$$+ \frac{1}{16} (\bar{\psi}_{\mu}\cdot\gamma\cdot\psi)^{2} + 3\frac{\Phi^{i}_{j}}{\Phi}\bar{\chi}_{Li}\not{D}\chi^{j}_{R} - 3\frac{\Phi^{ij}_{k}}{\Phi}\bar{\chi}_{Li}\not{D}z_{j}\chi^{k}_{R}$$

$$- 3\frac{\Phi^{i}_{j}}{\Phi}\bar{\psi}_{\mu L}\not{D}z^{*j}\gamma_{\mu}\chi_{Li} + \frac{4\Phi^{i}}{\Phi}\chi_{Li}\sigma^{\mu\nu}D_{\mu}\psi_{\nu}$$

$$- \frac{3\Phi^{i}}{\Phi}\bar{\chi}_{Li}\gamma^{\nu}\gamma^{\mu}\psi_{L\nu}D_{\mu}\sigma + \frac{3}{8} e^{-1}\ \varepsilon^{\mu\nu\rho\sigma}\psi_{\mu}\gamma_{\nu}\psi_{\rho}$$

$$\times\left(\frac{\Phi^{i}}{\Phi}D_{\sigma}z_{i} + \frac{1}{2}\frac{\Phi^{j}_{i}}{\Phi}\bar{\chi}^{i}_{R}\gamma_{\sigma}\chi_{Lj}\right) + \frac{3}{2}\frac{\Phi^{i}_{j}}{\Phi}\psi_{\mu L}\bar{\chi}_{Li}\bar{\psi}_{\mu R}\chi^{j}_{R}$$

$$- \frac{3}{2} \frac{\Phi^{ij}_{k}}{\Phi} \bar{\chi}_{Li} \chi_{Lj} \bar{\chi}^{k}_{R} \chi^{\ell}_{R} - \frac{1}{2} \Phi^{i} \bar{\chi}_{Li}$$

$$\times \left(\chi_{\mu L} \bar{\psi} \cdot \gamma \psi^{\mu} + \frac{1}{2} \sigma_{\mu\nu} \psi_{\rho L} \bar{\psi}^{\nu} \gamma^{\rho} \psi^{\mu} + \sigma^{\mu\nu} \psi_{\nu L} \bar{\psi}_{\mu} \gamma \cdot \psi \right)$$

$$- \frac{3}{2} i\tilde{g} \frac{\Phi^{i}}{\Phi} T^{\alpha j}_{i} z_{j} \bar{\psi}_{L} \cdot \gamma \lambda^{\alpha}_{R} + 6i\tilde{g} \frac{\Phi^{j}_{i}}{\Phi} T^{\alpha k}_{j} z_{k} \bar{\lambda}^{\alpha}_{R} \chi^{i}_{R}$$

$$- \frac{1}{2} g^{ij} e^{3\sigma} \bar{\chi}_{Li} \chi_{Lj} + \frac{1}{2} g^{i} e^{3\sigma} \psi_{R} \cdot \gamma \chi_{Li}$$

$$+ \frac{1}{2} g \, e^{3\sigma} \bar{\psi}_{\mu R} \sigma^{\mu\nu} \psi_{\nu R} + \frac{1}{2} \mathrm{Re}\, f_{\alpha\beta} \Big[- \frac{1}{2} \bar{\lambda}^{\alpha} \not{D} \lambda^{\beta} + \frac{1}{2} \bar{\lambda}^{\alpha} \gamma^{\mu} \sigma^{\rho\sigma} \psi_{\mu}$$

$$\times \left(F^{\beta}_{\rho\sigma} + \frac{1}{2} \bar{\psi}_{\rho} \gamma_{\sigma} \lambda^{\beta} \right) - \frac{1}{32} \bar{\lambda}^{\alpha} \gamma_{5} \gamma_{\nu} \lambda^{\beta} e^{-1} \varepsilon^{\mu\nu\rho\sigma} \bar{\psi}_{\mu} \gamma_{\rho} \psi_{\sigma} \Big]$$

$$- \frac{1}{4} i \,\mathrm{Im}\, f_{\alpha\beta} D^{\mu} (\bar{\lambda}^{\alpha} \gamma^{\mu} \lambda^{\beta}_{R}) + \frac{1}{2} f^{i}_{\alpha\beta} \Big[- \bar{\chi}_{Li} \sigma^{\mu\nu} \lambda^{\beta}_{L}$$

$$\times \left(F^{-\alpha}_{\mu\nu} - \frac{1}{2} \bar{\psi}_{\rho L} \sigma_{\mu\nu} \gamma^{\rho} \lambda^{\alpha}_{R} + \frac{1}{2} \bar{\psi}_{R} \cdot \gamma \sigma_{\mu\nu} \lambda^{\alpha}_{L} \right)$$

$$- \frac{1}{2} \bar{\psi}_{R} \cdot \gamma \chi_{Li} \bar{\lambda}^{\alpha}_{L} \lambda^{\beta}_{L} \Big] + \frac{1}{4} f^{ij}_{\alpha\beta} \bar{\chi}_{Li} \chi_{Lj} \bar{\lambda}^{\alpha}_{L} \lambda^{\beta}_{L} + \text{h.c.} \qquad (4.13)$$

The kinetic part of Eq. (4.13) can be further diagonalized through the redefinition

$$\psi_{\mu L} \to \psi_{\mu L} - \gamma_{\mu} \frac{\Phi_{i}}{\Phi} \chi^{i}_{R} \,. \qquad (4.14)$$

If, in addition, we perform the following chiral redefinitions on the fermion fields,

$$\psi_{\mu L} \to \left(\frac{g}{g^{*}}\right)^{\frac{1}{4}} \psi_{\mu L} \,, \quad \chi_{Li} \to \left(\frac{g}{g^{*}}\right)^{\frac{1}{4}} \chi_{Li} \,, \quad \lambda^{\alpha}_{L} \to \left(\frac{g}{g^{*}}\right)^{\frac{1}{4}} \lambda^{\alpha}_{L} \qquad (4.15)$$

and we define the function

$$G = J - \log \frac{1}{4} |g|^{2} \qquad (4.16)$$

so that

$$G_i = J^i - \frac{g^i}{g} , \quad G^i_j = J^i_j , \tag{4.17}$$

the final Lagrangian becomes

$$L = L_B + L_{FK} + L_{FM} + L_{(4)F} , \tag{4.18}$$

where

$$e^{-1} L_B = -\frac{1}{2} R + G^i_j D_\mu z_i D^\mu z^{*j} - \frac{1}{4} \text{Re } f_{\alpha\beta} F^\alpha_{\mu\nu} F^\beta_{\mu\nu}$$

$$+ \frac{1}{4} i \text{ Im } f_{\alpha\beta} F^\alpha_{\mu\nu} \tilde{F}^\beta_{\mu\nu} + e^{-G}(3 + G_k G^{-1k}_\ell G^\ell)$$

$$+ \frac{1}{2} \tilde{g}^2 \text{ Re } f^{-1}_{\alpha\beta} (G^i T^{\alpha j}_i z_j)(G^k T^{\beta\ell}_k z_\ell) ; \tag{4.19}$$

L_{FK} is the curved generalization of the fermionic terms; L_{FM} is the quadratic part in the fermion fields without derivative terms; $L_{(4)F}$ is the part of the Lagrangian which contains four-fermion interaction terms [not coming from the spin-3/2 torsion contribution already contained in the Lorentz connection $\omega_{\mu mn}(e,\psi)$] and also other interactions of dimension greater than four containing covariant derivatives of boson fields:

$$e^{-1} L_{FK} = -\frac{1}{4} \text{Re } f_{\alpha\beta} \bar{\lambda}^\alpha \not{D} \lambda^\beta - \frac{1}{4} e^{-1} \bar{\psi}_\mu \gamma_5 \gamma_\nu D_\rho \psi_\sigma \varepsilon^{\mu\nu\rho\sigma}$$

$$+ G^i_j \bar{\chi}_{Li} \not{D} \chi^j_R - \frac{i}{8} \text{Im } f_{\alpha\beta} \, e^{-1} D_\mu (e \bar{\lambda}^\alpha \gamma_5 \gamma_\mu \lambda^\beta) + \text{h.c.} , \tag{4.20}$$

$$e^{-1} L_{F,M} = e^{-G/2} \bar{\psi}_{\mu R} \sigma^{\mu\nu} \psi_{\nu R} - \bar{\psi}_R \cdot \gamma \Big(e^{-G/2} G^i \chi_{Li}$$

$$- \frac{i}{2} \tilde{g} G^i T^{\alpha j}_i z_j \lambda^\alpha_L \Big) + e^{-G/2} (G^{ij} - G^i G^j$$

$$- G^\ell G^{-1k}_\ell G^{ij}_k) \bar{\chi}_{Li} \chi_{Lj} - 2i\tilde{g} G^k_i z^{*j} T^{\alpha i}_j \bar{\lambda}^\alpha_L \chi_{Lk}$$

$$+ \frac{1}{2} f^k_{\alpha\beta} \Big(\frac{1}{2} e^{-G/2} G_\ell G^{-1\ell}_k \bar{\lambda}^\alpha_L + i\tilde{g} \text{ Re } f^{-1}_{\alpha\gamma} G^i T^{\gamma j}_i z_j \bar{\chi}_{Lk} \Big) \lambda^\beta_L + \text{h.c.} , \tag{4.21}$$

$$e^{-1} L_{(4)F} = \frac{1}{2}\,\mathrm{Re}\, f_{\alpha\beta}\left(\frac{i}{2}\,\bar{\lambda}^{\alpha}\gamma^{\mu}\sigma^{\rho\sigma}\psi_{\mu}F^{\beta}_{\rho\sigma} - \frac{1}{2}\,\bar{\lambda}^{\alpha}_{L}\gamma_{\mu}\lambda^{\beta}_{R}G^{i}D_{\mu}z_{i}\right)$$

$$- \frac{1}{2}\, f^{i}_{\alpha\beta}\bar{\chi}_{Li}\sigma\cdot F^{\alpha}\lambda^{\beta}_{L} + \frac{1}{8}\, e^{-1}\,\varepsilon^{\mu\nu\rho\sigma}\bar{\psi}_{\mu}\gamma_{\nu}\psi_{\rho}G^{i}D_{\sigma}z_{i}$$

$$- G^{j}_{i}\bar{\psi}_{\mu L}\not{\partial} z^{*i}\gamma^{\mu}\chi_{Lj} - \bar{\chi}_{Li}\not{\partial} z_{j}\chi^{k}_{R}\left(G^{ij}_{k} + \frac{1}{2}\,G^{i}_{k}G^{j}\right)$$

$$+ \frac{1}{32}\,G^{-1k}_{\ell} f^{\ell}_{\alpha\beta} f^{*}_{\gamma\delta k}\bar{\lambda}^{\alpha}_{L}\lambda^{\beta}_{L}\bar{\lambda}^{\gamma}_{R}\lambda^{\delta}_{R} + \frac{3}{32}(\mathrm{Re}\, f_{\alpha\beta}\bar{\lambda}^{\alpha}_{L}\gamma_{m}\gamma^{\beta}_{R})^{2}$$

$$+ \frac{1}{8}\,\mathrm{Re}\, f_{\alpha\beta}\bar{\lambda}^{\alpha}\gamma^{\mu}\sigma^{\rho\sigma}\psi_{\mu}\bar{\psi}_{\rho}\gamma_{\sigma}\lambda^{\beta} + \frac{1}{2}\, f^{i}_{\alpha\beta}\left(\bar{\chi}_{Li}\sigma^{\mu\nu}\lambda^{\alpha}\cdot\bar{\psi}_{\nu L}\gamma_{\mu}\lambda^{\beta}_{R}\right.$$

$$\left.+ \frac{1}{4}\,\bar{\psi}_{R}\cdot\gamma\chi_{Li}\bar{\lambda}^{\alpha}_{L}\lambda^{\beta}_{L}\right) + \frac{1}{8}\,G^{j}_{i}\bar{\chi}^{i}_{k}\gamma_{d}\chi_{Lj}(\varepsilon^{abcd}\bar{\psi}_{a}\gamma_{b}\psi_{c}$$

$$- \bar{\psi}_{a}\gamma_{5}\gamma^{d}\psi_{a}) + \frac{1}{16}\,\bar{\chi}_{Li}\gamma^{\mu}\chi^{j}_{R}\bar{\lambda}^{\delta}_{R}\gamma_{\mu}\lambda^{\gamma}_{L}(-2G^{i}_{j}\,\mathrm{Re}\, f_{\gamma\delta}$$

$$+ \mathrm{Re}\, f^{-1}_{\alpha\beta} f^{i}_{\alpha\gamma} f^{*}_{\beta\delta j}) - \frac{1}{16}\,\bar{\chi}_{Li}\sigma_{\mu\nu}\chi_{Lj}\bar{\lambda}^{\gamma}_{L}\sigma^{\mu\nu}\lambda^{\delta}_{L}$$

$$\times\,\mathrm{Re}\, f^{-1}_{\alpha\beta} f^{i}_{\alpha\gamma} f^{j}_{\beta\delta} + \frac{1}{16}\,\bar{\chi}_{Li}\chi_{Lj}\bar{\lambda}^{\gamma}_{L}\lambda^{\delta}_{L}$$

$$\times\,(-4G^{ij}_{k}G^{-1k}_{\ell} f^{\ell}_{\gamma\delta} + 4f^{ij}_{\gamma\delta} - \mathrm{Re}\, f^{-1}_{\alpha\beta} f^{i}_{\alpha\gamma} f^{j}_{\beta\delta})$$

$$+ \left(-\frac{1}{2}\,G^{ij}_{k\ell} + \frac{1}{2}\,G^{ij}_{m}G^{-1m}_{n}G^{n}_{k\ell} - \frac{1}{4}\,G^{i}_{k}G^{j}_{\ell}\right)$$

$$\times\,\bar{\chi}_{Li}\chi_{Lj}\bar{\chi}^{k}_{R}\chi^{\ell}_{R} + \mathrm{h.c.} \quad (4.22$$

The final form of the Lagrangian, as given by Eq. (4.18), is invariant under the following local supersymmetry transformations:

$$\delta e^{m}_{\mu} = \frac{1}{2}\,\bar{\varepsilon}_{L}\gamma^{m}\psi_{\mu R} + \mathrm{h.c.}\,, \quad (4.23a$$

$$\delta\psi_{\mu L} = \left[\partial_{\mu} + \frac{1}{2}\,\omega_{\mu mn}(e,\psi)\sigma^{mn}\right]\varepsilon_{L} + \frac{1}{2}\,\sigma_{\mu\nu}\varepsilon_{L}G^{j}_{i}\bar{\chi}^{i}_{R}\gamma^{\nu}\chi_{Lj}$$

$$+ \frac{1}{2}\,\gamma_{\mu}\varepsilon_{R}\, e^{-G/2} + \frac{1}{4}\,\psi_{\mu L}(G^{i}\bar{\varepsilon}_{L}\chi_{Li} - G_{i}\bar{\varepsilon}_{R}\chi^{i}_{R})$$

$$- \frac{1}{4}\,\varepsilon_{L}(G^{i}D_{\mu}z_{i} - G_{i}D_{\mu}z^{*i}) + \frac{1}{4}\,(\sigma_{\nu\mu} + g_{\mu\nu})\varepsilon_{L}\bar{\lambda}^{*}_{L}\gamma^{\nu}\lambda^{\beta}_{R}\,\mathrm{Re}\, f_{\alpha\beta}\,,$$

(4.23b

$$\delta B_\mu^\alpha = -\frac{1}{2}\bar{\varepsilon}_L\gamma_\mu\lambda_R^\alpha + \text{h.c.} \ , \tag{4.23c}$$

$$\delta\lambda_R^\alpha = \frac{1}{2}\sigma^{\mu\nu}\hat{F}_{\mu\nu}^\alpha\varepsilon_R + \frac{1}{2}i\varepsilon_R \ \text{Re}\ f_{\alpha\beta}^{-1}\Big(-\tilde{g}G^i T_i^{\beta j} z_j + \frac{1}{2}if_{\beta\gamma}^i\bar{\chi}_{Li}\lambda_L^\gamma - \frac{1}{2}if_{\beta\gamma i}^*\bar{\chi}_R^i\lambda_R^\gamma\Big) - \frac{1}{4}\lambda_R^\alpha(G^i\bar{\varepsilon}_L\chi_{Li} - G_i\bar{\varepsilon}_R\chi_R^i) \ , \tag{4.23d}$$

$$\delta z_i = \bar{\varepsilon}_L\chi_{Li} \ , \tag{4.23e}$$

$$\delta\chi_{Li} = \frac{1}{2}\hat{\not{\partial}}z_i\varepsilon_R - \frac{1}{2}\varepsilon_L\, e^{-G/2}\, G_i^{-1j}G_j - \frac{1}{8}\varepsilon_L\bar{\lambda}_R^\alpha\lambda_R^\beta G^{-1k}f_{\alpha\beta k}^* + \frac{1}{2}\varepsilon_L G_i^{-1k}G_k^{j\ell}\bar{\chi}_{Lj}\chi_{L\ell} + \frac{1}{4}\chi_{Li}(G_j\bar{\varepsilon}_R\chi_R^j - G^j\bar{\varepsilon}_L\chi_{Lj}) \ . \tag{4.23f}$$

The transformations (4.23) are directly obtained from the original transformations (3.25), with the substitution of Eqs. (4.5) in the fermionic field variations and with the redefinitions (4.8), (4.14), and (4.15) for the vierbein field and for the fermionic fields. In addition, the local supersymmetry parameter in Eqs. (4.23) has also been redefined, compared with Eqs. (3.25), as follows:

$$\varepsilon_L \to e^{-\frac{1}{2}J}\left(\frac{g}{g^*}\right)^{\frac{1}{4}}\varepsilon_L \ . \tag{4.24}$$

We note that, by using conformal tensor calculus, we could avoid the redefinitions given by Eqs. (4.8), (4.9), (4.14), (4.15), and (4.24), by suitably choosing a special conformal gauge. This has been shown elsewhere[17].

Finally, we discuss the case in which the gauge group is $G \times U_R(1)$, corresponding to the superspace action formulae given in Eqs. (3.11) and (3.12). As a consequence of Eq. (3.35), the presence of a gauged R-symmetry is most easily discussed in the final form of the Lagrangian and transformation laws as given by Eqs. (4.18) and (4.23).

In fact, the Lagrangian has precisely the same expression as in Eq. (4.18) with suitable gauge covariantization with respect to the $U_R(1)$ gauge field. The only difference now is the fact that the D-term of the R-multiplet, defined as

$$D_R = g_R G^i n_i z_i , \tag{4.25}$$

can have a constant term, owing to the non-invariance of the super-potential under R-transformations:

$$g^i(z) n_i g_R z_i = 3 g_R g(z) , \tag{4.26}$$

where $n_i g_R = q_i$ are the U(1) charges of the scalar fields z_i.

It is obvious that these considerations are no longer valid for a vanishing superpotential function $g(z) = 0$. In this case the final form of the Lagrangian and transformation laws comes directly from Eqs. (4.18) and (4.25) with the following substitution rules[11]:

i) e^{-G}, $e^{-G/2} \rightarrow 0$,

$$G^i \nabla_m z_i \Rightarrow J^i \nabla_m z_i + \frac{3}{2} i g_R V^R_m ,$$

ii) $\left(G_i \nabla_m z^{*i} \Rightarrow J_i \nabla_m z^{*i} - \frac{3}{2} i g_R V^R_m \right)$,

iii) $\tilde{g}_\alpha G^i T^{\alpha j}_i z_j \Rightarrow \tilde{g}_\alpha J^i T^{\alpha j}_i z_j - \frac{3}{2} g_R \delta^\alpha_R$,

iv) for other derivatives of G not in the form given by (ii) and (iii), replace $G \rightarrow J$.

We note, in particular, that the substitution rule (ii) produces new minimal gauge couplings of the R-gauge field to the gravitino, gauginos, and chiral fermions (but not to the scalars) which are usually eliminated through the chiral redefinitions (4.15) when $g(z) \neq 0$.

It is important to note that some models discussed in the literature fall within this class[19]. However, these models can only have spontaneously broken supersymmetry with a non-vanishing cosmo-logical constant.

In the more interesting physical situation of spontaneously broken R-symmetry and supersymmetry with a vanishing cosmological constant, the superpotential cannot vanish and the final form of the Lagrangian as given by Eq. (4.18) applies.

THE SUPER-HIGGS EFFECT IN N = 1 SUPERGRAVITY

Spontaneous breaking of local supersymmetry implies the super-Higgs effect, i.e. the gravitino mass generation and the disappearance of the massless goldstino which is absorbed by the 'massive' spin-3/2 gravitino.

This phenomenon may or may not occur with a vanishing cosmological constant, i.e. in a Minkowski or de Sitter space-time.

The goldstino field is uniquely defined by the spin-½ fermion which couples to the spin-3/2 gravitino in $L_{F,M}$ given by Eq. (4.21):

$$-\bar{\psi}_R \cdot \gamma \eta_L \ , \qquad \eta_L = e^{-G/2}\, G^i \chi_{Li} - \frac{i}{2} D^\alpha \lambda_L^\alpha \ , \tag{5.1}$$

where

$$D^\alpha = \tilde{g} G^i T^{\alpha j} z_j \ . \tag{5.2}$$

A necessary and sufficient condition for spontaneous supersymmetry breaking therefore implies that one of the quantities[6],

$$e^{-G/2}\, G^i, \qquad D^\alpha \ , \tag{5.3}$$

is different from zero at the minimum of the potential given in Eq. (4.19):

$$V(z,z^*) = -e^{-G}(3 + G^i G_i^{-1j} G_j) + \frac{1}{2} \mathrm{Re}\, f^{-1}_{\alpha\beta} D^\alpha D^\beta \ . \tag{5.4}$$

In the more interesting situation of a Minkowski space-time, we also demand the vanishing of the cosmological constant

$$V(z_0, z_0^*) = 0 \quad \text{at} \quad \left.\frac{\partial V}{\partial z}\right|_{z=z_0} = 0 \ . \tag{5.5}$$

Under condition (5.5), the gravitino mass has a precise meaning, and this is given by the universal formula[5]:

$$m_{3/2} = e^{-G\,2} . \tag{5.6}$$

We now discuss the particularly interesting case of 'minimal' coupling of the Yang-Mills system to supergravity. This is defined by th two conditions

$$G^i_j = -\frac{1}{2}\,\delta^i_j , \quad f_{\alpha\beta} = \delta_{\alpha\beta} , \tag{5.7}$$

in which case all the kinetic terms are canonical and the Lagrangian depends only on the superpotential g(z). In this case we have

$$G(z,z^*) = -\frac{1}{2}|z|^2 - \log\frac{1}{4}|g(z)|^2 . \tag{5.8}$$

The scalar potential (5.4) becomes

$$V(z,z^*) = e^{-G}(2G^iG_i - 3) + \frac{1}{2}(D^\alpha)^2 , \tag{5.9}$$

with

$$D^\alpha = -\frac{1}{2}\,\tilde{g}_\alpha z^{*i}T^{\alpha j}_i z_j + \xi\delta^\alpha_R .$$

The term $L_{F,M}$ given by Eq. (4.21), which determines the fermion mass matrix is

$$e^{-1}\,L_{F,M} = e^{-G/2}\left(\bar{\psi}_{\mu R}\sigma^{\mu\nu}\psi_{\nu R} - \bar{\psi}_R\cdot\gamma\tilde{\eta}_L - \frac{2}{3}\,\bar{\tilde{\eta}}_L\tilde{\eta}_L\right) + \bar{\chi}_{Li}M^{ij}\chi_{Lj} + 2\bar{\chi}_{Li}M^{i\alpha}\lambda^\alpha_L + \bar{\lambda}^\alpha_L M^{\alpha\beta}\lambda^\beta_L + \text{h.c.} \tag{5.10}$$

where

$$\tilde{\eta}_L = G^i\chi_{Li} - \frac{i}{2}\,e^{G/2}\,D^\alpha\lambda^\alpha_L . \tag{5.11}$$

The spin-½ fermion mass matrix has the form

$$M^{ij} = e^{-G/2}\left(G^{ij} - \frac{1}{3}\,G^iG^j\right) , \tag{5.12a}$$

$$M^{i\alpha} = -\frac{1}{3}\,iG^iD^\alpha - ID^{\alpha i} , \tag{5.12b}$$

$$M^{\alpha\beta} = -\frac{1}{6} e^{G/2} D^{\alpha} D^{\beta} . \tag{5.12c}$$

We can now compute the quadratic mass relation which generalizes the result of spontaneously broken, globally supersymmetric Yang-Mills theories. The scalar contribution to the trace of the square mass is

$$\begin{aligned} \mathrm{Tr}\, M_0^2 = & -(N+4) D^{\alpha} D^{\alpha} + 2G^{i} G_{i} D^{\alpha} D^{\alpha} + \eta\, e^{-G} (4G_{ij} G^{ij} + N) \\ & + 4D^{\alpha i} D^{\alpha}_{i} + 4D^{\alpha} D^{\alpha i}_{i} . \end{aligned} \tag{5.13}$$

The spin-1 contribution is

$$\mathrm{Tr}\, M_1^2 = 12 D^{\alpha}_{i} D^{i\alpha} . \tag{5.14}$$

The spin-½ contribution is

$$\begin{aligned} \mathrm{Tr}\, M_{\frac{1}{2}}^2 = & 8\, e^{-G} \left(G^{ij} - \frac{1}{3} G^{i} G^{j} \right) \left(G_{ij} - \frac{1}{3} G_{i} G_{j} \right) \\ & + 16 \left(D^{\alpha i} + \frac{1}{3} G^{i} D^{\alpha} \right) \left(D^{\alpha}_{i} + \frac{1}{3} G_{i} D^{\alpha} \right) + \frac{2}{g} e^{G} (D^{\alpha})^2 (D^{\beta})^2 \end{aligned} \tag{5.15}$$

and the gravitino contribution is

$$\mathrm{Tr}\, M_{3/2}^2 = 4\, e^{-G} . \tag{5.16}$$

Therefore we finally get the mass formula

$$\begin{aligned} \text{super-trace } M^2 = & \sum_{J=0}^{3/2} (-)^{2J} (2J+1) m_j^2 = (N-1)(2m_{3/2}^2 - K^2 D^{\alpha} D^{\alpha}) \\ & - 2\tilde{g}_{\alpha} D^{\alpha}\, \mathrm{Tr}\, T^{\alpha} . \end{aligned} \tag{5.17}$$

Equation (5.17) gives back the result of global supersymmetry[20] if we set $m_{3/2} = 0$, $k = 0$. It also reproduces the result of Ref. 11 for $N = 1$ and $\tilde{g} = 0$. We also note that the last term in Eq. (5.17) is only possible for Abelian U(1) factors of G with $\mathrm{Tr}\, T^{\alpha} \neq 0$. Equation (5.17) summarizes the mass relations of spontaneously broken supersymmetry when the simultaneous occurrence of the Higgs and super-Higgs effects takes place.

6. LOW-ENERGY LIMIT OF SPONTANEOUSLY BROKEN, LOCALLY SUPERSYMMETRIC LAGRANGIANS

In the practical current applications of spontaneously broken supergravity models to particle interactions, the Lagrangians described in the previous sections will contain, beyond the Planck scale m_P, some lower energy scales.

In particular they will contain the weak interaction scale m_W and other possible fundamental scales, such as the GUT scale $m_X = O(10^{15}-10^{16}\ \text{GeV})$.

Although our previous formalism is valid for any value of the gravitino mass, which is related to the supersymmetry-breaking scale M_{SB} as follows[4,5],

$$m_{3/2} = \sqrt{\frac{8\pi}{3}}\frac{M_{SB}^2}{m_P}\ , \tag{6.1}$$

we will consider the particular situation of an intermediate supersymmetry breaking scale

$$M_{SB} = \sqrt{\mu m_P}\ , \tag{6.2}$$

where μ is a scale parameter $O(m_W)$. Under this circumstance

$$m_{3/2} = \sqrt{\frac{8\pi}{3}} = O(m_W) \tag{6.3}$$

and supergravity corrections to low-energy physics play a crucial role. This is already implied by the modification of the mass spectrum through the quadratic mass relation given by Eq. (5.17).

The purpose of the present section is to derive all possible supergravity-induced breaking terms which survive in the low-energy limit defined by letting $m_P \to \infty$ with $m_{3/2}$ kept fixed.

The final result will be that in the limit $m_{3/2}/m_P \ll 1$, spontaneously broken supergravity will manifest itself as an explicitly softly broken global supersymmetric theory with well-defined renormalizable interactions[21].

In particular, the soft-breaking terms, at most of dimension three in the particle fields, are such that no quadratic divergence arises through radiative corrections if gravitational interactions are neglected[22]. This circumstance has the resulting effect of not destroying the characteristic feature of supersymmetric theories, i.e. of keeping radiative corrections of light particles, including Higgs, generally small, compared to some other large scale of the theory such as the GUT scale m_X of the Planck scale m_P.

The important point in obtaining the low-energy limit of a spontaneously broken supergravity model is to realize that the cancellation of the cosmological constant [see Eq. (5.5)] implies the presence, in the theory, of some scalar fields x_i acquiring v.e.v.'s of order m_P which break supersymmetry[5]. It is natural, although not compulsory, that these fields are singlets under the Yang-Mills group G introduced in the previous section, especially if we do not want the GUT scale $m_X < m_P$ to be related to the Planck scale. The full set of scalar fields z_i belonging to the chiral multiplets S^i of the theory therefore split into

$$z_i = (x_i, y_i) , \tag{6.4}$$

where $x_i = O(m_P)$ and $\langle y_i \rangle / \langle x_j \rangle \ll 1$.

The remaining fields y_i are called low-energy fields and generally transform non-trivially under G.

The requirement that the low-energy effective Lagrangian indeed describes light fields with renormalizable interactions can be fulfilled under the condition that a decoupling occurs between x-type and y-type multiplets as well as gauge vector multiplets. Furthermore, the gravitino mass $m_{3/2}$ must be finite for $m_P \to 0$ [21].

We define the dimensionful G function to be (dim G = 2):

$$G(z,z^*) = J(z,z^*) - \frac{1}{k^2} \log |k^3 g(z)|^2 . \tag{6.5}$$

The y-dependent part of the Kähler potential is

$$J\left(\frac{1}{k}\zeta, \frac{1}{k}\zeta^*; y,y^*\right) = -\frac{1}{2} y_i \Lambda^i_j(\zeta,\zeta^*) y^{*j} + B^i(\zeta,\zeta^*) y_i + B^*_i(\zeta,\zeta^*) y^{*i} , \quad (6.6)$$

where $\zeta_i = kx_i$. Analogously, the y-dependent part of the superpotential $g(z) = g(x,y)$ must be a finite polynomial, at most of degree three in the y-variables with ζ-dependent coefficients.

Equation (3.5) defines the Kähler potential $J(z,z^*)$ up to a Kähler transformation

$$J(z,z^*) \to J(z,z^*) + f(z) + f^*(z^*) , \quad (6.7a)$$

$$g(z) \to g(z) \exp k^2 f(z) . \quad (6.7b)$$

The behaviour of the y-dependent part of the superpotential g(z) for $m_P \to \infty$, consistent with Eq. (6.6), must be

$$g(\zeta,0) = O(m_P^2) , \quad k \to 0 . \quad (6.8)$$

In addition, the dimensionless function $f_{\alpha\rho}(z)$ must behave as

$$f_{\alpha\beta}(z) = \delta_{\alpha\beta}(1 + i\theta_\alpha) , \quad k \to 0 \quad (6.9)$$

for $m_P \to \infty$ and x_i, y_i fixed.

Let us now consider the general form of the Lagrangian (4.18) in the limit $m_P \to \infty$ $(k \to 0)$ in the y_i fields, after the x_i fields have acquired v.e.v.'s of $O(m_P)$ which break supersymmetry.

Equation (6.8) implies a finite gravitino mass,

$$m_{3/2} = \lim_{k\to 0} k^2 |\langle g\rangle| \exp - \frac{k^2}{2} \langle J\rangle . \quad (6.10)$$

The low-energy fields also include (coming from the x-type fields) the goldstino, which provides the missing degrees of freedom of the massive spin-$3/2$ gravitino and two G-singlet scalar fields, the scalar partners of the would-be goldstino, with masses $O(m_{3/2})$[6]. However, these fields are not included in the low-energy Lagrangian of the y fields, because in the limit $m_P \to \infty$ they decouple from the y-type multiplets.

The gravitino mass defined by Eq. (6.10) provides the scale of all explicit supersymmetry-breaking terms appearing in the limiting Lagrangian for the low-energy fields y_i. This has the general form[23]

$$L = L(\text{SUSY};f) + m_{3/2}\,(c_\alpha \bar{\lambda}^\alpha_L \lambda^\alpha_L + \text{h.c.}) - m_{3/2}\left[h(y) + h^*(y^*)\right] - \frac{1}{2}\, m^2_{3/2}\, y^{*j} \Lambda^i_j y_i \;. \tag{6.11}$$

The first term is a globally supersymmetric Lagrangian with superpotential

$$f(y) = \lim_{k\to 0} g(\langle x\rangle, y)\, \exp\left(-\frac{1}{2}\, k^2 \langle J\rangle\right). \tag{6.12}$$

The function h(y) is another analytic polynomial, invariant under G, of (at most) degree three in the y fields; the $c_{\alpha,s}$ in the gaugino mass terms are complex dimensionless coefficients of order 1; and Λ^i_j is a numerical Hermitian matrix which depends on the Kähler potential J defined by Eq. (6.6).

In addition, the D-term of the supersymmetric part of Eq. (6.11) can contain dimensionful constants (Fayet-Iliopoulos terms), corresponding to U(1) factors of G:

$$D^2_{U(1)} = (\tilde{g} y_i Q^i_j y^{*j} + \xi)^2 \;, \tag{6.13}$$

this even if the original Lagrangian given by Eq. (4.18) does not contain any gauged R-symmetry, as defined in Section 3. The reason is that if the scalar fields x_i have v.e.v.'s $\langle x_i\rangle = O(m_P)$, then we can get a purely gravitationally induced effective D-term coming from terms linear in the U(1) gauge multiplet which couples to the x-fields*),

$$k^2 D^\alpha W_\alpha \Psi(S,\bar{S})|_D \, , \tag{6.14}$$

producing an effective interaction,

$$D_{U(1)} \Psi^j_i h^i h^*_j \to c D_{U(1)} m^2_{3/2} \, , \tag{6.15}$$

for $\Psi^j_i \sim \delta^j_i$ and $|h_x| = O(m_{3/2})$.

This term should not be confused with a possible intrinsic ξ term in the original Lagrangian (4.18) which could be introduced by the gauging of R-symmetry as explained in Section 3. The term in Eq. (6.13) rather produces a soft-breaking term of dimension two in the scalar fields, of the type

$$L_{soft} \simeq c m^2_{3/2} \; y_i (Q)^i_j y^{*j} \tag{6.16}$$

where Q is the U(1) charge matrix of the y fields and which looks li a Fayet-Iliopoulos term since it can be rewritten in the form (6.13)

It is interesting to note that the general form (6.11) can also be obtained by the coupling of a locally supersymmetric Lagrangian for the matter fields y_i, to a Volkov-Akulov field, i.e. the non-linearly realized goldstino field. This has recently been shown by Samuel and Wess[24].

The reason why this approach gives the same result as the linea realization is due to the decoupling which occurs, for $m_P \to \infty$, between the goldstino multiplet and the light-matter fields y_i. In particular, the scalar degrees of freedom of the goldstino multiplet which acquire v.e.v.'s of $O(m_P)$, do not play any role in the final softly broken Lagrangian we have considered. Then the entire

*) For a related argument, in the framework of non-linear realization, see Samuel and Wess[24].

goldstino multiplet can be simply replaced by a Volkov-Akulov field θ transforming, for constant ε, as[24]:

$$\delta_{\varepsilon}\theta(x) = M_{SB}^2\varepsilon + \frac{1}{M_{SB}^2}\,\xi^{\mu}(x)\partial_{\mu}\theta(x) \;, \tag{6.17}$$

with

$$\xi^{\mu}(x) = \frac{1}{2}\,\bar{\varepsilon}_L\gamma^{\mu}\theta_R(x) - \frac{1}{2}\,\bar{\theta}_L(x)\gamma^{\mu}\varepsilon_R \;. \tag{6.18}$$

The Lagrangian (6.11) shows that the low-energy effects due to spontaneously broken supergravity, with $m_{3/2} = O(m_W)$, result in explicit soft-breaking terms in the global limit which do not alter the ultraviolet behaviour of the purely globally supersymmetric part of the Lagrangian. In particular, they do not introduce quadratic divergences since these terms are soft in a supersymmetric sense, according to the analysis of Ref. 22.

To understand the origin of the various soft-breaking terms in Eq. (6.11), it is useful to consider special forms for the functions $G(z,z^*)$ and $f_{\alpha\beta}(z)$ which appear in the general form of the matter-Yang-Mills supergravity action given by Eq. (4.18).

In the case of minimal coupling, defined by Eq. (5.7), we get[21,25]

$$h(y) = (A-3)f(y) + \sum_{I} y_i\,\frac{\partial f}{\partial y_i} \;, \tag{6.19a}$$

$$c_{\alpha} = 0 \;, \tag{6.19b}$$

$$\Lambda_i^j = \delta_i^j \;, \tag{6.19c}$$

where A is a parameter depending on the particular way the x fields enter into the superpotential $g(z) = g(x,y)$.

For example, if

$$g(x,y) = g(x) + g(y) \;, \tag{6.20}$$

with

$$g(x) = ax + b \;, \tag{6.21}$$

then

$$A = 3 - \sqrt{3} \, . \tag{6.22}$$

In the minimal case we may obtain an arbitrary function h(y) for the soft-breaking term in Eq. (6.11), which is linear in $m_{3/2}$, by relaxing the property (Eq. 6.19)*).

If we introduce general mixing terms of the y fields to the x fields, then h(y) receives an additional contribution given by

$$\Delta h(y) = \lim_{k\to 0} \frac{\sqrt{3}}{k} \left.\frac{\langle \partial g(x,y)\rangle}{\partial x}\right|_{x=x_0} \exp\left(-\frac{k^2}{2}\langle J\rangle\right) , \tag{6.23}$$

where $\partial g/\partial x$ is the derivative of g in the direction of the would-be goldstino.

Note that in order to get Eq. (6.22), what is needed is to introduce a superpotential of the form**)

$$g(x,y) = g(x) + g(y) + xg'(y) + O\left(\frac{1}{m_P}\right) + \ldots \, . \tag{6.24}$$

The reason for introducing non-vanishing gaugino mass terms in Eq. (6.11) is because we can have an x-dependence in the function $f_{\alpha\beta}(z)$ which appears in the Yang-Mills part of the original Lagrangian[6]. The coefficients c_α in Eq. (6.11) are related to the first derivative of the function $f_{\alpha\beta}$ in the x-direction:

$$\begin{aligned} m_{3/2} c_\alpha \delta_{\alpha\beta} &= \lim_{k\to 0} \frac{1}{k} f^i_{\alpha\beta} G_j (-J^{-1})^j_i \exp\left(-\frac{k^2\langle G\rangle}{2}\right) \\ &= \lim_{k\to 0} \frac{1}{k} f^x_{\alpha\beta} G_x (-J^{-1})^x_x \exp\left(-\frac{k^2\langle G\rangle}{2}\right). \end{aligned} \tag{6.25}$$

---

*) An example of the soft-breaking term given by an analytic function h(y) is obtained by integrating out the heavy degrees of freedom of GUTs, which in general modify the soft-breaking term of light fields[26].

**) An example which is a particular case of Eq. (6.23) has been considered by Cremmer et al.[27].

Finally, to get mass terms for the y fields of the form $y_i \Lambda^i_j y^{*j}$ with $\Lambda^i_j \neq \delta^i_j$, we need a non-minimal Kähler potential[23]:

$$J(z,z^*) = -\frac{1}{2}|x_i|^2 - \frac{1}{2}|y_i|^2 - \frac{1}{2}k^2|x_i|^2 y^{*\ell} A^m_\ell y_m \ , \qquad (6.26)$$

where A^m_ℓ is a numerical Hermitian matrix.

Then one gets in this case the matrix relation

$$\Lambda = \mathbb{1} + 3A \ . \qquad (6.27)$$

It turns out that Eqs. (4.24) and (4.26) give a correction to the general y-independent mass sum rule for minimal coupling regarding the y multiplets[6]:

$$\sum_{J=0}^{1} (-)^{2J}(2J+1)m_J^2 = 2(\dim T)m_{3/2}^2 - 2\tilde{g}^\alpha D^\alpha \operatorname{Tr} T^\alpha . \qquad (6.28)$$

This is related to the modification of the general quadratic mass sum rule in the case of non-minimal coupling[16].

For instance, a non-minimal Kähler potential gives an additional contribution to Eq. (6.27) of the form[16]:

$$e^{-G}\, G_a (G^{-1})^a_b G^c (G^{-1})^d_c R^b_d \ , \qquad R^b_d = \partial^b \partial_d \log \operatorname{Det} G^m_n \ , \qquad (6.29)$$

when R^ℓ_i is the contracted Riemann tensor (Ricci tensor) constructed out of the Kähler metric G^j_i.

For the particular example given by Eq. (6.25), the term in Eq. (6.28) gives a contribution to the quadratic mass sum rule proportional to $3m_{3/2}^2 \operatorname{Tr} A$.

Further examples of a super-Higgs sector of the theory with non-trivial Kähler curvature and non-trivial mass relations will be given in the next section.

The Lagrangian given by Eq. (6.11) is a tree-level result. If we start with a minimal coupling of matter to supergravity, then $c_\alpha = 0$ and $\Lambda^j_i = \delta^j_i$. It is clear that for general values of Λ^j_i and

c_α the low-energy Lagrangian (6.11) is stable under radiative corrections. However, if we start with the minimal coupling [Eq. (5.7) then radiative corrections, including gravitational ones, will induce corrections which at most will be reabsorbed in definitions of the coefficients c_α and the matrix Λ_i^j as well as in a possible Fayet-Iliopoulos term for Abelian factors of G. These corrections, depending on the particular model under consideration, can turn out to be small. In this situation the low-energy spectrum can be described by Eq. (6.18), in terms of a few parameters, and we can build simple extensions of the standard model for weak interactions as wel as GUT models with well-defined predictions for particle phenomenolo in the 100 GeV to 1 TeV energy range.

Simple examples of the generalization of the Standard Model, based on low-energy supergravity, will be given in Section 8.

7. FURTHER EXAMPLES OF A SUPER-HIGGS SECTOR: FLAT POTENTIALS AND GAUGED R-SYMMETRY

In this section we will consider the super-Higgs effect in situations that are different from the flat Kähler potential situation

$$\mathcal{G}_i^j = -\frac{1}{2}\,\delta_i^j \tag{7.1}$$

as considered in Sections 5 and 6.

We again consider the situation of a single chiral multiplet (z,χ_L). The z-potential is given, in general, by the first term of Eq. (5.4) for i = 1:

$$V(z,z^*) = -e^{-G}(G_z G_{z^*} G^{-1}_{zz^*} + 3) \ . \tag{7.2}$$

The gravitino mass is given by Eq. (5.6), provided $V(z,z^*)$ vanishes at the point $V_z = 0$. The Kähler curvature, for a signle scalar fiel is given by[16]

$$R_{zz^*} = \partial_z \partial_{z^*} \log G_{zz^*} \tag{7.3}$$

and it vanishes for the 'minimal kinetic term' (7.1). This corresponds to the super-Higgs sector described by a Polony field[28]. For a non-trivial, non-constant metric G_{zz^*}, the mass-sum rule of the super-Higgs sector becomes[16]:

$$m_A^2 + m_B^2 - 4m_{3/2}^2 = -2\, e^{-G} \frac{G_z G_{z^*}}{(G_{zz^*})^2} R_{zz^*} \tag{7.4}$$

There are interesting cases where the scalar potential given by Eq. (7.2) is identically equal to zero, although the supersymmetry may be broken. Rewriting the scalar potential of Eq. (7.2) as[5]

$$V = -g\, e^{-(4/3)G}\, G^{-1}_{zz^*} \partial_z \partial_{z^*}\, e^{(1/3)G} , \tag{7.5}$$

it follows immediately that the potential is identically equal to zero[29] if*)

$$G = \frac{3}{2} \log \left[\phi(z) + \phi^*(z^*)\right]^2 . \tag{7.6}$$

---

*) To better emphasize the geometric relevance of global flatness it is worth noting that Eq. (7.6) is equivalent to the unique solution

$$G = \frac{3}{2} \log\, (z + z^*)$$

up to field redefinition $z \to f(z)$. The scalar Lagrangian takes the form

$$3\sqrt{-g}\, g^{\mu\nu}\, \partial_\mu z \partial_\nu z^* / (z + z^*)^2$$

This Lagrangian is the same as the scalar sector of the N = 4 supergravity theory[30] and it describes a non-linear σ-model with an SU(1,1) non-compact symmetry

$$z \to \alpha z + i\beta / i\gamma z + \delta ,$$

with α, β, γ, δ real, and $\alpha\beta + \beta\gamma = 1$.

We also note that through the particular redefinition

$$z \to z + \sqrt{3}/z - \sqrt{3}$$

the kinetic term takes the form of the conformal scalar coupling [15,31] with $J = +3 \log\, (1 - \tfrac{1}{3}\, zz^*)$.

Contrary to the minimal kinetic term case, the Kähler manifold is not flat ($R_{zz^*} = 0$), since using Eq. (7.3) it is found that

$$R_{zz^*} = -\frac{2}{3} G_{zz^*} . \qquad (7.7)$$

Therefore, this defines an Einstein-Kähler manifold. However, rela tion (7.7) is not sufficient to ensure the vanishing of the potenti

Note that in the positive energy domain defined by

$$\mathcal{D} : G_{zz^*} < 0 \qquad (7.8)$$

supersymmetry is spontaneously broken and the gravitino mass is different from zero:

$$m^2_{3/2} = e^{-G} = \frac{1}{|\phi(z) + \phi^*(z^*)|^3} , \quad \text{for all } z \in \mathcal{D} \qquad (7.9)$$

From the explicit expression for G_{zz^*}, if follows that

$$\mathcal{D} : |\phi_z| \neq 0, \quad (\phi + \phi^*)^2 < \infty . \qquad (7.10$$

So the 'gravitino mass' is zero only at the boundary points $(\phi + \phi^*) = \infty$. The fact that the scalar potential is degenerate ($V \equiv 0$) reflects an arbitrariness for the values of the gravitino mass. This ambiguity can easily be removed if we demand the flatness condition ($V \equiv 0$) only locally, around a point $z_0 \in \mathcal{D}$.

$$V_{z \to z_0} \simeq C|z - z_0|^{2n} , \; n > 2 . \qquad (7.11$$

In fact we can build a large class of positive definite potentials demanding the relation

$$\partial_z \partial_{z^*} e^{(1/3)G} = \Phi_{zz^*}(z,z^*) , \qquad (7.12$$

where the real function $\Phi(z,z^*)$ satisfies the conditions

$$\Phi_{zz^*}(z,z^*) > 0, \quad \text{for all } z \in \mathcal{D} , \qquad (7.13$$

with $\mathcal{D}$ the positive kinetic energy domain ($G_{zz^+} < 0$), and

$$\Phi_{zz^*}(z_0, z_0^*) = 0 \,.$$

The general solution of Eq. (7.12) is then

$$G = \frac{3}{2} \log (\phi + \phi^* + \Phi)^2 \,, \tag{7.15}$$

with

$$G_{zz^*} = 3 \, \frac{\Phi_{zz^*}(\phi + \phi^* + \Phi) - |\phi_z + \Phi_z|^2}{(\phi + \phi^* + \Phi)^2} < 0 \,. \tag{7.16}$$

Equation (7.16) defines the positive kinetic energy domain $\mathcal{D}$.

The corresponding scalar potential is positive definite in $\mathcal{D}$, provided that $\Phi_{zz^*} > 0$ and $\phi + \phi^+ + \Phi > 0$, as can easily be seen from its analytic expression

$$V_0 = 3 \, \frac{\Phi_{zz^*}(\phi + \phi^* + \Phi)}{|\phi + \phi^* + \Phi|^3 \left[|\phi_z + \Phi_z|^2 - \Phi_{zz^*}(\phi + \phi^* + \Phi) \right]} \,. \tag{7.17}$$

At the minimum ($z = z_0$) the potential vanishes identically and the gravitino mass is well defined. If the potential is locally flat at z_0, the fourth derivative of Φ at $z = z_0$ must also vanish:

$$\Phi_{zzz^*z^*}(z_0, z_0^*) = 0 \,. \tag{7.18}$$

In that case, the relation (7.7) between the curvature R_{zz^*}, and the Kähler metric G_{zz^*} is still valid locally ($z \sim z_0$), and the curvature contribution to the $(\text{mass})^2$ sum rule[16] is minus four times the gravitino mass squared:

$$\Delta m^2 = -2 \, e^{-G} \, \frac{G_z G_{z^*}}{(G_{zz^*})^2} \, R_{zz^*} \Big|_{z=z_0} = -4 m_{3/2}^2 \,. \tag{7.19}$$

We may understand how the $(\text{mass})^2$ sum rule can be satisfied with a non-vanishing gravitino mass (signal of supersymmetry breaking) and vanishing scalar masses, as is obvious from the local flatness of the potential at the minimum.

Let us now examine the interesting physical case where the usual matter fields are also coupled to supergravity. In general, the positivity properties of the potential are destroyed when the matter fields are coupled in an arbitrary way. The solution to thi problem is to assume that G is the sum of two uncoupled functions $G(z,z^*)$, as in Eq. (7.15), and $\bar{G}(y^i,y^*_i)$, so that the potential becomes

$$V = -e^{-G_T}\left[\frac{G_z G_{z^*}}{G_{zz^*}} + 3\right] - e^{-G_T}\,\bar{G}_i(\bar{G}^{-1})^i_j\bar{G}^j + \frac{1}{2}D^\alpha D^\alpha , \qquad (7.20)$$

with

$$G_T = G(z,z^*) + \bar{G}(y_i,y^{*i}) , \qquad (7.21)$$

and, therefore, remains positive definite provided that $-\bar{G}^i_j$ is a Hermitian positive matrix, the latter being a necessary condition for ensuring a meaningful kinetic term.

The vanishing minimum arises for $z = z_0$ as before, and

$$G_i(\bar{G}^{-1})^i_j\bar{G}^j = 0 , \qquad (7.22)$$

$$D^\alpha = 0 .$$

The choice of minimal kinetic terms for the matter sector (y^i) lead to the following positive definite potential:

$$V(z,y_i) = e^{-\bar{G}}V_0(z,z^*) + e^{-G_T}\left(\sum_i \left|h_i + y_i\right|^2\right) + \frac{1}{2}(D^\alpha)^2 , \qquad (7.23)$$

where $V_0(z,z^*)$ is given by Eq. (7.17). The total G_T function reads

$$G_T = \frac{3}{2}\log(\phi + \phi^* + \Phi)^2 - y_i y^{i*} - h(y_i) - h^*(y^{*i}) . \qquad (7.24)$$

The absolute minimum of the potential (7.23) is zero and occurs at $z = z_0$ and $y_i = -h^*_i$ (if there are solutions). Note that the condition $y_i = -h^*_i$ automatically implies the vanishing of D terms becaus of the invariance of h under the internal group[6]. The vanishing of the cosmological constant is thus automatically satisfied without any unnatural fine tuning. The gravitino mass, in that case, is given by

$$m^2_{3/2} = e^{-G_T(z,z^*,y_i,y^{i*})}\Big|_{\text{at the minimum}} \tag{7.25}$$

In the flat limit, the hidden sector decouples and the scalar particle z remains massless for locally flat potentials. The potential for the matter fields becomes

$$V(y_i,y^{*i}) = \sum_i \left| \frac{\partial g(y_i)}{\partial y_i} + m_{3/2}\, y^{*i} \right|^2 + \frac{1}{2} D^\alpha D^\alpha \,, \tag{7.26}$$

where we have rescaled $h(y^i) = (1/M^2 m_{3/2})g(y^i)$ in order to have a residual effect. The D term in that case is given by

$$D^\alpha = g^\alpha y^{*i} T^{\alpha j}_i y_j \,. \tag{7.27}$$

The form of the potential presented in Eq. (7.26) no longer depends on the specific form of the hidden sector, and gives rise to soft global supersymmetry-breaking terms of the same form as in the A = 3 case or the case of a factorized form of the superpotential[27], and therefore the supertrace formula for the matter fields remains the same as in the minimal case[6,27]

$$\sum_B m_B^2 - 2\sum_F m_F^2 = 2N_y m^2_{3/2} - 2r_i m^2_{\lambda_i} \,, \tag{7.28}$$

where N_y is the number of chiral multiplets in the y sector, r_i is the dimension of the internal gauge group, and m_{λ_i} are the gaugino masses coming from a non-minimal choice of f_{AB}[6,27]. In the gravity-hidden sector, the bosonic degrees of freedom are massless, although their fermionic partners obtain masses equal to the gravitino mass ($m_{3/2}$). To appreciate this fact better, let us generalize our results to the case of N chiral superfields (z_a,χ_a) in the hidden sector. In that case, the corresponding scalar potential of Eq. (7.2) becomes

$$V = -e^{-G}\left[G_a (G^{-1})^a_b G^b + 3\right] \tag{7.29}$$

and may be rewritten as

$$V = -\frac{g}{N^2}\, e^{-[(N+3)/N]G}\,((G^{-1})^a_b \partial_a \partial^b e^{(N/3)G} \,. \tag{7.30}$$

The flatness of the potential in this case implies a particular G, so that

$$(G^{-1})^a_b \partial_a \partial^b \; e^{(N/3)G} = 0 \,. \tag{7.31}$$

For simplicity, we examine only the case of chiral superfields which are singlets under the internal group. An obvious particular solution of Eq. (7.31) is

$$G = \sum_{a=1}^{N} \frac{3}{2N} \log\left[\phi_a(z_a) + \phi_a^*(z_a^*)\right] , \tag{7.32}$$

where $\phi_a(z^a)$ is a function of the z^a field only. The potential is identically zero, as in the one-superfield case. Here, also, we can realize the locally flat potential requirement ($z_a \sim z_a^0$) by modifying G to

$$G = \sum_{a=1}^{N} \frac{3}{2N} \log\left[\phi_a(z_a) + \phi_a^*(z_a^*) + \Phi_a(z_a, z_a^*)\right] , \tag{7.33}$$

with

$$\partial_a \partial^a \Phi_a > 0 \qquad \text{for all } z_a \in \mathcal{D} ,$$
$$\partial_a \partial^a \partial_a \partial^a \Phi_a = 0 \quad \text{at } z_a^0 \,. \tag{7.34}$$

All scalar fields are then massless, contrary to their fermionic partners and the gravitino whose masses satisfy the following (mass) formula[16]:

$$-2 \sum_{F=1}^{N-1} m_F^2 - 4m_{3/2}^2 = -2\, e^{-G}\, G_a (G^{-1})^a_b G^c (G^{-1})^d_c R^b_d + 2(N-1)m_{3/2}^2 \,, \tag{7.35}$$

where the curvature tensor R^b_a is given by[16]

$$R^b_a = \partial^b \partial_a \log \operatorname{Det} G^m_n \,. \tag{7.36}$$

The particular structure of the function G in Eq. (7.33) implies the following non-trivial R^b_a:

$$R^b_a = -\frac{2N}{3} G^b_a . \tag{7.37}$$

Here also the Kähler space is an Einstein manifold.

Using Eq. (7.37) we find

$$e^{-G} G_a (G^{-1})^a_b G^c (G^{-1})^d_c R^b_d = 2N m^2_{3/2} \tag{7.38}$$

and, using Eqs. (7.5) and (7.7), we finally obtain the following mass formula:

$$2 \sum_{F=1}^{N-1} m_F^2 = 2(N-1) m^2_{3/2} , \tag{7.39}$$

which means that every fermionic degree of freedom acquires a mass equal to the gravitino mass. Their bosonic partners are massless. Note the reverse role of bosonic and fermionic degrees of freedom. In fact, contrary to the minimal kinetic term case case[6], the extra mass contribution originating in the supersymmetry breaking is distributed among the fermionic degrees of freedom, whilst their bosonic partners remain massless.

Just as in the case of one chiral superfield in the hidden sector, the matter fields must be coupled in such a way that the properties of the hidden sector are not destroyed. We must choose the total G function as the sum of the hidden sector $G(z^a, z^*_a)$ and the matter one $G(y^i, y^*_i)$. The resulting potential for the y sector is the same as before.

In the second part of this section we consider a super-Higgs sector implemented by local R-symmetry[15]. The introduction of local R-symmetry in supergravity has been extensively explained in Section 6, and it allows us to generalize the Fayet-Iliopoulos mechanism[13] for spontaneous supersymmetry breaking to local supersymmetry.

We consider a class of 'minimal' models in which the super-Higgs sector is due to spin $(0^{\pm}, \frac{1}{2})$ chiral multiplet carrying non-trivial R-charge and a spin $(1, \frac{1}{2})$ vector multiplet gauging R-symmetry[32]. This is the minimal multiplet content for a super-Higgs sector with local R-symmetry if we demand the vanishing of the cosmological constant with positive definite potential. We exhibit a simple model in which the supersymmetry breaking receives equal contributions from the D-term due to the vector multiplet and the f-term due to the chiral multiplet.

The final mass spectrum consists of a real scalar, a chiral spinor, and a massive vector, all with the same mass $m = 2m_{3/2}$. In an R-symmetric theory with a single chiral multiplet coupled to N = 1 supergravity, the most general form of the scalar potential is[11]

$$V(z,z^*) = -e^{-G}(G_z G_{z^*} G^{-1}_{zz^*} + 3) + \frac{1}{2} g^2 D^2 , \tag{7.40}$$

where $G = G(z,z^*)$ and $D = -G_z z$; g is the R-gauge coupling constant. Using the identity

$$G_z = z^* G'(\rho) , \quad G_{z^*} = zG'(\rho) , \quad G_{zz^*} = G'(\rho) + \rho G''(\rho) =$$
$$= -D'(\rho) = -\frac{1}{\rho} D'(\omega) , \tag{7.41}$$

$$\rho = zz^* , \quad \omega = \log \rho ,$$

we can rewrite (7.40) as follows:

$$V(\omega) = e^{-G} \left(\frac{D^2(\omega)}{D'(\omega)}\right) - 3 + \frac{1}{2} g^2 D^2(\omega) , \tag{7.42}$$

with $D' = f(D)$, $dD/f(D) = d\omega$.

The G function is then given by the following integral:

$$G(\omega) = - \int d\omega D(\omega) = - \int dD \frac{D}{F(D)} . \tag{7.43}$$

f we now define

$$e^{-G} = \phi(D) \ , \qquad (7.44)$$

hen

$$-G' \ e^{-G} = \phi'(D)D' = \phi(D)D \ ; \qquad \frac{1}{D'} = \frac{\phi'(D)}{\phi(D)D} \ , \qquad (7.45)$$

o we finally get

$$\omega = \int \frac{dD}{D} \frac{\phi'(D)}{\phi(D)}$$

nd the potential can be rewritten in the D-variable as follows[32]:

$$V = \phi(D)\left(\frac{D^2\phi'(D)}{D\phi(D)} - 3\right) + \frac{1}{2} g^2 D^2$$

$$= -3\phi(D) + D\phi'(D) + \frac{1}{2} g^2 D^2 \ . \qquad (7.47)$$

quation (7.47) defines the most general R-symmetric potential in erms of the function $\phi(D)$, which is related to G and D through qs. (7.44) and (7.45).

Positive definite potentials correspond to different choices of $\phi(D)$, so that Eq. (7.47) is semipositive definite.

The simplest case is to take $\phi(D)$ linear in D[32]:

$$\phi(D) = \alpha D - \beta/3 \ . \qquad (7.48)$$

Then Eq. (7.47) becomes

$$V = \frac{1}{2} g^2 \left(D^2 - \frac{4\alpha}{g^2} D + \frac{2\beta}{g^2}\right) \ . \qquad (7.49)$$

Positivity means

$$V = \frac{1}{2} g^2 (D - \xi)^2 \ , \qquad (7.50)$$

which demands

$$\frac{2\alpha}{g^2} = \xi \ , \quad \frac{2\beta}{g^2} = \xi^2 \ ; \quad \phi(D) = \frac{1}{2} g^2 \xi D - \frac{1}{6} g^2 \xi^2 \ . \qquad (7.51)$$

At the minimum

$$D = \xi \ , \quad \phi(\xi) = \frac{1}{3} g^2\xi^2 \ , \quad m_{3/2}^2 = \frac{1}{3} g^2\xi^2 \ . \tag{7.52}$$

Using Eq. (7.46) we can rewrite the potential in terms of ρ:

$$D = \frac{\xi}{3} \frac{1}{1 - (\rho \ \rho_0)^{\xi/3}} \ . \tag{7.53}$$

So we finally get

$$V(\rho) = \frac{2}{3} m_{3/2}^2 \frac{\left[\left[1 - (\rho/\rho_{min})^{\xi/3}\right]^2\right]}{\left[1 - \frac{2}{3} (\rho/\rho_{min})^{\xi/3}\right]^2} \ , \quad \rho_{min} = \frac{2m_{3/2}^3}{g^2} \ , \tag{7.54}$$

where the following conditions have been used to compute ρ_{min} and ρ

$$D(\rho_{min}) = \xi \ , \quad D'(\rho_{min}) = 1 \ . \tag{7.55}$$

From Eqs. (7.55) we get

$$\rho_{min} = \frac{2}{3} \xi^2 \ , \quad (\rho_{min}/\rho_0)^{\xi/3} = \frac{2}{3} \ . \tag{7.56}$$

Note that the potential (7.54) becomes singular for $\rho/\rho_{min} = (3/2)^3$ so the variable ρ is constrained in the region $0 < \rho < \rho_{min}(3/2)^{3/\xi}$. From Eq. (7.47) we also get, at the minimum,

$$-e^{-G} G_z G_{z^*} G_{z^*} G^{-1}_{zz^*} = \frac{1}{2} g^2 D^2 = \frac{1}{2} g^2\xi^2 = \frac{2}{3} m_{3/2}^2 \ ,$$

which means that the gravitino mass gets an equal contribution from the 'D'-breaking and the 'f'-breaking terms. As we will see below, this means that the would-be goldstino is an equal mixture of the χ and λ spinors of the chiral and vector multiplets. It is now straight-forward to compute the three-level particle masses of the theory.

The scalar square-mass matrix is

$$M_B^2 = g^2\rho_{min}\begin{pmatrix} 1 & 1 \\ 1 & 1 \end{pmatrix} , \quad \text{i.e. } M_1^2 = 0, \quad M_2^2 = 2g^2\rho_{min} = 4m_{3/2}^2 \ . \tag{7.57}$$

The massless mode is the would-be Goldstone boson of spontaneously broken R-symmetry; the physical massive mode has mass $M_2 = 2m_{3/2}$.

The vector boson mass is

$$m_V^2 = 2g^2\rho_{min} = 4m_{3/2}^2 . \tag{7.58}$$

For the fermion mass matrix we notice that the $\chi\chi$ term of the N = 1 supergravity Lagrangian[6],

$$+\bar{\chi}_L\chi_L\left(G_{zz} - G_zG_z - G_zG_{zzz^*}G^{-1}_{zz^*}\right) , \tag{7.59}$$

is absent owing to the vanishing of the expression in brackets in our model.

The remaining term of the spin-½ mass matrix is

$$\lambda_L\chi_L(2igz^*) + \text{h.c.} , \tag{7.60}$$

The goldstino mode is defined by the term coupled to the spin-½ gauge field in the supergravity Lagrangian[6],

$$\bar{\psi}_R\cdot\gamma\eta_L + \text{h.c.}, \tag{7.61}$$

$$\eta_L = m_{3/2}G_z\chi_L - \frac{i}{2}\, gzG_z\lambda_L = i\sqrt{\frac{3}{2}}\, m_{3/2}\left(\frac{i\sqrt{2}\chi_L + \lambda_L}{\sqrt{2}}\right) . \tag{7.62}$$

Therefore, Eq. (7.60) can be rewritten as

$$\bar{\lambda}_L(i\sqrt{2}\chi_L)\sqrt{2}\, gz^* + \text{h.c.} , \qquad \sqrt{2}\, gz^*_{min} = 2m_{3/2} . \tag{7.63}$$

Equation (7.63) shows that the orthogonal combination to the would-be Goldstone fermion

$$\psi_L = \frac{i\sqrt{2}\cdot\chi_L - \lambda_L}{\sqrt{2}} \tag{7.64}$$

has mass $m_\psi = 2m_{3/2}$. Therefore, the physical spectrum consists of a massless spin-2 graviton, a massive spin-3/2 gravitino, and a scalar, a spinor, and a vector, all with the same mass:

$$m_S = m_V = m_\psi = 2m_{3/2} . \tag{7.65}$$

The square-mass sum rule gives in this case

$$StM^2 = \sum_{J=0}^{3/2} (-)^{2J} m_J^2 = 4m_{3/2}^2 \; . \qquad (7.66)$$

To understand this result, we have to apply the general mass formul in N = 1 supergravity coupled to matter in the presence of non-vanishing Kähler curvature and the non-vanishing D-term[16]:

$$StM^2 = 2(g^2 D + g^2 z\Gamma_z D) = 2R_{zz^*} \frac{G_z G_{z^*}}{(G_{zz^*})^2} e^{-G} \; , \qquad (7.67)$$

where $z\Gamma_z$ is the Kähler connection

$$z\Gamma_z = z\partial_z \log G_{zz^*} \; . \qquad (7.68)$$

In our case,

$$z\Gamma_z \Big|_{min} = z\partial_z \log D'(\rho)\Big|_{min} = \left(-1 + \frac{5}{3}\xi\right) , \qquad (7.69)$$

$$R_{zz^*} = -2G_{zz^*} \; , \qquad (7.70)$$

which at the minimum gives

$$R_{zz^*}\Big|_{min} = 2 \; . \qquad (7.70)$$

Therefore we get

$$StM^2 = 2\left[g^2\xi + g^2\xi\left(-1 + \frac{5}{3}\xi\right)\right] + 4m_{3/2}^2 \frac{G_z G_{z^*}}{G_{zz^*}}$$

$$= \frac{10}{3} g^2\xi^2 - 4m_{3/2}^2 \frac{\xi^2}{\rho} = 10m_{3/2}^2 - 4 \times \frac{3}{2} m_{3/2}^2 = 4m_{3/2}^2 \; , \qquad (7.71)$$

which agrees with the left-hand side given by Eq. (7.66). Inciden-tally, we note that Eq. (7.71) shows that the Kähler manifold is an Einstein space.

We now consider the potential given by Eq. (7.54) in the low-energy limit $m_P \to \infty$, with $m_{3/2}$ fixed[32]. By means of Eq. (7.52) we can express the dimensionless variable ξ as

$$\xi = \sqrt{24\pi}\,\frac{m_{3/2}}{g m_P} \tag{7.72}$$

so that two different limits are possible.

If $g \to 0$ with $m_P \to \infty$, then it is easy to see that

$$\underset{m_P\to\infty,\,g\to 0}{V} = 2m_{3/2}^2\,\phi^2 + O\left(\frac{1}{m_P}\right) , \tag{7.73}$$

where ϕ is the physical scalar degree of freedom. The vector, spinor, and scalar degrees of freedom decouple in this limit, and we get a free theory of massive particles of spin 0, ½, and 1, respectively. This situation is entirely analogous to the normal super-Higgs effect without local chiral symmetry when the scalar field of the hidden sector just decouples in the limit $m_P \to \infty$, $m_{3/2}$ fixed[21].

However, in the presence of a gauged R-symmetry a second non-trivial limit exists for the supergravity Lagrangian with $M_P \to \infty$, $m_{3/2}$ fixed, and g fixed[32]. In this case the gauge-R interaction is non-gravitational in the sense that g does not vanish with $M_P \to \infty$.

From a physical point of view, this second limit is less interesting: it corresponds to non-renormalizable interactions at low energy, since non-renormalizable terms in the effective Lagrangian appear, which are scaled by inverse powers of $m_{3/2}$.

REFERENCES

1. For a review see, for example, van Nieuwenhuizen, P. 1981, Phys Rep. 68 191.
2. For recent reviews see, for example, Barbieri, R. and Ferrara, S. 1983, Surveys in High-Energy Physics 4 33.
   Zumino, B. 1982, Proc. Solvay Conf., Austin, Texas [Phys. Rep. 104 113 (1984)].
   Nilles, H.-P. 1983, Proc. Conf. on Problems of Unification and Supergravity, La Jolla, Calif. (AIP Conference Proceedings No. 116, 1983), p. 109.
   Polchinski, J. 1983, Harvard preprint HUTP-83/A036.
   Ferrara, S. 1982, Proc. 4th Silarg Symp. on Gravitation, Caracas, ed. C. Aragone (World Scientific, Singapore, 1982), p. 11.
   Nanopoulos, D.V. 1983, CERN preprint TH.3699, to appear in Proc Europhysics Study Conference on Electroweak Effects at High Energies, Erice, 1983.
   Ellis, J. 1983, preprint CERN TH.3718, to appear in Proc. Int. Symp. on Lepton and Photon Interactions at High Energies, Cornell, 1983.
   Barbieri, R. 1983, Unconventional weak interactions, Univ. Pisa preprint, to appear in Proc. Int. Symp. on Lepton and Photon Interactions at High Energies, Cornell, 1983.
3. Barbieri, R., Ferrara, S. and Nanopoulos, D.V. 1982, Z. Phys. C13 267; Phys. Lett. 116B 6.
   Ellis, J., Ibanez, L. and Ross, G. 1982, Phys. Lett. 113B 283.
   Dimopoulos, S. and Raby, S. 1983, Nucl. Phys. B219 479.
   Polchinski, J. and Susskind, L. 1982, Phys. Rev. D26 3661.
4. Deser, S. and Zumino, B. 1977, Phys. Rev. Lett. 38 1433.
5. Cremmer, E., Julia, B., Scherk, J., Ferrara, S., Girardello, L. and van Nieuwenhuizen, P. 1978, Phys. Lett. 79B 231; 1979, Nucl. Phys. B147 105.
6. Cremmer, E., Ferrara, S., Girardello, L. and van Proeyen, A. 1982, Phys. Lett. 116B 231 and 1983, Nucl. Phys. B212 413.
7. Gildener, E. 1976, Phys. Rev. D14 1667.
   Gildener, E. and Weinberg, S. 1976, Phys. Rev. D15 3333.
   Maiani, L. 1979, Proc. Summer School on Weak Interactions, Gif-sur-Yvette (IN2P3, Paris, 1980), p. 3.
   Veltman, M. 1981, Acta Phys. Pol. B12 437.
   Witten, E. 1981, Nucl. Phys. B188 513.
   Dimopoulos, S. and Raby, S. 1981, Nucl. Phys. B199 353.
8. Ferrara, S. and van Nieuwenhuizen, P. 1978, Phys. Lett. 74B 333; 1978, Phys. Lett. 76B 404.
   Stelle, K.S. and West, P.C. 1978, Phys. Lett. 74B 330; 1978, 77B 376.
9. Kaku, M., Townsend, P.K. and van Nieuwenhuizen, P. 1978, Phys. Rev. D17 3179.

de Wit, B. in Supergravity 1982, eds. Ferrara, S., Taylor, J.G. and van Nieuwenhuizen, P. (World Scientific, Singapore, 1982), p. 85.
Kugo, T. and Uehara, S. 1983, Nucl. Phys. B222 125.
10. Sohnius, M.F. and West, P.C. 1981, Phys. Lett. 105B 353.
11. Ferrara, S., Girardello, L., Kugo, T. and van Proeyen, A. 1983, Nucl. Phys. B223 191.
12. See, for example, Wess, J. and Bagger, J. 1983, Supersymmetry and supergravity (Princeton Univ. Press, Princeton, NJ).
13. Fayet P. and Iliopoulos, J. 1974, Phys. Lett. 51B 461.
14. Stelle, K.S. and West, P.C. 1978, Nucl. Phys. B145 175.
15. Barbieri, R., Ferrara, S., Nanopoulos, D.V. and Stelle, K.S. 1982, Phys. Lett. 113B 219.
16. Grisaru, M.T., Rocek, M. and Karlhede, A. 1982, Phys. Lett. 120B 189.
17. Kugo, T. and Uehara, S. 1982, Kyoto preprint KUNS 646, to be published in Nuclear Physics B.
18. Bagger, J. 1983, Nucl. Phys. B211 302.
19. Freedman, D.Z. 1977, Phys. Rev. D15 1173.
de Wit, B. and van Nieuwenhuizen, P. 1979, Nucl. Phys. B139 531.
20. Ferrara, S., Girardello, L. and Palumbo, F. 1979, Phys. Rev. D20 403.
21. Barbieri, R., Ferrara, S. and Savoy, C.A. 1982, Phys. Lett. 119B 343.
22. Girardello L. and Grisaru M.T. 1982, Nucl. Phys. B194 65.
23. Soni, S.K. and Weldon, H.A. 1983, Phys. Lett. 126B 215.
Barbieri, R. and Ferrara, S., see Ref. 2.
24. Samuel S. and Wess, J. 1982, Columbia Univ. preprint CU-TP-258 and 1983, CU-TP-260.
25. Nilles, H.-P., Srednicki, M. and Wyler, D. 1983, Phys. Lett. 120B 346.
26. Hall, L., Likken, J. and Weinberg, S. 1983, Phys. Rev. D27 2359.
27. Cremmer, E., Fayet, P. and Girardello, L. 1983, Phys. Lett. 122B 41.
28. Polony, J., Budapest report KFKI-(1977) (unpublished).
29. Cremmer, E., Ferrara, S., Kounnas, C. and Nanopoulos, D.V. 1983, Phys. Lett. B113 61.
Ferrara, S. and van Proeyen, A. 1984, Phys. Lett. 138B 77.
30. Cremmer, E., Scherk, J. and Ferrara, S. 1978, Phys. Lett. 74B 61.
31. Ngee-Poug Chang, Ouvry, S. and Xizeng Wu, 1983, Phys. Rev. Lett. 51 327.
32. Cremmer, E., Ferrara, S., Girardello, L., Kounnas, C. and Masiero, A. 1984, Phys. Lett. 137B 62.

SUPERSPACE OF $N=1$ SUPERGRAVITY

W SIEGEL

1. GAUGE GROUPS: SUPERTRANSLATIONAL AND SUPERCONFORMAL

As for global supersymmetry, the use of superspace in formulating supergravity has not only great technical advantages (at both the classical and quantum levels) but also elucidates the conceptual differences from ordinary gravity. One fundamental difference is that conformal transformations, which play a somewhat trivial role in ordinary gravity, are essential in the understanding of supergravity.

1.1 Translations in gravity

In ordinary gravity, an analogy can be drawn between Yang-Mill groups and the group of general coordinate transformations by identifying partial derivatives (or any basis for translations as the (antihermitian) group generators. (This analogy also has not only the conceptual advantages of relating gravity to Yang-Mills theory, but in addition simplifies the computational algebra.) We therefore write an arbitrary group element as $e^{i\lambda}$, where $\lambda=\lambda^m(x)i\partial_m$ is an element of the corresponding Lie algebra (Gates, Grisaru, Roček, and Siegel, 1983 e). The algebraic properties are given by $[\lambda_1,\lambda_2] = i\lambda_{12}$, where $\lambda_{12}{}^n = \lambda_{[1}{}^m\partial_m\lambda_{2]}{}^n$. (See Appendix A for notational conventions. The simplest representation is the scalar field, which transforms as $\psi'(x) = e^{i\lambda}\psi(x)$. This transformation can be related directly to a general coordinate transformation by noting that its effect on ψ can be expressed totally by its action on ψ's argument: $\psi'(x') = \psi(x)$, where $x' = e^{-i\lambda}x = x + \lambda(x)+\cdots$. The group is not unimodular: Its Jacobian is given by $1\cdot e^{i\overleftarrow{\lambda}}$, where the $\overleftarrow{\partial}_m$ in $\overleftarrow{\lambda} = \lambda^m i\overleftarrow{\partial}_m$ acts on all objects to its left till it hits the 1. (Consider, e.g., a Taylor expansion of the exponential.) Invariants take the form $\int d^4x\, \mathcal{L}$, where $\mathcal{L}$ is a density, transforming as $\mathcal{L}' = \mathcal{L}e^{i\overleftarrow{\lambda}} = (1\cdot e^{i\overleftarrow{\lambda}})(e^{i\lambda}\mathcal{L}) = \mathcal{L}+$ (total derivative). (Thus, a scalar times a density is also a density.)

1.2 Supertranslations as superconformal

In supergravity, we again have $e^{i\Lambda}$ as a group element, but with $\Lambda = \Lambda^M(z)i\partial_M$ a function of all superspace coordinates $z^M = (x^m, \theta^\mu, \bar\theta^{\dot\mu})$ and expanded over all partial derivatives $\partial_M = \partial/\partial z^M$ (Siegel, 1977 a-d, 1978; Siegel and Gates, 1979; Gates *et al.*,1983 g). Irreducible representations of global supersymmetry can be represented by chiral superfields (Siegel and Gates, 1981; Gates *et al.*, 1983 b). Furthermore, the most general form of the scalar multiplet is represented by a chiral scalar superfield, which is also the gauge parameter for supersymmetric Yang-Mills theory. Thus, in order to find the supergravity group, it's sufficient to consider its action on the simplest representation, the chiral scalar. In the "chiral representation" a chiral superfield can be defined simply as a (complex) superfield $\phi(x,\theta)$ which is independent of $\bar\theta$. In order for the group to preserve this property, $\phi' = e^{i\Lambda}\phi$ must also be chiral, and thus

$$\Lambda=\Lambda^m(x,\theta)i\partial_m + \Lambda^\mu(x,\theta)i\partial_\mu + \Lambda^{\dot\mu}(x,\theta,\bar\theta)i\bar\partial_{\dot\mu}, \qquad (1.1)$$

where Λ is complex, Λ^m and Λ^μ are chiral, and $\Lambda^{\dot\mu}$ is general. Since $\bar\partial_{\dot\mu}\phi = 0$, $\Lambda^{\dot\mu}$ doesn't act on it, but this won't be true for more general representations. (The role of $\Lambda^{\dot\mu}$ will be clarified when we consider its action on the supergravity superfields themselves.) It's easy to show that the group with elements $e^{i\Lambda}$ closes when these chirality conditions are imposed (i.e., the chirality conditions define a subgroup of the original group).

In order to understand the relation of this group to the group of general coordinate transformations, we consider the action of some of its components on the components of the chiral scalar. In the similar case of the vector multiplet, which also has a chiral gauge parameter Λ (also acting as $\phi'=e^{i\Lambda}\phi$), it's only the hermitian part of the gauge parameter which survives choosing a Wess-Zumino gauge for the gauge superfield: $\Lambda(x,\theta)=\bar\Lambda(x,\bar\theta)=\lambda(x)$, the usual nonsupersymmetric gauge parameter. Making a similar analysis here, we write

$\Lambda = \lambda^m(x) i\partial_m + \Lambda^\mu(x,\theta) i\partial_\mu + \bar{\Lambda}^{\dot\mu}(x,\bar\theta) i\bar\partial_{\dot\mu}$, where now $\bar\Lambda^{\dot\mu}$ is the hermitian conjugate of Λ^μ. $\lambda^m(x)$ is the usual nonsupersymmetric general coordinate transformation parameter (it acts through $e^{i\Lambda}$ on any function of x just as before), so let's now consider just the action of the rest of the group, contained in $\Lambda^\mu(x,\theta)$. We expand in components as

$$\Lambda^\mu| = \varepsilon^\mu, \quad \partial_\mu \Lambda^\nu| = \omega_\mu{}^\nu - \tfrac{1}{2}\delta_\mu{}^\nu \sigma, \quad \partial^2 \Lambda^\mu| = \eta^\mu; \tag{1.2a}$$

$$\phi| = A, \quad \partial_\mu \phi| = \psi_\mu, \quad \partial^2 \phi| = F; \tag{1.2b}$$

where $\omega_\mu{}^\mu = 0$. ("|" means to evaluate at $\theta=0$.) The action of infinitesimal Λ^μ, $\delta\phi = -\Lambda^\mu \partial_\mu \phi$, is then given by

$$\delta A = -\varepsilon^\mu \psi_\mu, \quad \delta\psi_\mu = \varepsilon_\mu F - \omega_\mu{}^\nu \psi_\nu + \tfrac{1}{2}\sigma\psi_\mu, \quad \delta F = \sigma F - \eta^\mu \psi_\mu . \tag{1.3}$$

If we ignore the absence of spacetime-derivative terms (they contain the vierbein, which will be introduced later), the ε^μ transformations can be recognized as the local form of supersymmetry transformations. Furthermore, the σ transformations are simply local scale transformations (*Re* σ) plus local chiral (U(1), R-symmetry) transformations (*Im* σ), and $\omega_\mu{}^\nu$ represents local Lorentz transformations (it leaves ψ^2 invariant). η^μ can be identified as the local generalization of S-supersymmetry transformations (the "spin" part, as opposed to the x-dependent "orbital" part), which are part of conformal supersymmetry. Together with λ^m, these parameters describe local conformal supersymmetry (λ^m and *Re* σ alone describe the nonsupersymmetric local conformal group).

1.3 <u>Poincaré subgroups</u>

The group can be restricted to various subgroups, each describing local Poincaré supergravity, by the constraint

$$(3n+1)\bar\partial_{\dot\mu}\Lambda^{\dot\mu} = (n+1)(\partial_m \Lambda^m - \partial_\mu \Lambda^\mu), \tag{1.4}$$

for an arbitrary (finite, complex) parameter n. However, since this constraint is differential (when expressed in terms of superfields), it's preferable to avoid it by instead introducing extra fields whose elimination produces (1.4) as a (partial) gauge choice. The analog in ordinary gravity is that λ^m, when expanded in x, is found to include global conformal transformations. The volume-preserving constraint $\partial_m \lambda^m = 0$ (which sets the Jacobian $1 \cdot e^{i\overleftarrow{\lambda}} = 1$) eliminates those conformal transformations which aren't part of the global Poincaré group. Although this subgroup is sufficient to describe gravity, with the unit-determinant part of the metric as the gauge field, using the full group is more convenient for both classical and quantum applications. One can then restrict the group by choosing $\det g_{mn} = -1$ as a gauge condition, or choose other, more convenient gauges. (We'll return to these points in the next section.)

1.4 Vector representations

In order to relate local supersymmetry to global supersymmetry, and for purposes of quantization, it's more useful to express supertranslations in terms of the global supersymmetry derivatives $D_M = (\partial_m, D_\mu, \bar{D}_{\dot{\mu}})$ rather than partial derivatives ∂_M. To preserve the chirality condition $\bar{D}_{\dot{\mu}}\phi = 0$ (instead of $\bar{\partial}_{\dot{\mu}}\phi = 0$), $\Lambda = \Lambda^M iD_M$ must now satisfy

$$\bar{D}_{\dot{\mu}}\Lambda^\nu = 0, \quad \bar{D}_{\dot{\mu}}\Lambda^{\nu\dot{\nu}} = i\,\delta_{\dot{\mu}}{}^{\dot{\nu}}\Lambda^\nu, \tag{1.5}$$

with solution

$$\Lambda^\mu = \bar{D}^2 L^\mu, \qquad \Lambda^{\mu\dot{\mu}} = -\,i\bar{D}^{\dot{\mu}}L^\mu, \tag{1.6a}$$

which can be expressed as

$$\Lambda^m\partial_m + \Lambda^\mu D_\mu = \tfrac{1}{2}\,\{\bar{D}^{\dot{\mu}}, [\bar{D}_{\dot{\mu}}, L^\mu D_\mu]\}. \tag{1.6b}$$

$\Lambda^{\dot{\mu}}$ is again arbitary. The previous Λ^m, Λ^μ are the chiral terms in the expansion of L^μ in $\bar{\theta}$. (The lowest term drops out, since L always appears in (1.6a) with at least one $\bar{D}$.) This "vector-representation" Λ is related to the chiral-representation Λ used above by

$$\Lambda_{\text{vector}} = e^{U/2}\,\Lambda_{\text{chiral}}\,e^{-U/2}, \quad U = \theta^\alpha\bar{\theta}^{\dot{\alpha}} i\partial_{\alpha\dot{\alpha}}. \tag{1.7}$$

(1.4) is modified only by replacing $\partial_M \to D_M$. From now on we use the vector representation unless explicitly stated otherwise.

For purposes of more general geometries (with nontrivial global topologies), or background-field quantization, D_M may be replaced with covariant derivatives $\mathcal{D}_M$ representing a curved superspace (Grisaru and Siegel, 1981, 1982; Gates *et al.*, 1983 j).

2. PREPOTENTIALS: REPRESENTATIONS OF GLOBAL AND LOCAL SUPERSYMMETRY

2.1 On-shell/conformal field strength

Before studying the geometry of the full interacting supergravity theory, we first need to understand the free (linearized) theory. Just as the construction of ordinary free field theories involves the study of representations of the Poincaré group, the construction of free supergravity involves the study of representations of global supersymmetry. Consideration of such representations (Siegel, 1981; Gates *et al.*, 1983 f) shows that the physical polarizations of an N=1 supersymmetry multiplet of spins 2 and 3/2 are contained in a chiral superfield $W_{\alpha\beta\gamma}$, totally symmetric in its spinor indices (analogous to the chiral W_α of super-Yang-Mills) (Ferrara and Zumino, 1978). As for the analogous case of the Weyl tensor $W_{\alpha\beta\gamma\delta}$ in ordinary gravity, this superfield is the complete field strength of conformal supergravity (Kaku, Townsend, and van Nieuwenhuizen, 1978), but the nonconformal theory contains other field strengths which vanish on shell. This is due to the fact that off-shell theories are reducible representations of supersymmetry (or the Poincaré group): e.g., they contain all values of p^2. Even after restricting $p^2 = -m^2$, the representation may still be reducible, since general theories have field equations more complicated than just the Klein-Gordon equation. This is the case for Poincaré supergravity (as well as Poincaré gravity). However, conformal supergravity is a submultiplet of Poincaré supergravity (it has a larger gauge group, so more can be gauged away), and thus its field strengths are a subset of those of Poincaré supergravity. As a result, its only field strength is the one which contains on-shell spins 2 and 3/2. Thus, by first analyzing the representation which gives on-shell Poincaré supergravity, we are led to a description of off-shell conformal supergravity. Poincaré supergravity is then obtained by the introduction of multiplets which vanish on shell ("auxiliary" multiplets).

2.2 Conformal prepotential

The conditions of irreducibility on the field strength $W_{\alpha\beta\gamma}$ are solved by its expression (Ferrara and Zumino, 1978) in terms of a real vector gauge superfield "prepotential" H^a (Ferrara and Zumino, 1975):

$$W_{\alpha\beta\gamma} = -i\,\frac{1}{6}\,\bar{D}^2 D_{(\alpha}\partial_{\beta}{}^{\dot\beta} H_{\gamma)\dot\beta}, \tag{2.1}$$

where H^a has the gauge transformation

$$\delta H_{\alpha\dot\alpha} = D_\alpha \bar{L}_{\dot\alpha} - \bar{D}_{\dot\alpha} L_\alpha . \tag{2.2a}$$

By comparison with (1.6a), if we identify this spinor gauge parameter with the one there, we can rewrite this transformation as

$$\delta H^m = i(\bar{\Lambda}^m - \Lambda^m). \tag{2.2b}$$

In this form the relation to the gauge transformation $\delta V = i(\bar{\Lambda}-\Lambda)$ of super-Yang-Mills is clear, especially if we notice that Λ^m would be chiral in the chiral representation of (1.1). (However, the latter representation doesn't linearize to a globally supersymmetric form, and the nonlinear gauge transformations would be required to include the effects of Λ^μ.)

Again analyzing in terms of components, we choose a Wess-Zumino gauge where the remaining components of $H_{\alpha\dot\alpha}$ are the same as those for the vector multiplet V but with an extra vector index, and find the remaining components of L_α correspond to the hermitian Λ described above: The nonvanishing components in this gauge are (cf. (1.2a))

$$\tfrac{1}{2}(\Lambda^{\alpha\dot\alpha} + \bar{\Lambda}^{\alpha\dot\alpha})| = \lambda^{\alpha\dot\alpha},\quad \Lambda^\alpha| = \epsilon^\alpha,\quad D_\alpha \Lambda^\beta| = \omega_\alpha{}^\beta - \tfrac{1}{2}\delta_\alpha{}^\beta \sigma,\quad D^2\Lambda^\alpha| = \eta^\alpha; \tag{2.3a}$$

$$\tfrac{1}{2}[\bar{D}_{\dot\alpha},D_\alpha]\ H_{\beta\dot\beta}| = h_{\alpha\dot\alpha,\beta\dot\beta},\quad i\bar{D}^2 D_\alpha H_{\beta\dot\alpha}| = \psi_{\alpha\dot\alpha,\beta},$$

$$(-\tfrac{2}{3}\ D^\beta \bar{D}^2 D_\beta H_a + \tfrac{1}{6}\varepsilon_{abcd}\partial^b [\bar{D}^{\dot\gamma},D^\gamma] H^d)| = A_a; \tag{2.3b}$$

with Λ expressed in terms of L by (1.6a).
The remaining gauge invariance is

$$\delta h_{ab} = -\ \omega_{ab} + \eta_{ab}\ \mathit{Re}\,\sigma - \partial_a \lambda_b,$$

$$\delta\psi_{a\beta} = \partial_a \varepsilon_\beta + iC_{\beta\alpha}\ \bar{\eta}_{\dot\alpha},\quad \delta A_a = \frac{2}{3}\ \partial_a\ \mathit{Im}\ \sigma; \tag{2.4}$$

where $\omega_{ab} = C_{\dot\alpha\dot\beta}\ \omega_{\alpha\beta} + C_{\alpha\beta}\bar{\omega}_{\dot\alpha\dot\beta}$. From these transformations we can make the same identifications of the parameters as from (1.3) for (1.2). h_{ab} then has the usual gauge transformation of a (conformal) graviton, and $\psi_{a\beta}$ of a (conformal) gravitino, while A_a is the gauge field for local chiral transformations (Kaku *et al.*, 1978; Ferrara and Zumino, 1978).

2.3 Finite group transformations

In order to make H a gauge field for the full nonlinear group, it must contain all the group generators iD_M, and thus we generalize it to a (hermitian) supervector $H=H^M iD_M$. The linearized transformation (2.2b) then follows from the general transformation

$$e^{H'} = e^{i\bar\Lambda}\ e^H\ e^{-i\Lambda} \tag{2.5}$$

where Λ satisfies (1.6a) and $\bar\Lambda = \bar\Lambda^M iD_M$. We thus have a complete analogy with super-Yang-Mills theory. The purpose of the independent parameter $\Lambda^{\hat\mu}$ can now be explained: Besides giving (2.2b), linearization also gives $\delta H^\mu = i(\bar\Lambda^\mu - \Lambda^\mu) = i(\bar\Lambda^\mu - \bar{D}^2 L^\mu)$, and thus $\Lambda^{\hat\mu}$ can be used to gauge $H^\mu = 0$. In such a gauge, H^m is the only gauge field (for conformal supergravity) and L^μ the only gauge parameter.

2.4 Chiral representation

Alternatively, we could have used $i\partial_M$ as our group generators. The relation between the two choices can be seen easily if we note that part of the construction of covariant derivatives involves objects analogous to the covariant derivatives of super-Yang-Mills (see below): Consider the objects

$$\hat{E}_\alpha = e^{-H} D_\alpha e^{H}, \quad \hat{E}_{\dot\alpha} = \bar{D}_{\dot\alpha}. \qquad (2.6)$$

We now use the fact that the spinor derivatives can be written as

$$D_\alpha = e^{-U/2} \partial_\alpha e^{U/2}, \quad \bar{D}_{\dot\alpha} = e^{U/2} \bar{\partial}_{\dot\alpha} e^{-U/2}, \qquad (2.7)$$

and perform a change of representation (from vector to chiral)

$$(\hat{\underline{E}}_\alpha, \hat{\underline{E}}_{\dot\alpha}) = e^{-U/2} (\hat{E}_\alpha, \hat{E}_{\dot\alpha}) e^{U/2} \qquad (2.8)$$

to write

$$\hat{\underline{E}}_\alpha = e^{-\underline{H}} \partial_\alpha e^{\underline{H}}, \quad \hat{\underline{E}}_{\dot\alpha} = \bar{\partial}_{\dot\alpha}, \qquad (2.9)$$

where

$$e^{\underline{H}} = e^{U/2} e^{H} e^{U/2}, \qquad (2.10)$$

so that $\underline{H} = U + H + \cdots$. If we note that the only nonvanishing component of $U = \theta^\alpha \bar{\theta}^{\dot\alpha} i\partial_{\alpha\dot\alpha}$ corresponds to the vacuum vierbein part of H, we see that $\underline{H}$ contains the *full* vierbein $e_a{}^m = \delta_a{}^m + h_a{}^m$. Conversely, if we start with a formulation using (2.9) (using the basis $i\partial_M$ for $\underline{H} = \underline{H}^M i\partial_M$), we see that the flat-space part of $\underline{H}$ (i.e., U) gives the flat-space covariant derivatives. Thus, flat-space supersymmetry can be *derived* by considering the local symmetries of curved superspace. In the analogous case of super-Yang-Mills, $\underline{V}=U+V+\cdots$ differs from V mainly in that the vector component gauge field A_a is replaced with the vector component covariant derivative $i\nabla_a = i\partial_a + A_a$. (This could have been expected from the fact that $\underline{\Lambda}$ in $e^{\underline{V}'} = e^{i\bar{\underline{\Lambda}}} e^{\underline{V}} e^{-i\underline{\Lambda}}$ contains no derivative terms, since $\bar{\partial}_{\dot\alpha} \underline{\Lambda} = 0$ instead of $\bar{D}_{\dot\alpha} \Lambda = 0$.)

2.5 Compensator in Poincaré gravity

Now that we have the prepotentials for conformal supergravity and their complete gauge transformation laws, we can obtain the same results for Poincaré supergravity by the addition of appropriate fields. In the case of nonsupersymmetric conformal gravity, the analog of H_a is the linearized, symmetric, *traceless* tensor $\hat{h}_{ab}$, with linearized gauge transformation

$$\delta\hat{h}_{ab} = -\tfrac{1}{2}\partial_{(a}\lambda_{b)} + \tfrac{1}{4}\eta_{ab}\partial_c\lambda^c. \tag{2.11}$$

The corresponding *global* symmetry group consists of those $e^{i\lambda}$'s for which $\delta\hat{h}_{ab} = 0$: the conformal group, described by (even for finite transformations)

$$\lambda^a(x) = \zeta^a + x^b(\omega_b{}^a - \delta_b{}^a\sigma) + (x^a x^b\xi_b - \tfrac{1}{2}x^b x_b\xi^a), \tag{2.12}$$

where $\omega_{ab} = -\omega_{ba}$. (Note the simple form of λ, vs. the nonlinear form of $x'(x)$.) The global Poincaré group is a subgroup described by applying a constraint:

$$\partial_a\lambda^a = 0 \rightarrow \lambda^a(x) = \zeta^a + x^b\omega_b{}^a. \tag{2.13}$$

In order to obtain this constraint on general λ's, we introduce a *scalar* gauge field $\hat{h}$ with linearized gauge transformation

$$\delta\hat{h} = -\partial_a\lambda^a. \tag{2.14}$$

We can then choose a gauge $\hat{h} = 0$: The remaining gauge invariance is then restricted by $\partial_a\lambda^a = 0$. In particular, the vacuum of the theory $\hat{h}_{ab} = \hat{h} = 0$ is invariant under $\delta\hat{h}_{ab} = \delta\hat{h} = 0$, and thus has only global Poincaré invariance (2.13). We now note that $\hat{h}_{ab}$ and $\hat{h}$ can be combined conveniently into a symmetric tensor with nonvanishing trace:

$$h_{ab} = \hat{h}_{ab} + \tfrac{1}{4}\eta_{ab}\,\hat{h} \rightarrow \delta h_{ab} = -\tfrac{1}{2}\partial_{(a}\lambda_{b)}\,. \qquad (2.15)$$

The corresponding expression in the nonlinear case is

$$g_{mn} = (-g)^{\frac{1}{4}}\,\hat{g}_{mn}, \quad \det g_{mn} = g, \quad \det \hat{g}_{mn} = -1, \qquad (2.16)$$

and the constraint on λ follows from choosing a gauge by

$$[(-g)^{\frac{1}{2}}]' = [(-g)^{\frac{1}{2}}]\,e^{i\overleftarrow{\lambda}} = (1\cdot e^{i\overleftarrow{\lambda}})\,e^{i\lambda}\,[(-g)^{\frac{1}{2}}];$$

$$g = -1 \rightarrow 1\cdot e^{i\overleftarrow{\lambda}} = 1 \rightarrow 0 = 1\cdot i\overleftarrow{\lambda} = -\partial_m\lambda^m. \qquad (2.17)$$

2.6 Chiral compensator

The analogy for supergravity is : Compare the x-expansion of $\lambda^a(x)$ in (2.12) with the θ-expansion of $\Lambda^\mu(x,\theta)$ in (1.2a). (We can also write $\lambda^a| = \zeta^a$, etc., with "|" meaning "evaluate at x = 0".) Λ^μ can be considered the "square root" of λ^a in that supersymmetry (ε^μ) is the square root of translations (ζ^a), S-supersymmetry (η^μ) of conformal boosts (ξ^a), and Lorentz ($\omega_\mu{}^\nu, \omega_a{}^b$) and scale (real σ) transformations of themselves. (Explicitly, in terms of $[\Lambda_1,\Lambda_2] = i\Lambda_{12}$ for $\Lambda = \Lambda^M iD_M$, $\Lambda_{12}{}^{\dot\mu\mu} = \Lambda_{[1}{}^{\mu}\bar{\Lambda}_{2]}{}^{\dot\mu} + \ldots$ in the gauge $H^{\dot\mu} = 0$.) We then see that the analog of the constraint $\partial_m\lambda^m = 0$ is one of the form $\partial_\mu\Lambda^\mu + \ldots = 0$, where the "..." is chosen to preserve group properties, and is generally given by (1.4). (The analysis is performed also in the gauge $H^{\dot\mu} = 0$, where $\Lambda^{\dot\mu}$ is determined in terms of Λ^μ.)

H^M is the analog of $\hat{g}_{mn}$ (or $\hat{h}_{ab}$). We now introduce the analog of g (or $\hat{h}$), a chiral scalar "compensating" superfield χ (Siegel, 1977 c; Kaku and Townsend, 1978) with linearized gauge transformation

$$\delta\chi = \partial_\mu \Lambda^\mu - \partial_m \Lambda^m, \tag{2.18}$$

or, in the vector representation,

$$\delta\chi = D_\mu \Lambda^\mu - \partial_m \Lambda^m = \bar{D}^2 D_\mu L^\mu. \tag{2.19}$$

In the gauge $\chi=0$, the remaining gauge invariance is restricted by the case $n=-1/3$ of (1.4). Only in this case is the gauge condition chiral (since $\bar{D}_\mu \cdot \Lambda^\mu$ isn't chiral in general), and thus corresponds to a chiral compensator ($\bar{D}_\mu \cdot \chi=0$, or $\bar{\partial}_\mu \cdot \chi=0$ for (2.18)). (For other n, see Appendix B.) χ is the linearization of the compensator ϕ^3 about $<\phi>=1$ (as $\hat{h}$ is for $-g$ about $g=-1$), where

$$\phi'^3 = \phi^3 \exp(i\overleftarrow{\Lambda}_c),$$

$$\Lambda_c = \Lambda^m iD_m + \Lambda^\mu iD_\mu . \tag{2.20}$$

As a result, $\int d^4x d^2\theta \ \ \phi^3$ (the supersymmetrization of $\int d^4x \ \ \sqrt{-g}$) is the invariant cosmological term, so ϕ (or g) can't be gauged to 1 globally.

Although gravity and supergravity are thus closely analagous, there are three major differences: (a) Although g and $\hat{g}_{mn}$ can be combined algebraically to form g_{mn} as in (2.16), ϕ and H_a don't combine so simply. (They combine in a nontrivial way to form covariant derivatives, as explained below.) (b) H_a is the analog of $\hat{g}_{mn}$, but (in appropriate gauges) it contains both $\hat{g}_{mn}$ and g (as h_{ab}: see (2.4); however, gauges can be chosen where *Re* σ gauges g from H_a into ϕ). (c) The analog of g is not unique. (For choices other than ϕ, see Appendix B.)

In order to compare the field content of Poincaré supergravity to that of conformal supergravity, we expand (2.18) in components: Expanding χ as in (1.2b) (and using the "hermitian" Λ described there),

$$\delta A = -\partial_m \lambda^m - \sigma, \quad \delta\psi_\mu = \eta_\mu, \quad \delta F=0. \tag{2.21}$$

Thus, in the partially chosen gauge where H^a has remaining components transforming as in (2.4), we see that the additional fields of Poincaré supergravity (a) eliminate ("compensate for") the extra local invariances (in the gauge $A = \psi_\mu = 0$) of conformal supergravity (scale, chiral, and S-supersymmetry), and (b) introduce an additional (complex) auxiliary field F. In general, g can be moved algebraically between h_{ab} and A (and its supersymmetric partner between $\psi_{a\beta}$ and ψ_α), but F can be moved only by a differential transformation, and thus only locally ($\int d^4x\, F + \cdots$ is invariant), so we have chosen a gauge avoiding such problems.

Note that the gravity sector contains 2 compensators: g and *Re* A. The only difference is that one resides in g_{mn}, while the other is separate. This has the advantage of explicitly separating the conformal and nonconformal parts of the metric without placing restrictions on g_{mn}. We'll find the corresponding procedure in supergravity very useful, since it avoids the much more complicated separation of ϕ from the covariant derivatives (and allows a unified treament of compensators other than ϕ: see Appendix B).

3. TANGENT SPACES: LORENTZ AND R-SYMMETRY

At this point we could continue the construction of the previous section, constructing a linearized (Ogievetsky and Sokatchev, 1976, 1977) and full nonlinear (Siegel, 1977 b) action, and covariant derivatives, but we'll get a better understanding of the geometry (and in a way which can be applied to extended supergravity) by considering the opposite approach: beginning with a set of covariant derivatives (Akulov *et al.*, 1975) and applying appropriate constraints (Wess and Zumino, 1978) whose solution (Siegel, 1978) is in terms of the prepotentials of the previous section.

3.1 Covariant derivatives

We define general covariant derivatives as

$$\nabla_A = E_A{}^M D_M + \Gamma_A{}^i G_i, \tag{3.1}$$

where G_i are the (antihermitian) generators of some tangent-space group which act directly on the fields. $E_A{}^M$ is the vielbein (the supersymmetric generalization of the vierbein) and $\Gamma_A{}^i$ is the connection for G_i. The action of ∇ on an arbitrary superfield (and on ∇ itself) is determined by the superfield's transformation under the tangent-space group. In order that such an expression always be covariant, we work only with objects (superfields or ∇) whose explicit indices are all tangent-space indices. (From now on, we use letters from the beginning of the alphabets for tangent-space indices, and from the middle for base-space indices.) $E_A{}^M$ (and its inverse $E_M{}^A$) can be used to convert between flat (tangent-space) and curved (base-space) indices. (We could extend ∇ to act also on objects with curved indices by including the corresponding generators and defining $E_A{}^M$ to be covariantly constant under the extended ∇. However, this would be an unnecessary complication, since it would give the same result as converting to flat indices, using the above ∇, and converting back again.) The covariant field strengths are defined by

$$[\nabla_A, \nabla_B\} = T_{AB}{}^C \nabla_C + F_{AB}{}^i G_i, \tag{3.2}$$

where $T_{AB}{}^C$ is the torsion and $F_{AB}{}^i$ is the curvature for the tangent-space group. (These tensors could also be defined as in gravity by the use of the notation of differential forms, which are dual to the covariant derivatives, but in supergravity explicit calculations, such as solving constraints or Bianchi identities, are easier with covariant derivatives.)

These derivatives and field strengths are covariant with respect to gauge parameters expanded over the same set of generators:

$$K = K^M iD_M + K^i iG_i,$$
$$\nabla_A' = e^{iK} \nabla_A e^{-iK}. \tag{3.3}$$

In order to maintain the (anti)hermiticity of the covariant derivatives, K must be hermitian ($K^M = \bar{K}^M$, $K^i = \bar{K}^i$). Thus K is *not* to be identified with the nonhermitian, chiral Λ of the previous sections. (In particular, (2.5), unlike (3.3), isn't even a similarity transformation.) This is another major distinction between supersymmetric and nonsupersymmetric gauge theories. As in super-Yang-Mills, the potentials which appear explicitly in the covariant derivatives transform independently from the prepotentials from which they are constructed: the former by hermitian parameters, the latter by chiral ones. Upon solving the constraints on the potentials, the hermitian symmetry is found to be a redundant one which trivially gauges away a redundant prepotential (in the same way as $\Lambda^{\dot\mu}$ gauges away H^μ, sec. 2.3). In N=2 extended supergravity, even the chiral parameters are redundant, and the true gauge parameters are one more step removed from the naive ones (Gates and Siegel, 1982).

3.2 Choice of tangent-space group

We now consider the choice of generators G_i. In general, we want to choose the largest tangent space possible. From general considerations (Haag *et al.*, 1975), we know the largest possible symmetry of the vacuum (on-shell, or global, symmetry) in Poincaré supergravity is Poincaré symmetry plus supersymmetry plus R-symmetry. (For N=1, R-symmetry is just chiral U(1), but for N-extended supergravity it's U(N), or SU(8) for N=8, and includes chiral and nonchiral generators.) The translational part of Poincaré transformations, and supersymmetry transformations, is generated by D_M, so we choose G_i to consist of Lorentz and U(1) generators (Howe, 1982):

$$K=K^M iD_M + (K_\alpha{}^\beta iM_\beta{}^\alpha + \bar K_{\dot\alpha}{}^{\dot\beta} i\bar M_{\dot\beta}{}^{\dot\alpha}) + K_5 Y, \tag{3.4a}$$

$$\nabla_A = E_A{}^M D_M + (\Phi_{A\beta}{}^\gamma M_\gamma{}^\beta + \bar\Phi_{A\dot\beta}{}^{\dot\gamma}\bar M_{\dot\gamma}{}^{\dot\beta}) - i\Gamma_A Y, \tag{3.4b}$$

$$[\nabla_A,\nabla_B\} = T_{AB}{}^C\nabla_C + (R_{AB\gamma}{}^\delta M_\delta{}^\gamma + \bar R_{AB\dot\gamma}{}^{\dot\delta}\bar M_{\dot\delta}{}^{\dot\gamma}) - iF_{AB}Y. \tag{3.4c}$$

The antihermitian Lorentz generators M and the hermitian generator Y are defined by their action on the covariant derivatives:

$$[Y,\nabla_\alpha] = -\tfrac{1}{2}\nabla_\alpha, \quad [Y,\bar{\nabla}_{\dot\alpha}] = \tfrac{1}{2}\bar{\nabla}_{\dot\alpha}, \quad [Y,\nabla_a] = 0; \tag{3.5a}$$

$$[M_{\alpha\beta},\nabla_\gamma] = \tfrac{1}{2}C_{\gamma(\alpha}\nabla_{\beta)}, \quad [M_{\alpha\beta},\bar{\nabla}_{\dot\gamma}] = 0,$$

$$[M_{\alpha\beta}, \nabla_{\gamma\dot\gamma}] = \tfrac{1}{2}C_{\gamma(\alpha}\nabla_{\beta)\dot\gamma}; \tag{3.5b}$$

$$[\bar{M}_{\dot\alpha\dot\beta}, \nabla_\gamma] = 0, \quad [\bar{M}_{\dot\alpha\dot\beta},\bar{\nabla}_{\dot\gamma}] = \tfrac{1}{2}C_{\dot\gamma(\dot\alpha}\bar{\nabla}_{\dot\beta)},$$

$$[\bar{M}_{\dot\alpha\dot\beta}, \nabla_{\gamma\dot\gamma}] = \tfrac{1}{2}C_{\dot\gamma(\dot\alpha}\nabla_{\beta\dot\beta)}. \tag{3.5c}$$

($M_\alpha{}^\beta$ generates SL(2,C), and therefore transforms an undotted index arbitrarily, except for tracelessness: $M_\alpha{}^\alpha = 0$, or equivalently $M_{\alpha\beta} = M_{\beta\alpha}$. As for ω in (1.2) and (2.3-4), it thus also represents an antisymmetric tensor, generating SO(3,1).) As in gravity, the global symmetries (symmetries of the vacuum) are obtained by setting the covariant derivatives equal to their vacuum values ($\nabla_A = D_A$), and receive contributions from both K^M("orbital") and K^i ("spin").

3.3 Superconformal transformations

As in the case of gravity, where the (broken) conformal structure can be manifested by introducing local scale transformations (and a compensating scalar), we here introduce local scale transformations in superspace. However, also as in gravity, the covariant derivatives don't transform covariantly under them, since there is no true local scale invariance. The appropriate transformation can be defined from the transformation of the spinor vielbein:

$$E_\alpha{}' = e^{L/2}E_\alpha \tag{3.6}$$

in terms of a real scalar superfield L. (We use the notation $E_A \equiv E_A{}^M D_M$.) As we'll see in the next section, the constraints on the covariant derivatives define the construction of ∇_A from E_α, and thus determine the scale transformation of ∇.

4. COVARIANT DERIVATIVES: CONSTRAINTS AND TENSORS

As explained in sec. 2, the prepotentials for Poincaré supergravity naturally divide into two parts, conformal supergravity and compensator. We therefore make a parallel division for the covariant objects: Conformal supergravity is contained in the covariant derivatives (Howe, 1982), and the compensating mulitplet is contained in a separate scalar superfield (Gates *et al.*, 1983 i).

4.1 Conformal supergravity constraints

In order to determine the covariant derivatives in terms of the prepotential H^m, we impose the constraints

$$\nabla_{\alpha\dot{\alpha}} = -i\,\{\nabla_\alpha, \bar{\nabla}_{\dot{\alpha}}\}, \tag{4.1a}$$

$$T_{\alpha\beta}{}^{\gamma} = T_{\alpha,\beta(\dot{\beta}}{}^{\beta\dot{\gamma})} = T_{\alpha b}{}^{b} = 0, \tag{4.1b}$$

$$\nabla_\alpha \bar{\eta} = 0 \rightarrow \{\nabla_\alpha, \nabla_\beta\}\bar{\eta} = 0. \tag{4.1c}$$

Written completely in terms of field strengths, these take the form

$$T_{\alpha\dot{\beta}}{}^{\gamma\dot{\gamma}} = i\delta_\alpha{}^\gamma \delta_{\dot{\beta}}{}^{\dot{\gamma}}, \quad T_{\alpha\dot{\beta}}{}^{\gamma} = R_{\alpha\dot{\beta}\gamma}{}^{\delta} = F_{\alpha\dot{\beta}} = 0, \tag{4.2a}$$

$$T_{\alpha\beta}{}^{\gamma} = T_{\alpha,\beta(\dot{\beta}}{}^{\beta\dot{\gamma})} = T_{\alpha b}{}^{b} = 0, \tag{4.2b}$$

$$T_{\alpha\beta}{}^{c} = T_{\alpha\beta}{}^{\dot{\gamma}} = 0. \tag{4.2c}$$

(The hermitian conjugate equations are implied.) These constraints have been divided into three types: (a) "conventional" constraints that determine ∇_a in terms of ∇_α, as in super-Yang-Mills; (b) conventional constraints that determine the connections ($\Phi_{\alpha\beta}{}^{\gamma}$, $\Phi_{\alpha\dot{\beta}}{}^{\gamma}$, and Γ_α, respectively) in terms of the vielbein (E_α), as in gravity; (c) "representation-preserving" constraints which allow the existence of chiral superfields, also as in super-Yang-Mills. Thus, all constraints are directly analogous to those found in simpler theories (super-Yang-Mills or ordinary gravity).

4.2 Superscale invariance

These constraints are preserved under superscale transformations (in addition to their manifest superPoincaré ⊗ U(1) invariance). The extension of (3.6) consistent with (4.1) is

$$\delta\nabla_\alpha = \tfrac{1}{2} L\nabla_\alpha + 2(\nabla_\beta L)M_\alpha{}^\beta + 3(\nabla_\alpha L)Y,$$

$$\delta\nabla_{\alpha\dot\alpha} = L\nabla_{\alpha\dot\alpha} - 2i(\bar\nabla_{\dot\alpha}L)\nabla_\alpha - 2i(\nabla_\alpha L)\bar\nabla_{\dot\alpha}$$
$$- 2i(\bar\nabla_{\dot\alpha}\nabla_\beta L)M_\alpha{}^\beta - 2i(\nabla_\alpha \bar\nabla_{\dot\beta}L)\bar M_{\dot\alpha}{}^{\dot\beta} + 3i([\nabla_\alpha, \bar\nabla_{\dot\alpha}]L)Y. \quad (4.3)$$

These transformations, plus those in K of (3.3-4), are sufficient to gauge away all fields except those in H^m invariant under (1.6a), (2.5).

4.3 Solving the conventional constraints

(4.1a) is its own explicit solution. To solve (4.1b) directly in terms of E_α, we define

$$\check E_A \equiv (E_\alpha, E_{\dot\alpha}, -i\{E_\alpha, E_{\dot\alpha}\}), \quad (4.4a)$$

$$[\check E_A, \check E_B\} = \check C_{AB}{}^C \check E_C. \quad (4.4b)$$

Using the D_M part of (4.1a) we can then solve for E_a:

$$E_{\alpha\dot\alpha} = \check E_{\alpha\dot\alpha} - i(\Phi_{\alpha\dot\alpha}{}^{\dot\beta} - \tfrac{1}{2}i\delta_{\dot\alpha}{}^{\dot\beta}\Gamma_\alpha)E_{\dot\beta} - i(\Phi_{\dot\alpha\alpha}{}^\beta + \tfrac{1}{2}i\delta_\alpha{}^\beta\Gamma_{\dot\alpha})E_\beta. \quad (4.5)$$

Now, to solve the torsion constraints (4.1b) we express the torsion in terms of ∇_α using its definition (3.4c). $\Phi_{\alpha\beta}{}^\gamma$ is easily found from

$$0 = T_{\alpha\beta}{}^\gamma = \check C_{\alpha\beta}{}^\gamma + \Phi_{(\alpha\beta)}{}^\gamma + \tfrac{1}{2}i\delta_{(\alpha}{}^\gamma\Gamma_{\beta)}$$

$$\rightarrow \Phi_{\alpha\beta\gamma} = \tfrac{1}{2}(\check C_{\beta\gamma\alpha} - \check C_{\alpha(\beta\gamma)}) - \tfrac{1}{2}iC_{\alpha(\beta}\Gamma_{\gamma)}. \quad (4.6)$$

(We have used $\Phi_{\alpha\beta\gamma} = \Phi_{\alpha\gamma\beta}$.) To find $\Phi_{\alpha\dot\beta}{}^{\dot\gamma}$ and Γ_α, we use also (4.5) and $0 = T_{\alpha\beta}{}^c = \check C_{\alpha\beta}{}^c$ (from (4.1c)): we find

$$0 = T_{\alpha,\beta(\dot\beta}{}^{\beta\dot\gamma)} = \check C_{\alpha,\beta(\dot\beta}{}^{\beta\dot\gamma)} + 2\Phi_{\alpha\dot\beta}{}^{\dot\gamma}$$

$$\rightarrow \Phi_{\alpha\dot\beta}{}^{\dot\gamma} = -\tfrac{1}{2}\check C_{\alpha,\beta(\dot\beta}{}^{\beta\dot\gamma)}, \quad (4.7)$$

$$0 = T_{\alpha b}{}^{b} = \check{C}_{\alpha b}{}^{b} + i\Gamma_{\alpha}$$

$$\rightarrow \Gamma_{\alpha} = i\check{C}_{\alpha b}{}^{b}. \tag{4.8}$$

4.4 Solving the representation-preserving constraint

In (4.1c) we have defined a chiral scalar η by $\nabla_{\alpha}\bar{\eta} = 0$, and taking another derivative and symmetrizing found $\{\nabla_{\alpha},\nabla_{\beta}\}\bar{\eta} = 0$, which can be expressed as (4.2c) using (3.4c). However, since η is a scalar (and assuming for simplicity $Y\eta = 0$), we can write equivalently, and more directly,

$$E_{\alpha}\bar{\eta} = 0 \rightarrow \{E_{\alpha},E_{\beta}\}\bar{\eta} = 0 \rightarrow \{E_{\alpha},E_{\beta}\} = \check{C}_{\alpha\beta}{}^{\gamma}E_{\gamma}. \tag{4.9}$$

Geometrically, this says that the two "vectors" $E_{\alpha} = E_{\alpha}{}^{M}D_{M}$ form a basis of a subalgebra of the supertranslations (generated by E_{A} or D_{M}): All operators of the form $\lambda^{\alpha}E_{\alpha}$ generate complex translations with an algebra that closes, determining a two-dimensional subspace of the full superspace. We thus explicitly can express E_{α} in terms of derivatives with respect to the two coordinates of this subspace:

$$E_{\alpha} = A_{\alpha}{}^{\mu}\frac{\partial}{\partial\tau^{\mu}} = \bar{\Psi}N_{\alpha}{}^{\mu}e^{-\Omega}D_{\mu}e^{\Omega}, \tag{4.10}$$

where the "zweibein" $A_{\alpha}{}^{\mu}$ has been split into a scale $\otimes$ U(1) factor $\bar{\Psi}$ and Lorentz rotation $N_{\alpha}{}^{\mu}$ (with unit determinant), and the coordinates τ^{μ} have been expressed as a *complex* coordinate transformation of θ^{μ}: $\Omega = \Omega^{M}iD_{M} \neq \bar{\Omega}$. (Remember also that $D_{\mu} = e^{-U/2}\partial_{\mu}e^{U/2}$, so $\tau^{M} = e^{-\Omega}e^{-U/2}z^{M}$.) This is closely analagous to the super-Yang-Mills $\nabla_{\alpha} = e^{-\Omega}D_{\alpha}e^{\Omega}$, whose Ω can be interpreted as a complex translation in the group manifold. ($\bar{\Psi}$ corresponds to an Ω_{5} for Y; but $N_{\alpha}{}^{\mu}$ can't be considered a complex transformation because SL(2,C) is already complex, and in fact $N_{\alpha}{}^{\mu}$ can be gauged to $\delta_{\alpha}{}^{\mu}$.)

4.5 Chiral gauge group and representation

The solution (4.10) to the representation-preserving constraint introduces a new gauge invariance, as in super-Yang-Mills: Rewriting it as

$$E_{\alpha} = e^{-\Omega}\underline{A}_{\alpha}{}^{\mu}D_{\mu}e^{\Omega}, \tag{4.11}$$

the appropriate invariance is of the form

$$e^{\Omega'} = e^{i\bar{\Lambda}} e^{\Omega}, \tag{4.12a}$$

$$\underline{A}_\alpha{}^{\mu\prime} D_\mu = e^{i\bar{\Lambda}} \underline{A}_\alpha{}^\mu D_\mu e^{-i\bar{\Lambda}} = (e^{i\bar{\Lambda}} \underline{A}_\alpha{}^\mu) e^{i\bar{\Lambda}} D_\mu e^{-i\bar{\Lambda}} . \tag{4.12b}$$

The form of (4.12b) constains Λ: In infinitesimal form

$$[\bar{D}_{\dot{\mu}}, \Lambda] = \Lambda_{\dot{\mu}}{}^{\dot{\nu}} \bar{D}_{\dot{\nu}} \tag{4.13}$$

for some $\Lambda_{\dot{\mu}}{}^{\dot{\nu}}$. This constraint is in fact identical to the constraint on Λ in sec. 1: There we had $\bar{D}_{\dot{\mu}}\phi = 0$, $\delta\phi = i\Lambda\phi$, so $[\bar{D}_{\dot{\mu}}, \Lambda]\phi = 0$, implying (4.13). The solution to (4.13) is thus given by (1.6), with $\Lambda_{\dot{\mu}}{}^{\dot{\nu}}$ given in terms of arbitrary $\Lambda^{\dot{\mu}}$ by $\Lambda_{\dot{\mu}}{}^{\dot{\nu}} = \bar{D}_{\dot{\mu}}\Lambda^{\dot{\nu}}$. This matrix can be separated into its trace and traceless pieces, which transform the $\bar{\Psi}$ and $N_\alpha{}^\mu$ parts of $A_\alpha{}^\mu$, respectively.

We still have the original K invariance. In particular, the supertranslation part $K = K^M iD_M$ acts on Ω as $e^{\Omega'} = e^{\Omega} e^{-iK}$ (cf. (4.12a)), or infinitesimally as $\delta\Omega = -iK$. Since K is real, while Ω is complex, it can be used to gauge away the imaginary part of Ω, making it real. An equivalent but more useful procedure is to perform a nonunitarity similarity transformation on all covariant quantities F:

$$F^{(+)} \equiv e^{-\bar{\Omega}} F e^{\bar{\Omega}} . \tag{4.14}$$

This defines a chiral representation, analagous to the one used in sec. 2.4. In addition, we use $K_\alpha{}^\beta$ to choose the gauge $N_\alpha{}^\mu = \delta_\alpha{}^\mu$. As a result, we have

$$E^{(+)}_{\dot{\alpha}} = \Psi\bar{D}_{\dot{\alpha}}, \quad E^{(+)}_{\alpha} = e^{-H}\bar{\Psi} D_\alpha e^{H} \equiv \tilde{\Psi} e^{-H} D_\alpha e^{H}; \tag{4.15a}$$

$$e^{H} \equiv e^{\Omega} e^{\bar{\Omega}} . \tag{4.15b}$$

All other quantities can be constructed as in sec. 4.3. As in super-Yang-Mills, the net effects of this transformation are: (a) simplification of $E_{\dot{\alpha}}$, (b) replacement of the complex Ω with the real H, (c) replacement of K transformations (except K_5) with Λ transformations. In particular, H transforms as (2.5), so we can now make complete identification of the conformal sector. (We could also use a nonunitary U(1) transformation to eliminate the phase of Ψ, but this will be unnecessary, since we'll later introduce a compensator for U(1) and scale transformations. However, the opposite is true for n=0: see Appendix B.) An equivalent rule for obtaining chiral representation quantities from vector representation ones is to make the replacements $\Omega \to H, \bar{\Omega} \to 0$ (and $N_\alpha{}^\mu = \delta_\alpha{}^\mu$).

4.6 Compensator

At this point we have a description of the conformal part of supergravity: The solution of the constraints has reduced our set of fields $\nabla_A \to \nabla_\alpha \to E_\alpha \to H, \Psi$ *via* (4.1a,b,c) respectively. Furthermore, we could use the parameters K_5 and L (all that remain of the original invariances of ∇_A, the rest having been replaced by Λ) to gauge Ψ to 1. We would then be left with the conformal fields and invariances of sec. 2.2. Our next step is thus to introduce the covariant analog of the compensator of sec. 2.6. The obvious choice is to introduce a *covariantly* chiral scalar superfield:

$$\bar{\nabla}_{\dot{\alpha}} \Phi = 0. \tag{4.16}$$

The requirement that this constraint, as those of (4.1), be covariant with respect to local scale transformations (4.3) determines the chiral weight of Φ: Defining

$$\delta\Phi = L\Phi \tag{4.17}$$

under scale transformations (arbitrarily, since $\bar{\nabla}_{\dot{\alpha}} \Phi^m = 0$ and $\delta\Phi^m = mL\Phi^m$ for arbitrary m), we find

$$\begin{aligned} 0 &= \delta(\bar{\nabla}_{\dot{\alpha}}\Phi) = (\delta\bar{\nabla}_{\dot{\alpha}})\Phi + \bar{\nabla}_{\dot{\alpha}}(\delta\Phi) \\ &= \tfrac{1}{2}L\bar{\nabla}_{\dot{\alpha}}\Phi - 3(\bar{\nabla}_{\dot{\alpha}}L)Y\Phi + \bar{\nabla}_{\dot{\alpha}}(L\Phi) \\ &= (\bar{\nabla}_{\dot{\alpha}}L)(1 - 3Y)\Phi \\ &\to Y\Phi = \frac{1}{3}\Phi. \end{aligned} \tag{4.18}$$

4.7 Solving the compensator's constraint

To solve the constraint (4.16) (subject to (4.18)) on the

compensator, we need a more explicit form of Γ_α. To derive this form we make use of the identity

$$(-)^B \check{C}_{\alpha B}{}^B \equiv \check{C}_\alpha{}^B{}_B = E_\alpha \ell n E - 1\cdot\overleftarrow{E}_\alpha , \qquad (4.19)$$

where $(-)^B$ indicates the extra signs due to the reordering or raising and lowering of the B indices, and E is the superdeterminant of $E_A{}^M$ (see Appendix A). To derive this identity, it's easier to use the basis $\check{E}_A = \check{E}_A{}^M \partial_M$ (rather than $\check{E}_A{}^M D_M$) but the result is basis-independent (since $D_A = \mathring{E}_A{}^M \partial_M$, with $\mathring{E} = 1$). From the definition (4.4b), and defining $\check{E}_M{}^A$ as the matrix inverse of $\check{E}_A{}^M$,

$$\check{C}_{AB}{}^C = (\check{E}_{[A}\check{E}_{B)}{}^M)\check{E}_M{}^C . \qquad (4.20)$$

Temporarily ignoring grading signs such as $(-)^B$ in (4.19), which will automatically be fixed at the end of the calculation by appropriate ordering of indices, we then have

$$\begin{aligned}\check{C}_\alpha{}^B{}_B &= (E_\alpha \check{E}^{BM})\check{E}_{MB} - (\check{E}^B E_\alpha{}^M)\check{E}_{MB} \\ &= E_\alpha \ell n\, \check{E} - \partial_M E_\alpha{}^M \\ &= E_\alpha \ell n\, E - 1\cdot\overleftarrow{E}_\alpha , \end{aligned} \qquad (4.21)$$

where we have used $\check{E}=E$, as follows from $\check{E}_\alpha{}^M \equiv E_\alpha{}^M$ and (4.5). We can now evaluate (4.8) as

$$-i\Gamma_\alpha = \check{C}_{\alpha b}{}^b = \check{C}_\alpha{}^B{}_B - \check{C}_\alpha{}^\beta{}_\beta - \check{C}_\alpha{}^{\dot\beta}{}_{\dot\beta} = E_\alpha \ell n E - 1\cdot\overleftarrow{E}_\alpha + \check{C}_{\alpha\beta}{}^\beta , \qquad (4.22)$$

using $\check{C}_{\alpha\dot\beta}{}^{\dot\gamma} = 0$, as follows from the definition (4.4a). For simplicity we choose the Lorentz gauge $N_\alpha{}^\mu = \delta_\alpha{}^\mu$ (but the final result will be trivially generalized to arbitrary $N_\alpha{}^\mu$): Then, from (4.10) we easily calculate

$$\check{C}_{\alpha\beta}{}^\gamma = \delta_{(\alpha}{}^\gamma E_{\beta)} \ell n\, \bar\Psi \rightarrow \check{C}_{\alpha\beta}{}^\beta = 3E_\alpha \ell n\, \bar\Psi , \qquad (4.23a)$$

$$1\cdot\overleftarrow{E}_\alpha = E_\alpha \ell n \bar\Psi + \bar\Psi(1\cdot\overleftarrow{\hat{E}}_\alpha) , \qquad (4.23b)$$

where

$$\hat{E}_\alpha \equiv e^{-\Omega} D_\alpha e^{\Omega} \rightarrow \overleftarrow{\hat{E}}_\alpha = e^{\overleftarrow{\Omega}} D_\alpha e^{-\overleftarrow{\Omega}} \qquad (4.24)$$

(as seen, e.g., by expanding in commutators). We then find

$$0 = 1\cdot e^{-\overleftarrow{\Omega}} e^{\overleftarrow{\Omega}} D_\alpha e^{-\overleftarrow{\Omega}} = (1\cdot e^{-\overleftarrow{\Omega}})\overleftarrow{\hat{E}}_\alpha$$

$$= \hat{E}_\alpha (1\cdot e^{-\overleftarrow{\Omega}}) + (1\cdot e^{-\overleftarrow{\Omega}})(1\cdot \overleftarrow{\hat{E}}_\alpha)$$

$$\rightarrow 1\cdot \overleftarrow{\hat{E}}_\alpha = -\hat{E}_\alpha \ell n(1\cdot e^{-\overleftarrow{\Omega}}). \qquad (4.25)$$

The final result is (from (4.22,23,25))

$$-i\Gamma_\alpha = E_\alpha T, \quad e^{T} = E\bar{\Psi}^2 (1\cdot e^{-\overleftarrow{\Omega}}). \qquad (4.26)$$

The solution to the constraint (4.16,18) on Φ is thus

$$0 = \bar{\nabla}_{\dot{\alpha}}\Phi = (\bar{E}_{\dot{\alpha}} - \tfrac{1}{3}i\bar{\Gamma}_{\dot{\alpha}})\Phi = e^{\bar{T}/3}\bar{E}_{\dot{\alpha}} e^{-\bar{T}/3}\Phi$$

$$\rightarrow \Phi = e^{\bar{T}/3}\phi, \quad \bar{E}_{\dot{\alpha}}\phi = 0. \qquad (4.27)$$

Finally, to explicitly separate all Ψ and Ω dependence, we define

$$\hat{E}_A = (\hat{E}_\alpha, \hat{E}_{\dot{\alpha}}, -i\{\hat{E}_\alpha, E_{\dot{\alpha}}\}) \rightarrow E = \hat{E}\Psi^2\bar{\Psi}^2, \qquad (4.28)$$

so that

$$\Phi = [\hat{E}\Psi^4\bar{\Psi}^2 \ (1\cdot e^{\overleftarrow{\bar{\Omega}}})]^{1/3}\phi. \qquad (4.29)$$

Instead of the scale $\otimes$ U(1) gauge $\Psi=1$ mentioned above, it's also possible to choose the gauge

$$\Phi = 1 \rightarrow \Gamma_\alpha = 0, \quad \Psi = \phi^{-1}\bar{\phi}^{-1/2}\hat{E}^{-1/6}(1\cdot e^{-\overleftarrow{\Omega}})^{1/6}(1\cdot e^{\overleftarrow{\bar{\Omega}}})^{-1/3}. \qquad (4.30)$$

In this gauge, Poincaré supergravity is completely described by ∇_A, with only superPoincaré gauged ($\Gamma_A = F_{AB} = 0$; this isn't true for $n \neq -1/3$: see Appendix B).

4.8 Bianchi identities and independent tensors

Having expressed ∇_A completely in terms of $\Omega, \Psi, N_\alpha{}^\mu$, and ϕ

(or, in the chiral representation, after choosing scale, U(1), and $\Lambda^{\dot{\mu}}$ gauges, just H^m and ϕ), it's now possible to calculate all tensors in terms of these prepotentials. However, considerable information about these tensors can be gained without knowledge of the prepotentials by examining the Jacobi identities of the covariant derivatives (Bianchi identities) (Grimm, Wess, and Zumino, 1979; Wess and Bagger, 1983). We can simplify our analysis if we first notice a few more implications of chirality conditions: e.g., from (4.16,18) we have

$$0 = \{\bar{\nabla}_{\dot{\alpha}},\bar{\nabla}_{\dot{\beta}}\}\Phi = -i\tfrac{1}{3}F_{\dot{\alpha}\dot{\beta}}\Phi \rightarrow F_{\dot{\alpha}\dot{\beta}} = 0. \tag{4.31}$$

(This can also be directly computed from (4.26).) As a result of this and $F_{\alpha\dot{\beta}} = 0$ (in (4.2a)), Γ_α is the potential of a vector multiplet: The usual analysis of its Bianchi identities (i.e., those for F_{AB} alone) shows that is has a chiral spinor field strength, and thus

$$0 = \{\bar{\nabla}_{\dot{\alpha}},\bar{\nabla}_{\dot{\beta}}\}W_\alpha = R_{\dot{\alpha}\dot{\beta}\alpha}{}^{\beta}W_\beta \rightarrow R_{\dot{\alpha}\dot{\beta}\alpha}{}^{\beta} = 0. \tag{4.32}$$

These results, together with the constraint (4.1c) and the first part of (4.1b), can be summarized as

$$\{\nabla_\alpha,\nabla_\beta\} = R_{\alpha\beta\gamma}{}^{\delta}M_\delta{}^{\gamma}. \tag{4.33}$$

Next, we plug this expression into the Jacobi identity

$$0 = [\nabla_{(\alpha},\{\nabla_\beta,\nabla_{\gamma)}\}] = (\nabla_{(\alpha}R_{\beta\gamma)\delta}{}^{\varepsilon})M_\varepsilon{}^{\delta} - R_{(\alpha\beta\gamma)}{}^{\delta}\nabla_\delta$$

$$\rightarrow R_{\alpha\beta}{}^{\gamma\delta} = \delta_{(\alpha}{}^{\gamma}\delta_{\beta)}{}^{\delta}\,\bar{R}, \quad \nabla_\alpha\bar{R} = 0. \tag{4.34}$$

Thus (4.33) simplifies to

$$\{\nabla_\alpha,\nabla_\beta\} = -\,2\bar{R}M_{\alpha\beta}. \tag{4.35}$$

Combining this last identity with the constraint (4.1a), we find

$$[\bar\nabla_{(\dot\alpha}, i\nabla_{\beta\dot\beta)}] = [\bar\nabla_{(\dot\alpha}\ \{\bar\nabla_{\dot\beta)}, \nabla_\beta\}] = [\{\bar\nabla_{\dot\alpha}, \bar\nabla_{\dot\beta}\}, \nabla_\beta]$$

$$= [-2R\bar M_{\dot\alpha\dot\beta}, \nabla_\beta] = 2(\nabla_\beta R)\bar M_{\dot\alpha\dot\beta}$$

$$\rightarrow [\bar\nabla_{\dot\alpha},\ i\nabla_{\beta\dot\beta}] = C_{\dot\beta\dot\alpha}A_\beta + (\nabla_\beta R)\bar M_{\dot\alpha\dot\beta} \qquad (4.36)$$

in terms of some operator A_α. Furthermore, the remaining constraints (the last two parts of (4.1b)) simply state that the vector-derivative part of A_α vanishes:

$$A_\alpha = A_\alpha{}^\beta \nabla_\beta + A_\alpha{}^{\dot\beta}\bar\nabla_{\dot\beta} + A_{\alpha\beta}{}^\gamma M_\gamma{}^\beta + A_{\alpha\dot\beta}{}^{\dot\gamma}\bar M_{\dot\gamma}{}^{\dot\beta} - iA^5{}_\alpha Y. \qquad (4.37)$$

The next Bianchi identity to analyze is

$$0 = \{\bar\nabla_{(\dot\alpha},\ [\bar\nabla_{\dot\beta)},\ i\nabla_{\gamma\dot\gamma}]\} - [\{\bar\nabla_{\dot\alpha}, \bar\nabla_{\dot\beta}\}, i\nabla_{\gamma\dot\gamma}]$$

$$= C_{\dot\gamma(\dot\alpha}\{\bar\nabla_{\dot\beta)}, A_\gamma\} + \{\bar\nabla_{(\dot\alpha}, (\nabla_\gamma R)\bar M_{\dot\beta)\dot\gamma}\} - [-2R\bar M_{\dot\alpha\dot\beta}, i\nabla_{\gamma\dot\gamma}]$$

$$= C_{\dot\gamma(\dot\alpha}\{\bar\nabla_{\dot\beta)}, A_\gamma\} + (i\nabla_{\gamma(\dot\alpha}R)\bar M_{\dot\beta)\dot\gamma} - \tfrac{1}{2}(\nabla_\gamma R)C_{\dot\gamma(\dot\alpha}\bar\nabla_{\dot\beta)}$$

$$+ RC_{\dot\gamma(\dot\alpha}i\nabla_{\gamma\dot\beta)} - 2(i\nabla_{\gamma\dot\gamma}R)\bar M_{\dot\alpha\dot\beta}$$

$$= C_{\dot\gamma(\dot\alpha}B_{\gamma\dot\beta)},$$

$$B_{\alpha\dot\alpha} = \{\bar\nabla_{\dot\alpha}, A_\alpha\} - \tfrac{1}{2}(\nabla_\alpha R)\bar\nabla_{\dot\alpha} + Ri\nabla_{\alpha\dot\alpha} + (i\nabla_\alpha{}^{\dot\beta}R)\bar M_{\dot\alpha\dot\beta} = 0, \quad (4.38)$$

where we have used (4.1a, 34-36). Substituting (4.37), we find

$$A_\alpha{}^\beta = -\delta_\alpha{}^\beta R, \quad A_{\alpha\dot\alpha\dot\beta} = -\tfrac{1}{2}\bar\nabla_{(\dot\alpha}A_{\alpha\dot\beta)},$$

$$\bar\nabla^{\dot\alpha}A_{\alpha\dot\alpha} = -\nabla_\alpha R - iA^5{}_\alpha, \quad \bar\nabla_{\dot\alpha}A^5{}_\alpha = \bar\nabla_{\dot\alpha}A_{\alpha\beta\gamma} = 0. \qquad (4.39)$$

The final independent Bianchi identity is

$$0 = [i\nabla^{\alpha\dot\alpha},\ i\nabla_{\alpha\dot\alpha}] = [\{\nabla^\alpha, \bar\nabla^{\dot\alpha}\},\ i\nabla_{\alpha\dot\alpha}]$$

$$= \{\nabla^\alpha,\ [\bar\nabla^{\dot\alpha},\ i\nabla_{\alpha\dot\alpha}]\} + \{\bar\nabla^{\dot\alpha},\ [\nabla^\alpha,\ i\nabla_{\alpha\dot\alpha}]\}$$

$$= 2(\{\nabla^\alpha, A_\alpha\} - \{\bar{\nabla}^{\dot\alpha}, \bar{A}_{\dot\alpha}\}) . \qquad (4.40)$$

Again substituting (4.37), and using (4.39), we find

$$A_{\alpha\dot\alpha} = \bar{A}_{\alpha\dot\alpha}, \quad A^\alpha{}_{\alpha\beta} = \tfrac{1}{2}iA^5{}_\beta, \quad \nabla^\alpha A^5{}_\alpha + \bar{\nabla}^{\dot\alpha}\bar{A}^5{}_{\dot\alpha} = 0,$$

$$\nabla^\alpha A_{\alpha\beta\gamma} + \frac{1}{3}i\nabla_{(\beta}A^5{}_{\gamma)} = -\frac{1}{2}i\nabla_{(\beta}{}^{\dot\alpha}A_{\gamma)\dot\alpha} . \qquad (4.41)$$

Finally, we define

$$A_{\alpha\dot\alpha} = -G_{\alpha\dot\alpha}, \quad A^5{}_\alpha = W_\alpha, \quad A_{\alpha\beta\gamma} = W_{\alpha\beta\gamma} - \frac{1}{6}iC_{\alpha(\beta}W_{\gamma)}, \qquad (4.42)$$

where $W_{\alpha\beta\gamma}$ is totally symmetric, so that (4.36) becomes

$$[\bar{\nabla}_{\dot\alpha}, i\nabla_{\beta\dot\beta}] = C_{\dot\beta\dot\alpha}[-R\nabla_\beta - G_\beta{}^{\dot\gamma}\bar{\nabla}_{\dot\gamma} + (\bar{\nabla}^{\dot\gamma}G_{\beta\dot\delta})\bar{M}_{\dot\gamma}{}^{\dot\delta} - iW_\beta Y$$

$$- \frac{1}{3}iW_\gamma M_\beta{}^\gamma + W_{\beta\gamma}{}^\delta M_\delta{}^\gamma] + (\nabla_\beta R)\bar{M}_{\dot\alpha\dot\beta}, \qquad (4.43a)$$

where

$$G_{\alpha\dot\alpha} = \bar{G}_{\alpha\dot\alpha}, \quad \bar{\nabla}_{\dot\alpha}R = \bar{\nabla}_{\dot\alpha}W_{\alpha\beta\gamma} = \bar{\nabla}_{\dot\alpha}W_\alpha = 0,$$

$$\bar{\nabla}^{\dot\alpha}G_{\alpha\dot\alpha} = \nabla_\alpha R + iW_\alpha, \quad \nabla^\alpha W_\alpha + \bar{\nabla}^{\dot\alpha}\bar{W}_{\dot\alpha} = 0,$$

$$\nabla^\alpha W_{\alpha\beta\gamma} + \frac{1}{3}i\nabla_{(\beta}W_{\gamma)} = \frac{1}{2}i\nabla_{(\beta}{}^{\dot\alpha}G_{\gamma)\dot\alpha} . \qquad (4.43b)$$

The total solution of the Bianchis is given by (4.1a, 35,43). The final commutator can be directly computed by

$$[i\nabla_{\alpha\dot\alpha}, i\nabla_{\beta\dot\beta}] = \{\nabla_\alpha, [\bar{\nabla}_{\dot\alpha}, i\nabla_{\beta\dot\beta}]\} + \{\bar{\nabla}_{\dot\alpha}, [\nabla_\alpha, i\nabla_{\beta\dot\beta}]\}. \qquad (4.44)$$

Thus, R, $G_{\alpha\dot\alpha}$, and $W_{\alpha\beta\gamma}$ are the only independent tensors, satisfying the conditions (4.43b) (which also defines W_α). As discussed above, in the case $n = -\frac{1}{3}$ we can make further restrictions by a (scale $\otimes$U(1)) gauge choice: In the gauge $\Phi=1$, $W_\alpha = 0$ (since $\Gamma_A = F_{AB} = 0$).

5. ACTIONS

5.1 General measure

In gravity, the inverse determinant of $e_a{}^m$ acts as an appropriate density for constructing actions, since

$$(\delta e_a{}^m)\partial_m = [e_a{}^m\partial_m, \lambda^n\partial_n] = (e_a{}^n\partial_n\lambda^m - \lambda^n\partial_n e_a{}^m)\partial_m$$

$$\rightarrow \delta e^{-1} = - e^{-1}e_m{}^a\delta e_a{}^m$$

$$= -e^{-1}(\partial_m\lambda^m - e_m{}^a\lambda^n\partial_n e_a{}^m)$$

$$= -e^{-1}\ \partial_m\lambda^m - \lambda^m\partial_m e^{-1}$$

$$= e^{-1}i\overleftarrow{\lambda}$$

$$\rightarrow (e^{-1})' = e^{-1}e^{i\overleftarrow{\lambda}}, \tag{5.1}$$

where we have used det X = exp (tr ℓn X) $\rightarrow$ δ det X = det X tr $(X^{-1}\delta X)$, and iterated the infinitesimal transformation by $e^x = \lim_{n\to\infty}(1 + x/n)^n$. Thus, $\int d^4x\ e^{-1}\ \underline{L}$ is invariant (as explained in sec. 1.1) for any scalar $\underline{L}$, with $\underline{L}'=e^{i\overleftarrow{\lambda}}\underline{L}$. An analogous derivation to (5.1) gives $(E^{-1})'=E^{-1}e^{i\overleftarrow{K}}$, so

$$S = \int d^4x\ d^4\theta\ E^{-1}\ \underline{L} \tag{5.2}$$

is an invariant action for scalar $\underline{L}$, $\underline{L}' = e^{iK}\underline{L}$. In arbitrary scale gauges, $\underline{L}$ will depend on the scale compensator Φ in order to preserve local scale invariance, but actions can also be constructed without regard to scale invariance by choosing the scale ($\otimes$U(1)) gauge Φ=1.

5.2 Chiral measure

As the name indicates, actions integrated over chiral superspace are more conveniently written in the chiral representation. There, a scalar transforms as $\underline{L}' = e^{i\Lambda}\underline{L}$, where for chiral $\underline{L}$ ($\bar{\nabla}_\alpha\underline{L} = 0$ for $Y\underline{L} = 0$) Λ can be replaced with Λ_c of (2.20). (The transformation law (4.14) converts K into Λ due to $e^{-\Omega'} = e^{i\Lambda}e^{-\Omega}e^{-iK}$.) Then the transformation law (2.20) of ϕ^3 (as follows from (4.12,29)) implies that

$$S = \int d^4x\, d^2\theta \quad \phi^3 \underline{L} \tag{5.3}$$

is invariant (Siegel, 1978). This also holds in the gauge $\Phi=1$ for arbitrary U(1) weight (which fixes the scale weight in analogy to (4.18)), since then $\Gamma_A = 0$, or $\underline{L}$ can be made to satisfy $Y\underline{L} = 0$ by multiplying with an appropriate power of Φ.

Since in the chiral representation, and the gauge $\Phi=1(\rightarrow\Gamma_A=0)$

$$R = \bar{D}^2\Psi^2, \tag{5.4}$$

as follows from (4.6,15a,35), and also

$$E^{-1} = \Psi^2\phi^3, \tag{5.5}$$

as follows from (4.26,27) (with $\bar{\Omega} \rightarrow 0$), (5.3) can be rewritten as

$$S = \int d^4x\, d^4\theta\; E^{-1}R^{-1}\, \underline{L}, \tag{5.6}$$

since the $\int d^2\bar{\theta} = \bar{D}^2$ acts only on the Ψ^2 in the E^{-1}. This form can also be used in the vector representation (and in arbitrary scale gauges, with appropriate Φ-dependence in $\underline{L}$.)

5.3 Supergravity actions

The action for supergravity, by dimensional analysis, must take the simple form (Siegel, 1977b; Wess and Zumino, 1978)

$$S = -\,3\kappa^{-2}\int d^4x\, d^4\theta\; E^{-1}\,\bar{\Phi}\Phi, \tag{5.7}$$

where Φ-dependence is determined by local scale invariance (4.3,17). Using (5.4,5), this can be rewritten in the $\Phi=1$ gauge as

$$S = -\,3\kappa^{-2}\int d^4x\, d^2\theta\;\; \phi^3 R. \tag{5.8}$$

The latter form, which more closely resembles the nonsupersymmetric case, should also be compared with the cosmological term (Kaku and Townsend, 1978)

$$S_{cosmo} = \lambda\kappa^{-2} \int d^4x\, d^2\theta\; \phi^3 + \text{h.c.} \tag{5.9}$$

The action for conformal supergravity, from the arguments of sec. 2, or from demanding local scale invariance with Φ absent, is

$$S_{conf} = \int d^4x \; d^2\theta \quad \phi^3 \; (W_{\alpha\beta\gamma})^2 . \tag{5.10}$$

5.4 Matter actions

The simplest actions are for the scalar multiplet, defined by $\bar{\nabla}_{\dot\alpha} \eta = 0$. Without loss of generality, we can choose $Y\eta = 0$ (and thus scale weight = 0, in analogy to (4.18)), and find

$$S = \int d^4x \; d^4\theta \quad E^{-1}\{\bar{\Phi}\Phi A(\bar{\eta},\eta) + [R^{-1}\Phi^3 B(\eta) + h.c.]\}. \tag{5.11}$$

An interesting special case is the conformally invariant action. Φ-dependence can then be elininated by choosing $Y\eta = \frac{1}{3}\eta$ (as in (4.18)) (Siegel, 1977 c; Ferrara and van Nieuwenhuizen, 1978 b):

$$S = \int d^4x \; d^4\theta \quad E^{-1} \; [\bar{\eta}\eta + (R^{-1} \; \frac{1}{6}\lambda\eta^3 + h.c.)]. \tag{5.12}$$

With $\eta \to \Phi$, this action is proportional to that of supergravity with a cosmological term (see (5.7,9)).

Coupling to the vector multiplet is obtained by adding generators for the Yang-Mills group to the covariant derivatives (3.1). One then constrains $\underline{F}_{\alpha\beta} = \underline{F}_{\alpha\dot\beta} = 0$ as in flat superspace, and finds a chiral field strength

$$\bar{\nabla}_{\dot\alpha} \underline{W}_{\alpha} = 0, \quad Y\underline{W}_{\alpha} = \tfrac{1}{2} \underline{W}_{\alpha} . \tag{5.13}$$

The (conformally invariant, Φ-independent) action is thus

$$S = g^{-2} \mathrm{tr} \int d^4x \; d^4\theta \quad E^{-1}R^{-1}\underline{W}^2 . \tag{5.14}$$

By the same methods, more general actions can be written coupling scalar and vector mulitplets, directly generalizing the flat superspace results.

APPENDIX A: CONVENTIONS

We use the conventions of Gates *et al.* (1983 a) (except that we don't bother to underline vector indices, since there's no need to distinguish them from isospin indices, which don't occur here). All Lorentz indices are expressed in terms of two-component spinor indices (lower-case Greek letters): A Weyl spinor is written as ψ^α, and its hermitian conjugate as $(\psi^\alpha)^\dagger \equiv \bar{\psi}^{\dot\alpha}$. A Dirac spinor is the *sum* of these two representations, $(\psi^\alpha, \chi^{\dot\alpha})$, and a Majorana spinor is a Dirac spinor with $\chi^{\dot\alpha} = \bar{\psi}^{\dot\alpha}$ (and thus equivalent to a single Weyl spinor). A (four-)vector transforms as the *product* of a Weyl spinor and its hermitian conjugate, $V^{\alpha\dot\alpha}$. It can be thought of either as a hermitian 2x2 matrix ($V^{\alpha\dot\alpha} = \bar{V}^{\alpha\dot\alpha}$), or as a 4-component column vector V^a, where $a \equiv \alpha\dot\alpha$ takes 4 values (since α and $\dot\alpha$ each take 2). We thus use vector indices (lower-case Roman letters) interchangeably (and sometimes simultaneously) with dotted-undotted pairs of spinor indices. Supervector indices are upper-case Roman letters: e.g., the superspace coordinates are $z^M = (x^m, \theta^\mu, \bar\theta^{\dot\mu})$ and their partial derivatives $\partial_M = (\partial_m, \partial_\mu, \bar\partial_{\dot\mu})$, with $\partial_M z^N = \delta_M{}^N$.

Since the Lorentz group is SL(2,C), the only invariant tensor is the antisymmetric tensor $C_{\alpha\beta} = -C_{\beta\alpha} = -C^{\alpha\beta} = C_{\dot\alpha\dot\beta}$. Any antisymmetric tensor is proportional to it, and in particular

$$C_{\alpha\beta}C^{\gamma\delta} = \delta_{[\alpha}{}^\gamma \delta_{\beta]}{}^\delta \rightarrow \psi_{[\alpha}\chi_{\beta]} = C_{\alpha\beta}(C^{\gamma\delta}\psi_\gamma\chi_\delta), \tag{A.1}$$

where [ ] represents antisymmetrization and ( ) symmetrization (e.g., $\psi_{[\alpha}\chi_{\beta]} \equiv \psi_\alpha\chi_\beta - \psi_\beta\chi_\alpha$). C can also be used to raise, lower, and contract indices:

$$\psi^\alpha = C^{\alpha\beta}\psi_\beta, \quad \psi_\alpha = \psi^\beta C_{\beta\alpha}, \quad \bar\psi^{\dot\alpha} = C^{\dot\alpha\dot\beta}\bar\psi_{\dot\beta}, \quad \bar\psi_{\dot\alpha} = \bar\psi^{\dot\beta}C_{\dot\beta\dot\alpha};$$

$$\psi^2 = \tfrac{1}{2}\psi^\alpha\psi_\alpha, \quad \bar\psi^2 = \tfrac{1}{2}\bar\psi^{\dot\alpha}\bar\psi_{\dot\alpha}, \quad \psi_\alpha\psi_\beta = C_{\beta\alpha}\psi^2. \tag{A.2}$$

The invariant tensors of SO(3,1) ($\sim$SL(2,C)) can be expressed in terms of C:

$$\eta_{ab} = C_{\alpha\beta}C_{\dot\alpha\dot\beta}, \quad \epsilon_{abcd} = i(C_{\alpha\delta}C_{\beta\gamma}C_{\dot\alpha\dot\beta}C_{\dot\gamma\dot\delta} - C_{\alpha\beta}C_{\gamma\delta}C_{\dot\alpha\dot\delta}C_{\dot\beta\dot\gamma}). \tag{A.3}$$

(Remember $\eta_{ab} \equiv \eta_{\alpha\dot{\alpha}\beta\dot{\beta}}$, etc.) Irreducible representations of the Lorentz group are symmetric in all undotted spinor indices, and in all dotted ones: e.g., an arbitrary second-rank tensor can be decomposed as $\psi_{ab} = \psi_{(\alpha\beta)(\dot{\alpha}\dot{\beta})} + C_{\alpha\beta}\,\psi_{(\dot{\alpha}\dot{\beta})} + C_{\dot{\alpha}\dot{\beta}}\,\psi_{(\alpha\beta)} + C_{\alpha\beta}C_{\dot{\alpha}\dot{\beta}}\,\psi$.

Our supersymmetry-covariant derivatives are defined as

$$D_A = (\partial_a, D_\alpha, \bar{D}_{\dot{\alpha}}),\ D_\alpha = \partial_\alpha + \tfrac{1}{2}\bar{\theta}^{\dot{\alpha}} i\partial_{\alpha\dot{\alpha}},\ \bar{D}_{\dot{\alpha}} = \bar{\partial}_{\dot{\alpha}} + \tfrac{1}{2}\theta^{\alpha} i\partial_{\alpha\dot{\alpha}}; \qquad \text{(A.4)}$$

$$\{D_\alpha, \bar{D}_{\dot{\alpha}}\} = i\partial_{\alpha\dot{\alpha}},\ D^\alpha D_\beta = \delta_\beta{}^\alpha D^2,\ \bar{D}^{\dot{\alpha}}\bar{D}_{\dot{\beta}} = \delta_{\dot{\beta}}{}^{\dot{\alpha}}\bar{D}^2,\ \partial^{\alpha\dot{\gamma}}\partial_{\beta\dot{\gamma}} = \delta_\beta{}^\alpha\,\Box. \qquad \text{(A.5)}$$

θ-integration is performed by converting into derivatives:

$$\int d^4x\ d^2\theta = \int d^4x\ D^2,\quad \int d^4x\ d^4\theta = \int d^4x\ D^2\bar{D}^2. \qquad \text{(A.6)}$$

Commutators and antisymmetrization of indices are generalized to superspace by the graded commutator $[A,B\}$ (an anticommutator if both A and B are fermionic, commutator otherwise) and graded antisymmetrization $[\ )$ (between any pair of enclosed indices, symmetrization if both indices are spinorial, antisymmetrization otherwise). The determinant is also generalized: For example, one definition is to generalize the identity

$$\det M = \exp\,[\mathrm{tr}\ (\ell n\ M)]$$

$$\to \mathrm{sdet}\ M \equiv \exp\,[\mathrm{str}\ (\ell n\ M)]\ ,\ \mathrm{str}\ M_A{}^B \equiv M^A{}_A, \qquad \text{(A.7)}$$

where the raising and lowering of indices in str introduces an extra sign for the spinorial part of the trace.

Hermitian conjugation of a spinor ψ^α, vector $\psi^{\alpha\dot{\alpha}}$, and antisymmetric tensor (symmetric spinor) $\psi^{\alpha\beta}$ takes the form

$$(\psi^\alpha)^\dagger = \bar{\psi}^{\dot{\alpha}},\ (\psi_\alpha)^\dagger = -\bar{\psi}_{\dot{\alpha}},\ (\psi^\alpha\psi_\alpha)^\dagger = \bar{\psi}^{\dot{\alpha}}\bar{\psi}_{\dot{\alpha}};$$

$$(\psi^{\alpha\dot{\beta}})^\dagger = \bar{\psi}^{\beta\dot{\alpha}},\ (\psi_{\alpha\dot{\beta}})^\dagger = \bar{\psi}_{\beta\dot{\alpha}},\ (\psi_\alpha{}^{\dot{\beta}})^\dagger = -\bar{\psi}^\beta{}_{\dot{\alpha}};$$

$$(\psi^{\alpha\beta})^\dagger = -\bar{\psi}^{\dot{\alpha}\dot{\beta}},\ (\psi_{\alpha\beta})^\dagger = -\bar{\psi}_{\dot{\alpha}\dot{\beta}},\ (\psi_\alpha{}^\beta)^\dagger = \bar{\psi}_{\dot{\alpha}}{}^{\dot{\beta}}. \qquad \text{(A.8)}$$

The rules for all objects follow from those for the spinor: e.g., consider the cases $\psi^{\alpha\dot{\alpha}} = \psi^{\alpha}\bar{\chi}^{\dot{\alpha}}$, $\psi^{\alpha\beta} = \psi^{(\alpha}\chi^{\beta)}$. However, note that derivatives and Lorentz generators are defined to be *antihermitian:*

$$(D^{\alpha})^{\dagger} = -\bar{D}^{\dot{\alpha}}, \quad (\partial^{\alpha\dot{\beta}})^{\dagger} = -\partial^{\beta\dot{\alpha}}, \quad (M^{\alpha\beta})^{\dagger} = \bar{M}^{\dot{\alpha}\dot{\beta}},$$

$$(D_{\alpha})^{\dagger} = \bar{D}_{\dot{\alpha}}, \quad (\partial_{\alpha\dot{\beta}})^{\dagger} = -\partial_{\beta\dot{\alpha}}, \quad (M_{\alpha\beta})^{\dagger} = \bar{M}_{\dot{\alpha}\dot{\beta}} . \tag{A.9}$$

(I.e., the signs are exactly opposite those defined in (A.8).)

APPENDIX B: $n \neq -1/3$

All (finite) choices of the parameter n in the group restriction (1.4) lead to descriptions of Poincaré supergravity (Siegel, 1977 a-c; Siegel and Gates, 1979; Gates *et al.*, 1983 h,i). Different choices correspond to different compensators, and thus different auxiliary fields. The simplest auxiliary field structure is that of $n = -1/3$ (Siegel, 1977 a-c; Ferrara and van Nieuwenhuizen, 1978; Stelle and West, 1978), described in the main text.

An equally small set of fields, but a larger gauge group, is provided by the case n=0 (Sohnius and West, 1981). Here the chiral scalar compensator ϕ is replaced by a real, linear scalar $\hat{G}(\bar{D}^2\hat{G} = 0 \to \hat{G} = \frac{1}{2}(D_\alpha\phi^\alpha + \bar{D}_{\dot{\alpha}}\bar{\phi}^{\dot{\alpha}}),\ \bar{D}_{\dot{\alpha}}\phi_\alpha = 0)$. The corresponding covariant compensator G $((\bar{\nabla}^2 + R)G = 0)$, due to its reality, is invariant under U(1) $(YG = 0)$, and thus compensates only for scale transformations. As a result, in the gauge G=1 (where also R=0, but $W_\alpha \neq 0$), local U(1) invariance remains. This is due to the fact that $\hat{G}$ (or G) describes a tensor multiplet (Siegel, 1979; Gates *et al.*, 1983 c), whose components are: a (real) "physical" scalar, a "physical" (Weyl) spinor, and a "physical" antisymmetric tensor gauge field. Therefore, the auxiliary field structure differs from n=-1/3 in that: (a) the complex auxiliary scalar is replaced by the tensor gauge field, and (b) the auxiliary (axial) vector is a gauge field (as in the conformal case). The latter two fields appear in the lagrangian as $\varepsilon^{abcd}A_a\partial_b A_{cd}$.

All other n (Siegel, 1977 a-c; Gates and Siegel, 1980) (and in particular n=-1 (Breitenlohner, 1977)) have a larger set of auxiliary fields described by a complex linear compensator $\Upsilon(\bar{D}^2\Upsilon=0 \to \Upsilon=\bar{D}^{\dot{\alpha}}\bar{\psi}_{\dot{\alpha}})$, and a corresponding covariant compensator $\Sigma((\bar{\nabla}^2 + R)\Sigma=0)$. Now $Y\Sigma=-2n\,(3n+1)^{-1}\Sigma$, so again the gauge $\Sigma=1$ eliminates local U(1). In this gauge Γ_α (corresponding to the scale $\otimes$ U(1)-invariant $(\bar{\Sigma}\Sigma)^{-(3n+1)/8}\nabla_\alpha \ell n(\bar{\Sigma}^{3n-1}\Sigma^{3n+1})$) represents a tensor in addition to $R, G_a, W_{\alpha\beta\gamma}$, and W_α, corresponding to a spinor auxiliary field when evaluated at $\theta=0$, to which $W_\alpha|$ is the partner. (They appear in the lagrangian as $(\Gamma^\alpha W_\alpha)|$.) Altogether, the auxiliary fields additional to those of n=-1/3 are:

the above pair of spinors, a vector, and an additional axial vector. The compensator T represents a nonminimal scalar multiplet (Gates and Siegel, 1981; Gates *et al.*, 1983 d).

In general, any multiplet with a scalar field strength can serve as a compensator for supergravity. The only such multiplets are the above three forms of the scalar multiplet (and their first-order formulations). However, n = - 1/3 allows the most general couplings because for other n there is no chiral compensator Φ to compensate for the various U(1) weights in $\int d^4x\, d^2\theta$ terms, and as a result the only chiral action allowed for a physical chiral scalar η is $\int d^4x\, d^2\theta\, \eta^{2/w}$ (where $Y\eta = \frac{1}{2}w\eta$) (Siegel and Gates, 1979). For a similar reason (related to the allowed couplings for Pauli-Villars regulators), n=-1/3 is the only form of N=1 supergravity consistent at the quantum level (Gates, Grisaru, and Siegel, 1982; Gates *et al.*, 1983 k). The supergravity action for all n≠0, with all compensators gauged away, in the chiral representation, can be written as (Siegel, 1977 b)

$$S_n = \frac{1}{n\kappa^2} \int d^4x\, d^4\theta \quad \hat{E}^n \,(1\cdot e^{-\overleftarrow{H}})^{(n+1)/2}, \tag{B.1}$$

while for n=0 it takes the form (Gates, Roček, and Siegel, 1982)

$$S_0 = \frac{1}{\kappa^2} \int d^4x\, d^4\theta \quad \ln \hat{E} \tag{B.2}$$

in the $\dot{\Lambda}^{\mu}$-gauge

$$H^{\alpha} = -i\bar{D}_{\dot{\alpha}} H^{\alpha\dot{\alpha}} \rightarrow 1\cdot e^{-\overleftarrow{H}} = 1 \tag{B.3}$$

(which can also be used for n≠0 to simplify (B.1), whereupon $S_0 = \lim_{n\to 0} S_n$).

REFERENCES

Akulov, V.P., Volkov, D.V., and Soroka, V.A. 1975, JETP Lett. 22 187-188

Breitenlohner, P. 1977, Nucl. Phys. B124 500-510

Ferrara, S. and van Nieuwenhuizen, P. 1978a, Phys. Lett. 74B 333-335

Ferrara, S. and van Nieuwenhuizen, P. 1978b, Phys. Lett. 76B 404-408

Ferrara, S. and Zumino, B. 1975, Nucl. Phys. B87 207-220

Ferrara, S. and Zumino, B. 1978, Nucl. Phys. B134 301-326

Gates, S.J., Jr., Grisaru, M.T., Roček, M., and Siegel, W. 1983a, Superspace *or* One Thousand and One Lessons in Supersymmetry, Benjamin/Cummings, Reading, pp 55-61, 76,84-85, 97-100

Gates *et al.* 1983b, *ibid.*, pp 119-136

Gates *et al.* 1983c, *ibid.*, pp 181-193

Gates *et al.* 1983d, *ibid.*, pp 199-201

Gates *et al.* 1983e, *ibid.*, pp 232-244

Gates *et al.* 1983f, *ibid.*, pp 245-248

Gates *et al.* 1983g, *ibid.*, pp 248-249

Gates *et al.* 1983h, *ibid.*, pp 256-267

Gates *et al.* 1983i, *ibid.*, pp 268-315

Gates *et al.* 1983j, *ibid.*, pp 409-418

Gates *et al.* 1983k, *ibid.*, pp 489-495

Gates, S.J., Jr., Grisaru, M.T., and Siegel, W. 1982, Nucl. Phys. B203 189-204

Gates, S.J., Jr., Roček, M., and Siegel, W. 1982, Nucl. Phys. B198 113-118

Gates, S.J., Jr., and Siegel, W. 1980, Nucl. Phys. B163 519-545

Gates, S.J., Jr., and Siegel, W. 1981, Nucl Phys. B187 389-396

Gates, S.J., Jr., and Siegel, W. 1982, Nucl. Phys. B195 39-60

Grimm, R., Wess, J., and Zumino, B. 1979, Nucl. Phys. B152 255-265

Grisaru, M.T. and Siegel, W. 1981, Nucl. Phys. B187 149-183

Grisaru, M.T. and Siegel, W. 1982, Nucl. Phys. B201 292-314, 206 496-497

Haag, R., Łopuszański, J., and Sohnius, M.F. 1975, Nucl. Phys. B88 257-274

Howe, P. 1982, Nucl. Phys. B199 309-364

Kaku, M. and Townsend, P.K. 1978, Phys. Lett. 76B 54-58

Kaku, M., Townsend, P.K., and van Nieuwenhuizen, P. 1978, Phys. Rev. D17 3179-3187

Ogievetsky, V.I. and Sokatchev, E.S. 1976, Nonlocal, Nonlinear and Nonrenormalizable Field Theories, Proc. 4th Int. Symposium on Nonlocal Field Theories, JINR, Dubna, pp 183-203

Ogievetsky, V.I. and Sokatchev, E. 1977, Nucl. Phys. B124 309-316

Siegel, W. 1977a, Supergravity Superfields without a Supermetric, Harvard preprint HUTP-77/A068 (November), unpublished

Siegel, W. 1977b, The Superfield Supergravity Action, Harvard preprint HUTP-77/A080 (December), unpublished

Siegel, W. 1977c, A Polynomial Action for a Massive, Self-interacting Chiral Superfield Coupled to Supergravity, Harvard preprint HUTP-77/A077 (December), unpublished

Siegel, W. 1977d, A Derivation of the Supercurrent Superfield, Harvard preprint HUTP-77/A089 (December), unpublished
Siegel, W. 1978, Nucl. Phys. B142 301-305
Siegel, W. 1979, Phys. Lett. 85B 333-334
Siegel, W. 1981, Nucl. Phys. B177 325-332
Siegel, W. and Gates, S.J., Jr., 1979, Nucl. Phys. B147 77-104
Siegel, W. and Gates, S.J., Jr., 1981, Nucl. Phys. B189 295-316
Sohnius, M.F. and West, P.C. 1981, Phys. Lett. 105B 353-357
Stelle, K.S. and West, P.C. 1978, Phys. Lett. 74B 330-332
Wess, J. and Bagger, J. 1983, Supersymmetry and Supergravity, Princeton University, Princeton, pp 117-126
Wess, J. and Zumino, B. 1978, Phys. Lett. 74B 51-53

EXTENDED SUPERGRAVITIES IN COMPONENT FORMALISM

E CREMMER

1. INTRODUCTION

Extended supergravities are the minimal theories invariant under local extended supersymmetry

$$\{\bar{Q}^A, Q^B\} = \not{P}\,\delta^{AB} \qquad A = 1 \ldots N$$

The on-shell massless representations of these algebrasare well known. The particle content is

helicity	h_{MAX}	$h_{MAX}-\frac{1}{2}$	...	$h_{MAX}-\frac{K}{2}$	...	$h_{MAX}-\frac{N}{2}$	$= h_{MIN}$
multiplicity	1	N	....	$\frac{K!(N-K)!}{N!}$	......	1	

To have the field representation we must add the CPT conjugate multiplet unless it is already CPT self conjugate. If we limit ourselves to theories with spin $\leqslant 2$, since it seems that there is no consistent theory of massless field of spin > 2 coupled to gravitation, there is only a finite number of supergravity theories, namely for a $N \leqslant 8$ whose content is described in the following table

S \ N	1	2	3	4	5	6	7	8
2	1	1	1	1	1	1	1	1
3/2	1	2	3	4	5	6	7+1	8
1		1	3	6	10	15+1	21+7	28
1/2			1	4	10+1	20+6	35+21	56
0				1+1	5+5	15+15	35+35	70

We note that N=7 and N=8 supergravities have the same field content. N=8 supergravity is the maximal supergravity and is CPT self conjugate.

The content of these theories suggests that they could lead to a unified theory of all interactions and the reduction of ultraviolet divergences due to supersymmetry could lead to a quantum theory of gravitation. These remarks make worthwhile the study of these theories. Moreover, the number of vector fields is precisely equal to N(N-1)/2 (except for N=7 equivalent to N=8, and N=6 where there is an extra U(1)) which is the dimension of the adjoint representation of O(N); this suggests the possibility of gauging the group O(N). In fact, it is possible to include the O(N) transformations into the supersymmetry algebra provided we start from the algebra of super De Sitter instead of super Poincare, this means that the gauging of O(N) will generate a cosmological constant.

The extended supersymmetry algebra is obviously invariant under global O(N) transformation. By separating the supersymmetric charges Q^A, $\bar{Q}^B$ in left and right-handed parts, it is easy to see that this O(N) symmetry can be extended to U(N). Although the fields of the supergravity N do not fall in representation of U(N), we shall see that this symmetry is present in supergravity theories and realized in a very specific way. This will be seen in particular in section 2 for the simplest extended supergravity N=2 where the duality transformations appear. Section 3 will describe more generally the realization of duality invariance of a theory. In section 4 we shall study the supergravity N=4 in its various forms where there appear for the first time scalar fields which will be shown to be described by coset spaces SU(1,1)/U(1). This general structure of coset space related to non linear sigma-models for the description of self interaction of scalar fields will be studied in section 5. With the help of these properties we shall show how they lead to the construction of the maximal supergravity N=8 with its symmetry global $E_{7(+7)}$ and local SU(8) in section 6. The introduction of a gauge coupling constant for O(N) symmetry will be discussed in section 7. Sections 8 and 9 will be more speculative and devoted to the possible origins of these symmetries and to the possible physical interpretations of N=8 supergravity.

Before going into the details of the formulation of extended supergravities, let us give a few recipes which allow to simplify the direct determination of the Lagrangian and supersymmetry transformations.

(1) Use all implications of bosonic symmetries (if you know them or conjecture them) in particular the covariance of the supersymmetry transformation laws with respect to these symmetries,

(2) Use the knowledge of a sub-theory to determine in part some couplings,

(3) Ask for supercovariant field equations for the fermionic fields (this is useful because there are first order differential equations),

(4) Use the 1.5 order formalism to describe the Lorentz connexion $\hat{\omega}_{\mu ab}$. This will allow to take into account some fermionic couplings, in particular the quartic ones,

(5) Ask for the closure of the local supersymmetry algebra on the bosonic fields.

These requirements will be sufficient to determine completely the Lagrangian and supersymmetry transformations with a lot of double checks. It can eventually be necessary to verify the complete invariance of the Lagrangian (in particular the cancellation of some variations of quartic fermionic terms which have not been reabsorbed in supercovariant derivatives or eliminated by 1.5 order formalism).

2 N=2 SUPERGRAVITY

Let us first begin with the simplest extended supergravity, namely N=2. Besides exhibiting already new symmetries, it will allow us to precise our notations and conventions. In particular we use the metric η_{ab} = (+ - - -). The Dirac matrices satisfy

$$\{\gamma^a, \gamma^b\} = 2\,\eta^{ab}$$

γ_5 is defined by

$$\gamma_5 = i\,\gamma^0\gamma^1\gamma^2\gamma^3 \quad , \quad \gamma_5^2 = 1$$

ε^{abcd} is defined by

$$\varepsilon^{0123} = +1$$

$\bar{\Psi}$ is defined by

$$\bar{\Psi} = \Psi^{+}\gamma_0$$

Left and right-handed spinors are defined by

$$\Psi^R = \frac{1+\gamma_5}{2}\Psi \quad , \quad \bar{\Psi}^L = \bar{\Psi}\left(\frac{1+\gamma_5}{2}\right) = \overline{\Psi^L}$$

$$\Psi_L = \frac{1-\gamma_5}{2}\Psi \quad , \quad \bar{\Psi}_R = \bar{\Psi}\left(\frac{1-\gamma_5}{2}\right) = \overline{\Psi_R}$$

The supergravity N=2 describes 1 graviton (vierbein) $g_{\mu\nu}$ $(e_\mu{}^a)$, 2 gravitinos $\Psi_\mu{}^A$ and one abelian vector field A_μ. The Lagrangian is written as (Ferrara and Van Nieuwenhuizen, 1976)

$$e^{-1}\mathcal{L} = -\frac{1}{4K^2}R(\hat{\omega}) - \frac{e^{-1}}{2}\varepsilon^{\mu\nu\rho\sigma}\bar{\Psi}_\mu^A\gamma_5\gamma_\nu D_\rho(\hat{\omega})\Psi_\sigma^A + \frac{1}{4}F_{\mu\nu}\tilde{G}^{\mu\nu} - \frac{K}{4}\varepsilon^{AB}\bar{\Psi}_\mu^A(\hat{F}^{\mu\nu} - i\gamma_5\hat{\tilde{F}}^{\mu\nu})\Psi_\nu^B \tag{2.1}$$

where e is the determinant of the vierbien $e_\mu{}^a$

$$\tilde{G}_{\mu\nu} = -F_{\mu\nu} - K\varepsilon^{AB}\left(\bar{\Psi}_\mu^A\Psi_\nu^B - \frac{ie^{-1}}{2}\varepsilon_{\mu\nu}{}^{\rho\sigma}\bar{\Psi}_\rho^A\gamma_5\Psi_\sigma^B\right) \tag{2.2}$$

$$\hat{F}_{\mu\nu} = F_{\mu\nu} + K\varepsilon^{AB}\bar{\Psi}_\mu^A\Psi_\nu^B \tag{2.3}$$

$$\tilde{F}^{\mu\nu} = \frac{e^{-1}}{2}\varepsilon^{\mu\nu\rho\sigma}F_{\rho\sigma} \tag{2.4}$$

$$D_\rho(\hat{\omega}) = \partial_\rho + \frac{1}{4}\hat{\omega}_{\rho ab}\gamma^{ab} \tag{2.5}$$

$\hat{\omega}_{\mu ab}$ is considered as an independent field and replaced by the solution of its field equation

$$\hat{\omega}_{\mu ab} = \omega^0_{\mu ab}(e,\partial e) + \frac{iK^2}{2}\left(\bar{\Psi}_\mu^A\gamma_b\Psi_a^A - \bar{\Psi}_\mu^A\gamma_a\Psi_b^A + \bar{\Psi}_b^A\gamma_\mu\Psi_a^A\right) \tag{2.6}$$

$G_{\mu\nu}$ is defined such that the field equation of the vector field A_μ be

$$\partial_\mu(e\tilde{G}^{\mu\nu}) = 0 \tag{2.7}$$

$\mathcal{L}$ is invariant under the following supersymmetry transformations

$$\delta e_\mu{}^a = -iK\bar{\varepsilon}^A\gamma^a\Psi_\mu^A$$

$$\delta\psi_\mu^A = \frac{1}{\kappa} D_\mu^{AB} \epsilon^B$$
$$\delta A_\mu = - \varepsilon^{AB} \bar{\epsilon}^A \psi_\mu^B \tag{2.8}$$

with

$$D_\mu^{AB} = D_\mu(\hat{\omega})\delta^{AB} - \frac{i\kappa}{4} \varepsilon^{AB} \gamma^{\rho\sigma}\gamma_\mu \hat{F}_{\rho\sigma} \tag{2.9}$$

It is immediately checked that $\hat{F}$ and $\hat{\omega}$ are the supercovariant extensions of F and ω (they transform under supersymmetry without $\partial\epsilon$ terms). Defining the supercovariant field strength for the gravitino $\psi_{\mu\nu}^A$

$$\psi_{\mu\nu}^A = D_\mu^{AB}\psi_\nu^B - D_\nu^{AB}\psi_\mu^B \tag{2.10}$$

The gravitino field equation is very simple

$$\varepsilon^{\mu\nu\rho\sigma}\gamma_5\gamma_\nu \psi_{\rho\sigma}^A = 0 \tag{2.11}$$

Let us now come to the extra symmetries of N=2 supergravity. It has a built-in 0(2) symmetry which can be easily extended to SU(2) by chiral transformations on ψ_μ^A (Ferrara, Scherk and Zumino, 1977)

$$\delta\psi_\mu^A = (C^{AB} + i\gamma_5 D^{AB})\psi_\mu^B \tag{2.12}$$

where $\Lambda_A{}^B = C^{AB} + iD^{AB}$ is a traceless antihermitian matrix; $C^{AB} = -C^{BA}$ generating the O(2) transformations. For describing these transformations it is better to use left and right-handed spinors $\psi_{\mu A}^R$ and $\psi_{\mu L}^A$

$$\delta\psi_{\mu A}^R = \Lambda_A{}^B\psi_{\mu B}^R \,;\; \delta\psi_{\mu L}^A = (\Lambda_A{}^B)^*\psi_{\mu L}^B = -\Lambda_B{}^A\psi_{\mu L}^B \tag{2.13}$$

It is not possible to enlarge the symmetry of the Lagrangian, but we can define further U(1) chiral transformations on the field equations. Let us consider the following variations

$$\delta_{U(1)}\psi_\mu^A = -\frac{i}{2}\gamma_5\Lambda\psi_\mu^A \quad \left(\text{or } \delta_{U(1)}\psi_{\mu A}^R = -\frac{i}{2}\Lambda\psi_{\mu A}^R\right)$$

$$\delta_{U(1)} F_{\mu\nu} = -\Lambda G_{\mu\nu} \tag{2.14}$$

On-shell we have $\partial_\mu(e\tilde{G}^{\mu\nu}) = 0$ together with the Bianchi identity $\partial_\mu(e\tilde{F}^{\mu\nu}) = 0$ and the U(1) transformations can be transferred onto the vector field. Let us show now that the field equations are invariant. From the definition of $G_{\mu\nu}$ we deduce

$$\delta_{U(1)} G_{\mu\nu} = +\Lambda F_{\mu\nu} \tag{2.15}$$

and therefore the U(1) transformations exchange the A_μ field equation and the Bianchi identity. It can be rewritten as

$$\delta_{U(1)} \hat{F}_{\mu\nu} = \Lambda \tilde{\hat{F}}_{\mu\nu} \tag{2.16}$$

This shows immediately that

$$\delta_{U(1)}(\gamma^{\mu\nu}\hat{F}_{\mu\nu}) = \Lambda\gamma^{\mu\nu}\tilde{\hat{F}}_{\mu\nu} = -i\Lambda\gamma_5\gamma^{\mu\nu}\hat{F}_{\mu\nu} \tag{2.17}$$

and the gravitino field equation is also invariant. We can check also that the energy momentum tensor is invariant. This proves the complete invariance of the theory. Let us note that the Lagrangian although not invariant has the simple form

$$e^{-1}\mathcal{L} = \frac{1}{4} F_{\mu\nu}\tilde{G}^{\mu\nu} + e^{-1}\mathcal{L}_{inv} \tag{2.18}$$

where $\mathcal{L}_{inv}$ is invariant under the chiral U(1). This will be the starting point to the next section on generalized duality transformations.

3. GENERALIZED DUALITY TRANSFORMATIONS

We shall not discuss the complete formalism for describing the generalized duality transformations, but only the part directly relevant to supergravity theories. A more complete study of duality rotations for interacting fields has been discussed by Gaillard and Zumino (1981).

We consider theories with abelian gauge invariance in which the vector fields A_μ^i appear only through their field strengths $F_{\mu\nu}^i$ and we note by Φ^N the other fields of the theory (including the gravitational field $g_{\mu\nu}$)

$$\mathcal{L} = e\,L(\Phi^N, F_{\mu\nu}^i) \tag{3.1}$$

Under transformations $\Delta\phi^N$ and $\Delta F^i_{\mu\nu}$, the variations of L and $\delta L/\delta\phi^N$ are

$$\Delta L = \frac{\delta L}{\delta\phi^N}\Delta\phi^N + \frac{1}{2}\frac{\delta L}{\delta F^i_{\mu\nu}}\Delta F^i_{\mu\nu}$$

$$\Delta\left(\frac{\delta L}{\delta\phi^N}\right) = -\frac{\delta(\Delta\phi^M)}{\delta\phi^N}\frac{\delta L}{\delta\phi^M} + \frac{\delta}{\delta\phi^N}(\Delta L) - \frac{1}{2}\frac{\delta L}{\delta F^i_{\mu\nu}}\frac{\delta(\Delta F^i_{\mu\nu})}{\delta\phi^N} \tag{3.2}$$

where the factor 1/2 takes the antisymmetry in (μ,ν) into account.

Let us now apply these results to the specific form (2.18)

$$L = \frac{1}{4}F^i_{\mu\nu}\tilde{G}^{\mu\nu}_i + L_{inv}(\phi^N, F^i_{\mu\nu}) \tag{3.3}$$

where $\tilde{G}^{\mu\nu}_i$ is a function of ϕ^N and $F^i_{\mu\nu}$ given by

$$\tilde{G}^{\mu\nu}_i = \delta L/\delta F^i_{\mu\nu} \tag{3.4}$$

We shall show that if L_{inv} is invariant under some subgroup G of Sp(2n) (bigger than GL(n,R)) and the relation between and is covariant under G, then although the Lagrangian L is not invariant (because of the first term), the field equations are. Let us consider the Sp(2n) transformations on $F^i_{\mu\nu}$ and $G_{\mu\nu i}$

$$\delta F^i_{\mu\nu} = A^i{}_j F^j_{\mu\nu} + B^{ij}G_{\mu\nu j}$$

$$\delta G_{\mu\nu i} = C_{ij}F^j_{\mu\nu} - A^j{}_i G_{\mu\nu j} \tag{3.5}$$

with $C_{ij} = C_{ji}$, $B^{ij} = B^{ji}$

The coefficients A, B, C are further restricted by the condition that L_{inv} and the relation (3.4) should be invariant (with appropriate $\Delta\phi^N$). It is then obvious that the field equations and the Bianchi identities

$$\begin{cases} \partial_\mu(e\,\tilde{G}^{\mu\nu}_i) = 0 \\ \partial_\mu(e\,\tilde{F}^{\mu\nu i}) = 0 \end{cases} \tag{3.6}$$

are exchanged by transformations (3.5). When both equations are satisfied (<u>on-shell</u>) we can realize these transformations on the vectors A^i_μ and $B_{\mu i}$ defined by

$$G_{\mu\nu i} = \partial_\mu B_{\nu i} - \partial_\nu B_{\mu i} \tag{3.7}$$

Let us now consider the field equations of ϕ^N

$$\Delta L = \frac{1}{4}\left(G_{\mu\nu i} B^{ij} \tilde{G}^{\mu\nu}_{j} + F^{i}_{\mu\nu} C_{ij} \tilde{F}^{\mu\nu j}\right)$$

$$\frac{\delta(\Delta L)}{\delta\phi^N} = \frac{1}{2}\,\frac{\delta G_{\mu\nu i}}{\delta\phi^N}\, B^{ij}\,\tilde{G}^{\mu\nu}_{j}$$

$$\frac{\delta(\Delta F^{i}_{\mu\nu})}{\delta\phi^N} = B^{ij}\,\frac{\delta G_{\mu\nu j}}{\delta\phi^N}$$

therefore $\displaystyle \frac{\delta(\Delta L)}{\delta\phi^N} - \frac{1}{2}\,\frac{\delta L}{\delta F^{i}_{\mu\nu}}\,\frac{\delta(\Delta F^{i}_{\mu\nu})}{\delta\phi^N} = 0$

It follows from (3.2) that

$$\Delta\left(\frac{\delta L}{\delta\phi^N}\right) = -\,\frac{\delta(\Delta\phi^M)}{\delta\phi^N}\,\frac{\delta L}{\delta\phi^M} \tag{3.8}$$

and therefore the field equations of ϕ^N are invariant under G. These properties of supergravity theories will be a very powerful tool in the building of these theories as has been seen in the derivation of N=4 supergravity.

4. N=4 SUPERGRAVITIES

N=4 supergravity describes the interactions of 1 graviton $g_{\mu\nu}(e_\mu{}^a)$, 4 gravitinos ψ_μ^A , 6 vector fields $A_\mu^{AB} = -A_\mu^{BA}$ and 1 complex scalar field $A+iB$. There appears for the first time a scalar field with its related complications due to the fact that KA and KB , being without dimension , can appear non polynomially in the Lagrangian. This forbids the order by order in K determination of the Lagrangian and supersymmetric transformations. We must use functional techniques by parametrizing them with arbitrary functions of $z = K(A+iB)$ and $\bar{z}$. Although it is not necessary, in principle, an enormous simplification takes place if we conjecture that the theory is U(4) invariant in the same way that the N=2 supergravity is U(2) invariant. It is only an on-shell symmetry (invariance of equations of motion) and involves duality transformations discussed in the previous section.

O(4) formulation of N=4 supergravity

The conjecture of the U(4) invariance has allowed to determine with a minimum of complex calculation the Lagrangian and supersymmetry transformations of N=4 supergravity (Cremmer, Ferrara and Scherk, 1977; Cremmer and Scherk, 1977). In the 1.5 order formalism, the Lagrangian is written as

$$
\begin{aligned}
e^{-1}\mathcal{L} = & -\frac{1}{4K^2}R(\omega) - \frac{e^{-1}}{2}\varepsilon^{\lambda\mu\nu\rho}\bar{\psi}_\lambda^A\gamma_5\gamma_\mu\left(D_\nu(\omega) + \frac{iK^2}{2a}\gamma_5 A\overleftrightarrow{\partial}_\nu B\right)\psi_\rho^A \\
& + \frac{1}{8}F_{\mu\nu}^{AB}\tilde{G}^{\mu\nu AB} + \frac{i}{2}\bar{\chi}^A\gamma^\mu\left(D_\mu(\omega) + \frac{3iK^2}{2a}\gamma_5 A\overleftrightarrow{\partial}_\nu B\right)\chi^A \\
& + \frac{1}{2a^2K^2}\partial_\mu z\partial_\mu\bar{z} - \frac{1}{2\sqrt{2}}\frac{K}{a}\bar{\psi}_\mu^A\left(\partial_\nu(A-i\gamma_5 B) + \hat{D}_\nu(A-i\gamma_5 B)\right)\gamma^\nu\gamma^\mu\chi^A \\
& - \frac{1}{16}Ka^{-1/2}(\hat{F}+\hat{H})_{\mu\nu}^{AB}\left(P_{AB}^{\mu\nu} + 2\sqrt{2}Q_{AB}^{\mu\nu}\right)
\end{aligned}
\tag{4.1}
$$

where

$$
\begin{aligned}
\hat{D}_\mu A &= \partial_\mu A - \frac{K}{\sqrt{2}}a\bar{\psi}_\mu^A\chi^A \\
\hat{D}_\mu B &= \partial_\mu B + i\frac{K}{\sqrt{2}}a\bar{\psi}_\mu^A\gamma_5\chi^A
\end{aligned}
\tag{4.2}
$$

$$
\begin{aligned}
\hat{F}_{\mu\nu}^{AB} = & F_{\mu\nu}^{AB} - \frac{iK}{\sqrt{2a}}\left(\varepsilon^{ABCD}\bar{\psi}_{[\mu}^C\gamma_{\nu]}\chi^D - K\bar{\psi}_{[\mu}^{[A}\gamma_{\nu]}(A-iB\gamma_5)\chi^{B]}\right) \\
& - \frac{K}{\sqrt{a}}\left(\bar{\psi}_{[\mu}^{[B}\psi_{\nu]}^{A]} + K\varepsilon^{ABCD}\bar{\psi}_\mu^C(A-iB\gamma_5)\psi_\nu^D\right)
\end{aligned}
\tag{4.3}
$$

$$
\begin{aligned}
P_{\mu\nu}^{AB} &= 2\bar{\psi}_{[\mu}^A\psi_{\nu]}^B - ie^{-1}\varepsilon_{\mu\nu\rho\sigma}\bar{\psi}^{A\rho}\gamma_5\psi^{B\sigma} \\
Q_{\mu\nu}^{AB} &= \frac{i}{4}\varepsilon^{ABCD}\bar{\psi}_\lambda^C\gamma_{\mu\nu}\gamma^\lambda\chi^D
\end{aligned}
\tag{4.4}
$$

(4.5)

$$
F_{\mu\nu}^{AB}\tilde{G}^{AB\mu\nu} = -F_{\mu\nu}^{AB}H^{\mu\nu AB} - (F_{\mu\nu}^{AB} + H_{\mu\nu}^{AB})\frac{K\sqrt{a}}{2}(P^{\mu\nu AB} + 2\sqrt{2}Q^{\mu\nu AB})
\tag{4.6}
$$

with $H_{\mu\nu}^{AB}$ defined by

$$
H_{\mu\nu}^{AB} - i\tilde{H}_{\mu\nu}^{AB} = \frac{1+z^2}{1-z^2}(F_{\mu\nu}^{AB} - i\tilde{F}_{\mu\nu}^{AB}) + \frac{z}{1-z^2}\varepsilon^{ABCD}(F_{\mu\nu}^{CD} - i\tilde{F}_{\mu\nu}^{CD})
\tag{4.7}
$$

$$a = 1 - z\bar{z} \tag{4.8}$$

This Lagrangian is invariant under the following supersymmetry transformations

$$\delta e_\mu{}^a = -i\kappa\,\bar{\epsilon}^A \gamma^a \psi_\mu^A$$

$$\delta A_\mu^{AB} = \frac{i}{\sqrt{2a}}\left[\varepsilon^{ABCD}\,\bar{\epsilon}^C \gamma_\mu \chi^D - \kappa\,\bar{\epsilon}^{[A}\gamma_\mu (A - iB\gamma_5)\chi^{B]}\right]$$
$$+ \frac{1}{\sqrt{a}}\left[\bar{\epsilon}^{[B}\psi_\mu^{A]} + \kappa\,\varepsilon^{ABCD}\,\bar{\epsilon}^C (A - iB\gamma_5)\psi_\mu^D\right] \tag{4.9}$$

$$\delta A = \frac{a}{\sqrt{2}}\,\bar{\epsilon}^A \chi^A \quad , \quad \delta B = -\frac{ia}{\sqrt{2}}\,\bar{\epsilon}^A \gamma_5 \chi^A$$

$$\delta \bar{\chi}^A = \frac{i}{a\sqrt{2}}\,\bar{\epsilon}^A \hat{D}_\mu (A - iB\gamma_5)\gamma^\mu + \frac{1}{4\sqrt{2}}\,a^{1/2}\varepsilon^{ABCD}(\hat{F}_{\mu\nu}^{CD} + \hat{H}_{\mu\nu}^{CD})\bar{\epsilon}^B \gamma^{\mu\nu}$$
$$- \frac{3}{2\sqrt{2}}\kappa^2\,\bar{\epsilon}^B \gamma_5 (A - iB\gamma_5)\chi^B\,\bar{\chi}^A \gamma_5$$

$$\delta \bar{\psi}_\mu^A = \frac{1}{\kappa}\bar{\epsilon}^B(\overleftarrow{D}_\mu \delta^{AB} + \Omega_\mu^{AB}) + \frac{\kappa}{2\sqrt{2}}\varepsilon^{ABCD}(\bar{\psi}_\mu^C \gamma^a \chi^D \bar{\epsilon}^B \gamma_a + \bar{\psi}_\mu^C \gamma_5 \gamma^a \chi^D \bar{\epsilon}^B \gamma_5 \gamma_a)$$
$$+ \frac{1}{2\sqrt{2}}\kappa^2\left[(\bar{\chi}^B \gamma_5 (A - iB\gamma_5)\psi_\mu^B)\bar{\epsilon}^A \gamma_5 - (\bar{\chi}^B \gamma_5 (A - iB\gamma_5)\epsilon^B\,\bar{\psi}_\mu^A \gamma_5\right]$$

with

$$\Omega_\mu^{AB} = \frac{i\kappa^2}{2a}\gamma_5 A \overleftrightarrow{\hat{D}}_\mu B\,\delta^{AB} - \frac{i\kappa}{8}a^{-1/2}(\hat{F} + \hat{H})_{\rho\sigma}^{AB}\gamma^{\rho\sigma}\gamma_\mu$$
$$+ \frac{i\kappa^2}{4}\left[(\bar{\chi}^C \gamma_5 \gamma^a \chi^C)\gamma_5 \gamma_a \gamma_\mu \delta^{AB} + (\bar{\chi}^C \gamma^a \chi^C)\gamma_a \gamma_\mu \delta^{AB} - (\bar{\chi}^A \gamma_5 \gamma^a \chi^B)\gamma_5 \gamma_a \gamma_\mu\right] \tag{4.10}$$

We verify immediately that $\hat{D}_\mu A$, $\hat{D}_\mu B$, $\hat{F}_{\mu\nu}^{AB}$ are supercovariant derivatives. We can also define the supercovariant derivative of χ^A and supercovariant field strength of ψ_μ^A

$$\hat{D}_\mu \chi^A = D_\mu(\omega)\chi^A + \frac{i\kappa}{a\sqrt{2}}\gamma^\rho \hat{D}_\rho (A - iB\gamma_5)\psi_\mu^A$$
$$+ \frac{\kappa}{4\sqrt{2}}a^{-1/2}\varepsilon^{ABCD}(\hat{F} + \hat{H})_{\rho\sigma}^{CD}\gamma^{\rho\sigma}\psi_\mu^B + \frac{3\kappa^3}{2\sqrt{2}}(\bar{\psi}_\mu^B \gamma_5 (A - iB\gamma_5)\chi^B)\gamma_5 \chi^A \tag{4.11}$$

$$\psi_{\mu\nu}^{A} = D_{[\mu}\psi_{\nu]}^{A} + \Omega_{[\mu}^{AB}\psi_{\nu]}^{B} - \frac{\kappa^2}{\sqrt{2}}\varepsilon^{ABCD}[(\bar{\psi}_\mu^B\psi_\nu^C)\chi^D - (\bar{\psi}_\mu^B\gamma_5\psi_\nu^C)\gamma_5\chi^D]$$
$$+ \frac{\kappa^3}{2\sqrt{2}}(\bar{\chi}^B\gamma_5(A - iB\gamma_5)\psi_{[\nu}^B)\gamma_5\psi_{\mu]}^A \tag{4.12}$$

These covariant objects allow to write the fermionic field equations under a simple form

$$e^{-1}\varepsilon^{\lambda\mu\nu\rho}\gamma_5\gamma_\mu\psi_{\nu\rho}^A + \frac{\kappa\sqrt{2}}{a}\hat{D}_\rho(A - iB\gamma_5)\gamma^\rho\gamma^\lambda\chi^A$$
$$+ \frac{i\kappa a^{-1/2}}{2\sqrt{2}}\gamma^{\mu\nu}\gamma^\lambda\chi^B\varepsilon^{ABCD}(\hat{F} + \hat{H})_{\mu\nu}^{CD} = 0 \tag{4.13}$$

$$i\gamma^\mu\hat{D}_\mu\chi^A + \frac{3\kappa^2}{2a}\gamma_5\gamma^\mu\chi^A A\overleftrightarrow{\hat{D}_\mu}B = 0 \tag{4.14}$$

Using the results of the previous sections, we verify immediately that the theory is invariant under the following U(4) transformation

$$\delta_U e_\mu{}^a = 0$$
$$\delta_U A = -B\Lambda_A{}^A \quad , \quad \delta_U B = A\Lambda_A{}^A$$
$$\delta_U\psi_\mu^A = i\gamma_5\Lambda^{AB}\psi_\mu^B \quad , \quad \delta_U\chi^A = -i\gamma_5(\Lambda^{AB} - \delta^{AB}\Lambda_C^C)\chi^B$$
$$\delta_U F_{\mu\nu}^{AB} = -\Lambda^{AC}G_{\mu\nu}^{CB} - \Lambda^{BC}G_{\mu\nu}^{AC} \tag{4.15}$$

with $\Lambda^{AB} = \Lambda^{BA}$ (we have not written the pure O(4) part of the transformations). The last transformations can be rewritten in a supercovariant form

$$\delta_U\hat{F}_{\mu\nu}^{AB} = -\Lambda^{AC}\tilde{\hat{H}}_{\mu\nu}^{CB} - \Lambda^{BC}\tilde{\hat{H}}_{\mu\nu}^{AC} \tag{4.16}$$

The relation (4.7) ensures immediately that

$$\delta_U\tilde{\hat{H}}_{\mu\nu}^{AB} = \Lambda^{AC}\hat{F}_{\mu\nu}^{CB} + \Lambda^{BC}\hat{F}_{\mu\nu}^{AC} \tag{4.17}$$

and that $\hat{F}^{AB}_{\mu\nu} + \hat{H}^{AB}_{\mu\nu}$ transforms into itself by U(4) transformations. As has been mentioned previously, it can be useful to use left-handed and right-handed spinors as well as the combination $F \pm i\tilde{F}$ to define, in a simple way, the U(4) transformations. We note also that the vector fields which appear only through $F^{A}_{\mu\nu}$ are coupled to the fermion always via $\gamma^{\mu\nu}$ which satisfies self duality condition

$$\tilde{\gamma}_{\mu\nu} = -i\gamma_5\gamma_{\mu\nu} \tag{4.18}$$

SU(1,1) invariance of O(4) supergravity

We have constructed the theory by conjecturing the U(4) invariance. In fact if we study carefully the Lagrangian we discover with surprise that there is a bigger symmetry due to the presence of scalar fields. For pedagogical reasons let us concentrate on the bosonic part of the Lagrangian

$$e^{-1}L_B = -\frac{1}{4K^2}R - \frac{1}{8}F^{AB}_{\mu\nu}H^{\mu\nu}_{AB} + \frac{1}{2K^2}\frac{\partial_\mu z\,\partial_\mu\bar{z}}{(1-z\bar{z})^2} \tag{4.19}$$

with

$$H_{\mu\nu AB} - i\tilde{H}_{\mu\nu AB} = \mathcal{M}_{AB,CD}(F^{CD}_{\mu\nu} - i\tilde{F}^{CD}_{\mu\nu}) \tag{4.20}$$

Using contraction on pairs of antisymmetric indices for matrix multiplication with the unity defined by

$$\mathbb{1}_{AB,CD} = \frac{1}{2}(\delta_{AC}\delta_{BD} - \delta_{AD}\delta_{BC}) = \delta^{CD}_{AB} \tag{4.21}$$

and $$\mathbb{Z}_{AB,CD} = \frac{1}{2}\varepsilon_{ABCD}\,z \tag{4.22}$$

$\mathcal{M}$ can be written, from (4.7), as

$$\mathcal{M} = (\mathbb{1}+\mathbb{Z})(\mathbb{1}-\mathbb{Z})^{-1} \tag{4.23}$$

By using self and antiselfdual combinations in the indices AB we can further simplify these formulas and use only the scalar field z . The kinetic energy term of the scalar field z is well known in non-linear sigma model and is invariant under the following transformations

$$z \longrightarrow \frac{az+b}{cz+d} \quad ; \quad |a|^2 - |b|^2 = 1 \tag{4.24}$$

This generates the non-compact group SU(1,1). The scalar fields parametrize the coset space SU(1,1)/U(1). This structure will be discussed in detail in the next section. We must now show that we can extend these transformations onto F and $\tilde{H}$ so that the relation (4.7) be preserved. From the previous section this will ensure the SU(1,1) invariance of the theory. Writing $a = r + it$, $b = s + iq$ this is realized by the following transformations

$$F^{AB}_{\mu\nu} \rightarrow (r\,\delta^{AB}_{CD} - \frac{s}{2}\,\varepsilon^{ABCD})\,F^{CD}_{\mu\nu} - (t\,\delta^{AB}_{CD} + \frac{q}{2}\,\varepsilon^{ABCD})\,\tilde{H}^{CD}_{\mu\nu}$$

$$\tilde{H}^{AB}_{\mu\nu} \rightarrow (t\,\delta^{AB}_{CD} - \frac{q}{2}\,\varepsilon^{ABCD})\,F^{CD}_{\mu\nu} + (r\,\delta^{AB}_{CD} + \frac{s}{2}\,\varepsilon^{ABCD})\,\tilde{H}^{CD}_{\mu\nu} \tag{4.25}$$

This can be extended to fermions by chiral transformations and (4.25) has to be modified to include them. This is simply done by replacing F and $\tilde{H}$ by their supercovariant extension $\hat{F}$ and $\hat{\tilde{H}}$. This SU(1,1) contains the U(1) factor of U(4) discussed previously (it corresponds to $b = 0$).

SU(4) supergravity

The N=1 supergravity in 10 dimensions with field content $e_\mu{}^r$, $A_{\mu\nu}$, ϕ , ψ_μ , χ can be reduced to 4 dimensions by demanding that the fields do not depend on the 6 extra coordinates and we get an N=4 supergravity coupled to 6 N=4 Yang-Mills multiplets. The N=4 supergravity obtained in this way has a manifest O(6) invariance. Since O(6) is isomorphic to SU(4) this suggests the existence of a version of N=4 supergravity with a manifest SU(4) invariance of the Lagrangian (Gliozzi, Scherk and Olive (1977)). The spectrum of this theory consists of 1 graviton $e_\mu{}^r$, 4 Majorana gravitino ψ'^A_μ, 3 vector fields A^n_μ, 3 axial fields B^n_μ, 4 Majorana spinors χ'^A, one scalar A' and one pseudoscalar B'. This pseudoscalar field is obtained from the antisymmetric tensor $A_{\mu\nu}$ from 10 dimensions after a duality transformation and it can be conjectured that it will appear only through its derivative $\partial_\mu B'$. The complete theory has been written by Cremmer, Ferrara and Scherk (1978). For simplicity we shall write only the bosonic part of the Lagrangian. Let us note

$$z' = K(A' + iB')$$

$$e^{-1}\mathcal{L}_B = -\frac{1}{4K^2}R(\omega) + \frac{1}{2K^2}g^{\mu\nu}\frac{\partial_\mu z' \partial_\nu \bar{z}'}{(1-z'-\bar{z}')^2}$$
$$-\frac{1}{4}\left[(A^n_{\mu\nu}A^{\mu\nu n} + B^n_{\mu\nu}B^{\mu\nu n})(1-2KA') - 2KB'(A^n_{\mu\nu}\tilde{A}^{\mu\nu n} + B^n_{\mu\nu}\tilde{B}^{\mu\nu n})\right]$$

(4.26)

The SU(4) transformations act only on A^n_μ and B^n_μ

$$\delta A^n_\mu = \varepsilon^{nmr}\Lambda_r A^m_\mu - \Lambda''_{nm}B^m_\mu$$
$$\delta B^n_\mu = \varepsilon^{nmr}\Lambda'_r B^m_\mu - \Lambda''_{mn}A^m_\mu$$

(4.27)

Λ_r, Λ'_r, Λ_{nm} are 15 real parameters. As expected B' appears only through its derivative $\partial_\mu B'$ since $e\,A^n_{\mu\nu}\tilde{A}^{\mu\nu n}$ is a total derivative. There is still an invariance SU(1,1) of the field equations. In fact the two theories are equivalent on-shell: they have the same equations of motion after field redefinition. We must perform a duality transformation on the axial fields to turn them into vector fields. It is easy to find the redefinition of the scalar fields: it is a SL(2,C) transformation defined by

$$(1-z)(1-z') = 1$$

(4.28)

The vector fields A^n_μ are algebraically related to A^{AB}_μ.

The axial fields B_μ^n are related to A_μ^{AB} by duality transformations which exchange field equations and Bianchi identity and are valid only on-shell. Let us introduce the 4x4 matrices α^n_{AB} and β^n_{AB} respectively selfdual and antiselfdual for the indices AB which connect SO(4) to SU(2)xSU(2). The relations for vector fields are

$$\alpha^n_{AB} A^n_{\mu\nu} = \frac{1}{\sqrt{2}} \left(F^{AB}_{\mu\nu} + \frac{1}{2} \varepsilon^{ABCD} F^{CD}_{\mu\nu} \right)$$

$$\beta^n_{AB} B^n_{\mu\nu} = -\frac{1}{\sqrt{2}} \left(\tilde{H}^{AB}_{\mu\nu} - \frac{1}{2} \varepsilon^{ABCD} \tilde{H}^{CD}_{\mu\nu} \right)$$

$$\beta^n_{AB} H^{Bn}_{\mu\nu} = -\frac{1}{\sqrt{2}} \left(\tilde{F}^{AB}_{\mu\nu} - \frac{1}{2} \varepsilon^{ABCD} \tilde{F}^{CD}_{\mu\nu} \right)$$

with

$$H^{Bn}_{\mu\nu} = (1-2KA') B^n_{\mu\nu} - 2KB' \tilde{B}^n_{\mu\nu} \qquad (4.29)$$

When fermionic fields are included we must also perform chiral transformations on these fields (using left and right-handed spinors)

$$\psi'^R_{\mu A} = \exp i\left(\frac{\theta}{2} + \frac{\pi}{2}\right) \psi^R_{\mu A}$$

$$\chi'^R_A = \exp i\left(\frac{3\theta}{2} + \frac{\pi}{2}\right) \chi^R_A \qquad (4.30)$$

with

$$\exp 2i\theta = \frac{1-z}{1-\bar{z}} = \frac{1-\bar{z}'}{1-z'} \qquad (4.31)$$

In (4.25) we must also replace the field strength by its supercovariant generalization.

5. NON-COMPACT COSET SPACES AND NON-LINEAR SIGMA MODELS

The non-linear sigma models associated with a group G are described by the Lagrangian

$$\mathcal{L} = \mathrm{Tr}(\mathcal{V}^{-1}\partial_\mu \mathcal{V})^2 = - \mathrm{Tr}(\partial_\mu \mathcal{V} \partial_\mu \mathcal{V}^{-1}) \tag{5.1}$$

where $\mathcal{V}$ is an element of the group G, $\mathcal{V}^{-1}\partial_\mu \mathcal{V}$ being therefore an element of the Lie algebra of G. This Lagrangian is invariant under global transformations generating a group isomorph to GxG

$$\mathcal{V} \rightarrow g_1 \mathcal{V} g_2 \quad ; \quad g_1, g_2 \in G \tag{5.2}$$

If G is non-compact the Killing metric contains positive <u>and</u> negative signs, and the hamiltonian of such a theory is not positive definite.
The problem is solved as usual by introducing a gauge field such that the theory has an extra gauge invariance which eliminates the ghost fields. Let H be the maximal compact subgroup of G. It is easy to write a Lagrangian which is invariant under global G x local H

$$\mathcal{V} \rightarrow h(x)\, \mathcal{V}\, g$$

$$h \in H \quad , \quad g \in G \tag{5.3}$$

Denoting the gauge field for the local H invariance by Ω_μ which is an element of the Lie algebra of H, the Lagrangian is written

$$\mathcal{L} \simeq \mathrm{Tr}(\mathcal{V}^{-1} D_\mu \mathcal{V})^2 \tag{5.4}$$

with $D_\mu \mathcal{V} = \partial_\mu \mathcal{V} - \Omega_\mu \mathcal{V}$ (5.5)

The field Ω_μ has no kinetic terms and has an algebraic field equation and therefore can be eliminated from the Lagrangian.
The Lie algebra of G $\mathcal{L}(G)$ can be separated into two parts: the Lie algebra of H $\mathcal{L}(H)$ and its orthogonal part with respect to the Killing metric K. If $q \in \mathcal{L}(H)$ and $p \in K$ we have the following commutation relations which mean that H is a maximal subgroup of G

$$\begin{aligned} [q_1, q_2] &\in \mathcal{L}(H) \\ [q, p] &\in K \\ [p_1, p_2] &\in \mathcal{L}(H) \end{aligned} \tag{5.6}$$

Any element e of $\mathcal{L}(G)$ can be written in a unique way as

$$e = p + q = e_\perp + e_{//} \tag{5.7}$$

in particular

$$\partial_\mu \mathcal{V} \mathcal{V}^{-1} = (\partial_\mu \mathcal{V} \mathcal{V}^{-1})_{//} + (\partial_\mu \mathcal{V} \mathcal{V}^{-1})_\perp \equiv Q_\mu + P_\mu \tag{5.8}$$

The field equation of Ω_μ is

$$\Omega_\mu = Q_\mu \tag{5.9}$$

and the Lagrangian becomes

$$\mathcal{L} = \mathrm{Tr}\ (P_\mu)^2 \tag{5.10}$$

The fact that H is the maximal compact subgroup of G ensures that $\mathcal{L}$ is now definite positive. This Lagrangian is still gauge invariant although there is no longer gauge fields. It is easy to show that if

$$\mathcal{V} \to h(x)\, \mathcal{V}\, g$$

then

$$P_\mu \to h\, P_\mu\, h^{-1}$$

$$Q_\mu \to h\, Q_\mu\, h^{-1} + \partial_\mu h\, h^{-1} \tag{5.11}$$

$\mathcal{L}$ has been written in terms of P_μ which is invariant under G and covariant under H. It is also possible to write it in terms of the metric on the coset space G/H which is covariant for G and invariant under H.

We shall not discuss here the most general case of coset spaces or their classification and shall assume that H is a unitary group or a subgroup of a unitary group. This is not essential but will simplify the discussion. More precisely we assume that the elements of $\mathcal{L}(H)$ are anti-hermitian and those of K hermitian. The metric on the coset space is then definite positive and given by

$$g = \mathcal{V}^+ \mathcal{V} \tag{5.12}$$

which is obviously invariant under the local group H. It is easy to check that

$$\mathcal{L} = -\frac{1}{4}\,\mathrm{Tr}\,(\partial_\mu g\, \partial_\mu g^{-1}) \tag{5.13}$$

If the number of parameters of G is N and those of H M, the theory describes N-M scalar fields. Due to the gauge invariance there is an arbitrariness in the choice of parametrization. A particularly useful parametrization is the hermitian one. Since $\mathcal{V}^{+}\mathcal{V}$ is positive definite we can always define its square root as a positive hermitian matrix $(\mathcal{V}^{+}\mathcal{V})^{1/2}$, writing

$$\mathcal{V} = h\,(\mathcal{V}^{+}\mathcal{V})^{1/2} \tag{5.14}$$

We see immediately that h is a unitary matrix, therefore by a gauge transformation we can always take $\mathcal{V}$ hermitian. It can then be written as

$$\mathcal{V} = \mathcal{V}^{+} = \exp p \quad ; \quad p \in K \tag{5.15}$$

The maximal subgroup which acts linearly on p is the diagonal subgroup isomorph to H defined by

$$\mathcal{V} \rightarrow h\,\mathcal{V}\,h^{-1}$$
$$p \rightarrow h\,p\,h^{-1}$$

with h a constant element of H.

Let us assume now for definiteness that we can write an element of $\mathcal{L}(G)$ as

$$\begin{pmatrix} U & W \\ W^{+} & V \end{pmatrix} \tag{5.16}$$

with $\begin{pmatrix} U & 0 \\ 0 & V \end{pmatrix} \in \mathcal{L}(H)$ antihermitian ; $\begin{pmatrix} 0 & W \\ W^{+} & 0 \end{pmatrix} \in K$

Following a general method for parametrizing coset spaces we can introduce the following matrix variable

$$z = w \frac{\operatorname{th}\sqrt{w^+ w}}{\sqrt{w^+ w}} \tag{5.17}$$

n the hermitian gauge $\mathcal{V}$ is then simply written as

$$\mathcal{V} = \begin{pmatrix} \frac{1}{\sqrt{1-zz^+}} & ; & z\frac{1}{\sqrt{1-z^+z}} \\ z^+\frac{1}{\sqrt{1-zz^+}} & ; & \frac{1}{\sqrt{1-z^+z}} \end{pmatrix} \tag{5.18}$$

5.17) ensures the hermitian matrices $1-zz^+$ and $1-z^+z$ are positive definite. The action of G on Z is very simple, and efined by

$$\mathcal{V}(z)\, g = h(z,g)\, \mathcal{V}(z') \tag{5.19}$$

ith $h \in H$

riting $g = \begin{pmatrix} A & B \\ C & D \end{pmatrix}$, $h = \begin{pmatrix} U & 0 \\ 0 & V \end{pmatrix}$ (5.20)

e get immediately

$$z' = (A + zC)^{-1}(B + zD) \tag{5.21}$$

he z 's are called the inhomogeneous coordinates of the coset /H. Finally in terms of z, P_μ is written in a simple form

$$P_\mu = \begin{pmatrix} 0 & ; & \frac{1}{\sqrt{1-zz^+}}\,\partial_\mu z\,\frac{1}{\sqrt{1-z^+z}} \\ \frac{1}{\sqrt{1-z^+z}}\,\partial_\mu z^+\,\frac{1}{\sqrt{1-zz^+}} & ; & 0 \end{pmatrix} \tag{5.22}$$

nd the Lagrangian becomes

$$\mathcal{L} = 2\,\mathrm{Tr}\left(\frac{1}{1-ZZ^{+}}\,\partial_\mu Z\,\frac{1}{1-Z^{+}Z}\,\partial_\mu Z^{+}\right) \tag{5.23}$$

Before ending this section let us apply these results to N=4 supergravity, whose scalar fields are described by the coset space SU(1,1)/U(1).

$$\mathcal{V} = \begin{pmatrix} \varphi_0 , & \varphi_1 \\ \bar{\varphi}_1 , & \bar{\varphi}_0 \end{pmatrix}$$

with $|\varphi_0|^2 - |\varphi_1|^2 = 1$ is an element of SU(1,1). The subgroup U(1) acts on the left of $\mathcal{V}$ and consists of the matrices

$$U = \begin{pmatrix} e^{i\theta} & 0 \\ 0 & e^{-i\theta} \end{pmatrix}$$

The inhomogeneous coordinate is $z = \varphi_1/\varphi_0$ and the Lagrangian is

$$\mathcal{L} \simeq \frac{\partial_\mu z\,\partial_\mu \bar{z}}{(1-z\bar{z})^2} \tag{5.24}$$

A consequence of supersymmetry is the relation between the kinetic term of the scalar field and the Noether coupling

$$K\,\bar{\Psi}_\mu^{A}\,\frac{\partial_\nu (A - iB\gamma_5)}{1-z\bar{z}}\,\gamma^\nu\gamma^\mu\chi^A \tag{5.25}$$

This is a general feature of extended supergravity. In this case we can easily compute Q_μ in the hermitian gauge

$$Q_\mu \simeq \frac{A\overleftrightarrow{\partial_\mu}B}{1-z\bar{z}} \tag{5.26}$$

Q_μ appears in the N=4 Lagrangian together with the derivative of fermionic fields

$$\left[D_\mu(\omega) + i\kappa^2 \gamma_5 \frac{A \overleftrightarrow{\partial_\mu} B}{2(1-z\bar{z})} \right] \psi_\rho^A$$

$$\left[D_\mu(\omega) + 3i\kappa^2 \gamma_5 \frac{A \overleftrightarrow{\partial_\mu} B}{2(1-z\bar{z})} \right] \chi^A \qquad (5.27)$$

This suggests that the structure SU(1,1) global x U(1) local can be completely extended to the N=4 supergravity. SU(1,1) acts on the vector fields (on-shell) and the scalar fields, local U(1) acts on the scalar fields and the spin 1/2 field. This is easily verified.

6. N=8 SUPERGRAVITY

The N=8 supergravity describes the interactions of 1 graviton $g_{\mu\nu}$, 8 Majorana gravitinos $\psi_{\mu A}$, 28 vector particles $A_\mu^{MN} = - A_\mu^{NM}$, 56 Majorana spin 1/2 particles λ_{ABC} (anti symmetric in A,B,C) and 70 scalar fields $W_{ABCD} = \frac{y}{24} \varepsilon_{ABCDEFGH} \bar{W}^{EFGH}$ $(y = \pm 1)$ (indices running over 1...8). The first invariances the theory must have are:

-reparametrization invariance in 4 dimensions
-local Lorentz invariance SO(3,1)
-8 local supersymmetries
-28 abelian gauge invariances.

As in N=4 supergravity we can also conjecture an SU(8) global invariance on-shell (U(8) is forbidden by the relation for the scalar fields between W and $\bar{W}$). However this has not been sufficient to build directly the N=8 supergravity (de Wit and Freedman (1977), de Wit (1979)). The starting point has been the N=1 supergravity in 11 dimensions (Cremmer, Julia and Scherk (1978)) from which we can derive the N=8 supergravity by dimensional reduction (Cremmer and Julia (1978), (1979)). This has shown that the theory has a much bigger invariance generalizing the case of N=4 supergravity. By conjecturing these symmetries we can build the N=8 supergravity directly. We have seen that the 2 scalar fields of N=4 supergravity are described by a non-linear sigma model based on the coset space SU(1,1)/U(1). In N=8 supergravity we have 70 scalar fields which spanned a representation 70 of SU(8). From the previous section, the maximal linear group is isomorphic to the local group $H \approx SU(8)$. This implies that if we can describe these scalars by a coset space G/H, G must have 70+63=133 parameters. There is only one semi-simple group with this number of generators, namely E_7. In order to have a positive definite theory, SU(8) must be the maximal compact subgroup of E_7. We can check immediately that there exists precisely a non-compact form of E_7 (denoted by $E_{7(+7)}$) which has SU(8) as a maximal compact subgroup (+7= 70-63 is the signature of the non-compact group). E_7 must be the group of duality transformations, since there are 28 vector fields, there must be a representation 56 of E_7: it is in fact the fondamental representation of E_7. We can therefore conjecture that there are a global E_7 group which acts on scalar fields and generates duality transformations and a local SU(8) group acting on the scalar and spinor fields. The scalar fields will be represented by a 56x56 matrix $\mathcal{V}$ element of E_7 which can be interpreted as a 56-bein between E_7 and SU(8) like the vierbein $e_\mu{}^r$ between the local Lorentz group SO(3,1) and the global SL(4,R) (subgroup of the reparametrization group).

Let us first describe $E_{7(+7)}$. It will be defined by its infinitesimal transformations in the fundamental representation of dimension 56 acting on the 56-vector Z:

$$z_{MN} = -z_{NM} \ , \quad \bar{z}^{MN} = (z_{MN})^* \quad ; M,N = 1 \dots 8$$

$$\delta z_{MN} = \Lambda_M{}^P z_{PN} + \Lambda_N{}^P z_{MP} + \Sigma_{MNPQ}\, \bar{z}^{PQ}$$

$$\delta \bar{z}^{MN} = \bar{\Lambda}^M{}_P \bar{z}^{PN} + \bar{\Lambda}^N{}_P \bar{z}^{PM} + \bar{\Sigma}^{MNPQ} z_{PQ} \qquad (6.1)$$

$\Lambda_M{}^N$ is a traceless antihermitian 8x8 matrix

$\bar{\Lambda}^M{}_N = (\Lambda_M{}^N)^*$

Σ_{MNPQ} is totally antisymmetric in M,N,P,O

$$\bar{\Sigma}^{MNPQ} = (\Sigma_{MNPQ})^{*}$$

and we have the constraints

$$\Sigma_{MNPQ} = \frac{1}{24} \varepsilon_{MNPQRSTU} \bar{\Sigma}^{RSTU} \tag{6.2}$$

We shall write an element $\mathcal{V}$ of E_7 in the representation 56 as

$$\mathcal{V} = \begin{pmatrix} U_{AB}{}^{MN} , & V_{ABMN} \\ \bar{V}^{ABMN} , & \bar{U}^{AB}{}_{MN} \end{pmatrix} \tag{6.3}$$

These matrices preserve the quadratic symplectic invariant

$$\bar{z}_2^{t} \, \Omega \, z_1 \qquad ; \qquad \Omega = i \begin{pmatrix} \mathbb{1}_{28} & 0 \\ 0 & -\mathbb{1}_{28} \end{pmatrix}$$

or $\quad i\,(\bar{z}_2{}^{MN} z_{1MN} - \bar{z}_1{}^{MN} z_{2MN})$

$$\tag{6.4}$$

It follows that the inverse of $\mathcal{V}$ is easily computed

$$\mathcal{V}^{-1} = - \Omega \, \mathcal{V}^{+} \Omega \tag{6.5}$$

The $\Lambda_M{}^N$'s generate the compact subgroup SU(8). The elements X of the subgroup SU(8) satisfy

$$[X, \Omega] = 0$$
$$X^{+} X = \mathbb{1} \tag{6.6}$$

We can compute the trace of the product of two generators T_1 and T_2 defined by (6.1). This defines the Killing metric

$$\mathrm{Tr}\,(T_1 T_2) = 2 \, \Sigma_{1MNPQ} (\Sigma_{2MNPQ})^{*} - 12 \, \Lambda_{1M}{}^{N} (\Lambda_{2M}{}^{N})^{*} \tag{6.7}$$

and therefore there are 63 compact generators ($\Lambda_P{}^Q$ generating SU(8)) and 70 non-compact generators (Σ_{MNPQ}).

Let us now show how we can determine, up to a few numerical coefficients and quartic fermionic terms, the Lagrangian and supersymmetry transformations by conjecturing the global $E_{7(+7)}$ on-shell symmetry and local SU(8) symmetry. The structure of the kinetic terms of the scalar fields is very easily determined from the discussion of the previous section. Let us write $\partial_\mu \mathcal{V}\mathcal{V}^{-1}$ which, beeing an element of $\mathcal{L}(E_7)$, has the form given by (6.1).

$$\partial_\mu \mathcal{V}\mathcal{V}^{-1} \quad \begin{pmatrix} 2\,Q_{\mu[A}{}^{[C}\,\delta^{D]}_{B]} & , \; P_{\mu ABCD} \\ \bar{P}_\mu{}^{ABCD} & , \; 2\,\bar{Q}_\mu{}^{[A}{}_{[C}\,\delta^{B]}_{D]} \end{pmatrix} \tag{6.8}$$

$P_{\mu ABCD}$ satisfies the constraint (6.2) and $Q_{\mu A}{}^{B}$ is antihermitian and traceless. Both are invariant under the global $E_{7(+7)}$. $P_{\mu ABCD}$ is covariant for the local SU(8) and $Q_{\mu A}{}^{B}$ transforms like a gauge field for SU(8). The kinetic term of scalar fields is then (we put the gravitational coupling constant equal to 1)

$$\mathcal{L}_{KS} \simeq g^{\mu\nu}\, P_{\mu ABCD}\, \bar{P}_\nu{}^{ABCD} \tag{6.9}$$

This will also determine the coupling of the scalar fields to the fermions. The Noether coupling is given by P_μ :

$$\mathcal{L}_N \simeq \bar{\Psi}_{\mu A}\, \gamma^\nu \gamma^\mu\, \bar{P}_\nu{}^{ABCD}\, \lambda_{BCD} \tag{6.10}$$

(with the convention that the multiplication of complex function f by Majorana spinors is made with $\mathrm{Re} f + i\gamma_5 \,\mathrm{Im} f$). From the local SU(8) invariance $Q_{\mu A}{}^{B}$ must appear together with the derivative of spinor fields

$$\mathcal{D}_\mu \Psi_{\nu A} = \left(D_\mu(\omega)\, \delta_A^B - Q_{\mu A}{}^{B} \right) \Psi_{\nu B}$$

$$\mathcal{D}_\mu \lambda_{ABC} = \left(D_\mu(\omega)\, \delta_{[A}{}^{D} - 3\, \varphi_{\mu\,[A}{}^{D}\right) \lambda_{BC]D} \tag{6.11}$$

Let us come now to the vector fields A_μ^{MN}. They appear only through their field strengths $F_{\mu\nu}^{MN}$. Let us define $G_{\mu\nu\,MN}$ by

$$G_{\mu\nu\,MN} = -\frac{4}{e}\, \widetilde{\frac{\delta \mathcal{L}}{\delta F_{\mu\nu}^{MN}}} \tag{6.12}$$

Following the discussion of section 3, the Lagrangian will be written as

$$e^{-1}\mathcal{L} = \frac{1}{8} F_{\mu\nu}^{MN}\, \tilde{G}^{\mu\nu}_{MN} + \mathcal{L}_{inv} \tag{6.13}$$

We shall require that $\mathcal{L}_{inv}$ should be invariant under E_7, the relation between G and F covariant under E_7, and that E_7 should exchange the vector field equation of A_μ and the Bianchi identity. Let us first define the complex 56-vector $\mathcal{F}_{\mu\nu}$

$$\mathcal{F}_{\mu\nu} = \frac{1}{\sqrt{2}} \begin{pmatrix} F_{\mu\nu}^{MN} + i\, G_{\mu\nu\,MN} \\ F_{\mu\nu}^{MN} - i\, G_{\mu\nu\,MN} \end{pmatrix} \equiv \begin{pmatrix} \mathcal{F}_{\mu\nu\,MN} \\ \bar{\mathcal{F}}_{\mu\nu}{}^{MN} \end{pmatrix} \tag{6.14}$$

E_7 will act on $\mathcal{F}$ as in (6.1)

$$\delta \mathcal{F}_{\mu\nu} = \begin{pmatrix} 2\, \Lambda_{[M}{}^{P}\, \delta_{N]}{}^{Q} & , \Sigma_{MNPQ} \\ \bar{\Sigma}^{MNPQ} & , 2\, \bar{\Lambda}^{[M}{}_{P}\, \delta^{N]}{}_{Q} \end{pmatrix} \begin{pmatrix} \mathcal{F}_{\mu\nu\,PQ} \\ \bar{\mathcal{F}}_{\mu\nu}{}^{PQ} \end{pmatrix} \tag{6.15}$$

and on $\mathcal{V}$ as

$$\delta \mathcal{V} = -\mathcal{V} \begin{pmatrix} 2\Lambda_{[M}{}^{[P}\delta_{N]}{}^{Q]} & , \Sigma_{MNPQ} \\ \bar{\Sigma}^{MNPQ} & , 2\bar{\Lambda}^{[M}{}_{[P}\delta^{N]}{}_{Q]} \end{pmatrix} \tag{6.16}$$

The vector field equation and Bianchi identity can be written as

$$\partial_\mu (e\, \tilde{\mathcal{F}}^{\mu\nu}) = 0 \tag{6.17}$$

This is obviously invariant under E_7 (subgroup of Sp(56)). The spinor fields must be coupled to the vector field in an E_7 invariant way; since they are singlets of E_7, they must couple to the E_7 invariant combination

$$\mathcal{V} \mathcal{F}_{\mu\nu} = \begin{pmatrix} \mathcal{F}_{\mu\nu\, AB} \\ \bar{\mathcal{F}}_{\mu\nu}{}^{AB} \end{pmatrix} \tag{6.18}$$

We have the possibility of adding to (6.18) bilinear fermionic terms ($X_{\mu\nu AB}$, $\bar{X}_{\mu\nu}{}^{AB}$) which do not depend on the scalar fields; this will define $\hat{\mathcal{F}}_{\mu\nu AB}$ which will effectively couple to fermions in $\mathcal{L}_{inv}$. Generalizing the results of N=2,4 we conjecture that the couplings occur only through $\gamma^{\mu\nu}\hat{\mathcal{F}}_{\mu\nu AB}$. Let us use the duality properties of $\gamma^{\mu\nu}$ for $\gamma^{\mu\nu}(X_{\mu\nu}+i\gamma_5 Y_{\mu\nu})$

$$\begin{aligned} \gamma^{\mu\nu}(X_{\mu\nu}+i\gamma_5 Y_{\mu\nu}) &= i\gamma_5 \tilde{\gamma}^{\mu\nu}(X_{\mu\nu}+i\gamma_5 Y_{\mu\nu}) \\ &= i\gamma_5 \gamma^{\mu\nu}(\tilde{X}_{\mu\nu}+i\gamma_5 \tilde{Y}_{\mu\nu}) \end{aligned} \tag{6.19}$$

and therefore only the part which satisfies

$$X_{\mu\nu} + i Y_{\mu\nu} = i(\tilde{X}_{\mu\nu} + i\tilde{Y}_{\mu\nu}) \tag{6.20}$$

is coupled though $\gamma_{\mu\nu}$.
The local SU(8) invariance then fixes the various fermionic couplings

$$e\left[a\,\bar{\Psi}_{\mu A}\gamma^{\nu}\hat{\mathcal{F}}_{AB}\gamma^{\mu}\psi_{\nu B} + ib\,\bar{\Psi}_{\mu C}\hat{\mathcal{F}}_{AB}\gamma^{\mu}\lambda_{ABC} + c\,\varepsilon^{ABCDEFGH}\bar{\lambda}_{ABC}\hat{\mathcal{F}}_{DE}\lambda_{FGH}\right] \tag{6.21}$$

with

$$\hat{\mathcal{F}}_{AB} = \gamma^{\mu\nu}\hat{\mathcal{F}}_{\mu\nu AB} \tag{6.22}$$

or using chiral spinor they can be written as

$$e\left[a\,\bar{\Psi}^{A}_{\mu(R)}\gamma^{\nu}\hat{\mathcal{F}}_{AB}\gamma^{\mu}\psi^{B}_{\nu(L)} + ib\,\bar{\Psi}^{(L)}_{\mu C}\hat{\mathcal{F}}_{AB}\gamma^{\mu}\lambda^{ABC}_{(L)} + c\,\varepsilon^{ABCDEFGH}\bar{\lambda}^{(L)}_{ABC}\hat{\mathcal{F}}_{DE}\lambda^{(R)}_{FGH}\right] + h.c. \tag{6.23}$$

Let us remember that $\chi^{(R)}_{A}$ transforms like 8 of SU(8) and $\chi^{A}_{(L)}$ like 8 of SU(8). We can conjecture the constraint (6.20) on $\hat{\mathcal{F}}_{\mu\nu AB}$:

$$\hat{\mathcal{F}}_{\mu\nu AB} = i\,\tilde{\hat{\mathcal{F}}}_{\mu\nu AB} \tag{6.24}$$

Ignoring the fermions in (6.24) this determines uniquely the kinetic term of the vector fields, namely $G^{(0)}_{\mu\nu MN}$ as a function of $F^{MN}_{\mu\nu}$ and of the scalar fields

$$\mathcal{V}\,\mathcal{F}^{(0)}_{\mu\nu} = \Omega\,\mathcal{V}\,\tilde{\mathcal{F}}^{(0)}_{\mu\nu} \tag{6.25}$$

or

$$\bar{\mathcal{F}}^{(0)}_{\mu\nu} = \Omega\,\mathcal{V}^{+}\mathcal{V}\,\tilde{\mathcal{F}}^{(0)}_{\mu\nu} \tag{6.26}$$

it depends only on the scalar fields through the metric on the coset space $\mathcal{V}^{+}\mathcal{V}$ which is invariant under the local group SU(8). From the definition of $G_{\mu\nu MN}$ it follows that the quadratic fermionic terms of 1/8 $F\tilde{G}$ are equal to those of $\mathcal{L}_{inv}$ coupled to $\hat{\mathcal{F}}$. The same arguments of covariance can apply to the supersymmetry transformations. The check of supersymmetry of the Lagrangian determines the few coefficients undetermined. The recipes given in section 2 allow us to determine the remaining quartic fermionic terms.

Let us now write down the Lagrangian of N=8 supergravity with $E_{7(+7)}$ symmetry of the equations of motion and local SU(8) symmetry.

$$\begin{aligned} e^{-1}\mathcal{L} = & -\frac{1}{4}R(\omega,e) + \frac{e^{-1}}{2}\varepsilon^{\mu\nu\rho\sigma}\bar{\psi}_{\mu A}\gamma_\sigma\gamma_5\left(\delta_A{}^B D_\nu(\omega) - \mathcal{Q}_{\nu A}{}^B\right)\psi_{\rho B} \\ & + \frac{1}{8}F_{\mu\nu}^{MN}\tilde{G}^{\mu\nu}_{MN}(A_\mu, \mathcal{V}, \psi_{\mu A}, \lambda_{ABC}) + \frac{i}{12}\bar{\lambda}_{ABC}\gamma^\mu\left(\delta_A{}^D D_\mu(\omega) - 3\mathcal{Q}_{\nu A}{}^B\right)\lambda_{BCD} \\ & + \frac{1}{24}P_{\mu ABCD}\bar{P}^{\mu ABCD} + \frac{1}{6\sqrt{2}}\bar{\psi}_{\mu A}\gamma^\nu\gamma^\mu\left(\hat{\bar{P}}_\nu^{ABCD} + \hat{\bar{P}}^{ABCD}\right)\lambda_{BCD} \\ & + \frac{1}{8\sqrt{2}}\Big\{\bar{\psi}_{\mu A}\gamma^\nu\hat{\mathcal{F}}_{AB}\gamma^\mu\psi_{\nu B} - \frac{i}{\sqrt{2}}\bar{\psi}_{\mu C}\hat{\mathcal{F}}_{AB}\gamma^\mu\lambda_{ABC} \\ & \qquad - \frac{\eta}{72}\varepsilon^{ABCDEFGH}\bar{\lambda}_{ABC}\hat{\mathcal{F}}_{DE}\lambda_{FGH}\Big\} + \mathcal{L}_4 \end{aligned} \tag{6.27}$$

where $\mathcal{L}_4$ is the remaining quartic fermionic terms not absorbed in the supercovariant objects ω , $\hat{P}_\mu$ and $\hat{\mathcal{F}}$ defined by

$$\hat{P}_{\mu ABCD} = P_{\mu ABCD} + 2\sqrt{2}\left[\bar{\psi}^{(L)}_{\mu[A}\lambda_{BCD]} + \frac{\eta}{24}\varepsilon_{ABCDEFGH}\bar{\psi}^{E}_{\mu(R)}\lambda^{FGH}_{(L)}\right]$$

$$\hat{\mathcal{F}}_{AB} = \gamma^{\mu\nu}\hat{\mathcal{F}}_{\mu\nu AB}$$

$$\begin{aligned} \hat{\mathcal{F}}_{\mu\nu AB} = \mathcal{F}_{\mu\nu AB} + 2\sqrt{2}\Big[&\bar{\psi}^{(L)}_{\underline{\mu}[A}\psi^{(R)}_{\underline{\nu}B]} - \frac{i}{\sqrt{2}}\bar{\psi}^{C}_{[\mu(R)}\gamma_{\nu]}\lambda^{(R)}_{ABC} \\ & + \frac{\eta}{288}\varepsilon_{ABCDEFGH}\bar{\lambda}^{CDE}_{(R)}\gamma_{\mu\nu}\lambda^{FGH}_{(L)}\Big] \end{aligned} \tag{6.28}$$

ω is defined by its field equation. $\tilde{G}^{\mu\nu}_{MN}$ is determined from the relation (6.18) and (6.24)

$$\hat{\mathcal{F}}_{\mu\nu AB} = i\,\hat{\tilde{\mathcal{F}}}_{\mu\nu AB}$$

$P_{\mu ABCD}$ and $Q_{\mu A}{}^{B}$ are defined by (6.8) and $\mathcal{V}$ is written as in (6.3). It is invariant under the following supersymmetry transformations

$$\delta_s e_\mu{}^r = -i\,\bar{\epsilon}_A \gamma^r \psi_{\mu A} \tag{6.29}$$

$$\frac{-1}{2\sqrt{2}}\delta_s \mathcal{V}\mathcal{V}^{-1} = \begin{pmatrix} 0 & ;\ \bar{\epsilon}^{(L)}_{[A}\lambda^{(R)}_{BCD]} + \frac{\eta}{24}\varepsilon_{ABCDEFGH}\bar{\epsilon}^{E}_{(R)}\lambda^{FGH}_{(L)} \\ \bar{\epsilon}^{[A}_{(R)}\lambda^{BCD]}_{(L)} + \frac{\eta}{24}\varepsilon^{ABCDEFGH}\bar{\epsilon}^{(L)}_{E}\lambda^{(R)}_{FGH}; & 0 \end{pmatrix} \tag{6.30}$$

$$\delta_s A_\mu^{MN} = 4\,\mathrm{Re}\left[(\bar{V}^{ABMN} - \bar{U}^{AB}{}_{MN})\left(\bar{\epsilon}^{(L)}_A \psi^{(R)}_{\mu B} - \frac{i\sqrt{2}}{4}\bar{\epsilon}^{C}_{(R)}\gamma_\mu \lambda^{(R)}_{ABC}\right)\right] \tag{6.31}$$

$$\begin{aligned}\delta_s \psi^{(R)}_{\mu A} = {}& \left(D_\mu(\omega)\delta_A^B - Q_{\mu A}{}^{B}\right)\epsilon^{(R)}_B - \frac{i}{4\sqrt{2}}\hat{\mathcal{F}}_{AB}\gamma_\mu \epsilon^{B}_{(L)} - \frac{1}{\sqrt{2}}\gamma_\alpha \epsilon^{C}_{(L)}\bar{\psi}^{B}_{\mu(R)}\gamma^\alpha \lambda^{(R)}_{ABC} \\ & + \frac{i}{4}\gamma_\alpha\gamma_\mu \epsilon^{(R)}_D \bar{\lambda}^{(L)}_{ABC}\gamma^\alpha \lambda^{DBC}_{(L)} - \frac{i\eta}{4\times 144}\varepsilon_{ABCDEFGH}\gamma_\mu\gamma_{\alpha\beta}\epsilon^{B}_{(L)}\bar{\lambda}^{CDE}_{(R)}\gamma^{\alpha\beta}\lambda^{FGH}_{(L)}\end{aligned} \tag{6.32}$$

$$\begin{aligned}\delta_s \lambda^{(R)}_{ABC} = {}& -i\sqrt{2}\,\hat{P}_{\mu ABCD}\gamma^\mu \epsilon^{D}_{(L)} + \frac{3}{4}\hat{\mathcal{F}}_{[AB}\epsilon^{(R)}_{C]} \\ & - \frac{\sqrt{2}}{24}\eta\,\varepsilon_{ABCDEFGH}\bar{\lambda}^{DEF}_{(R)}\lambda^{GHI}_{(L)}\epsilon^{(R)}_I\end{aligned} \tag{6.33}$$

we check immediately that $\hat{P}_\mu$ and $\hat{\mathcal{F}}$ are supercovariant. From (6.30) we see that $Q_{\mu A}{}^{B}$ is supercovariant by itself.

Up to now we have not fixed the gauge of the local SU(8). Fixing the gauge corresponds to using a specific parametrization of the matrix $\mathcal{V}$ in terms of 70 parameters. It is sufficient to use it in the Lagrangian. We must be careful with the supersymmetry transformation because the gauge condition will not be preserved by the supersymmetry transformation (6.30). It will be necessary to add to it a local SU(8) transformation which will bring back the supersymmetry variation in the chosen gauge. This local SU(8) transformation will act on spinor fields also. Let us now apply the result of section 5 to write down the theory in the hermitian gauge. The resulting theory will no longer have the local SU(8) symmetry but will still have the E_7 symmetry. In the hermitian gauge $\mathcal{V}$ is written as

$$\mathcal{V} = \exp \begin{pmatrix} 0 , & W_{ABCD} \\ \bar{W}^{ABCD} , & 0 \end{pmatrix} \tag{6.34}$$

where W_{ABCD} is totally antisymmetric and η self dual. We define the inhomogeneous coordinates

$$z_{AB,\,CD} = \left(W \frac{\operatorname{th}\sqrt{\bar{W}W}}{\sqrt{\bar{W}W}} \right)_{AB,\,CD} \tag{6.35}$$

There is no simple characterization which would show directly that there is only 70 independent z 's (like the W 's). Although W_{ABCD} can take any numerical value, it is not true fo $z_{AB,\,CD}$, (6.35) implies that the hermitian matrices $(\mathbb{1} - z\bar{z})_{AB}{}^{CD}$ and $(\mathbb{1} - \bar{z}z)^{AB}{}_{CD}$ are definite positive (we could define a norm such that $\|z\| < 1$). We have immediately

$$P_{\mu\,ABCD} = \left(\frac{1}{\sqrt{1 - z\bar{z}}}\, \partial_\mu z\, \frac{1}{\sqrt{1 - \bar{z}z}} \right)_{AB\,CD} \tag{6.36}$$

which is automatically antisymmetric and η self dual like W_{ABCD} . The constraints which define the couplings of the vector fields are easily solved

$$\mathcal{F}^{(0)}_{\mu\nu\,AB} = \frac{1}{\sqrt{2}} \left(\frac{1}{\sqrt{1 - z\bar{z}}} \right)_{AB}{}^{CD} \left(F_{\mu\nu}^{CD} + i\, G^{(0)}_{\mu\nu\,CD} + z_{CDEF} \left(F_{\mu\nu}^{EF} - i\, G^{(0)}_{\mu\nu\,EF} \right) \right) \tag{6.37}$$

and (6.25) becomes

$$G^{(0)}_{\mu\nu\,AB} - i\,\tilde{G}^{(0)}_{\mu\nu\,AB} = i\left(\frac{1+Z}{1-Z}\right)_{AB,\,CD} \left(F^{CD}_{\mu\nu} - i\,\tilde{F}^{CD}_{\mu\nu}\right) \tag{6.38}$$

and $\mathcal{F}^{(0)}_{\mu\nu\,AB}$ is now simply written

$$\mathcal{F}^{(0)}_{\mu\nu\,AB} = \frac{1}{\sqrt{2}}\left(\sqrt{1-Z\bar{Z}}\;\frac{1}{1-\bar{Z}}\right)_{AB\,CD} \left(F^{CD}_{\mu\nu} + i\,\tilde{F}^{CD}_{\mu\nu}\right) \tag{6.39}$$

The limitation $\|Z\| < 1$ ensures that the kinetic terms for vector fields are positive definite.

From N=8 supergravity we can rederive all other extended supergravities by consistent truncation. They all have of course the same pattern of global and local symmetry. We get

	Global on-shell symmetry	local symmetry
N=7,8	$E_{7(+7)}$	SU(8)
N=6	$SO^*(12)$	U(6)
N=5	SU(5,1)	U(5)
N=4	SU(4)xSU(1,1)	U(4)

7. GAUGED SUPERGRAVITIES

So far, the extended supergravity theories we have discussed had only one coupling constant K and the vector fields were abelian. As we mentioned in the introduction for N≠6,7 there are exactly N(N-1)/2 vector fields, namely just the right number to gauge O(N). We can therefore try to gauge the O(N) group by introducing a new coupling constant g and replacing $F_{\mu\nu}^{AB}$ by the Yang-Mills curvature and derivatives by the covariant derivatives. These minimal requirements are not compatible with supersymmetry and it will be necessary to add more coupling in g and g^2 to the Lagrangian and modify the supersymmetry transformations.

<u>O(2) supergravity</u>

In N=2 (Das and Freedman (1977)) we can gauge O(2) by replacin

$$D_\mu(\hat\omega)\,\psi_\rho^{\ A} \quad \text{by} \quad \left(D_\mu(\hat\omega)\,\delta^A_{\ B} + g\,\varepsilon^{AB} A_\mu\right)\psi_\rho^{\ B} \tag{7.1}$$

It is necessary to add the following terms to the Lagrangian

$$e^{-1}\Delta\mathcal{L} = \frac{g}{2\kappa}\,\overline{\psi}_\mu^{\ A}\,\gamma^{\mu\nu}\,\psi_\nu^{\ A} + \frac{3}{2}\,\frac{g^2}{\kappa^4} \tag{7.2}$$

and to add to $\delta\,\overline{\psi}_\mu^{\ A}$

$$\Delta\delta\,\overline{\psi}_\mu^{\ A} = \frac{i}{2\kappa^2}\,g\,\overline{\varepsilon}^A\gamma_\mu \tag{7.3}$$

There is in (7.2) a mass-like term for the gravitino and a cosmological constant. The relation between the two is that required to describe an SO(3,2) de Sitter universe. There is n breaking of supersymmetry and the gravitino is massless in the sense of de Sitter space. This theory can be viewed as the gauging of the super de Sitter group. Since A_ν appears explicitly in (7.1), we can't have the U(1) duality symmetry of the ungauged case any longer . The SU(2) symmetry is also brok by the gauging (ε^{AB} is not preserved by SU(2) transformations).

N=4 gauged supergravity

We can gauge SO(4) ∼ SO(3) x SO(3) and further problems arise because of the presence of scalar fields. The gauging can be made both in the O(4) and SU(4) versions and the results will not be equivalent. It is possible to introduce two coupling constants, one for each SO(3).

In the O(4) formulation, the complete gauged theory has been written by Gates and Zwiebach (1983). Let us call g_1 and g_2 the two coupling constants and use the matrices α^n_{AB} and β^n_{AB} which relate SO(4) to SO(3) x SO(3). The covariant derivatives and Yang-Mills field strength are defined by

$$\mathcal{D}_\mu \psi_\nu^A = D_\mu(\omega)\psi_\nu^A + [(g_1+g_2)A_\mu^{AB} + (g_1-g_2)\tfrac{1}{2}\varepsilon^{ABCD}A_\mu^{CD}]\psi_\nu^B$$

$$F_{\mu\nu}^{AB} = \partial_\mu A_\nu^{AB} - \partial_\nu A_\mu^{AB} + \frac{1}{4}\varepsilon^{mnp}(g_1\alpha_{AB}^m \varphi_\mu^n \varphi_\nu^p + g_2\beta_{AB}^m \varphi_\mu'^n \varphi_\nu'^p)$$

(7.4)

with

$$\varphi_\mu^n = \alpha_{AB}^n A_\mu^{AB} \quad ; \quad \varphi_\mu'^n = \beta_{AB}^n A_\mu^{AB}$$

except the obvious covariantization with respect to the local group SO(3) x SO(3), it is found that we must add to the original Lagrangian

$$\begin{aligned} e^{-1}\Delta\mathcal{L} = &\frac{1}{2K(1-z\bar{z})^{1/2}} \bar{\psi}_\mu^A \gamma^{\mu\nu}(g_+ - K(A-iB\gamma_5)g_-)\psi_\nu^A \\ &+ \frac{i}{K\sqrt{2}(1-z\bar{z})^{1/2}} \bar{\psi}_\mu^A \gamma^\mu (K(A-iB\gamma_5)g_+ - g_-)\chi^A \\ &+ \frac{1}{2K^4}\left[g_+^2 \frac{3-z\bar{z}}{1-z\bar{z}} - g_-^2 \frac{1-3z\bar{z}}{1-z\bar{z}} - 4g_+g_- \frac{KA}{1-z\bar{z}}\right] \end{aligned}$$

(7.5)

with

$$g_\pm = g_1 \pm g_2$$

The supersymmetry transformations must also be modified

$$\Delta\delta\bar{\psi}_\mu^A = \frac{i}{2K^2} \frac{\bar{\varepsilon}^A \gamma_\mu (g_+ - K(A+iB\gamma_5)g_-)}{(1-z\bar{z})^{1/2}}$$

$$\Delta\delta\bar{\chi}^A = \frac{1}{\sqrt{2}K^2} \frac{\bar{\varepsilon}^A (K(A+iB\gamma_5)g_+ - g_-)}{(1-z\bar{z})^{1/2}}$$

(7.6)

It is easy to show that the gauged theory has a "pseudo" invariance SO(1,1)

$$z \longrightarrow \frac{rz+s}{sz+r}$$

$$A_\mu^{AB} \longrightarrow r A_\mu^{AB} - s\,\tfrac{1}{2}\varepsilon^{ABCD}A_\mu^{CD}$$

$$g_+ \longrightarrow r g_+ + s g_-$$
$$g_- \longrightarrow s g_+ + r g_- \qquad (7.7)$$

with

$$r^2 - s^2 = 1$$

This means that the theory will depend only on the invariant

$$g_+^2 - g_-^2 = 4 g_1 g_2 \qquad (7.8)$$

This pseudo symmetry can be easily understood if we remark that for $g_1 = g_2 = 0$, it is the subgroup of the on-shell SU(1,1) invariance which leaves the Lagrangian invariant. The potential is always unbounded from below (for $|z| \to 1$). More detailed properties will depend on the value of g_1 g_2 and using the invariance (7.7), we have 3 generic cases

$$\begin{cases} g_1 = g_2 & ; \quad g_1 = -g_2 \quad ; \quad g_1 g_2 = 0 \\ g_- = 0 & ; \quad g_+ = 0 \quad ; \quad |g_+| = |g_-| \end{cases} \qquad (7.9)$$

(a) $g_1 = g_2$ The maximum of the potential occurs for z=0 with a negative cosmological constant $\Lambda = -\frac{3}{2K^4} g_+^2$ and N=4 supersymmetry is unbroken;

(b) $g_1 = -g_2$ The maximum of the potential still occurs for z=0 but with a positive cosmological constant $\Lambda = g_-^2/2K^4$ and all 4 supersymmetries are broken $(\delta \bar{\chi}^A = -(g_-/\sqrt{2}K^2)\bar{\epsilon}^A)$

(c) $g_2 = 0$ There is no critical points of the potential and supersymmetry is always broken.

The 3 cases can be related by analytic continuation or limiting process. Writing $r = \text{ch}\,t$, $s = \text{sh}\,t$, case (b) can be obtained from case (a) by analytic continuation of t to $i\pi/2$ and $g^2 = g_+^2 - g_-^2$ to $-g^2$. The potential (a) $-g^2(3 - z\bar{z})/(1 - z\bar{z})$ becomes $g^2(1 - 3z\bar{z})/(1 - z\bar{z})$, the potential (b) by $z \to 1/z$ and $g^2 \to -g^2$. The case (c) is obtained by a limiting process $t \to \infty$ where $g_1(A_\mu^{AB} + \frac{1}{2}\epsilon^{ABCD} A_\mu^{CD})$ remains finite but $g_2(A_\mu^{AB} - \frac{1}{2}\epsilon^{ABCD} A_\mu^{CD})$ goes to zero.

In the SU(4) version, the complete gauged theory has been written by Freedman and Schwarz (1978). The A_μ^n fields are proportional to φ_μ^n and the B_μ fields correspond to the dual fields of φ'^n_μ. Defining the covariant derivatives by

$$\mathcal{D}_\mu \psi_\nu^A = D_\mu(\omega)\psi_\nu^A - \frac{1}{2}(e_A A_\mu^n \alpha^n_{AB} + e_B B_\mu^n \beta^n_{AB})\psi_\nu^B$$

$$A_{\mu\nu}^n = \partial_\mu A_\nu^n - \partial_\nu A_\mu^n + e_A \varepsilon^{nmp} A_\mu^m A_\nu^p$$

$$B_{\mu\nu}^n = \partial_\mu B_\nu^n - \partial_\nu B_\mu^n + e_B \varepsilon^{nmp} B_\mu^m B_\nu^p \tag{7.10}$$

local supersymmetry requires the new terms in the Lagrangian

$$e^{-1}\Delta\mathcal{L} = -\frac{i}{4K}\frac{1}{(1-z'-\bar{z}')^{1/2}}\bar{\chi}^A(e_A - ie_B\gamma_5)\gamma^\mu\psi_\mu^A + \frac{\sqrt{2}}{8K}\frac{1}{(1-z'-\bar{z}')^{1/2}}\bar{\psi}_\mu^A(e_A + ie_B\gamma_5)\gamma^{\mu\nu}\psi_\nu^A \tag{7.11}$$

and the supersymmetry transformations have to be modified

$$\Delta\delta\bar{\psi}_\mu^A = \frac{i\sqrt{2}}{8K^2}\frac{1}{(1-z'-\bar{z}')^{1/2}}\bar{\epsilon}^A(e_A + i\gamma_5 e_B)\gamma_\mu$$

$$\Delta\delta\bar{\chi}^A = \frac{1}{4K^2}\frac{1}{(1-z'-\bar{z}')^{1/2}}\bar{\epsilon}^A(e_A + i\gamma_5 e_B) \tag{7.12}$$

There is no critical point of the potential and supersymmetry is always broken. In fact when $e_B = 0$, since the $B_{\mu\nu}^n$ are abelian, it is possible to dualize them and get an O(4) theory with $g_2 = 0$. This is immediately checked by the equality of the potentials

$$\frac{1}{1-z'-\bar{z}'} = \frac{1+z\bar{z}-z-\bar{z}}{1-z\bar{z}} \tag{7.13}$$

with $(1-z)(1-z') = 1$ according to Sect. 4. The previous SO(1,1) pseudo invariance becomes

$$(e_A, e_B) \rightarrow \alpha(e_A, e_B)\ ; \ (A_\mu^n, B_\mu^n) \rightarrow \frac{1}{\alpha}(A_\mu^n, B_\mu^n)$$

$$1-2z' \rightarrow \alpha^2(1-2z') \tag{7.14}$$

There is no pseudo invariance associated with the SU(4) invariance of the ungauged Lagrangian.

O(8) supergravity

We have seen in section 4 that the ungauged theory N=8 has an on-shell global $E_7(+7)$ invariance and a local SU(8) invariance of the Lagrangian. There is a global subgroup of $E_7(+7)$ which is an invariance of the Lagrangian. It is defined as a subgroup which does not mix $F_{\mu\nu}^{MN}$ and $G_{\mu\nu MN}$, namely, from (6.15), the subgroup corresponding to real coefficients $\Lambda_M{}^N$ and Σ_{MNPQ}. It can be realized on the vector fields A_μ^{MN} themselves

$$\delta A_\mu^{MN} = \Lambda^M{}_P A_\mu^{PN} + \Lambda^N{}_P A_\mu^{MP} + \Sigma_{MNPQ} A_\mu^{PQ}$$

with

$$\Lambda^M{}_P = -\Lambda^P{}_M \; ; \quad \Sigma_{MNPQ} = \frac{\eta}{24} \varepsilon_{MNPQRSTU} \Sigma_{RSTU} \qquad (7.15)$$

$\Sigma = 0$ corresponds to the subgroup O(8). Let us use the redefinition

$$A_\mu^{MN} = \frac{1}{4} (\Gamma_{ij})_{MN} A_\mu^{ij} \qquad (7.16)$$

where the $(\Gamma_{ij})_{MN}$ are 8x8 antisymmetrical matrices realizing the SO(8) Lie algebra which can be built from the SO(7) Clifford algebra (Cremmer and Julia, 1979). In this basis it is easy to see that $\Lambda_M{}^N$ and Σ_{MNPQ} generate the subgroup SL(8,R)

$$\delta A_\mu^{ij} = \Lambda^i{}_k A_\mu^{kj} + \Lambda^j{}_k A_\mu^{ik} \qquad (7.17)$$

where $\Lambda^i{}_k$ is a traceless 8x8 real matrix. We can therefore try to gauge any 28-dimensional subgroup of SL(8,R) keeping all the previous structure, in particular the local SU(8) invariance. It is not necessary to limit ourselves to a compact group because the kinetic term of the vector fields (invariant under SL(8,R)) is positive definite.

Let us concentrate on the gauging of SO(8) (De Wit and Nicolai, 1981,1982). Only the vector fields and the scalar fields transform under this SO(8); the spinors are singlet. The Yang-Mills field strength is defined by

$$F_{\mu\nu}^{MN} = 2\, \partial_{[\mu} A_{\nu]}^{MN} - 2g\, A_{[\mu}^{MK} A_{\nu]}^{KN} \tag{7.18}$$

and the covariant derivative of the 56-bein of scalars by

$$\mathcal{D}_\mu \mathcal{V} = \partial_\mu \mathcal{V} - 2g\, \mathcal{V} \begin{pmatrix} A_\mu{}^{[M}{}_{[K}\, \delta^{N]}_{L]} \;; & 0 \\ 0 \;; & A_{\mu[M}{}^{[K}\, \delta^{L]}_{N]} \end{pmatrix} \tag{7.19}$$

This will define the covariant $\mathcal{P}_{\mu\, ABCD}$ and $\mathcal{Q}_{\mu\, A}{}^{B}$

$$\mathcal{D}_\mu \mathcal{V}\, \mathcal{V}^{-1} = \begin{pmatrix} 2\, \mathcal{Q}_{\mu[A}{}^{[C}\, \delta^{D]}_{B]} \;; & \mathcal{P}_{\mu\, ABCD} \\ \bar{\mathcal{P}}_\mu{}^{ABCD} \;; & 2\, \bar{\mathcal{Q}}_\mu{}^{[A}{}_{[C}\, \delta^{B]}_{D]} \end{pmatrix} \tag{7.20}$$

The gauge couplings will occur also for fermions indirectly through $\mathcal{Q}_{\mu A}{}^{B}$ in the SU(8) covariant derivative. As for N=2 and 4 these modifications are not compatible with supersymmetry and it is necessary to add new terms in the Lagrangian and the supersymmetry transformations of the fermion fields

$$\begin{aligned} e^{-1}\Delta\mathcal{L} = \Big\{ & -\frac{g}{2} A_{1AB}\, \bar{\Psi}^{A}_{\mu(R)} \gamma^{\mu\nu} \Psi^{B}_{\nu(L)} - \frac{i g}{6\sqrt{2}} A_{2D}{}^{ABC}\, \bar{\Psi}^{D}_{\mu(R)} \gamma^{\mu} \lambda^{(R)}_{ABC} \\ & + \frac{g}{144}\, \eta\, \varepsilon^{ABCDA'B'C'D'} A_{2}{}^{E'}{}_{DA'B'}\, \bar{\lambda}^{(L)}_{ABC}\, \lambda^{(R)}_{E'C'D'} + h.c \Big\} \\ & + \frac{g^2}{16} \left(3\, |A_1^{AB}|^2 - \frac{1}{6} |A_2{}^{D}{}_{ABC}|^2 \right) \end{aligned} \tag{7.21}$$

$$\Delta\, \delta \bar{\Psi}^{A}_{\mu(R)} = \frac{i g}{2}\, \bar{\epsilon}^{(L)}_{B} \gamma_\mu A_1^{BA}$$

$$\Delta\, \delta \bar{\chi}^{ABC}_{(R)} = -\frac{g}{\sqrt{2}}\, \bar{\epsilon}^{D}_{(R)} A_{2D}{}^{ABC} \tag{7.22}$$

where the SU(8) covariant tensor A_1^{AB} and $A_{2D}{}^{ABC}$ are defined by

$$A_1^{AB} = \frac{4}{21} T_C^{ACB} \quad ; \quad A_{2D}^{ABC} = -\frac{4}{3} T_D^{[ABC]}$$ (7.23)

The tensor T_D^{ABC} is defined in terms of the 28x28 matrices U and V which are the elements of $\mathcal{V}$ following (6.3)

$$T_D^{ABC} = (\bar{U}^{BC}{}_{IJ} + \bar{V}^{BCIJ})(U_{DE}{}^{JK}\bar{U}^{AE}{}_{KI} - V_{DEJK}\bar{V}^{AEKI})$$ (7.24)

This tensor is no longer invariant under $E_{7(+7)}$ (acting on the pair of indices IJ) but only under SO(8). T_D^{ABC} is uniquely defined by A_1 and A_2 and satisfies

$$T_D^{ABC} + \frac{3}{2} A_1^{A[B} \delta_D^{C]} + \frac{3}{4} A_{2D}^{ABC} = 0$$ (7.25)

The potential depends now on 42 variables (70-28) and is difficult to analyze. We shall not discuss here the problem of finding its critical points which is not yet completely solved (Warner, 1983;1984). By truncation of N=8 to N=4, it is seen that this potential is also unbounded from below.
It is also possible to gauge non-compact versions of SO(8), namely SO(p,q) (Hull, 1984; Hull and Warner, 1984). These gaugings can be either directly constructed using the same method as for SO(8), or using the SL (8,R) invariance of the ungauged Lagrangian and making an analytic continuation in the parameter of the SL(8,R) transformation together with

$$g^8 \rightarrow -g^8$$ (7.26)

each branch of this mapping will give a different SO(p,q). By a limiting process where a parameter is going to ∞, we can obtain a gauging of the contraction of these groups.

8. HIDDEN SYMMETRY IN EXTENDED SUPERGRAVITIES

We have seen that the extended supergravities admit a non-compact global invariance realized non-linearly and an associated local invariance whose gauge fields are not elementary but composite fields. However there is no deep understanding yet of the reasons why these symmetries appear.
The first sign for the existence of these hidden symmetries comes from dimensional reduction. It allows, from a theory in D+N dimensions, to derive a theory in D dimensions with a finite number of degrees of freedom. The simplest dimensional reduction consists in assuming that all fields do not depend on the N extra coordinates. When the theory in D+N dimensions is invariant under reparametrization, it is easily shown that, as a result of the simplest dimensional reduction, this reparametrization invariance generates the product of reparametrization in D dimensions, local $U(1)^N$ and global SL(N,R) in D dimensions. The local Lorentz invariance SO(D+N -1,1) becomes local Lorentz invariance SO(D-1,1) in D dimensions times local SO(N) invariance. A detailed analysis shows that the scalar fields coming from the tensor metric in D+N dimensions (but one) are described by a non-linear sigma-model based on the coset space SL(N,R)/SO(N). The vector fields coming from the metric are abelian and transform linearly under SL(N,R), but are inert under local SO(N). The spinor fields are inert under SL(N,R) but transform linearly under local SO(N). This is the minimal symmetry that the theory can have (if we do not make any truncation on the reduced theory). In particular we have seen for N=8 supergravity which can be derived from 11-dimension supergravity, that SL(7,R) can be extended to SL (8,R) and by duality transformation to an $E_{7(+7)}$ invariance of the field equations.
Therefore we can conjecture the following symmetries for pure extended supergravity in D dimensions. We ask that the scalar fields be described by non-linear sigma models based on the coset space G/H. The maximal subgroup which acts linearly on the field is H_D isomorph to H and is expected to be the group of invariance of extended supersymmetry in D dimensions. The

dimension of G must be the dimension of H_D plus the number of scalar fields. To avoid ghosts, H must be the maximal compact subgroup of G. This invariance cannot be realized on vector fields in 4 dimensions, it is only realized on-shell. We therefore expect that H_D and G exchange the vector field equations and the Bianchi identity; thus G must have a representation of dimension twice the number of vector fields. In other even dimensions 2d, the same thing happens not for vector fields, but for antisymmetric gauge tensor fields of rank d-1; the tensor fields of rank different from d-1 transform linearly under G. In odd dimensions, there are no duality transformations, and G is a symmetry of the Lagrangian. All these counting arguments fix quite uniquely G and H. This can be applied, in particular, to all maximal extended supergravities in D dimensions (≤ 11) which can all be derived by dimensional reduction from 11-dimension supergravity and we find

D	G global	x	H local
9	GL(2,R)		SO(2)
8	$E_{3(+3)}$ =SL(3,R)xSL(2,R)		SO(3)xSO(2)
7	$E_{4(+4)}$ =SL(5,R)		SO(5)
6	$E_{5(+5)}$ =SO(5,5)		SO(5)xSO(5)
5	$E_{6(+6)}$		USp(8)
4	$E_{7(+7)}$		SU(8)
3	$E_{8(+8)}$		SO(16)

In two dimensions Julia (1982) has conjectured that the group G will have an infinite number of generators and should be the affine group $E_8^{(1)}$.

It has been suggested that the global invariance in 4 dimensions will be better described by means of geometrical objects if we introduce extra bosonic coordinates; for example 56 additional coordinates for $E_{7(+7)}$ $Z_{IJ} = - Z_{JI}$ and $\bar{Z}^{IJ}$ (I/J = 1...8) (Howe and Lindström (1981), Kallosh (1982)). This will in particular allow us to describe at the same time the ordinary vierbein as well as the 56-bein associated with the scalar fields and, moreover, it seems natural to include the vector fields too in this matrix $\mathcal{E}$

$$\mathcal{E} = \begin{pmatrix} e_\mu{}^r & v(C_\mu{}^{MN}, \bar{C}_{\mu MN}) \\ 0 & v \end{pmatrix} \tag{8.1}$$

where C_μ^{MN} are complex vector fields whose real parts are the usual abelian vector fields A_μ^{MN} and the imaginary parts are the dual fields B_μ^{MN} defined only on-shell

$$G_{\mu\nu}^{MN} = \partial_\mu B_\nu^{MN} - \partial_\nu B_\mu^{MN} \tag{8.2}$$

If, instead of real ordinary space time, we use the complex space of Ogievestky and Sokatchev (1978) we are led to consider a space of 32 complex dimensions. We could therefore look if there exists a dimensional reduction scheme such that the reparametrization in 32 complex dimensions reduces the reparametrization in 4 complex coordinates x local $U(1)^{56}$ x $E_{7(+7)}$ global : this should not be the trivial dimensional reduction scheme since it would lead to an SL(56,R) global symmetry. Further constraints should be needed to reduce the theory to 4 real space time with only 28 real vector fields. In this spirit Galperin et al. (1984) have recently succeeded in describing the N=3 super Yang-Mills with the help of 6 extra bosonic coordinates associated with the coset space $SU(3)/U(1)^2$ which allows to define harmonic superspace. For N=8 we can conjecture that the extra coordinates could be associated with the coset space $SU(8)/U(1)^7$. It is possible to enlarge the global supersymmetry algebra by the addition of central charges X^{AB}, Y^{AB}

$$\{Q_\alpha^A, Q_\beta^B\} = (\gamma_0 P)_{\alpha\beta}\delta^{AB} + (\gamma_0)_{\alpha\beta} X^{AB} + i(\gamma_5\gamma_0)_{\alpha\beta} Y^{AB} \tag{8.3}$$

where X^{AB} and Y^{AB} are real and antisymmetric in A and B and commute with all the other generators of the algebra. It has been suggested that these central charges could arise from a dimensional reduction in 56 coordinates (Taylor, 1982). Another striking feature has been found by Gunaydin and Saçlioglu (1982) who have shown that the Lie algebra of the non-compact groups for extended supergravities in 4 dimensions can be realized using dibosonic creation and annihilation operators transforming like the antisymmetric 2-tensor representation of SU(N). For example for N=8 we start from the creation and annihilation operators a^{+AB} and a_{AB} satisfying

$$[a_{AB}, a^{+CD}] = \delta_A^C \delta_B^D - \delta_A^D \delta_B^C$$

$$[a_{AB}, a_{CD}] = [a^{+AB}, a^{+CD}] = 0 \tag{8.4}$$

The SU(8) generators $T^A{}_B$ are defined by

$$T^A{}_B = a^{+AC} a_{BC} - \frac{1}{8} \delta^A_B (a^{+CD} a_{CD}) \tag{8.5}$$

The 70 extra generators of $E_{7(+7)}$ V_{ABCD} are written as

$$4 V_{ABCD} = a_{[AB} a_{CD]} + \frac{1}{4!} \varepsilon_{ABCDEFGH} a^{+EF} a^{+GH} \tag{8.6}$$

and satisfy obviously

$$V_{ABCD} = \frac{1}{4!} \varepsilon_{ABCDEFGH} V^{+EFGH} \tag{8.7}$$

and we can check immediately that

$$[\Lambda_A{}^B T^A{}_B + \Sigma^{ABCD} V_{ABCD}, a_{EF}] =$$

$$= \Lambda_E{}^B a_{AF} + \Lambda_F{}^B a_{EB} + \Sigma_{EFGH} a^{+GH} \tag{8.8}$$

We may wonder if these operators a_{AB} and a^{+AB} could be related to some transformations in the 56 space of extra coordinates.

9. PHYSICAL CONTENT OF EXTENDED SUPERGRAVITIES

The relation between extended supergravity and particle physics depends on the question of renormalizability of the theory.

Theories including gravity are not renormalizable in the ordinary sense since the coupling constant K has the dimension (M^{-1}); however the theory could be finite due to miraculous cancellations implied by local supersymmetry. This hope is at most valid for N=8 supergravity. The $N \leq 4$ supergravities can be coupled to matter (spin ≤ 1) and they can be considered as phenomenological theories in which matter is added by hand. This program has been pursued in details for N=1 supergravity by studying the superHiggs effect (spontaneous breakdown of local supersymmetry) and the low energy limit of these theories (decoupling of gravity). It has been shown that this is in agreement with present low energy physics but not very predictive. It is hoped that the extension of such analysis to the coupling of N=2 supergravity to matter fields will provide more restriction on soft supersymmetry breaking terms induced by supergravity at low energy. The coupling of N=4 supergravity to N=4 super Yang-Mills can be viewed as a 4-dimensional low enery phenomenological Lagrangian of the superstrings in 10 dimensions (Schwartz, 1982; Green, 1983).

N=8 supergravity is the maximal extended supergravity and it is not possible to add any matter. Before any attempt at making contact with physics, the first conjecture to make is that we can give a sense to the theory. The most conservative conjecture is that N=8 supergravity is finite. It implies that we cannot adjust parameters by renormalization procedure, K is not really a parameter but only a mass scale. The N=8 ungauged (gauged) has no (only one) parameter. It is now well known that a straightforward identification of the elementary fields of supergravity with the particle spectrum of conventional gauge theories is not possible. One of the reasons is that the maximal conventional gauge group SO(8) does not contain SU(3) x SU(2) x U(1). Therefore, in analogy with the CP^{N-1} models in 2 dimensions (in which they are renormalizable) where the local symmetry is realized also by composite gauge fields, we conjecture (Cremmer and Julia, 1979):

At the quantum level, the local symmetry SU(8) becomes dynamical. In particular the SU(8) gauge fields can propagate (acquire a kinetic term) and have a Yang-Mills type of coupling with a

gauge constant g_0 which should be predictable (eventually as a function of the SO(8) gauge coupling g which has a beta function equal to zero at least at one loop).
It was subsequently conjectured (Ellis et al, 1980) that the field spectrum of conventional gauge theories (quarks, leptons, Higgs,...) should arise also as supergravity bound states. The first problem to solve is to determine if there is a finite N=8 supermultiplet to which the composite gauge field belongs. We would also like to know the behaviour of the component fields under both local SU(8) and global $E_{7(+7)}$ (or gauged SO(8)) like for the fundamental N=8 multiplet. We could then make further conjectures on the particle effectively present at low energy, for example: only the singlets of $E_{7(+7)}$ are present in the ungauged case or only the singlets of gauged O(8) are not preconfined. In fact in the ungauged case, we expect that the bosonic fields, except the scalar fields which are non singlet for $E_{7(+7)}$, will have non minimal couplings like the magnetic type of coupling of A_μ^{MN} and will decouple in the effective theory when we decouple gravity (keeping g_0 the SU(8) gauge coupling constant fixed). The higher spin states, if they are described also by an effective Lagrangian, should be massive in order to avoid inconsistency in the coupling to gravity.
Always in analogy with CP^{N-1} models it has been conjectured (Ellis, Gaillard and Zumino (1981)) that the global symmetry $E_{7(+7)}$ is realized linearly on the bound states. Since $E_{7(+7)}$ is non-compact, this implies that the unitary representations are infinite dimensional. There have been arguments in favour of an infinite spectrum based on the possible Regge behaviour of amplitudes in N=8 supergravity (Grisaru and Schnitzer, 1981).

All these remarks show that a lot of work remains to be done concerning the structure of N=8 supergravity as well as the dynamics of the theory. We are led to a new concept of superunification: Supergravity is fundamental. It should be viewed as a preon theory whose spectrum we should compute. In this spirit the graviton should be fundamental but the photon should be composite.
We could also imagine scenarios where for example SU(3) x U(1) is included in the gauged SO(8) and SU(2) should be included

in the composite SU(8). They have been considered in particular in the Kaluza Klein approach based on 11 dimensional supergravity. It seems in fact difficult to obtain SU(3) x SU(2) x U(1) as the isometry group of a compactified 7 dimensional space together with chiral fermions and a local chiral SU(8) seems to appear independently of a specific compactification.

REFERENCES

Cremmer, E., Ferrara, S., Scherk, J., 1977, Phys. Lett. 68B 234-238

Cremmer, E., Scherk, J., 1977, Nucl. Phys. B127 259-268

Cremmer, E., Ferrara, S., Scherk, J., 1978, Phys. Lett. 74B 61-64

Cremmer, E., Julia, B., Scherk, J. 1978, Phys. Lett. 76B 409

Cremmer, E., Julia, B., 1978, Phys. Lett. 80B 48-51

Cremmer, E., Julia, B., 1979,Nucl. Phys. B159 141-212

Das, A., Freedman, D.Z., 1977, Nucl. Phys. B120 221-230

De Wit, B., Freedman, D.Z., 1977, Nucl. Phys. B130 105-113

De Wit, B., 1979, Nucl. Phys. B158 189-212

De Wit, B., Nicolai, H., 1981, Phys. Lett. 108B 285-290

De Wit, B., Nicolai, H., 1982, Nucl. Phys. B208 323-364

Ellis, J., Gaillard, M.K., Maiani, L., Zumino, B., 1980, Unification of the fundamental particle interactions, ed. Ferrara, S., Ellis, J., Van Nieuwenhuizen, P., Plenum Press, New York, pp. 69-88

Ellis, J., Gaillard, M.K., Zumino, B., 1981, CERN preprint TH 3152

Ferrara, S., Van Nieuwenhuizen, P., 1976, Phys. Rev. Lett. 37 1669-1671

Ferrara, S., Scherk, J., Zumino, B., 1977, Nucl. Phys. B121 393-402

Freedman, D.Z., Schwarz, J.H., 1978, Nucl. Phys. B137 333-339

Gaillard, M.K., Zumino, B., 1981, Nucl. Phys. B193 221-244

Galperin, A. Ivanov, E., Kalitzin, S., Ogievetsky, V., Sokatchev, E., 1984, Preprint DUBNA

Gates, J., Zwiebach, B., 1983, Phys. Lett. 123B 200-204

Gliozzi, F., Olive, D., Scherk, J., 1977, Nucl. Phys. B122 253-290

Green, M., 1983, Surveys in High Energy Physics 3 127

Grisaru, M., Schnitzer, H., 1981, Phys. Lett. 107B 196-200

Gunaydin, M., Saçlioglu, C., 1982, Phys. Lett. 108B 180-186

Howe, P., Lindström, U., 1981, Nucl. Phys. B181 487-501

Hull, C.M., 1984, More gauging of N = 8 supergravity, MIT preprint

Hull, C.M., Warner, N.P., 1984, The structure of the gauged N = 8 supergravity theories, CALTECH Preprint 68-1177

Julia, B., 1982, Lecture notes in Phys. 180 Springer Verlag p. 214

Kallosh, R., 1982, Supergravity 81, ed. Ferrara, S., Taylor, J.G., Cambridge University Press, Cambridge, pp. 397-420

Ogievetsky, V., Sokatchev, E., 1978, Phys. Lett. 79B 222-224

Schwarz, J.H., 1982, Phys. Reports 89 223-322

Taylor, J.G., 1982, Supergravity 81, ed. Ferrara, S., Taylor, J.G., Cambridge University Press, Cambridge, pp. 17-46

Warner, N.P., 1983, Phys. Lett. 128B 169-173
Warner, N.P., 1984, Nucl. Phys. B231 250-268

EXTENDED SUPERGRAVITY IN SUPERSPACE

P S HOWE

1. INTRODUCTION

A superspace (supermanifold) is a generalization of an ordinary differentiable manifold in which the local co-ordinates consist of both even and odd elements of a Grassmann algebra. Although superspaces are radically different in some ways from ordinary manifolds, some features of ordinary geometry are readily generalized to the super case. In particular it is straightforward to set up tensor geometry in superspace.

Superspaces of various dimensions provide the natural setting for supersymmetric field theories, which are based on representations of the Poincaré superalgebras. In the case of rigid (space-time independent) supersymmetry, the appropriate superspaces are the homogeneous superspaces formed by factoring the super Poincaré group by SO(1,d - 1) × G where d is the dimension of spacetime and G the internal symmetry group corresponding to the internal symmetry subalgebra of the Poincaré superalgebra. These spaces are therefore extensions of ordinary Minkowski spacetime which is just the Poincaré group factored by the Lorentz group. In the case of locally (space-time dependent) supersymmetric theories, i.e. supergravity theories, the appropriate superspaces are curved supermanifolds, the equations of notion taking the form of restrictions on the supertorsion and supercurvature tensors.

Both types of theory can be formulated in terms of fields on superspace (superfields) and because of the anti-commuting nature of the odd co-ordinates, any superfield can be expanded as a power series in the odd co-ordinates with co-efficients which depend on the even co-ordinates. These co-efficient fields are component fields on spacetime, although not all of these components need correspond to physical fields.

The superspace approach to supersymmetry has several attractive features: firstly, the formalism is manifestly co-variant and provides a geometrical interpretation of super-gravity theories; secondly it shows how the space-time notions of local supersymmetry transformations, supercovariant derivatives and supercovariant derivatives have geometrical origins in superspace and thirdly, it opens up the possibility of manifestly covariant quantization of supersymmetric field theories. This last feature is perhaps the most important but it is not straightforward to implement in all theories of interest. This is because not all supersymmetric theories admit auxiliary fields (Rivelles and Taylor 1981) which are a necessary prerequisite for a superspace perturbation theory based on unconstrained superfields.

In this article we shall discuss extended supergravity theories in superspace at the level of the equations of motion. In the next section the basic formalism of superspace differential geometry is introduced; in particular the super-torsion and supercurvature tensors are defined and the Bianchi Identities which they satisfy derived. A particularly inter-esting feature is that the tangent space group is taken to be SO(1, d - 1) × G which implies that the tangent spaces have a direct sum structure under the action of this group. As a consequence it turns out that the supercurvature tensor is expressible in terms of the supertorsion tensor and (co-variant) derivatives thereof. Hence, in contrast to ordinary Einstein gravity it is the supertorsion tensor which plays the central rôle in supergravity theories. Indeed if the super-torsion tensor vanishes so does the supercurvature tensor and there is no supersymmetry at all, i.e. even rigid superspace is not flat. In this section we also show how local super-symmetry transformations in spacetime are defined in terms of general co-ordinate transformations in superspace and how one identifies the physical component fields as components of the various geometrical fields in superspace. We also define supercovariant derivatives and supercovariant field strength tensors.

In section 3 the question of constraints is discussed. The equations of motion for supergravity theories take the form of constraints on the superspace torsion tensor and these constraints depend on the theory in question. A simple method of finding these constraints is outlined and a brief discussion of the complications which arise when scalar fields are present is given. Briefly it turns out that, given the representation of supersymmetry one is seeking to describe, the full non-linear equations of motion may be derived straightforwardly, and shown to be consistent.

In sections 4 and 5 the formalism is applied to two specific examples, N = 1 supergravity in d = 11 and the chiral N = 2 supergravity in d = 10. These theories are the two maximal supergravity theories which both correspond to N = 8 d = 4 supergravity under conventional dimensional reduction. However it is not possible to derive the N = 2, d = 10 theory from the N = 1 d = 11 theory and the two theories are completely inequivalent from the point of view of reduction schemes of the Kaluza-Klein type. On-shell supergravity theories in d = 4 can be obtained using the same methods and are described in some detail by Howe (1982).

2. GENERAL FORMALISM

2.1 Co-ordinates and differential forms

Superspaces of the type we shall use were first introduced by Salam and Strathdee (1975) in the context of d = 4 N = 1 rigid supersymmetry. The curved superspace formalism suitable for describing supergravity theories was developed by Wess and Zumino (1977, 1978a). It is possible to make the notion of a supermanifold mathematically precise (see e.g. Rogers 1980, de Witt 1984) but we shall not go into such details here. We shall consider superspaces with d even co-ordinates x^m, corresponding to d-dimensional spacetime and d' odd co-ordinates θ^μ, where d' depends on the theory under consideration. For example, the superspaces appropriate for maximal supergravity theories have 32 (real) odd co-ordinates irrespective of the dimension of spacetime ($d \leqslant 11$) in which they are formulated. The entire set of co-ordinates is denoted by z^M,

$$z^M = (x^m, \theta^\mu) \tag{2.1}$$

and the Grassmann properties of the z's are summarized by the formula

$$z^M z^N = (-1)^{MN} z^N z^M \tag{2.2}$$

$$(-1)^{MN} = \begin{cases} +1 & \text{if either or both indices are even} \\ -1 & \text{if both indices are odd} \end{cases}$$

Because of the anticommuting nature of the θ's any local function f(z) of the superspace co-ordinates can be expanded in a finite power series in θ with x-dependent co-efficients

$$f(z) = \sum_{n=o}^{d'} f_{\mu_1 \ldots \mu_n}(x)\, \theta^{\mu_1} \ldots \theta^{\mu_n} \tag{2.3}$$

where the co-efficient functions are antisymmetric on the indices $\mu_1 \ldots \mu_n$. Thus a superfield is equivalent to a set of space-time fields which are alternately of bosonic and fermionic type in the θ-expansion.

Concepts from ordinary tensor geometry are readily generalized to supermanifolds. For example a vector field X may be written

$$X = X^M \frac{\partial}{\partial z^M} \tag{2.4}$$

and a differential form ω can be written

$$\omega = dz^M \omega_M \tag{2.5}$$

The generalisations to tensors of arbitrary type are straightforward although care must be taken with the ordering of factors because of the anticommuting quantities present. Particularly useful tensors from the point of view of supergravity are differential forms in superspace; the wedge product is defined by (Zumino 1975)

$$dz^M \wedge dz^N = -(-1)^{MN} dz^N \wedge dz^M \tag{2.6}$$

A p-form ϕ_p can be written

$$\phi_p = dz^{M_p} \wedge \ldots \wedge dz^{M_1} \; \phi_{M_1 \ldots M_p} \tag{2.7}$$

here $\phi_{M_1 \ldots M_p}$ is generalised antisymmetric, i.e.

$$\phi_{M_1 M_2 \ldots M_p} = -(-1)^{M_1 M_2} \phi_{M_2 M_1 \ldots M_p}\text{, etc.} \tag{2.8}$$

n the following we shall not explicitly include the wedge ymbol in formulae involving differential forms.

As in ordinary space one has

$$\phi_p \, \phi_q = (-1)^{pq} \, \phi_q \, \phi_p \tag{2.9}$$

The exterior derivative operator d takes a p-form ϕ_p into (p+1)-form $d\phi_p$ defined by

$$d\phi_p = dz^{M_p} \ldots dz^{M_1} \, dz^N \, \partial_N \, \phi_{M_1 \ldots M_p} \tag{2.10}$$

nd it is straightforward to verify that

$$d^2 = 0. \tag{2.11}$$

The d operator starts from the right so that

$$d(\phi_p \, \phi_q) = \phi_p (d\phi_q) + (-1)^q \, (d\phi_p) \, \phi_q \tag{2.12}$$

Superspace differential forms differ from ordinary differntial forms in that the definition (2.6) allows there to be ifferential forms of arbitrarily high degree in superspace; urthermore there is no direct generalization of the interation of forms in ordinary geometry to the supermanifold ase.

.2 Structure Equations

t each point in superspace it is assumed that there is a referred set of frames E^A which are related to the co-ordinte basis one-forms dz^M by the supervielbein matrix $E_M{}^A$:

$$E^A = dz^M \, E_M{}^A \tag{2.13}$$

$$E^A = (E^a, E^\alpha); \quad \begin{cases} a = 0, 1, \ldots d-1 \\ \alpha = 1, \ldots \; d' \end{cases}$$

he matrix $E_M{}^A$ has an inverse $E_A{}^M$:

$$E_A{}^M \, E_M{}^B = \delta_A{}^B \tag{2.14}$$

Under a tangent space transformation the frame E^A

transforms into a new preferred frame. Infinitesimally we have,

$$\delta E^A = E^B L_B{}^A \tag{2.15}$$

In ordinary Lorentzian geometry the preferred frames are the orthonormal ones determined by the metric and the tangent space group is just the Lorentz group. In the superspaces we wish to consider a straightforward generalization of the metric does not exist and the tangent space group is taken to be SO(1, d - 1) × G where G is the internal symmetry group corresponding to the internal Lie subalgebra of the relevant Poincaré superalgebra. Therefore we always have

$$L_{ab} = L_a{}^c \eta_{cb} = - L_{ba} \tag{2.16}$$

where η_{ab} is the Minkowski spacetime metric, and

$$L_\alpha{}^b = L_a{}^\beta = 0 \tag{2.17}$$

i.e. there is no tangent space supersymmetry. The precise form of $L_\alpha{}^\beta$ depends on the theory under consideration; for simple supergravity theories G is trivial and $L_\alpha{}^\beta$ is just the Lorentz transformation (2.16) in the spinor representation. The choice of the internal symmetry group factor is a matter of taste to some extent. For example, in d = 4 N = 8 supergravity it is convenient to include SU(8) in the tangent space group even though the SU(8) gauge fields are composite; furthermore, it is even convenient to keep SU(8) in the case of gauged N = 8 supergravity and to introduce the SO(8) gauge fields separately (Howe and Nicolai, 1982).

In order to define covariant exterior differentiation we introduce the connection one-form $\Omega_A{}^B$ which has the same symmetry properties as $L_A{}^B$. Under (2.16) $\Omega_A{}^B$ transforms inhomogeneously

$$\delta\Omega_A{}^B = - dL_A{}^B + \Omega_A{}^C L_C{}^B - L_A{}^C \Omega_C{}^B \tag{2.18}$$

With the aid of $\Omega_A{}^B$ a covariant exterior derivative, D, can be defined; for example, for a p-form $\phi_A{}^B$

$$D\phi_A{}^B = d\phi_A{}^B - (1)^P \Omega_A{}^C \phi_C{}^B + \phi_A{}^C \Omega_C{}^B \tag{2.19}$$

The supertorsion two-form T^A and the supercurvature

two-form $R_A{}^B$ are defined in the usual way

$$T^A = DE^A = \frac{1}{2} E^C E^B T_{BC}{}^A \tag{2.20}$$

$$R_A{}^B = \frac{1}{2} E^D E^C R_{CD,A}{}^B = d\Omega_A{}^B + \Omega_A{}^C \Omega_C{}^B \tag{2.21}$$

Using the definitions and $d^2 = 0$ one derives the Ricci Identity

$$D^2 \phi_A{}^B = - R_A{}^C \phi_C{}^B + \phi_A{}^C R_C{}^B \tag{2.22}$$

and the structure equations (Bianchi Identities)

$$DT^A - E^B R_B{}^A = I^{(1)A} = 0 \tag{2.23}$$

$$DR_A{}^B = I^{(2)}{}_A{}^B = 0 \tag{2.24}$$

In index notation

$$I^{(1)}_{ABC,}{}^D = \sum_{(\widehat{ABC})} \{D_A T_{BC}{}^D + T_{AB}{}^E T_{EC}{}^D - R_{AB,C}{}^D\} \tag{2.25}$$

$$I^{(2)}_{ABC,D}{}^E = \sum_{(\widehat{ABC})} \{D_A R_{BC,D}{}^E + T_{AB}{}^F R_{FCD}{}^E\} \tag{2.26}$$

where $\sum_{(\widehat{ABC})}$ denotes the graded cyclic sum:

$$\sum_{(\widehat{ABC})} X_{AB,C} = X_{AB,C} + (-1)^{C(A+B)} X_{CA,B} + (-1)^{A(B+C)} X_{BC,A} \tag{2.27}$$

As a consequence of the structure of the tangent space group one can prove two useful theorems (Dragon, 1979): firstly, the supercurvature tensor can be expressed in terms of $T_{AB}{}^C$ and covariant derivatives thereof and secondly, the identities $I^{(2)}{}_A{}^B$ are not independent. That is, given $I^{(1)A} = 0$ and the Ricci Identity, one can show that $I^{(2)}{}_A{}^B = 0$. This is a useful result from the point of view of supergravity in superspace where the Bianchi Identities play a crucial rôle. We shall not prove these theorems here, but indicate how the first one follows from the form of $L_A{}^B$ in the case of simple supergravity. Considering $I^{(1)}_{\alpha\beta c}{}^d$ one has

$$D_\alpha T_{\beta c}{}^d + T_{\alpha\beta}{}^E T_{Ec}{}^d + \text{cyclic terms} = R_{\alpha\beta,c}{}^d \tag{2.28}$$

since

$$R_{c\alpha,\beta}{}^d = R_{\beta c,\alpha}{}^d = 0. \tag{2.29}$$

Hence $R_{\alpha\beta,c}{}^{d}$ is immediately expressed in terms of T and since $R_{\gamma}{}^{\delta}$ is related to $R_{c}{}^{d}$, one can also find $R_{\alpha\beta,\gamma}{}^{\delta}$. The remaining parts of R may be expressed in terms of T in a similar fashion.

2.3 Additional gauge fields

In the component formalism, extended supergravity theories have additional gauge fields over and above the vielbein and the spin $3/_2$ fields. These latter fields are components of the supervielbein but the other gauge fields are not. This can be seen by the following argument: under a superspace general co-ordinate transformation one has

$$\delta E_M{}^A = \xi^N \partial_N E_M{}^A + \partial_M \xi^N E_N{}^A \tag{2.30}$$

Hence

$$\delta E_\mu{}^A = \partial_\mu \xi^A + \ldots \tag{2.31}$$

where

$$\xi^A = \xi^M E_M{}^A \tag{2.32}$$

and, similarly,

$$\delta\Omega_{\mu,A}{}^B = - \partial_\mu L_A{}^B + \ldots \tag{2.33}$$

Expanding ξ^A and $L_A{}^B$ in a θ power series, we see from (2.31) and (2.32) that all the components of these parameters except the leading (θ^o) ones act as shift transformations on components of $E_\mu{}^A$ and $\Omega_{\mu A}{}^B$ and can therefore be used to transform these components to zero. In this gauge (the Wess-Zumino gauge) the only remaining gauge transformations are $\xi^A|_{\theta=0}$ and $L_A{}^B|_{\theta=0}$, and these correspond to x-space general co-ordinate, local supersymmetry, local Lorentz and internal symmetry transformations. Thus, for example, in d = 11 supergravity the three index antisymmetric tensor gauge field cannot be a component of the super vielbein which implies that it must be located elsewhere. To accommodate such fields it is convenient to introduce further gauge fields into the superspace formalism. Typically these fields are p-forms so we introduce p-form gauge fields $A_{(p)}$

$$A_{(p)} = \frac{1}{p!} E^{A_p} \ldots E^{A_1} A_{A_1 \ldots A_p} \tag{2.34}$$

and corresponding (p + 1)-form field strengths F,

$$F = dA_{(p)} \tag{2.35}$$

which are invariant under the transformations

$$A_p \rightarrow A_p + d\Lambda_{p-1} \tag{2.36}$$

Such field strengths obey Bianchi Identities

$$dF = 0 \tag{2.37}$$

A complication may arise if we have a set of F's which transform under a representation of the internal symmetry group. In this case simple identities of the form (2.37) would be inconsistent with the local internal symmetry. These covariant F's must therefore be related to a new set of F's which do satisfy (2.37) but which do not transform under the local internal symmetry group. As we shall see the transformation which accomplishes this is related to the presence of scalar fields and an associated rigid symmetry group.

2.4 Superspace and components

To conclude this section we outline the relation of the superspace formalism to the component formalism (Wess and Zumino 1978b). Let ϕ be a 0-form superfield with a set of unspecified tangent space indices. Under an infinitesimal superspace general co-ordinate transformation one has

$$\delta\phi = \xi^M \partial_M \phi \tag{2.38}$$

This can be rewritten in the form,

$$\delta\phi = \xi^A D_A \phi \tag{2.39}$$

with ξ^A given by (2.32) and where we have dropped an Ω-dependent tangent space transformation. For the supervielbein we find

$$\delta E_M{}^A = D_M \xi^A + \xi^B T_{BM}{}^A \tag{2.40}$$

Instead of defining the components of a superfield by a straightforward θ-expression as in equation (2.3), we define them instead by evaluating products of D_α's on the superfield at $\theta = 0$. Thus the components of ϕ are

$$\phi\big|_{\theta=0}, \quad D_\alpha \phi\big|_{\theta=0}, \quad D_{[\alpha} D_{\beta]} \phi\big|_{\theta=0}, \ \cdots \tag{2.41}$$

Clearly this set of fields is in one-to-one correspondence with those of the θ-expansion. For the $(D_\alpha)^2$ and higher terms in the expansion there is some freedom in the definition of the components. For example, instead of using $D_{[\alpha} D_{\beta]} \phi\big|_{\theta=0}$ we could use

$$D_{[\alpha} D_{\beta]} \phi + S_{\alpha\beta} \phi\big|_{\theta=0} \tag{2.42}$$

where $S_{\alpha\beta}$ is another covariant superfield with appropriate dimensions. Of course we should also like to define the components so that they transform irreducibly under the tangent space group and this is straightforward to do. As far as the enumeration of components is concerned it is not necessary to antisymmetrize on the indices of products of D_α's since in practice

$$\{D_\alpha, D_\beta\} = -\, i(\Gamma^c)_{\alpha\beta}\, D_c + \ldots \tag{2.43}$$

where $(\Gamma^c)_{\alpha\beta}$ is a numerical tensor, and $D_c\, \phi\big|_{\theta=0}$ is not a new component.

Because of the tangent space group structure, $D_\alpha\, \phi$ is again a superfield which transforms covariantly. Hence it will also transform under general co-ordinate transformations as ϕ does (equation 2.38). We now define the x-space local supersymmetry parameter $\zeta^\alpha(x)$ by

$$\zeta^\alpha(x) = \xi^\alpha(z)\big|_{\theta=0} \tag{2.44}$$

The local supersymmetry transformations of the components of ϕ can therefore be read off using (2.39), e.g.

$$\delta\phi\big|_{\theta=0} = \zeta^\alpha(x)\, D_\alpha\, \phi\big|_{\theta=0} \tag{2.45}$$

The covariant superfields ϕ will be parts of $T_{AB}{}^C$ which are left over after constraints have been imposed. The components of ϕ will therefore correspond to covariant fields in space-time, i.e. spinor fields and gauge field strengths. After having imposed the constraints all parts of $T_{AB}{}^C$ and $R_{ABC}{}^D$ will be expressible in terms of a set of ϕ's which will be related to one another through the Bianchi Identities. In general this set of ϕ's will contain relatively few components so that the sequence of fields defined by (2.41) will terminate rather rapidly in the sense that all higher components will be expressible in terms of derivatives and products of the independent components. Therefore, providing that the Bianchi Identities are satisfied, i.e. that we have not imposed constraints which lead to inconsistencies or only have trivial solutions, it will suffice to determine the independent components of the ϕ's, the higher order terms being guaranteed to take care of themselves.

For the x-space gauge fields themselves will be components of the supervielbein and any other p-form gauge fields we have introduced. For the supervielbein it is always possible to choose the gauge in which

$$\begin{aligned} E_m{}^a &= e_m{}^a(x) + O(\theta) \\ E_m{}^\alpha &= \psi_m{}^\alpha(x) + O(\theta) \\ E_\mu{}^a &= 0 + O(\theta) \\ E_\mu{}^\alpha &= \delta_\mu{}^\alpha + O(\theta) \end{aligned} \tag{2.46}$$

where $e_m{}^a(x)$ is the x-space vielbein and $\psi_m{}^\alpha(x)$ the x-space Rarita-Schwinger fields. Similarly, for the connection we may choose

$$\begin{aligned} \Omega_{ma}{}^b &= \omega_{ma}{}^b(x) + O(\theta) \\ \Omega_{\mu A}{}^B &= O(\theta) \\ \Omega_{m\alpha}{}^\beta &= \omega_{m\alpha}{}^\beta(x) + O(\theta) \end{aligned} \tag{2.47}$$

where $\omega_{ma}{}^b(x)$ is the x-space Lorentz connection and $\omega_{m\alpha}{}^\beta(x)$ is the x-space Lorentz connection in the spinor representation together with any internal symmetry connection. The $\omega_{mA}{}^B$'s are composite fields so that we do not need to find their variations although this can be done straightforwardly. The variations of $e_m{}^a$ and $\psi_m{}^\alpha$ are found by using (2.46) in conjunction with (2.40).

Now suppose we have additional gauge fields. For simplicity let us consider a single one-form field A. Using the superspace Λ-gauge transformations (2.36) we may choose a gauge in which

$$A_m = a_m(x) + O(\theta)$$
$$A_\mu = O(\theta) \tag{2.48}$$

where $a_m(x)$ is to be identified as the x-space gauge field. Under superspace general co-ordinate transformations we have

$$\delta A_M = \xi^N \partial_N A_M + \partial_M \xi^N A_N \tag{2.49}$$

which can be written in the form

$$\delta A_M = \xi^A F_{AM} \tag{2.50}$$

where we have dropped a field-dependent Λ-gauge transformation. Using (2.50) we can find δa_m if we have determined F_{AB}.

Lastly, we briefly discuss supercovariant derivatives and field strengths. Given a covariant superfield ϕ we can construct the covariant superfield $D_a \phi$. Evaluating this superfield at $\theta = 0$ in the Wess-Zumino gauge gives the supercovariant derivative of $\phi|_{\theta=0}$ (= $\hat{\phi}(x)$), i.e.

$$(D_a \phi)|_{\theta=0} = \hat{D}_a(\phi|_{\theta=0}) = \hat{D}_a \hat{\phi} \tag{2.51}$$

defines the x-space supercovariant derivative $\hat{D}_a$. Now

$$D_a \phi = E_a{}^M D_M \phi \tag{2.52}$$

so at $\theta = 0$

$$\hat{D}_a \hat{\phi} = e_a{}^m \nabla_m \hat{\phi} + E_a{}^\mu D_\mu \phi|_{\theta=0} \tag{2.53}$$

where ∇_m is the x-space SO(1,d - 1) × G covariant derivative. For the second term we have

$$D_\mu \phi|_{\theta=0} = \delta_\mu{}^\alpha (D_\alpha \phi)_{\theta=0}$$

and

$$E_a{}^\mu|_{\theta=0} = -\psi_a{}^\beta \delta_\beta{}^\mu \text{ where } \psi_a{}^\beta = e_a{}^m(x) \psi_m{}^\beta \tag{2.54}$$

Hence

$$\hat{D}_a \hat{\phi} = e_a{}^m \nabla_m \hat{\phi} - \psi_a{}^\alpha (D_\alpha \phi)_{\theta=0} \tag{2.55}$$

or

$$\hat{D}_m \hat{\phi} = e_m{}^a \hat{D}_a \hat{\phi} = \nabla_m \hat{\phi} - \psi_m{}^\alpha (D_\alpha \phi)_{\theta=0} \tag{2.56}$$

Now consider the field strength F_{AB} corresponding to a 1-form gauge field. We define the x-space supercovariant field strength $\hat{f}_{ab}(x)$ by

$$\hat{f}_{ab}(x) = F_{ab}|_{\theta=0} \tag{2.57}$$

but

$$F_{ab} = E_b{}^N E_a{}^M F_{MN} \text{ and } F_{MN} = E_N{}^B E_M{}^A F_{AB} (-1)^{M(B+N)} \tag{2.58}$$

So at $\theta = 0$ in the Wess-Zumino gauge, we find

$$\hat{f}_{ab} = e_a{}^m e_b{}^n f_{mn} + 2\psi_{[a}{}^\gamma F_{b]\gamma}|_{\theta=0} + \psi_a{}^\alpha \psi_b{}^\beta F_{\alpha\beta}|_{\theta=0} \tag{2.59}$$

or

$$\hat{f}_{mn} = f_{mn} + 2\psi_{[m}{}^\beta e_{n]}{}^a F_{a\beta}|_{\theta=0} + \psi_m{}^\alpha \psi_n{}^\beta F_{\alpha\beta}|_{\theta=0} \tag{2.60}$$

thus expressing the supercovariantized x-space field strength in terms of the ordinary x-space field strength f_{mn} ($= \partial_m a_n - \partial_n a_m$) together with additional pieces which depend on the Rarita-Schwinger field. We remark that the above applies to components of the supercurvature and supertorsion tensors. Thus, $T_{ab}{}^c|_{\theta=0}$ is the supercovariantized x-space torsion tensor, $R_{ab,c}{}^d|_{\theta=0}$ is the x-space supercovariantized Riemann tensor and $T_{ab}{}^\gamma|_{\theta=0}$ is the supercovariantized x-space Rarita-Schwinger field strength. For this last tensor we have

$$T_{ab}{}^\alpha|_{\theta=0} = e_a{}^m e_b{}^n T_{mn}{}^\alpha|_{\theta=0} + \ldots \tag{2.61}$$

but

$$T_{mn}{}^\alpha|_{\theta=0} = \nabla_m \psi_n{}^\alpha - \nabla_n \psi_m{}^\alpha \tag{2.62}$$

so that

$$\hat{t}_{mn}{}^\alpha = e_m{}^a e_n{}^b T_{ab}{}^\alpha|_{\theta=0} = 2\nabla_{[m} \psi_{n]}{}^\alpha + \ldots \tag{2.63}$$

3. CONSTRAINTS

3.1 Necessity of constraints

A scalar superfield in a superspace appropriate for a maximally extended supergravity theory has 2^{32} components whereas the number of on-shell degrees of freedom in such a theory is only 128 + 128 (bosonic plus fermionic). It is therefore evident that the supervielbein contains many more components than are required to describe the theory. This is also true for supergravity theories which have off-shell extensions, i.e. auxiliary fields. The basic reason for this is that in rigid superspace the supersymmetry generators Q_α anticommute with the covariant derivatives D_α so that a general superfield defines a reducible representation of supersymmetry. The central problem of superspace supergravity is therefore to find the constraints which determine an irreducible representation corresponding either to fields (off-shell) or particle states (on-shell).

Clearly some of the redundant degrees of freedom correspond to the Wess-Zumino gauge transformations discussed in section 2. However it would be extremely tedious to expand the supervielbein in a θ-expansion, fix Wess-Zumino gauges and then try to guess the required constraints on the remaining components. This problem is avoided by imposing the constraints covariantly on the supertorsion tensor, so that the Wess-Zumino gauges need not be fixed in the analysis. As a result of imposing constraints all parts of $T_{AB}{}^{C}$ will be expressible in terms of a set of covariant superfields W having various tangent space indices. In addition the W's will be related to each other by the Bianchi Identities, typically by relations involving the spinorial derivative D_α. One can then investigate the independent components of the W's which are defined by the method outlined in section 2. The enforcement of the Bianchi Identities thus allows an enumeration of the number of independent components and ensures that the constraints that have been chosen are consistent in the sense that they do not lead to a trivial system. In this sense we may speak of a solution to the Bianchi Identities. In the case that the constraints are on-shell the Bianchi Identities also provide a check on the consistency of the dynamical equations. In the case that some of the W's have tangent space symmetry properties which could correspond to (x-space) gauge field strengths the question arises as to how many degrees of freedom $W|_{\theta=0}$ has. For example, in all supergravity theories $R_{ab,cd|\theta=0}$ occurs as a higher component of one or more W's.

However, the x-space Riemann tensor is the field strength

corresponding to vielbein and hence corresponds to $\frac{d(d-1)}{2}$ degrees of freedom if the representation is off-shell and $\frac{d(d-3)}{2}$ degrees of freedom on-shell (not counting gauge degrees of freedom). Similar reasoning can be applied to $T_{ab}{}^{\alpha}\big|_{\theta=0}$, but there may be other W's with no obvious geometrical interpretation such as tensors which could correspond to anti-symmetric tensor gauge fields. If W corresponds to a gauge field strength then it will satisfy a covariantized Bianchi Identity of the form

$$D_{[a} W_{bc]} + \ldots = 0 \tag{3.1}$$

where we have taken W to be a second rank anti-symmetric tensor for simplicity. If an equation such as (3.1) arises as a consequence of the Bianchi Identities (2.25) then W_{ab} must be counted as a gauge vector field. If no such identity is true then the number of degrees of freedom is equal to the number of components of W. Such non-gauge antisymmetric tensors can occur in off-shell representations of supergravity. If a W does correspond to a gauge field then the analysis is simplified by introducing additional superspace p-form potentials A_p as described in section 2.

3.2 Conventional constraints

Given an ordinary manifold with a metric g and a metric connection ω there is a unique torsion-free connection. This is determined by imposing the constraint

$$t_{ab}{}^{c} = 0 \tag{3.2}$$

from which $\omega_{m,a}{}^{b}$ can be solved for algebraically in terms of $e_m{}^a$. Imposing the constraint (3.2) can be made without loss of generality since the dynamics of the system are determined by the Lagrangian. For example for gravity coupled to spinor matter one can choose to use minimal coupling and treat $\omega_{ma}{}^{b}$ as an independent variable, or one can impose (3.2) but have the same physical system by including non-minimal terms in the Lagrangian.

In a similar fashion one can choose conventional constraints in superspace to eliminate Ω as an independent superfield. Because of the tangent space group structure it is not possible to choose $T_{AB}{}^{C} = 0$ as there are many more parts of $T_{AB}{}^{C}$ than there are of $\Omega_{AB}{}^{C} = E_A{}^M \Omega_{M,B}{}^{C}$. Furthermore, in superspace, one may choose additional conventional constraints corresponding to parts of the supervielbein. This is most easily seen at the linearized level. Let

$$E_M{}^A = E_M^{(o)A} + H_M{}^A; \quad H_A{}^B = E_A^{(o)M} H_M{}^B$$

and

$$\Omega_{MA}{}^B = \Omega_{MA}^{(o)B} + \phi_{MA}{}^B; \quad \phi_{A,B}{}^C = E_A^{(o)M} \phi_{MB}{}^C \tag{3.3}$$

where $E_M^{(o)A}$ and $\Omega_{MA}^{(o)B}$ correspond to rigid superspace, i.e.

$$T_{AB}^{(o)C} = 0 \text{ except for } T_{\alpha\beta}^{(o)C} = -i(\Gamma^c)_{\alpha\beta} \tag{3.4}$$

so that

$$\{D_\alpha^{(o)}, D_\beta^{(o)}\} = i(\Gamma^c)_{\alpha\beta} D_c^{(o)}$$

$$[D_\alpha^{(o)}, D_b^{(o)}] = [D_\alpha^{(o)}, D_b^{(o)}] = 0 \tag{3.5}$$

Then

$$T_{AB}^{(1)C} = D_A^{(o)} H_B{}^C - (-1)^{AB} D_B^{(o)} H_A{}^C + T_{AB}^{(o)D} H_D{}^C$$
$$- H_A{}^D T_{DB}^{(o)C} + (-1)^{AB} H_B{}^D T_{DA}^{(o)C} + \phi_{AB}{}^C - (-1)^{AB} \phi_{BA}{}^C \tag{3.6}$$

is the linearized torsion tensor, i.e.

$$T_{AB}{}^C = T_{AB}^{(o)C} + T_{AB}^{(1)C} \tag{3.7}$$

Consider equation (3.6) for N = 1, d = 4. We have

$$T_{\alpha\beta}^{(1)c} = D_\alpha^{(o)} H_\beta{}^c + D_\beta^{(o)} H_\alpha{}^c - i(\gamma^d)_{\alpha\beta} H_d{}^c$$
$$+ i H_\alpha{}^\delta (\gamma^c)_{\delta\beta} + i H_\beta{}^\delta (\gamma^c)_{\delta\alpha} \tag{3.8}$$

Now redefine $H_d{}^c$,

$$H_d{}^c \to H_d{}^c + K_d{}^c = H_d'^{\,c} \tag{3.9}$$

then we can choose K such that the redefined torsion satisfies

$$(\gamma^d)^{\alpha\beta} (T^{(1)}{}_{\alpha\beta}{}^c)' = 0 \tag{3.10}$$

We shall not go into the details of precisely which conventional constraints can be chosen, since this is not necessary in practice when one is interested in on-shell representations. However, we remark that as N increases (in d = 4), so the relative number of conventional constraints decreases.

3.3 Kinematic and dynamical constraints

After the conventional constraints have been imposed there are still a vast number of component fields left in superspace compared to the dimensions of the representations one wishes to describe. The remaining constraints can be subdivided into two classes: kinematical and dynamical. The imposition of kinematical constraints leads to a representation of supersymmetry on fields, usually taken to be irreducible, but which does not imply any equations of motion for the component fields. The imposition of further dynamical constraints then eliminates the auxiliary fields so that the remaining physical fields are on-shell, i.e., obey the appropriate equations of motion. If one wishes to derive the equations of motion from a superspace Lagrangian, it is necessary to make this division since superspace Lagrangians only exist for off-shell representations. If one is content with deriving the field equations alone, then it is not necessary to make the distinction, one merely imposes the constraints that lead directly to the equations of motion.

In the case of N = 1 and 2 supergravity in d = 4 it is possible to adduce arguments for which kinematical constraints should be imposed. We shall not follow this approach here, but refer the interested reader to the literature (Stelle and West, 1979; Gates, 1979; Siegel 1979; Gates and Siegel, 1980; Gates, Stelle and West, 1980).

Suppose instead that we know which representation we are looking for. This is always the case for on-shell representations which are determined by the representations of the super-Poincaré algebra on states. It is also partially true for off-shell representations where one can obtain irreducible representations from the study of supercurrent multiplets.

For d = 4 $N \leq 4$ this approach yields the conformal supergravity multiplets which can be represented within the standard superspace framework of section 2 (Bergshoett, de Roo and de Wit, 1981; Howe, 1981). These conformal representations can then be used as a basis for building off-shell Poincaré supergravity theories by introducing compensating multiplets (Kaku and Townsend, 1978; Das, Kaku and Townsend, 1978; de Wit, Van Holten and Van Proeyen, 1980). In superspace this corresponds to the reduction of the superconformal invariance of the constraints.

In either case, given the representation, we can determine the constraints by simple dimensional analysis. Each covariant component field in the representation (i.e. spinor fields, gauge field strengths, etc.) can be thought of as the $\theta = 0$ component of a covariant superfield W. Furthermore they will have definite dimensions and symmetry properties with respect to the tangent space group. We then set to zero all parts of $T_{AB}{}^{C}$ which cannot correspond to such W's and place the W's in $T_{AB}{}^{C}$ wherever they are allowed on symmetry and dimensional grounds. There will then be some arbitrary numerical co-efficients to determine and this can be done by using the Bianchi Identities. In practice one starts with the lowest (mass) dimension parts of $T_{AB}{}^{C}$ and determines the higher dimensional parts by the Bianchi Identities thus avoiding problems with possible non-linear terms. That is to say, given the representation at the linearized level one may use the Bianchi Identities to determine the full non-linear theory and, using the methods of section 2, the x-space supersymmetry transformations. In the case of on-shell representations one can go on to deduce the component equations of motion again using the Bianchi Identities.

To do the dimensional analysis it is convenient to adopt geometrical rather than canonical dimensions since the former are the same for all d. Thus

$$\dim(E^{a}) = -1$$
$$\dim(E^{\alpha}) = -\tfrac{1}{2} \qquad (3.11)$$

so that

$$\dim T_{AB}{}^{C} = [A] + [B] - [C] \qquad (3.12)$$

where

$$[A] = \begin{cases} 1 & \text{if } A = a \\ \frac{1}{2} & \text{if } A = \alpha \end{cases} \quad (3.13)$$

For a p-form gauge field $A_{(p)}$, we choose

$$\dim A_{(p)} = -p \quad (3.14)$$

so that

$$\dim A_{a_1 \ldots a_p} = 0. \quad (3.15)$$

Then

$$\dim F_{A_1 \ldots A_{p+1}} = \sum_{i=1}^{p+1} [A_i] - p. \quad (3.16)$$

Canonical dimension fields are obtained by expanding about rigid superspace,

$$E_M{}^A = \overset{o}{E}_M{}^A + \kappa H_K{}^A \quad (3.17)$$

where κ is the gravitational coupling constant which has dimension $-\frac{1}{2}(d-2)$.

3.4 Scalar fields

The above method works perfectly well when there are no dimension zero covariant fields, i.e. scalar fields, in the problem. When these are present then the method as it stands is not very tractable since we could have in principle arbitrary functions of the scalar fields appearing in $T_{AB}{}^C$ rather than mere numerical co-efficients. Fortunately, when more than one scalar field is present then there is always an associated non-linear rigid symmetry group G' (Cremmer, Scherk and Ferrara, 1978; Cremmer and Julia, 1979), i.e., the scalar sector of extended supergravity theories is of σ-model type.

We recall that for a $G'/_G$ σ-model in space-time the scalars may be described by a field $V(x)$ which is an element of G' (Coleman, Wess and Zumino, 1969; Callam, Coleman, Wess and Zumino, 1969). The transformation of V under G' and G is given by

$$V(x) \to g' \, v \, g \quad (3.18)$$

where $g' \in G'$ is space-time independent and $g \in G$ is space-time dependent. If $\{X_i\}$ are the generators of the Lie algebra of G and $\{Y_r\}$ the remaining generators of the Lie algebra of G'

then we can use g(x) to choose a gauge in which

$$V = \exp \phi_r Y_r \tag{3.19}$$

where the ϕ_r's are n fields, n = dim G′ - dim G. The transformation (3.18) then defines a non-linear realization of G′ acting on the fields ϕ_r with the subgroup G acting linearly.

In supergravity theories G is the internal symmetry group contained in the super-Poincaré group and G′ is a non-compact group with dim G′ = dim G + number of scalar fields in the model. For example in N = 8 d = 4 supergravity $G' = E_7$, G = SU(8).

To accommodate the scalar fields in superspace we introduce a superfield V(z) which is an element of G′. Introducing the pure gauge connection $\tilde{\Omega}$ taking values in the Lie algebra of G′ we have

$$\tilde{\Omega} = V^{-1}\, dV = Q + P = Q^i X_i + P^r Y_r \tag{3.20}$$

The associated curvature $\tilde{R}$ vanishes,

$$\tilde{R} = d\tilde{\Omega} - \tilde{\Omega}\,\tilde{\Omega} = 0 \tag{3.21}$$

which gives

$$R^i X_i = (P \wedge P)^i X_i \tag{3.22}$$

and

$$(DP)^r Y_r = + (P \wedge P)^r Y_r \tag{3.23}$$

where R^i is the G curvature, and the one-form P transforms covariantly under G but is invariant under G′. Equations (3.22) and (3.23) may be taken as supplementary equations to the Bianchi Identities (2.25) and (2.26). Furthermore (3.20) determines the G-connection in terms of the scalar fields, and this connection is to be identified with the internal symmetry part of $\Omega_A{}^B$. The advantage of this approach is now clear: all parts of $T_{AB}{}^C$, etc., are invariant under G′ so that the scalar fields can only enter via the covariant one-form P. Hence, given that the scalar sector has a coset space structure, we may use dimensional methods to determine $T_{AB}{}^C$ straightforwardly.

4. N = 1 d = 11 SUPERGRAVITY

4.1 Conventions and component fields

In d = 11 it is convenient to use a space-time with signature + 9 so that the γ-matrices may be chosen to be real. In this representation d = 11 Majorana spinors have 32 real components, so that d = 11 is the maximum number of dimensions in which one can have a supersymmetric theory which reduces to a four-dimensional theory with spins ≤ 2. A basis for the 32 × 32 component matrices is given by the $\gamma^{a_1 \ldots a_n}$'s where

$$\gamma^{a_1 \ldots a_n} = \gamma^{[a_1} \gamma^{a_2} \ldots \gamma^{a_n]} \qquad (4.1)$$

An independent set consists of C $(= \gamma^o)$, γ^a, γ^{ab}, γ^{abc}, γ^{abcd} and γ^{abcde}. For any γ we have

$$\gamma_{\alpha\beta} = \gamma_\alpha{}^\beta \, C_{\beta\alpha} \qquad (4.2)$$

and $(\gamma^a)_{\alpha\beta}$, $(\gamma^{ab})_{\alpha\beta}$, $(\gamma^{abcde})_{\alpha\beta}$ are symmetric while $C_{\alpha\beta}$ $(\gamma^{abc})_{\alpha\beta}$ and $(\gamma^{abcd})_{\alpha\beta}$ are skew-symmetric. Spinorial indices are raised and lowered by the C-matrix,

$$\psi^\alpha = C^{\alpha\beta} \psi_\beta ; \quad \psi_\alpha = \psi^\beta C_{\beta\alpha} ; \quad C^{\alpha\beta} C_{\alpha\gamma} = \delta^\beta_\gamma \qquad (4.3)$$

The co-ordinates of superspace are

$$z^M = (x^m, \theta^\mu); \qquad m = 0, 1, \ldots, 10$$
$$\mu = 1, \ldots, 32 \qquad (4.4)$$

where the θ^μ are real. We use the summation convention

$$U^A V_A = U^a V_a + U^\alpha V_\alpha \qquad (4.5)$$

so that if U^α is real then V_α is imaginary and vice versa since complex conjugation is taken to reverse the order of odd elements of the Grassmann algebra. In particular E^α is real so that, from

$$D = E^a D_a + E^\alpha D_\alpha \qquad (4.6)$$

D_α is imaginary.

The component fields of on-shell d = 11 supergravity are the elfbein $e_m{}^a$ (44 degrees of freedom), an antisymmetric tensor gauge field a_{mnr} (84) and the Rarita-Schwinger field $\psi_m{}^\alpha$ (128). The theory was first constructed by Cremmer,

Julia and Scherk (1978) and subsequently recast into superspace language by Cremmer and Ferrara (1980) and independently by Brink and Howe (1980).

As we have remarked in section 2, a_{mnr} cannot be a component of the supervielbein and it is convenient to introduce a three-form superspace gauge field A together with its corresponding four-form field strength F,

$$A = \frac{1}{3!} E^C E^B E^A A_{ABC}$$

$$F = dA = \frac{1}{4!} E^D E^C E^B E^A F_{ABCD} \qquad (4.7)$$

Evidently F is invariant under the gauge transformation

$$A \to A + d\Lambda \qquad (4.8)$$

where Λ is a two-form gauge parameter. The Bianchi Identity for F is

$$dF = 0 \qquad (4.9)$$

or in index notation,

$$D_{[A} F_{BCDE]} + 2\, T_{[AB}{}^{G} F_{|G|CDE]} = 0 \qquad (4.10)$$

where the brackets denote generalized antisymmetrization and the notation $|G|$ indicates that the index G is not included in the antisymmetrization.

4.2 <u>The theory in superspace</u>

The tangent space group in d = 11 superspace is just the Lorentz group SO(1,10), so we have

$$L_{ab} = - L_{ba}$$

$$L_{\alpha}{}^{b} = L_{a}{}^{\beta} = 0$$

$$L_{\alpha\beta} = \frac{1}{4} (\gamma^{ab})_{\alpha\beta} L_{ab} \qquad (4.11)$$

and similarly for $\Omega_A{}^B$ and $R_A{}^B$.

All the component fields are gauge fields so that the only covariant components are the field strengths. These correspond to F_{abcd}, $T_{ab}{}^{\alpha}$ and $R_{ab,cd}$ evaluated at $\theta = 0$. Hence it is almost trivial to write down the theory in superspace. Since F_{abcd} has dimension one, we have immediately for $T_{AB}{}^{C}$, dim 0 $\quad T_{\alpha\beta}{}^{c} = -\, i\, (\gamma^{c})_{\alpha\beta}$

dim $\frac{1}{2}$ $\quad T_{\alpha b}{}^{c} = T_{\alpha\beta}{}^{\gamma} = 0$

dim 1 $\quad T_{ab}{}^{c} = 0$ (conventional constraint)

$$T_{a\beta\gamma} = k_1\, F_{abcd}\, (\gamma^{bcd})_{\beta\gamma} + k_2\, (\gamma_{abcde})_{\beta\gamma}\, F^{bcde} \qquad (4.12)$$

Similarly the only non-vanishing parts of F_{ABCD} are the dimension zero and dimension one parts. The latter is just F_{abcd} while for the former we have

$$F_{\alpha\beta cd} = i\, (\gamma_{cd})_{\alpha\beta} \qquad (4.13)$$

The non-vanishing of $T_{\alpha\beta}{}^{c}$ is fixed by the requirement that rigid superspace be a solution of the equations of motion while $F_{\alpha\beta cd} \neq 0$ is forced by the dimension zero Bianchi Identity (4.10). The choice of numerical factors for these two tensors is conventional.

We now have to check the Bianchi Identities. At dimension one these imply

$$T_{a\beta\gamma} = \frac{1}{36}\, F_{abcd}\, (\gamma^{bcd})_{\beta\gamma} + \frac{1}{288}\, (\gamma_{abcde})_{\beta\gamma}\, F^{bcde} \qquad (4.14)$$

while at dimension $\frac{3}{2}$ we find

$$T_{ab\alpha} = \frac{i}{42}\, (\gamma^{cd})_{\alpha}{}^{\beta}\, D_{\beta}\, F_{abcd} \qquad (4.15)$$

and

$$D_{\alpha}\, F_{bcde} = -\, \frac{1}{7}\, (\gamma_{[bc}\, \gamma^{fg})_{\alpha}{}^{\beta}\, D_{\beta}\, F_{de]fg} \qquad (4.16)$$

together with

$$(\gamma^{abc})_{\alpha}{}^{\beta}\, T_{bc\beta} = 0 \qquad (4.17)$$

Hence the component $\hat{f}_{abcd} = F_{abcd}|_{\theta=0}$ varies into $\hat{t}_{ab\alpha} = T_{ab\alpha}|_{\theta=0}$ which is the covariantized Rarita-Schwinger field strength.

At dimension two we have the Bianchi Identity

$$R_{ab,\gamma\delta} = D_a\, T_{b\gamma\delta} - D_b\, T_{a\gamma\delta} + D_{\gamma}\, T_{ab\delta} + T_{a\gamma}{}^{\varepsilon} T_{b\varepsilon}{}^{\delta} - T_{b\gamma}{}^{\varepsilon}\, T_{a\varepsilon}{}^{\delta} \qquad (4.18)$$

so that the second variation of F_{abcd} yields the Riemann tensor and no new components as required. It is straightforward to check that nothing new is generated at dimension $\frac{5}{2}$

so that we have a complete solution of the Bianchi Identities in which all parts of $T_{AB}{}^{C}$ are expressed in terms of the superfield F_{abcd} and covariant derivatives thereof.

4.3 Equations of motion

It is straightforward to derive the x-space equations of motion from the Bianchi Identities in supercovariantized form.

Define

$$\hat{f}_{abcd} = F_{abcd}|_{\theta=0}$$

$$\hat{t}_{ab\alpha} = T_{ab\alpha}|_{\theta=0}$$

$$\hat{r}_{ab,cd} = R_{ab,cd}|_{\theta=0} \tag{4.19}$$

and $\hat{r}_{ab} = \hat{r}_{ac,b}{}^{c}$, $\hat{r} = r_{a}{}^{a}$.

Then from (4.17) we have the Rarita-Schwinger field equation

$$(\gamma^{abc})_{\alpha}{}^{\beta}\,\hat{t}_{bc\alpha} = 0 \tag{4.20}$$

Contracting (4.18) with $(\gamma^{cd})^{\gamma\delta}$ we find $R_{ab,cd}$ and hence the graviton equation of motion

$$\hat{r}_{ab} - \frac{1}{2}\eta_{ab}\,\hat{r} = -\frac{1}{48}\left(4\,\hat{f}_{acde}\,\hat{f}_{b}{}^{cde} - \frac{1}{2}\eta_{ab}\,\hat{f}_{cdef}\,\hat{f}^{cdef}\right) \tag{4.21}$$

Contracting (4.18) with $(\gamma_{c})^{\gamma\delta}$ we find

$$\hat{D}^{a}\,\hat{f}_{abcd} = -\frac{1}{36.48}\,\varepsilon_{bcd\,e_1\ldots e_4\,f_1\ldots f_4}\,\hat{f}^{e_1\ldots e_4}\,\hat{f}^{f_1\ldots f_4} \tag{4.22}$$

which is the supercovariant form of the a_{mnr} equation of motion.

4.4 Component fields

In the Wess-Zumino gauge we have

$$E_{m}{}^{a}|_{\theta=0} = e_{m}{}^{a} \qquad E_{\mu}{}^{a}|_{\theta=0} = 0$$

$$E_{m}{}^{\alpha}|_{\theta=0} = \psi_{m}{}^{\alpha} \qquad E_{\mu}{}^{\alpha}|_{\theta=0} = \delta_{\mu}{}^{\alpha} \tag{4.23}$$

and

$$A_{mnr}|_{\theta=0} = a_{mnr}, \quad A_{\mu\nu\rho} = A_{\mu\nu r} = A_{\mu nr}|_{\theta=0} = 0 \tag{4.24}$$

as well as

$$\Omega_{m,a}{}^{b}\Big|_{\theta=0} = \omega_{m,a}{}^{b} \tag{4.25}$$

To find the x-space torsion we use (2.59) to write

$$T_{ab}{}^{c}\Big|_{\theta=0} = \hat{t}_{ab}{}^{c} = e_{a}{}^{m}\, e_{b}{}^{n}\, t_{mn}{}^{c} + 2\psi_{[a}{}^{\gamma}\, T_{b]\gamma}{}^{c}\Big|_{\theta=0}$$

$$+ \psi_{a}{}^{\alpha}\, \psi_{b}{}^{\beta}\, T_{\alpha\beta}{}^{c} \tag{4.26}$$

but

$$T_{ab}{}^{c} = T_{\alpha b}{}^{c} = 0$$

so that

$$t_{mn}{}^{c} = -\, i\, \bar{\psi}_{m}\, \gamma^{c}\, \psi_{n} \tag{4.27}$$

from which one can determine $\omega_{m,a}{}^{b}$. The variations of the component fields can be found from (2.40) and the analogue of (2.50). For the elfbein

$$\delta e_{m}{}^{a} = -\, i\, \bar{\zeta}\, \gamma^{a}\, \psi_{m} \tag{4.28}$$

while for the Rarita-Schwinger field

$$\delta\psi_{m} = \nabla_{m}\zeta - \frac{1}{36}\, \hat{f}_{mabc}\, \gamma^{abc}\, \zeta + \frac{1}{288}\, \hat{f}^{abcd}\, \gamma_{mabcd}\, \zeta \tag{4.29}$$

where

$$\psi_{m} \equiv \psi_{m\alpha}, \quad \bar{\zeta}\, \gamma^{a} \psi_{m} = \zeta^{\alpha} (\gamma^{a})_{\alpha}{}^{\beta}\, \psi_{m\beta} \tag{4.30}$$

Finally, for a_{mur}, we have

$$\delta a_{mur} = 3i\, \bar{\zeta}\, \gamma_{[mn}\, \psi_{r]} \tag{4.31}$$

In both (4.29) and (4.31) the curved indices on the γ-matrices are formed using $e_{m}{}^{a}$.

5. N = 2 d = 10 CHIRAL SUPERGRAVITY

5.1 Component fields

In ten dimensional space-time Dirac spinors again have 32 components but these can be split into two 16 component Weyl spinors. Furthermore ten-dimensional spinors can satisfy a Majorana condition and the Weyl condition simultaneously, i.e. we can have

$$\psi = C\bar{\psi}^{T} \text{ and } \gamma^{11}\psi = \pm\, \psi \tag{5.1}$$

where C is the charge conjugation matrix and $\gamma^{11} = \gamma^{0}\, \gamma^{1} \ldots \gamma^{9}$.

Hence the basic ten-dimensional spinor is a Majorana-Weyl spinor which has sixteen real components. As a consequence there are two N = 2 d = 10 supersymmetry algebras: one has one left-handed and one right-handed Majorana supersymmetry generators, while the second has two left (or right)-handed Majorana spinor generators, or equivalently one complex left-handed Weyl generator. Corresponding to the two supersymmetry algebras there are two N = 2 d = 10 supergravity theories. The first can be obtained by dimensional reduction from the d = 11 theory described in the last section (Campbell and West, 1984), but the second cannot be so obtained. It is therefore a distinct maximal supergravity theory, a feature which may be of interest from the point of view of Kaluza-Klein theory, although as yet no compactifying solution of the form d = 4 space-time X six-dimensional internal compact space has been found (apart from the six-torus).

The possibility that N = 1 chiral supergravity might exist was pointed out some time ago by Nahm (1978); partial steps towards constructing the theory were taken by Green and Schwarz (1982, 1983a) and by Schwarz and West (1983), the full theory being constructed by the method we shall now describe by Howe and West (1984). The equations of motion to linear order in the fermion fields were also obtained by Schwarz (1983) using a different technique.

The component fields of the theory are as follows:

$$e_m{}^a,\ a,\ a_{mn},\ b_{murs};\ \psi_m,\ \ \lambda\ . \qquad (5.2)$$

$e_m{}^a$ is the zehnbein (35 degrees of freedom on-shell), a is a complex scalar (2), a_{mn} is a complex antisymmetric gauge tensor (56), b_{mnrs} is a real antisymmetric tensor gauge field with a self-dual field strength (35), ψ_m is a complex left-handed vector-spinor field (112) and λ a complex right-handed spinor (16). Thus there are 128 + 128 degrees of freedom as in d = 11 supergravity.

The theory has some interesting features. Firstly it is not possible to write down a Lorentz covariant Lagrangian which gives rise to a self-dual antisymmetric tensor gauge field (Marcus and Schwarz, 1982), although Siegel (198) has shown that this can be circumvented at the linearized level by using a non-linear Lagrangian. This fact makes it difficult to write down an x-space Lagrangian for the on-shell supergravity fields and consequently the equations of motion have to be derived by other means. The superspace method is perfectly suited to this purpose. A second feature is that a theory with Weyl spinors and self-dual antisymmetric tensors in d = 2 mod 4 dimensions can have gravitational anomalies (Alvarez-Gaumé and Witten, 1983); however, in the same paper it was shown that they cancel in the theory under consideration. Thirdly, ten dimensions is of interest because it is the dimension of space-time in which superstring theories are believed to be consistent (Green and Schwarz, 1983b; Green, Schwarz and Brink, 1983), and indeed, it is thought that ten-dimensional supergravity theories can be obtained as zero-slope limits of superstring theories.

The N = 2 chiral theory is the most complicated supergravity theory to construct. It cannot be obtained from a higher dimensional field theory with a simpler field structure; it has scalars and hence a non-linear σ-model structure; the spinor field λ has dimension one-half so that there are quartic fermion terms even in supercovariantized tensors at dimension two, and, lastly, one has to contend with the ten-dimensional γ-matrices.

5.2 Conventions

We use a metric with signature - 8 and

$$\varepsilon^{0123456789} = +1 \tag{5.3}$$

For a five-index antisymmetric tensor G_{abcde}, we define its dual $*G_{abcde}$ by

$$*G^{abcde} = \frac{1}{5!}\,\varepsilon^{abcdefghij}\,G_{fghij} \tag{5.4}$$

we can write

$$G_{abcde} = G^{+}_{abcde} + G^{-}_{abcde} \tag{5.5}$$

where

$$G^{+} = \frac{1}{2}\,(G + *G); \quad G^{-} = \frac{1}{2}(G - *G) \tag{5.6}$$

A ten dimensional Dirac spinor can be written in the form

$$\psi = \begin{pmatrix} \phi_\alpha \\ \bar{\chi}^{\dot{\alpha}} \end{pmatrix} \tag{5.7}$$

where ϕ_α is a 16-component right-handed spinor and $\bar{\chi}^{\dot{\alpha}}$ a 16-component left-handed spinor.

$$\psi = \psi_R + \psi_L$$

$$\psi_R = \tfrac{1}{2}(1 + \gamma_{11})\psi = \begin{pmatrix} \phi \\ 0 \end{pmatrix}$$

$$\psi_L = \tfrac{1}{2}(1 - \gamma_{11})\psi = \begin{pmatrix} 0 \\ \bar{\chi} \end{pmatrix} \tag{5.8}$$

with

$$\gamma_{11} = \begin{pmatrix} 1 & 0 \\ 0 & -1 \end{pmatrix} \tag{5.9}$$

In ten dimensions dotted and undotted indices are equivalent and are related by a matrix $B_{\dot{\alpha}}{}^{\alpha}$ with

$$B = B^T, \quad \bar{B}B = 1 \tag{5.10}$$

We can define

$$\phi_{\dot{\alpha}} = B_{\dot{\alpha}}{}^{\alpha}\, \phi_\alpha$$

and

$$\bar{\chi}^\alpha = \bar{\chi}^{\dot{\alpha}}\, B_{\dot{\alpha}}{}^{\alpha} \tag{5.11}$$

It is in fact consistent to set $\bar{\phi}_{\dot{\alpha}} = B_{\dot{\alpha}}{}^{\alpha}\, \phi_\alpha$ in which case we have a Majorana-Weyl spinor, but we shall use complex spinors rather than two real Majorana-Weyl spinors.

The γ-matrices are

$$\gamma^a = \begin{pmatrix} 0 & (\sigma^c)_{\alpha\dot{\beta}} \\ (\hat{\sigma}^a)^{\dot{\alpha}\beta} & 0 \end{pmatrix} \tag{5.12}$$

with

$$(\sigma^a)_{\dot{\alpha}\beta} = (1, \sigma^i), \quad i = 1, \ldots 9$$

$$(\hat{\sigma}^a)^{\alpha\beta} = (1, -\sigma^i) \tag{5.13}$$

$$\{\sigma^i, \sigma^j\} = 2\delta^{ij}$$

the σ-matrices being hermitian.

Using the B-matrix we can convert dotted indices to undotted ones:

$$(\sigma^a)_{\alpha\beta} = \bar{B}_\alpha{}^{\dot\beta} (\sigma^a)_{\alpha\dot\beta}$$
$$(\hat\sigma^a)^{\alpha\beta} = (\hat\sigma^a)^{\alpha\dot\beta} B_{\dot\alpha}{}^{\alpha} \tag{5.14}$$

For a general multispinor we define a conjugation by

$$X_{\alpha\ldots}{}^{\beta\ldots} \rightarrow \bar{X}_{\alpha\ldots}{}^{\beta\ldots} = B_\alpha{}^{\dot\alpha} \ldots \bar{X}_{\dot\alpha\ldots}{}^{\dot\beta\ldots} B_{\dot\beta}{}^{\beta} \ldots \tag{5.15}$$

So

$$(\bar\sigma^a)_{\alpha\beta} = (\sigma^a)_{\alpha\beta} \tag{5.16}$$

Antisymmetric products of γ-matrices have the form

$$\gamma^{ab} = \begin{pmatrix} \sigma^{ab} & 0 \\ 0 & \hat\sigma^{ab} \end{pmatrix}, \quad \gamma^{abc} = \begin{pmatrix} 0 & \sigma^{abc} \\ \hat\sigma^{abc} & 0 \end{pmatrix}, \text{ etc.} \tag{5.17}$$

where

$$(\sigma^{ab})_\alpha{}^\beta = \sigma^{[a}_{\alpha\gamma} \hat\sigma^{b]\gamma\beta}, \quad (\hat\sigma^{ab})^\alpha{}_\beta = \hat\sigma^{[a\,\alpha\gamma} \sigma^{b]}_{\alpha\beta}$$

$$(\sigma^{abc})_{\alpha\beta} = \sigma^{[ab}{}_a{}^\gamma \sigma^{c]}_{\alpha\beta}, \quad (\hat\sigma^{abc})^{\alpha\beta} = \hat\sigma^{[ab\alpha}{}_\gamma \hat\sigma^{c]\gamma\beta} \tag{5.18}$$

Because the indices on the σ's and the $\hat\sigma$'s are in different positions we shall not need to retain the carets. In any product involving spinors and σ-matrices there is no ambiguity since it is not possible to raise and lower spinor indices, e.g.

$$\bar\lambda \, \sigma_a \, \lambda = \bar\lambda_\alpha (\sigma_a)^{\alpha\beta} \lambda_\beta \tag{5.19}$$

σ_a and σ_{abcde} are symmetric, σ_{abc} is antisymmetric; $(\sigma_{abcde})_{\alpha\beta}$ is self-dual while $(\sigma_{abcde})^{\alpha\beta}$ is anti-self-dual. Also $(\sigma_{ab})^\alpha{}_\beta = -(\sigma_{ab})_\beta{}^\alpha$ and $(\sigma_{abcd})^\alpha{}_\beta = (\sigma_{abcd})_\beta{}^\alpha$.

Superspace co-ordinates are

$$z^M = (x^m, \theta^\mu, \bar\theta^\mu) \quad m = 0, 1, \ldots 9 \quad \mu = 1, \ldots 16 \tag{5.20}$$

and tangent space indices are; e.g. on the frames

$$E^A = (E^a, E^\alpha, \bar{E}^\alpha) \quad (E^a, E^\alpha, E^{\bar\alpha}) \; \alpha = 1, \ldots 16. \tag{5.21}$$

The summation convention is

$$U^A V_A = U^a V_a + U^\alpha V_\alpha - U^{\bar\alpha} V_{\bar\alpha} \tag{5.22}$$

The tangent space group is taken to be $SO(1,9) \times U(1)$:

$$\delta E^A = E^B \hat{L}_B{}^A$$

$$\hat{L}_{ab} = L_{ab} = - L_{ba}$$

$$\hat{L}_\alpha{}^\beta = L_\alpha{}^\beta + i\delta_\alpha{}^\beta K, \quad \hat{L}_{\bar{\alpha}}{}^{\bar{\beta}} = - L_\alpha{}^\beta + i\delta_\alpha{}^\beta K$$

$$K = \bar{K}$$

$$L_\alpha{}^\beta = \frac{1}{4} (\sigma^{ab})_\alpha{}^\beta L_{ab} = \bar{L}_\alpha{}^\beta$$

$$\hat{L}_\alpha{}^{\bar{\beta}} = \hat{L}_{\bar{\alpha}}{}^\beta = \hat{L}_\alpha{}^b = \hat{L}_{\bar{\alpha}}{}^b = \hat{L}_a{}^\beta = \hat{L}_a{}^{\bar{\beta}} = 0 \tag{5.23}$$

Similar algebraic relations hold for the connection $\hat{\Omega}_A{}^B$ and the curvature $\hat{R}_A{}^B$.

In particular

$$\hat{\Omega}_\alpha{}^\beta = \Omega_\alpha{}^\beta + i\delta_\alpha{}^\beta Q$$

and

$$\hat{R}_\alpha{}^\beta = R_\alpha{}^\beta + i\delta_\alpha{}^\beta M \tag{5.24}$$

where Q and M are the U(1) connection and field strength tensor respectively while $\Omega_\alpha{}^\beta$ and $R_\alpha{}^\beta$ are the corresponding quantities for SO(1,9).

Tensors such as $T_{AB}{}^C$ have well-defined U(1) weights which can be read off from their indices, i.e. an upper (lower) α index has weight 1 (-1) and an upper (lower) $\bar{\alpha}$ index has weight -1 (+1). For example $T_{\alpha\beta}{}^{\bar{\gamma}}$ has U(1) weight -3. The only field with U(1) weight -3 in the theory is $\bar{\Lambda}_\alpha = \bar{\lambda}_\alpha(x) + O(\theta)$, so we have

$$T_{\alpha\beta}{}^{\bar{\gamma}} = c\ (\sigma^a)_{\alpha\beta}\ (\sigma_a)^{\gamma\delta}\ \bar{\Lambda}_\delta + d\ \delta^\gamma_{(\alpha}\ \bar{\Lambda}_{\beta)} \tag{5.25}$$

where c and d are constants. Note that barred and unbarred indices transform in the same way under SO(1,9) and only in tensors such as $T_{AB}{}^C$ do they automatically correspond to a definite U(1) weight. Otherwise, fields such as Λ_α have to have their U(1) weights specified, but we shall always take the covariant derivative D to be covariant with respect to SO(1,9) × U(1), e.g.

$$D\Lambda_\alpha = d\Lambda_\alpha - \Omega_\alpha{}^\beta \Lambda_\beta + 3iQ\ \Lambda_\alpha \tag{5.26}$$

5.3 Superspace tensors

As we have emphasized, the constraint problem is simplified if one introduces from the start superspace p-forms corresponding to the p-form component gauge fields. In addition the N = 2 chiral theory has 2 real scalars and in the non-linear theory they lie in the coset space SU(1,1)/U(1). Furthermore at the linearized level the fields have the following U(1) weights:

$$\begin{array}{lccccccc} \text{field} & e_m{}^a & a & a_{mn} & b_{mnrs} & ; & \psi_m & \lambda \\ U(1) & 0 & 4 & 2 & 0 & & 1 & 3 \end{array} \tag{5.27}$$

We wish to introduce a complex two-form gauge field corresponding to a_{mn} but it cannot transform under the local U(1) and still satisfy an identity of the form dF = 0. This problem is resolved by the scalar fields. Let $\mathcal{V}$ be a superfield which is an element of SU(1,1):

$$\mathcal{V} = \begin{pmatrix} U & V \\ \bar{V} & \bar{U} \end{pmatrix} \tag{5.28}$$

$$U\bar{U} - V\bar{V} = 1 \tag{5.29}$$

Define $\tilde{\Omega}$ by

$$\tilde{\Omega} = \mathcal{V}^{-1}\, d\mathcal{V} = \begin{pmatrix} 2iQ & P \\ \bar{P} & -2iQ \end{pmatrix}, \quad Q = \bar{Q} \tag{5.30}$$

where we have anticipated the identification of the U(1) part of $\tilde{\Omega}$ with the U(1) part of the superspace connection $\hat{\Omega}_A{}^B$. Under rigid SU(1,1) and local U(1) transformations we have

$$\delta\mathcal{V} = X\mathcal{V} + \mathcal{V}\begin{pmatrix} -2iK & 0 \\ 0 & 2iK \end{pmatrix} \tag{5.31}$$

where X is an infinitesimal SU(1,1) transformation,

$$X = \begin{pmatrix} iB & C \\ \bar{C} & -iB \end{pmatrix}, \quad B = \bar{B} \tag{5.32}$$

$\tilde{\Omega}$ and hence P and Q are inert under SU(1,1), while under U(1) we have

$$\delta Q = -dK, \quad \delta P = 4iKP \tag{5.33}$$

The SU(1,1) curvature $\tilde{R} = d\tilde{\Omega} - \tilde{\Omega}\tilde{\Omega}$ vanishes, so we have

$$DP = 0 \tag{5.34}$$

$$dQ = M = -\frac{i}{2} P\bar{P} \tag{5.35}$$

The one-form P has the expansion

$$P = E^a P_a + E^\alpha P_\alpha - E^{\bar{\alpha}} P_{\bar{\alpha}} \tag{5.36}$$

with the U(1) weights of P_a, P_α, $P_{\bar{\alpha}}$ being 4, 3 and 5 respectively.

Although the two-form gauge field cannot transform under U(1) it can transform under SU(1,1) as this is a rigid symmet group. We introduce therefore a complex two-form A and form from it the row vector $(\bar{A},A)$ which transforms under SU(1,1) b

$$\delta(\bar{A},A) = -(\bar{A},A)X \tag{5.37}$$

where X is the matrix given in (5.32). The field strength $\mathcal{F}$ is defined in the usual way

$$\mathcal{F} = dA \tag{5.38}$$

and satisfies the Bianchi Identity

$$d\mathcal{F} = 0 \tag{5.39}$$

We now introduce the three-form F by

$$(\bar{\mathcal{F}},\mathcal{F})V = (\bar{F},F) \tag{5.40}$$

F is an SU(1,1) singlet while under U(1),

$$\delta F = 2iKF \tag{5.41}$$

In terms of F, (5.39) takes the form

$$DF - \bar{F}P = 0 \tag{5.42}$$

We now turn to the real four-form gauge field. There is no problem here defining a four-form field B in superspace with five-form field strength G:

$$G = dB, \quad dG = 0 \tag{5.43}$$

The trouble with (5.43) is that the Bianchi Identity does not admit a solution. This is resolved by redefining G to be

$$G = dB - 2i(\bar{A}\mathcal{F} - A\bar{\mathcal{F}}) \tag{5.44}$$

with the corresponding Bianchi Identity

$$dG = 2\bar{F}F \tag{5.45}$$

The gauge transformations for B and A are then

$$\delta A = dX$$

$$\delta B = dY - 2i(\bar{X}F - X\bar{F}) \tag{5.46}$$

where X is a one-form and Y a three-form.

To summarize, the covariant superspace forms are the supertorsion T^A, the supercurvature $\hat{R}_A{}^B$, the complex 2-form F, the real five-form G and the complex one-form P satisfying the Identities (2.25), (2.26), (5.42), (5.45), (5.34) and (5.35) and the problem is to demonstrate that constraints can be imposed on the parts of these tensors in such a way that the identities are satisfied and that the independent components are those listed in (5.2).

5.4 The solution to the Bianchi Identities

The Bianchi Identities are "solved" by the method described in sections 3 and 4. We shall not reproduce the details of the calculation here referring the interested reader to Howe and West (1984). The main point is that the superspace tensors T^A, F, G, P are singlets under SU(1,1) so that the parts of these tensors will be expressed in terms of the following superfields,

$$\begin{aligned} \Lambda_\alpha &= \lambda_\alpha(x) + O(\theta) \\ F_{abc} &= \hat{f}_{abc}(x) + O(\theta) \\ G_{abcde} &= \hat{g}_{abcde}(x) + O(\theta) \\ P_a &= \hat{p}_a(x). \end{aligned} \tag{5.47}$$

P_a, F_{abc} and G_{abcde} having dimension one. $\hat{f}_{abc}$, $\hat{g}_{abcde}$ and $\hat{p}_a$ are suitable supercovariant x-space field strengths corresponding to a_{mn}, b_{mnrs} and the scalar fields respectively. (At the linearized level $\hat{p}_a$ is just the derivative of the complex scalar a if the U(1) gauge is fixed appropriately).

The superspace torsion tensor is given by:

dim 0: $T_{\alpha\beta}{}^{c} = 0$

$$T_{\alpha\bar{\beta}}{}^{c} = - i(\sigma^{c})_{\alpha\beta}$$

dim $\frac{1}{2}$: $T_{\alpha b}{}^{c} = T_{\alpha\beta}{}^{\gamma} = T_{\alpha\bar{\beta}}{}^{\bar{\gamma}} = 0$

$$T_{\alpha\beta}{}^{\bar{\gamma}} = (\sigma^{a})_{\alpha\beta}\, (\sigma_{a})^{\gamma\delta}\, \bar{\Lambda}_{\delta} - 2\delta^{\gamma}_{(\alpha}\, \bar{\Lambda}_{\beta)}$$

dim 1: $T_{ab}{}^{c} = 0$

$$T_{a\beta}{}^{\bar{\gamma}} = - \frac{3}{16}\, (\sigma^{bc})_{\beta}{}^{\gamma}\, \bar{F}_{abc} - \frac{1}{48}(\sigma_{abcd})_{\beta}{}^{\gamma}\, \bar{F}^{bcd}$$

$$T_{\alpha\beta}{}^{\gamma} = \frac{21i}{2}\, X_{a}\, \delta_{\beta}{}^{\gamma} + \frac{3i}{2}\, (\sigma_{ab})_{\beta}{}^{\gamma}\, X^{b} + \frac{5i}{4}\, (\sigma^{bc})_{\beta}{}^{\gamma}\, X_{abc} + \frac{i}{4}\, (\sigma_{abcd})_{\beta}{}^{\gamma}\, X^{bcd} + i(\sigma^{bcde})_{\beta}{}^{\gamma}\, (\frac{1}{192}\, G_{abcde} + \frac{1}{16}\, X_{abcde})$$

where

$$X_{r} = \frac{1}{16}\, \bar{\Lambda}\, \sigma_{r}\, \Lambda; \quad \sigma_{r} = (\sigma_{a}, \sigma_{abc}, \sigma_{abcde}) \tag{5.48}$$

The dimension $\frac{3}{2}$ torsion $T_{ab}{}^{\gamma}$ evaluated at $\theta = 0$ is the covariantized Rarita-Schwinger field strength.

The SO(1,9) supercurvature tensor is

dim 1: $R_{\alpha\beta,ab} = \frac{3i}{4}\, (\sigma^{c})_{\alpha\beta}\, \bar{F}_{abc} + \frac{i}{24}\, (\sigma_{abcde})_{\alpha\beta}\, \bar{F}^{cde}$

$$R_{\alpha\bar{\beta},ab} = - 3(\sigma_{abc})_{\alpha\beta}\, X^{c} - 5(\sigma^{c})_{\alpha\beta}\, X_{abc} - \frac{1}{2}(\sigma_{abcde})_{\alpha\beta}\, X^{cde} - \frac{1}{2}(\sigma^{cde})_{\alpha\beta}\, (\frac{1}{12}\, G_{abcde} + X_{abcde})$$

dim $\frac{3}{2}$

$$R_{\alpha b,cd} = - \frac{i}{2}\, [(\sigma_{b})_{\alpha\beta}\, T_{bc}{}^{\bar{\beta}} + (\sigma_{c})_{\alpha\beta}\, T_{bd}{}^{\bar{\beta}} - (\sigma_{d})_{\alpha\beta}\, T_{bc}{}^{\bar{\beta}}] \tag{5.49}$$

The U(1) field strength M is

dim 1 $M_{\alpha\beta} = 0$

$$M_{\alpha\bar{\beta}} = 2i\, \Lambda_{\alpha}\, \bar{\Lambda}_{\beta}$$

dim $\frac{3}{2}$ $M_{\alpha b} = - i\, P_{b}\, \Lambda_{\alpha}$

dim 2

$$M_{ab} = - i\, \bar{P}_{[a}\, P_{b]} \tag{5.50}$$

The one-form P is

$$\dim \tfrac{1}{2} \qquad P_\alpha = -2\Lambda_\alpha, \quad P_{\bar\alpha} = 0$$

$$\dim 1 \qquad P_a \tag{5.51}$$

The three form F is:

dim $-\frac{1}{2}$ all zero

dim 0 zero except for $F_{\alpha\beta\gamma} = -i\,(\sigma_a)_{\beta\gamma}$

$$\dim \tfrac{1}{2} \qquad F_{ab\gamma} = 0$$

$$F_{ab\bar\gamma} = -(\sigma_{ab})_\gamma{}^\delta\,\Lambda_\delta \tag{5.52}$$

$$\dim 1 \qquad F_{abc}$$

While for G we have

dim $-\frac{3}{2}, -1, -\frac{1}{2}$ all zero

$$\dim 0: \qquad G_{abc\alpha\bar\beta} = (\sigma_{abc})_{\alpha\beta}$$

$$G_{abc\delta\varepsilon} = 0$$

$$\dim \tfrac{1}{2}: \qquad G_{\alpha bcde} = 0$$

$$\dim 1: \qquad G_{abcde} = G^+_{abcde} - 8\,X_{abcde} \tag{5.53}$$

It has been verified that the field strengths listed above satisfy all the Bianchi Identities (Howe and West, 1984).

The superfields Λ, P, F and G are not independent. For example from the Bianchi Identities, we have

$$D_\alpha\,\Lambda_\beta = -\frac{i}{24}\,(\sigma^{abc})_{\alpha\beta}\,F_{abc}$$

$$\bar D_\alpha\,\Lambda_\beta = -\frac{i}{2}\,(\sigma^a)_{\alpha\beta}\,P_a \tag{5.54}$$

and Λ_α is itself related to the superfield V via equation (5.30). Note also that G_{abcde} is self-dual up to bilinear terms in the Fermion fields so that the degree of freedom count is correct.

.4 Equations of motion and components

he equations of motion of this system are somewhat harder to dentify than for N = 1 d = 11, but they can be obtained onetheless using the Bianchi Identities and the Ricci dentities. The relevant superfield equations are

$$\sigma^a D_a \bar{\Lambda} = -3i \bar{Y}_{abc} \sigma^{abc} \Lambda + \frac{i}{5.192} \sigma^{abcde} G_{abcde} \Lambda \quad (5.55)$$

$$D^a \bar{P}_a = -36i X_a \bar{P}^a + 6i Y_{abc} F^{abc} + \frac{1}{6} \bar{F}_{abc} \bar{F}^{abc} \quad (5.56)$$

$$D^c \bar{F}_{abc} = -20i F_{abc} X^c - 4i F_{[a}{}^{cd} X_{b]cd} + \frac{2i}{3} X_{abcde} \bar{F}^{cde} - F_{abc} \bar{P}^c + \frac{i}{6} G^+_{abcde} \bar{F}^{cde} + 96i Y_{abc} P^c - \bar{\Lambda}_\alpha T_{bc}{}^\alpha \quad (5.57)$$

$$G_{abcde} = {}^*G_{abcde} - 16 X_{abcde} \quad (5.58)$$

$$(\sigma^a)_{\alpha\beta} T_{ab}{}^{\bar{\beta}} = -\frac{3i}{8} F_{bcd} (\sigma^{cd})_\alpha{}^\beta \bar{\Lambda}_\beta + \frac{i}{24} F^{cde} (\sigma_{bcde})_\alpha{}^\beta \bar{\Lambda}_\beta - 2i\Lambda_\alpha \bar{P}_b \quad (5.59)$$

$$R'_{ab} = -2\bar{P}_{(a} P_{b)} - \bar{F}_{(a}{}^{cd} F_{b)cd} + \frac{1}{12} \eta_{ab} \bar{F}_{cde} F^{cde} - \frac{1}{96} G_a^{+cdef} G_{bcdef} - 18.80 Y_{(a}{}^{cd} \bar{Y}_{b)cd} + 8.18 \eta_{ab} Y_{cde} \bar{Y}^{cde} \quad (5.60)$$

Here

$$R'_{ab} = R_{acb}{}^c \qquad Y_{abc} = \frac{1}{3!16} \Lambda \sigma_{abc} \Lambda \quad (5.61)$$

The $\theta = 0$ components of equations (5.55) to (5.60) are the x-space field equations for λ, a, a_{mn}, b_{mnrs}, ψ_m and $e_m{}^a$ respectively. Note that the b_{murs} equation is first-order; as is well-known a self-dual gauge field automatically satisfies the standard second-order equations of motion by virtue of the Bianchi identity.

The component fields are defined in the usual way,

$$e_m{}^a = E_m{}^a\big|_{\theta=0}, \quad \psi_m{}^\alpha = E_m{}^\alpha\big|_{\theta=0}$$

$$a_{mn} = A_{mn}\big|_{\theta=0}, \quad b_{mnrs} = B_{mnrs}\big|_{\theta=0} \quad (5.62)$$

while for the scalar fields we set

$$V\big|_{\theta=0} = v = \begin{pmatrix} u & v \\ \bar{v} & \bar{u} \end{pmatrix} \quad (5.63)$$

which, in the symmetric gauge takes the form,

$$\upsilon = \exp \begin{pmatrix} 0 & a \\ \bar{a} & 0 \end{pmatrix} \tag{5.64}$$

we have

$$\upsilon^{-1} \partial_m \upsilon = \begin{pmatrix} 2iq_m & P_m \\ \bar{P}_m & -2iq_m \end{pmatrix} \tag{5.65}$$

where q_m is the x-space U(1) connection.

The supercovariant quantities $\hat{f}$, $\hat{g}$, $\hat{p}$ are given by

$$\hat{p}_a = p_a + 2\psi_a \lambda$$

$$\hat{f}_{abc} = f_{abc} - 3\bar{\psi}_{[a} \sigma_{bc]} \lambda + 3i\, \psi_{[a} \sigma_b \psi_{c]}$$

$$\hat{g}_{abcde} = g_{abcde} - 20\, \bar{\psi}_{[a} \sigma_{bcd} \psi_{e]} \tag{5.66}$$

where

$$p_a = e_a{}^m p_m, \quad \psi_a = e_a{}^m \psi_m, \text{ etc.} \tag{5.67}$$

We also have

$$\begin{aligned} T_{ab}{}^{\bar{\alpha}}\big|_{\theta=0} = {} & 2e_a{}^m e_b{}^n \nabla_{[m} \psi_{n]}{}^{\bar{\alpha}} + 2\psi_{[a}{}^{\beta} T_{b]\beta}{}^{\bar{\alpha}}\big|_{\theta=0} \\ & - 2\bar{\psi}_{[a}{}^{\bar{\beta}} T_{b]\bar{\beta}}{}^{\bar{\alpha}}\big|_{\theta=0} + \psi_a \sigma^c \psi_b (\sigma^c \bar{\lambda})^{\alpha} \\ & - 2\psi_{[a}{}^{\alpha} \psi_{b]}{}^{\beta} \bar{\lambda}_{\beta} \end{aligned} \tag{5.68}$$

and

$$\begin{aligned} R_{ab,c}{}^d\big|_{\theta=0} = {} & r_{abc}{}^d + 2\psi_{[a}{}^{\alpha} R_{b]\alpha c}{}^d\big|_{\theta=0} \\ & - 2\bar{\psi}_{[a}{}^{\bar{\alpha}} R_{b]\bar{\alpha}c}{}^d\big|_{\theta=0} + \psi_a{}^{\alpha} \psi_p{}^{\beta} R_{\alpha\beta c}{}^d\big|_{\theta=0} \\ & + \bar{\psi}_a{}^{\alpha} \bar{\psi}_b{}^{\beta} R_{\bar{\alpha}\bar{\beta}c}{}^d\big|_{\theta=0} - 2\bar{\psi}_{[a}{}^{\alpha} \psi_{b]}{}^{\beta} R_{\bar{\alpha}\beta c}{}^d\big|_{\theta=0} \end{aligned} \tag{5.69}$$

The derivative ∇_m in (5.68) is covariant with respect to SO(1,9) × U(1). These formulae together with the results for the supertorsion and supercurvature tensors allow one to rewrite the equations of motion directly in terms of the component fields. The x-space torsion tensor is found using $T_{ab}{}^c = 0$,

$$t_{mn}{}^{a}(x) = -2i\,\bar{\psi}_{[m}\,\sigma^{a}\,\psi_{n]} \tag{5.70}$$

from which one can find the x-space Lorentz connection

$$\omega_{m,ab} = \omega_{m,ab}(e) + k_{mab} \tag{5.71}$$

$$k_{a,bc} = \frac{1}{2}\,(t_{ab,c} + t_{c,ab} - t_{bc,a})$$

where $\omega_{m,ab}(e)$ is the usual Christoffel connection.

Finally, using the techniques of section 2 one finds the local supersymmetric variations of the component fields to be

$$\delta e_{m}{}^{a} = -i\bar{\zeta}\,\sigma^{a}\,\psi_{m} - i\zeta\,\sigma^{a}\,\bar{\psi}_{m}$$

$$\begin{aligned}\delta\psi_{m} = {} & \nabla_{m}\,\zeta - \frac{3}{16}\,\hat{f}_{mab}\,\sigma^{ab}\,\bar{\zeta} + \frac{1}{48}\,\hat{f}^{bcd}\,\sigma_{mbcd}\,\bar{\zeta} \\ & - \frac{i}{192}\,\hat{g}_{mbcde}\,\sigma^{bcde}\,\zeta \\ & + \frac{i}{16}\Big(-\frac{21}{2}\,\bar{\lambda}\,\sigma_{m}\,\lambda + \frac{3}{2}\,\bar{\lambda}\,\sigma_{a}\,\lambda\,\sigma_{m}{}^{a} + \frac{5}{4}\,\bar{\lambda}\,\sigma_{mab}\,\lambda\,\sigma^{ab} \\ & - \frac{1}{4}\,\bar{\lambda}\,\sigma^{abc}\,\lambda\,\sigma_{mabc} - \frac{1}{16}\,\bar{\lambda}\,\sigma_{mabcd}\,\lambda\,\sigma^{abcd}\Big)\,\zeta\end{aligned}$$

$$\delta\upsilon = 2\upsilon\begin{pmatrix} 0 & -\zeta\lambda \\ \bar{\zeta}\bar{\lambda} & 0 \end{pmatrix}$$

$$\delta\lambda = \frac{i}{24}\,\hat{f}_{abc}\,\sigma^{abc}\,\zeta + \frac{i}{2}\,\hat{P}_{a}\,\sigma^{a}\,\zeta$$

$$\begin{aligned}\delta(\bar{a}_{mn},\,a_{mn}) = {} & -(\zeta\,\sigma_{mn}\,\lambda + 2i\,\bar{\zeta}\,\sigma_{[m}\,\lambda_{n]},\; -\bar{\zeta}\,\sigma_{mn}\,\lambda \\ & + 2i\zeta\,\sigma_{[m}\,\psi_{n]})\,\upsilon^{-1}\end{aligned}$$

$$\begin{aligned}\delta b_{mnrs} = {} & -4\zeta\,\sigma_{[mnr}\,\bar{\psi}_{s]} + 4\bar{\zeta}\,\sigma_{[mnr}\,\psi_{s]} \\ & + 12i\,(a_{[mn}\,\delta\,\bar{a}_{rs]} - \bar{a}_{[mn}\,\delta a_{rs]})\end{aligned} \tag{5.72}$$

REFERENCES

Alvarez-Gaume, L. and Witten, E. 1983. Nucl. Phys. B234, 269.
Bergshoeff, E., de Roo, M. and de Wit, B. 1981, Nucl. Phys. B182, 173.
Brink, L. and Howe, P. 1980, Phys. Lett 91B, 153.
Callan, C., Coleman, S., Wess, J. and Zumino, B. 1969, Phys. Rev. 177, 2247.
Campbell, I.C.G. and West, P.C. 1984, Nucl. Phys. B243, 112.
Coleman, S., Wess, J. and Zumino, B. 1969, Phys. Rev. 177, 2239.
Cremmer, E. and Ferrara, S. 1980, Phys. Lett. 91B, 61.
Cremmer, E. and Julia, B. 1979, Nucl. Phys. B159, 141.
Cremmer, E., Julia, B. and Scherk, J. 1978, Phys. Lett. 76B, 409.
Cremmer, E., Scherk, J. and Ferrara, S. 1978, Phys. Lett. 74B, 61.
Das, A., Kaku, M. and Townsend, P. 1978, Phys. Rev. Lett. 40, 1215.
de Wit, B., Van Holten, J. and Van Proeyen, A. 1980, Phys. Lett. 95B, 51.
de Witt, B. 1984, Supermanifolds, Cambridge Univ. Press.
Dragon, N. 1979, Z. Phys. G2, 29.
Gates, S.J. 1979, In Supergravity, eds. P. Van Nieuwenhuizen and D.Z. Freedman, North-Holland.
Gates, S.J. and Siegel, W. 1980, Nucl. Phys. B163, 519.
Gates, S.J., Stelle, K. and West, P. 1980, Nucl. Phys. B169, 347.
Green, M. and Schwarz, J. 1982, Phys. Lett. 109B, 444.
Green, M. and Schwarz, J. 1983a, Phys. Lett. 122B, 143.
Green, M. and Schwarz, J. 1983b, Nucl. Phys. B218, 43.
Green, M., Schwarz, J. and Brink, L. 1983, Nucl. Phys. B219, 437.
Howe, P. 1981, Phys. Lett. 100B, 389.
Howe, P. 1982, Nucl. Phys. B199, 309.
Howe, P. and Nicolai, H. 1982, Phys. Lett. 109B, 269.
Howe, P. and West, P. 1984, Nucl. Phys. B238, 181.
Kaku, M. and Townsend, P. 1978, Phys. Lett. 76B, 54.
Marcus, M. and Schwarz, J. 1982, Phys. Lett. 115B, 111.
Nahm, W. 1978, Nucl. Phys. B135, 149.
Rivelles, V.O. and Taylor, J.G. 1981, Phys. Lett. 104B, 131.
Rogers, A. 1980, J. Math. Phys. 21, 1352.
Salam, A. and Strathdee, J. 1975, Phys. Rev. D11, 1521.
Schwarz, J. 1983, Nucl. Phys. B226, 269.
Schwarz, J. and West, P. 1983, Phys. Lett. 126B, 301.
Siegel, W. 1979, In Supergravity, eds. P. Van Nieuwenhuizen and D.Z. Freedman, North-Holland.
Siegel, W. 198
Stelle, K. and West, P. 1978, In Supergravity, eds. P. Van Nieuwenhuizen and D.Z. Freedman, North-Holland.
Wess, J. and Zumino, B. 1977, Phys. Lett. 66B, 361.
Wess, J. and Zumino, B. 1978a, Phys. Lett. 74B, 51.
Wess, J. and Zumino, B. 1978b, Phys. Lett. 79B, 394.
Zumino, B. 1975 in Proc. Northern Univ. Conf. eds. R. Arnowitt and P. Nath.

REALISTIC MODELS OF SUPERSYMMETRY

G G ROSS

1. INTRODUCTION

Gauge theory models have been remarkably successful in describing the strong, weak and electromagnetic interactions. However, the standard model, based on the gauge group SU(3) x SU(1) x U(1), falls short of what one might hope for in the ultimate theory: particles are assigned by hand to multiplets of the gauge group and there are a large number of parameters, gauge couplings and masses, needed to specify the theory. Grand Unified Theories (GUTs) seek to improve this situation by embedding the standard gauge group in a simple (grand unified) gauge group with a single gauge coupling constant. The multiplet structure for a given particle spin can also be simplified in GUTs, but there is still no symmetry relating different spins. Only the vector bosons are uniquely specified,by the local gauge principle, to belong to a definite representation of the gauge group (the adjoint).

Supersymmetry is the only symmetry known which can relate different spins, and one may hope it supplies the missing ingredient for constructing the ultimate theory by relating the representation content of scalars and fermions to those of vectors and by relating the scalar and Yukawa couplings to the gauge coupling. Moreover, local supersymmetry necessarily includes the theory of gravity, leading to the unification of gravity with the other interactions. Unfortunately, these supergravity theories do not have the correct spectrum to describe the known particles. For example, even the largest (N=8) supergravity includes only the gauge group SU(3) x U(1) x U(1) and not the standard model. Thus supergravity theories may be the fundamental theory, but only as a theory for constituents in which the $W^{\pm}$ bosons, for example, are composite.

Thus, in building supersymmetric models, we must give up the hope that the extended supersymmetry models contain the spectrum of the observed states in a single irreducible representation. This leads to models based on the direct product structure G x [N extended supersymmetry]. At first sight, this appears to lose many of the potential benefits of supersymmetry, and indeed in this type of theory there may remain many arbitrary parameters. However, such models can solve the hierarchy problem common to Grand Unified Theories, and for this reason they have been extensively studied. In this review, we will mainly discuss simplest such theories based on the group G x [N=1 supersymmetry]. Higher N theories are possible, but they all require mirror fermions and, since there is no evidence yet for such states, it seems best to start with a structure capable of accommodating the known chiral fermion structure.

1.1 Rules for the construction of N=1 globally supersymmetric Lagrangians

It is straightforward to construct Lagrangian densities which are globally supersymmetric using the standard superspace rules.(1) The gauge invariant supersymmetric matter kinetic energy term is of the form

$$L_{kin} = [\Phi^{+} e^{2gV} \Phi]_D \tag{1}$$

where D denotes the $\theta\theta\bar{\theta}\bar{\theta}$ projection. The superfield Φ is a (left handed) chiral superfield with components

$$\Phi = \phi + \sqrt{2}\,\psi\,\theta + F\theta\theta \tag{2}$$

where ϕ is a complex scalar field, ψ is a two component Weyl fermion field and F is an auxiliary field. V is a vector superfield defined by

$$V \equiv \{-\theta\,\sigma^{\mu}\,\bar{\theta}\,\nu^{a}_{\mu} + i\theta\theta\bar{\theta}\,\bar{\lambda}^{a} - i\bar{\theta}\bar{\theta}\theta\,\lambda^{a} + \tfrac{1}{2}\,\theta\theta\bar{\theta}\bar{\theta}\,D^{a}\}T^{a} \tag{3}$$

where ν^{a}_{μ}, λ^{a} are gauge bosons and their fermion partners the gauginos respectively. The index a refers to the local gauge group, the fields ν^{a}_{μ} and λ^{a} belonging to its adjoint representation. The matrices T^{a} are representations of the group acting on the matter fields Φ.

Eq.(1) uniquely specifies the couplings of the gauge bosons and their gaugino partners to matter fields once their representation is specified. For example, if the gauge group is SU(2) with $\binom{\nu}{e}$ a left-handed fermion doublet component of eq.(2), the interaction with the charged gauge field, W^+ is given by the usual form

$$\frac{g}{2\sqrt{2}}\,\bar{e}\,\gamma_\mu(1-\gamma_5)\nu\, W^{\mu-} + \text{h.c.} \tag{4}$$

while the gaugino coupling contains the piece

$$\frac{g}{\sqrt{2}}\,(\bar{e}\,\tilde{\nu}\,\lambda_W^- + \tilde{e}^*\,\nu\,\lambda_W^-) + \text{h.c.} \tag{5}$$

in which $\tilde{\nu}$ and $\tilde{e}$ are the scalar neutrino and electron respectively, and λ_W is the W gaugino, the Wino. We will give a more complete form for the standard model in the next section. Thus, to define a supersymmetric gauge theory, it is necessary to choose the gauge group and the matter representations. This specifies the gauge boson and gaugino couplings. In addition, we must specify the non-gauge interactions between the components of the chiral (matter) supermultiplets. The form of these interactions follows from the allowed supersymmetric interaction term which may be written[1] as

$$L^{int} : [P(\Phi)]_F + \text{h.c.} \tag{6}$$

where F denotes the $\theta\theta$ projection and the superpotential, P, is a general function of the chiral superfields Φ_i, but not of their hermitian conjugates. If L^{int} is to have maximum field dimension of 4, necessary for a renormalisable theory, $P(\Phi)$ should be at most cubic in Φ. In terms of the component fields, L^{int} may be written as[1] $L^{int} = L^{Fermion} + L^{scalar}$, where

$$L^{Fermion} = \sum_{i,j} \left.\frac{\partial^2 P}{\partial\Phi_i\,\partial\Phi_j}\right|_{\Phi=\phi} \psi_i\psi_j$$

$$L^{scalar} = \sum_i \left|\frac{\partial P}{\partial\Phi_i}\right|^2_{\Phi=\phi} + \sum_{\alpha,i} |g_\alpha\, \phi_i^+\, T^{\alpha,i}\, \phi_i|^2 \tag{7}$$

where $T^{\alpha,i}$ are representations of the gauge group, α, acting on the representation ϕ_i and g_α is the associated gauge coupling. The last term in L^{scalar} comes from eliminating the (non-propagating) D^a auxiliary field in eq. (3) while the first term in L^{scalar} comes from eliminating the F auxiliary field in eq.(2). They are often referred to as D and F terms respectively.

Once P is specified, the model is complete and in the next section I will discuss reasonable forms for P which can generate realistic fermion masses through $L^{Fermion}$ and acceptable patterns of spontaneous symmetry breaking via L^{scalar}.

1.2 A supersymmetric version of the standard model(2,3)

We turn now to the construction of a supersymmetric version of the standard model based on the gauge group SU(3) x SU(2) x U(1). Although our discussion is in terms of a globally supersymmetric theory, the same considerations apply to locally supersymmetric theories and the multiplet structure and non-gravitational interactions are common to both. In section (3.3)we will discuss some of the new features of the locally supersymmetric theories which arise as a result of gravitational strength couplings.

In the standard model the vector supermultiplet transforms as the adjoint under the group giving the particle content shown in Table 1. Unfortunately, we cannot identify any of the new fermions, the gauginos, with the quarks and leptons as the latter do not transform as the adjoint representation. We are therefore forced to introduce chiral supermultiplets to accommodate the quarks and leptons as in Table 1. (Note that, for a theory with renormalisable gauge interactions, the vector supermultiplets must be in the adjoint of the gauge group). Once again we are forced to double the number of states this time introducing (complex) scalar fields to partner the known fermions. Finally, we must include Higgs scalar fields, needed to spontaneously break the gauge group SU(3) x SU(2) x U(1) to SU(3) x $U(1)_{em}$. Our original hope was that the Higgs sector would be simplified, by assigning the Higgs scalar to the same supermultiplets as the known fermions. An obvious possibility is to identify the Higgs SU(2) doublet as a partner of a lepton doublet. However, this is not possible, for such an assignment in supersymmetry does not give an acceptable pattern of fermion masses. The reason is that supersymmetry restricts the possible forms of Yukawa couplings and the couplings necessary to give down quarks and charged leptons a mass are not present. To see this, note that Yukawa couplings are simple described in terms of the superpotential P via eq.(7). The Yukawa couplings needed to give all charged fermions a mass has a superpotential of the form

$$P = \sum_{j,i=d,s,b} m_{ij}^{(d)} \varepsilon^{\alpha\beta} \psi_{i,a}^{(q)} H_{2\beta} \psi_j^{(d^c)} + \sum_{i,j=u,c.t} m_{ij}^{(u)} \varepsilon^{\alpha\beta} \psi_{i,\alpha}^{(q)} H_{1\beta} \psi_j^{(u^c)}$$

$$+ \sum_{i=e,\mu,\tau} m_i \varepsilon^{\alpha\beta} \psi_{i,\alpha}^{(1)} H_{2\beta} \psi_i^{(1^c)} \qquad (8)$$

where $H_{1\alpha}$ and $H_{2\alpha}$ are chiral supermultiplets transforming as doublets under SU(2), but with U(1) charge ±1 respectively, so that their charge states are $[H_1^+, H_1^o]$ and $[H_2^o, H_2^-]$ respectively. In eq.(8) the indices i and j are family indices and $m_{ij}^{(d)}$, $m_{ij}^{(u)}$ are the mass matrices for the up and down quark masses. The lepton supermultiplet doublets, $\psi_{i,\beta}^{(1)}$, have the correct charges to be identified with $H_{2\beta}$. In the non-supersymmetric standard model $H_{2\beta} = \varepsilon_{\beta\gamma} H_1^{c\gamma}$, where the superscript c denotes the charge conjugate state, but, in the supersymmetric case, the rules of section (1.2) require that P only be formed using products of (left handed chiral) supermultiplets and not their (right handed chiral) charge conjugates. Thus, in eq.(8), $H_{1\beta}$ must be identified with a completely new chiral supermultiplet. In addition, more states are needed, for $H_{1\beta}$ contains new, charged, Weyl fermions and we must add, further, charged fermions to allow for the construction of Dirac masses to ensure that the final theory has no massless charged states (we cannot give charged fermions a Majorana mass without violating charge conservation) and to avoid anomalies. The simplest solution is to introduce another new SU(2) doublet chiral superfield which is usually identified with $H_{2\beta}$. In constructing simple grand unified generalisation of the standard model this, in fact, is the only possibility, for, if we insist on identifying $H_{2\beta}$ with a lepton doublet, we find proton decay proceeds too fast because the colour triplet components of H_2, now the multiplet partners of the leptons, are necessarily light. They will then mediate proton decay at an unacceptably fast rate (see section 2.2).

Thus the final multiplet structure for a supersymmetric version of the standard model includes two new chiral supermultiplets whose scalar partners are to be identified with the Higgs scalars needed to break the SU(3) x SU(2) x U(1) to SU(3) x $U(1)_{em}$ and to give all charged fermions a mass. The full multiplet ctructure is given in Table (1).

Although the SU(3) x SU(2) x U(1) x N=1 supersymmetry structure fails to simplify the multiplet structure of the original model (indeed it more than doubles the spectrum!) it does have a redeeming property that has caused it to be studied intensively, recently, as a possible theory for the strong, weak and electromagnetic interactions - it solves the hierarchy problem. In the next sections, we will construct the Lagrangian for this theory and show how this solution is achieved.

Table 1

Multiplet structure for the minimal supersymmetric SU(3) x SU(2) x U(1) theory.

Vector supermultiplets		Name	Spin J	SU(3) x SU(2) x U(1) transformations
V_G	$g_\mu^{a=1...8}$	Gluons	1	(8,1,0)
	$\tilde{g}^{a=1...8}$	Gluinos	½	(8,1,0)
V_W	$W_\mu^\pm$, Z_μ	W, Z bosons	1	(1,3,0)
	$\tilde{W}^\pm$, $\tilde{Z}$	Winos, Zino	½	(1,3,0)
V_γ	A_μ	Photon	1	(1,1,0)
	$\tilde{A}$	Photino	½	(1,1,0)
Chiral Supermultiplets				
$\psi_{i,\alpha}^{(q)}$	q_L, $q_L^{(c)q}$	Quarks	½	$(3,2,\frac{1}{3})$, $(\bar{3},1,\frac{-4}{3}$ or $\frac{+2}{3})$
$\psi_i^{u^c}$, $\psi_i^{d^c}$	$\tilde{q}$, $\tilde{q}^c$	Scalar quarks	0	$(3,2,\frac{1}{3})$, $(\bar{3},1,\frac{-4}{3}$ or $\frac{+2}{3})$
$\psi_{i,d}^{(\ell)}$	ℓ_L, $\ell_L^{(c)}$	Leptons	½	(1,2,-1),(1,1,0 or 2)
ψ_i^{1c}	$\tilde{\ell}$, $\tilde{\ell}^c$	Scalar leptons	0	(1,2,-1),(1,1,0 or 2)
	H_1, H_2	Fermionic Higgs	½	(1,2,1), (1,2,-1)
	H_1, H_2	Higgs doublets	0	(1,2,1), (1,2,-1)

1.3 The SU(3) x SU(2) x U(1) supersymmetric Lagrangian

Once the transformation properties of the supermultiplets under the gauge group are specified and the superpotential is given, the Lagrangian density may be immediately constructed using the result of section (1.2).

$$\begin{aligned} L^G = & -\frac{1}{4}\ \mathrm{Tr}\{W^{\mu\nu}W_{\mu\nu}\} - i\mathrm{Tr}\{\lambda\sigma^m D_m\bar{\lambda}\} \\ & + \sum_j \phi_j D_m D^m \phi^*_j + i\sum_j D_m\bar{\psi}_j\bar{\sigma}^m\psi_j \\ & - \frac{i}{\sqrt{2}}\sum_{j,a} g_a(\phi_j\tau^a\bar{\psi}_j - \phi^*_j\tau^a\psi_j)\lambda^a \\ & - \frac{1}{8}\left|\sum_a g_a \sum_j \phi^+_j \tau^a \phi_j\right|^2 \end{aligned} \tag{9}$$

In this, λ represents the gaugino, the trace implies a sum over all the gauge indices, a, of SU(3) x SU(2) x U(1) and the sum over j is over all the chiral superfields of Table 1. D_m are the usual gauge covariant derivatives.

The Yukawa couplings may be read immediately from the superpotential using the form of eq.(7). They are of the form given in the standard model, except that the Higgs, H_1, couples only to the right handed up quarks and H_2 couples only to the right handed down quarks and charged leptons. The scalar interactions in L_{int} also follow from P using eq.(7).

The two component form for the interactions, used here, is simply related to the more familiar four component form by noting, in the Weyl basis, that

$$\lambda\sigma^\mu\psi = \Lambda\gamma^\mu\left(\frac{1-\gamma_5}{2}\right)\Psi \tag{10}$$

where λ and ψ are 2 2 component Weyl spinors related to the usual four component Dirac spinors Λ and Ψ by

$$\Psi = \begin{bmatrix}\psi \\ \text{any}\end{bmatrix} \qquad \Lambda = \begin{bmatrix}\Lambda \\ \text{any}\end{bmatrix} \tag{11}$$

We may also write this vertex in the form

$$\bar{\lambda}\bar{\sigma}^\mu\psi = -\bar{\Lambda}\ \gamma^\mu\left(\frac{1+\gamma_5}{2}\right)\psi' \tag{12}$$

where

$$\Psi' = \begin{bmatrix} \text{any} \\ - \\ \psi \end{bmatrix}, \qquad \Lambda' = \begin{bmatrix} \text{any} \\ - \\ \lambda \end{bmatrix} \tag{13}$$

(This follows from the result $\bar{\lambda}\bar{\sigma}^{\mu}\psi = -\psi\sigma^{\mu}\bar{\lambda}$ which applies to anti-commuting operators.)

For scalar couplings $\lambda\psi = \psi\lambda$ so

$$\lambda\psi = \bar{\Lambda}'_R\Psi_L = \bar{\Psi}'_R\Lambda_L \tag{14}$$

The multiplet structure and interactions discussed above are common to all attempts to supersymmetrise the standard model. Even in the local supersymmetric models this form persists, although additional terms may appear (see section 3.3). As a result many of the phenomenological implications are model independent and we will discuss these in the next four sections. The main diversity in models proves to be the mechanisms driving spontaneous symmetry breaking of the SU(2) x U(1) gauge group and of supersymmetry itself and in sections (3.1) to (3.3) we will discuss the alternatives that have been explored paying particular regard to the possibility that the supersymmetry be local.

1.4 Symmetries of the standard SUSY model

In the non-supersymmetric standard model there are various symmetries which occur. Baryon and lepton numbers are automatically conserved, as is lepton type (e, μ, τ). Strangeness is conserved by the strong and electromagnetic interactions and, due to the GIM mechanism, by the neutral weak currents at tree level. C and P are conserved by the strong and electromagnetic interactions, but not by the weak. CP is violated if there are three or more generations. Neutrinos have zero mass.

Let us consider the situation in the supersymmetric case. This time, because there are new supersymmetric partners carrying the same quantum numbers as their partners, there are new possibilities for symmetry violation.

Baryon and lepton numbers are conserved by the supersymmetric Lagrangian (cf. eqs.(7)-(9)). However, unlike the standard model, this does not automatically follow[(4)(5)] for we have omitted allowed supersymmetric terms in the Lagrangian which would have violated B and L. For example the terms

$$(\psi^{\ell}H_1)_F, \quad (\psi^{q}\psi^{q}\psi^{q})_F \tag{15}$$

are SU(3) x SU(2) x U(1) symmetric and supersymmetric. Thus they could be added to the original Lagrangian, but as they violate Lepton and Baryon numbers, respectively, they were excluded from the original Lagrangian. An interesting property of supersymmetric theories is that if a term is not included in the original superpotential, it will not be generated in any order of perturbation theory(6), so that, even if there is no symmetry forbidding terms such as in eq.(15), they will not occur if not present at tree level. This remarkable fact can give rise to 'supernatural' or accidental symmetries. However, for many people, this is an unsatisfactory origin for a symmetry as basic as lepton or baryon number and as we will discuss shortly, terms such as in eq.(15) can be forbidden by new symmetries known as R symmetries.

Strangeness in the standard model is violated because the quark mass matrix splits the degeneracy between d and s quarks and so the charged currents, when expressed in terms of the quark mass eigenstates, have a strangeness changing component. However, in the lepton sector, because neutrinos are massless, the charged currents do not change τ, μ or e lepton number, as the neutrino current eigenstates can be taken as the physical neutrinos leaving the charged currents lepton flavour diagonal.

In the supersymmetric version of the standard model of course these features persist in the quark and lepton sector. However, in addition, strangeness may be violated in the squark sector giving rise to new sources of strangeness violation. In the lepton sector there is an even more interesting change for the sneutrinos are not massless and, when the charged currents are expressed in terms of the neutrino mass eigenstates, they have a lepton flavour violating component, which can induce processes such as $\mu \to e\gamma$, $\mu \to 3e$, etc. Finally, the squark and slepton sectors may introduce new sources of CP violation as discussed below.

R symmetry and R parity.(11) As mentioned above in addition to the usual symmetries of gauge theories their supersymmetric versions have new "R" symmetries. Note that the supersymmetric Lagrangian was formed by either a $\theta\theta\bar{\theta}\bar{\theta}$ projection, eq.(1), or a $\theta\theta$ projection, eq.(6). The first is invariant under the "R symmetry" characterised by the angle β given by

$$\theta \rightarrow e^{-i\beta}\theta$$
$$P(\Phi) \rightarrow e^{2i\beta}P(\Phi) \tag{16}$$
$$V \rightarrow V$$

The form of the superpotential P determines whether there is an R symmetry for a general β, other than the discrete values $\beta = 0,\pi$. The value $\beta = \pi$ is always a symmetry of the Lagrangian and corresponds to fermion number conservation since (cf. eqs.(2) and (3)) under such a rotation fermion fermion fields change sign. L will be invariant under a more general R symmetry, characterised by β, provided we can find a choice of phase rotations for the chiral fields Φ_i

$$\Phi_i \rightarrow e^{in_i\beta}\Phi_i \tag{17}$$

under which all components of $P(\Phi)$ have the transformation of eq.(16). The transformation of the component fields under R symmetry are [11]

$$V_\mu{}^a \rightarrow V_\mu{}^a \; ; \; \lambda_\alpha{}^a \rightarrow e^{i\beta}\lambda_\alpha{}^a ; \; D^a \rightarrow D^a \tag{18}$$

$$\phi^i \rightarrow e^{in_i\beta}\phi^i \; ; \; \psi^i \rightarrow e^{i(n_i-1)\beta}\psi^i \; ; \; F \rightarrow e^{i(n_i-2)\beta}F$$

Clearly, any R symmetry forbids a gaugino mass term of the form $\lambda^a\lambda^a$. However, it is likely that gluinos, if massless, would have been observed and so, to build a viable theory, it is necessary to break all R symmetries, either explicitly or spontaneously. In the next sections we discuss how this may happen. However, even if R symmetry is broken, there may remain unbroken discrete symmetries. For example, the superpotential of eq.(8) is invariant under "matter parity".

$$\psi^{(i)} \rightarrow -\psi^{(i)} \; ; \; i = q, u^c, d^c, \ell, \ell^c \tag{19}$$
$$H_{1,2} \rightarrow H_{1,2}$$

Such a symmetry is clearly desirable as it forbids terms such as in eq.(15) which could violate lepton and baryon number and for this reason almost all supersymmetric models are matter parity invariant. Combining matter parity and the residual, unbroken, R symmetry with $\beta = \pi \equiv$ fermion number gives a symmetry known as "R parity" under which

$$\nu_\mu, q, \ell, H^o \xrightarrow{R_\pi} \nu_\mu, q, \ell, H^o_{1,2}$$

$$\lambda, \tilde{q}, \tilde{\ell}, \tilde{H}_{1,2} \xrightarrow{R_\pi} -\lambda, -\tilde{q}, -\tilde{\ell}, -\tilde{H}^o_{1,2} \tag{20}$$

Thus all the new supersymmetric states are odd under R parity. This has a profound effect on the phenomenology of such theories for, if R parity is conserved, the new states introduced to supersymmetrise the theory will only be produced and decay in pairs.

Although most models do have an exact R parity, it should be emphasised that this is not inevitable. R parity can be broken by the judicious inclusion of terms(5) such as given in eq.(15), although one must be sure they do not violate baryon or lepton number with rates in conflict with experiment. Perhaps more plausibly R parity may be spontaneously broken through a scalar field with odd R parity acquiring a vacuum expectation value(8) (v.e.v.) A glance at eq.(20) shows that the only neutral scalar fields with odd R parity are the sneutrinos. If a sneutrino acquires a v.e.v. it breaks both the R parity and the lepton number carried by that sneutrino. We will discuss the phenomenological implications of such models in section (1.9).

1.5 Mixing angles and CP violation in the supersymmetric sector

Let us discuss the structure of the charged and neutral currents appearing in eq.(9) in some more detail(9). Expressed in terms of "current" squark and quark eigenstates the charged weak current is diagonal and is simply given by

$$L_{cc} = \frac{g_2}{\sqrt{2}} \tilde{u}_i^{+L} \overleftrightarrow{\partial}_\mu \tilde{d}_i^L W_\mu^+ + \text{h.c.} + \frac{g_2}{\sqrt{2}} \bar{u}_i^L \gamma_\mu d_i^L W^\mu \tag{21}$$

where the superscript L refers to the helicity of the fermion in the supermultiplet.

The mass eigenstates, $\tilde{u}_i^m$ and $\tilde{d}_i^m$, (u_i, d_i) are related to the current eigenstates by unitary matrices $\tilde{X}$, $\tilde{Y}$ (X, Y)

$$\tilde{u}_i^m = \tilde{X}_{ij}\tilde{u}_j, \ \tilde{d}_i^m = \tilde{Y}_{ij}\tilde{d}_j \ : \ u_i^m = X_{ij}u_j, \ d_i^m = Y_{ij}d_j \tag{22}$$

and hence the charged current becomes

$$L_{cc} = \frac{g_2}{\sqrt{2}} \tilde{u}^{m+L}_{i} \partial^{\mu} \tilde{U}_{KMij} \tilde{d}^{mL}_{j} W^{+}_{\mu} + \text{h.c.} + \frac{g_2}{\sqrt{2}} u^{mL}_{i} \gamma_{\mu} U_{KMij} d^{m}_{j} W^{+}_{\mu} \tag{23}$$

where

$$\tilde{U}_{KM} = \tilde{X}_L \tilde{Y}_L^{\,+} \quad ; \quad U_{KM} = X_L Y_L^{\,+} \tag{24}$$

and, clearly, $\tilde{U}_{KM}$ and U_{KM} are unitary matrices. Since supersymmetry is broken there is no need for the squark mass eigenstates to correspond to the quark eigenstates and so $\tilde{U}_{KM}$ need not be equal to U_{KM}. $\tilde{U}_{KM}$ may be written in the same form as the Kobayashi Maskawa mixing matrix, U_{KM}, in the standard model, but with different angles and phase. The neutral currents involving both squarks and quarks coupled to the Z, the γ or the gluons remain invariant under this rotation - the supersymmetric generalisation of the GIM mechanism. This is because they always involve the combinations $\tilde{X}\tilde{X}^+$ or $\tilde{Y}\tilde{Y}^+$ (XX^+ or YY^+) which are unity.

The situation is quite different for interactions involving the gauginos. Consider, for example, the interaction of gluinos which is (suppressing colour indices on the quarks and squarks)

$$L_g = \sqrt{2} i g_3 \bar{\lambda}^{\alpha} [\tilde{u}^{+}_{L_i} T^{\alpha} u_{L_i} + \tilde{d}^{+}_{L_i} T^{\alpha} d_{L_i}] + \text{h.c.} \tag{25}$$

The interaction between mass eigenstates is

$$L_g = \sqrt{2} i g_3 \bar{\lambda}^{\alpha} [\tilde{u}^{m+}_{L_i} V^{u}_{L} T^{\alpha}_{ij} u^{m}_{L} + \tilde{d}^{+m}_{L} V^{d}_{L} T^{\alpha}_{ij} d^{m}_{l_j}] + \text{h.c.} \tag{26}$$

where

$$V^{u}_{L} = \tilde{X}_L X^{+}_{L} \quad \text{etc.} \tag{27}$$

and i, j are family indices. There is a similar form for the right handed components.

V^u and V^d can again be written in the standard Kobayashi Maskawa form, but in general involving new mixing angles and phases. Their values depend on the mechanism giving squarks and sleptons mass, a subject discussed further in sections (3.2 - 3.4). In the minimal version of the model[10]

$$\tilde{U}_{KM} \simeq I$$

$$V_L^u \simeq V_R^u \simeq I$$

$$V_L^d \simeq U_{KM} \qquad V_R^d \approx I \tag{28}$$

where U_{KM} is the Kobayashi Maskawa mixing matrix of the standard model. Thus gluinos will change flavour principally where coupled to the left-handed down doublets of quarks. Non-minimal versions, with enlarged Higgs structure giving good fermion mass predictions, give $V^u \neq I$ and gluinos appreciable flavour changing components when coupled to top quarks and squarks[11].

The lepton sector may similarly be analysed and again we find new mixing angles are needed when expressing the current in terms of mass eigenstates. In the lepton sector the masslessness of the neutrinos meant these angles could be rotated away, but in the slepton sector, the sneutrinos are massive and these mixing angles cannot be rotated to zero and give rise to lepton number violating currents described in terms of mixing matrices $\tilde{U}^{\ell}$, V^{e+}, V^{e^o} in complete analogy with the quark case. Thus supersymmetric theories differ from the standard model in that they predict lepton number violating processes.

In addition to the CP violation introduced in the mixing matrix, there are further new sources of CP violation arising from the gaugino mass terms (see section (1.8)). To be consistent with observed limits on CP violation for the neutron dipole electric moment, the imaginary part of the gluino mass must satisfy[12].

$$\mathrm{Im}[m_{\tilde{g}}] < 10^{-3} \frac{m^3{}_{3/2}}{(1000\mathrm{GeV})^2} \tag{29}$$

In non-minimally coupled supergravity models, the gauginos acquire mass of $O(m_{3/2})$ at tree level and eq.(29) gives a constraint which may be difficult to satisfy.

1.6 Flavour changing neutral currents[9, 13]

The supersymmetric partners we have introduced, contribute, via loops, to flavour changing neutral currents involving light hadrons and leptons. For example, the graph of Fig.(1a) gives (assuming $m_{\tilde{q}} > m_{\tilde{W}}$) the $\Delta S = 2$ operator contributing to the $K_L - K_S$ mass difference:

$$\frac{g_2^4}{256\pi^2 \, m_{\tilde{q}}^6} \, U^+_{KM_{s_i}} \, \Delta m^2_{\tilde{q}_i} \, U_{KM_{id}} \, (\bar{s}_L \gamma_\mu d_L)^2 \tag{30}$$

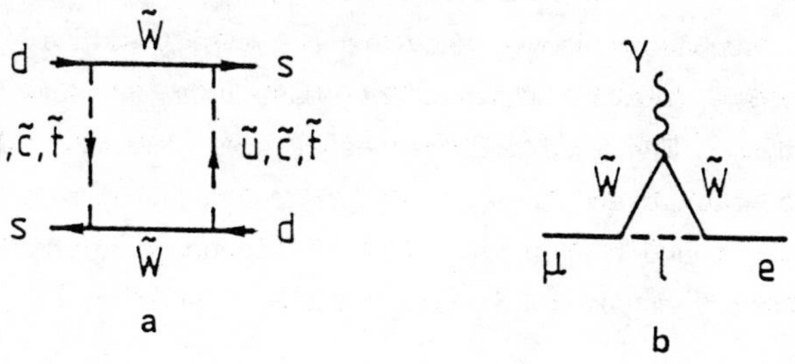

Fig. 1
(a) Graph contributing to the ΔS=2, ΔQ=0 amplitude via super-symmetric virtual states. (b) Graph generating $\mu \to e\gamma$

$m_{\tilde{q}}$ is the mean up squark mass and $\Delta m^2_{\tilde{q}i} \equiv m^2_{\tilde{q}i} - m^2_{\tilde{q}}$ is the up squark mass difference. The experimental measurement of the $K_L - K_S$ mass difference puts strong constraints on the size of this term, and assuming the mixing angles of the $\tilde{u}$ and $\tilde{d}$ squarks are smaller or comparable to the Cabibbo angle, gives the bound

$$\frac{\Delta \tilde{m}^2_{q_{u,c}}}{m^2_{\tilde{q}}} < 0\left[\frac{1}{30} \frac{(m_{\tilde{q}})^2}{M_W^2}\right] \tag{31}$$

An even stronger bound comes from the contribution of Fig.(1a) to the imaginary part of the K_L, K_S mass difference and the associated CP violation in the decays of the kaon (assuming phases of order one).

$$\frac{\Delta m^2_{q_{u,c}}}{m^2_{\tilde{q}}} < 0(10^{-3}) \left(\frac{m^2_{\tilde{q}}}{M_W^2}\right) \tag{32}$$

For the top squark the bound may be weaker because we expect the associate mixing angle to be smaller.

In Fig.(1a) we obtain non-zero contribuations if we replace the $W^{\pm}$ by $\tilde{W}^o$, $\tilde{B}^o$ and $\tilde{g}$ exchange with $\tilde{d}$ and $\tilde{s}$ squark exchange along the horizontal for, as discussed in section (1.6), in this case the GIM mechanism does not prevent flavour changing neutral currents. Evaluation of these graphs lead to similar constraints for the $\tilde{d}$ and $\tilde{s}$ squark mass differences, slightly weakened by the additional mixing angle factors.

For values of $m_{\tilde{q}} \lesssim M_W$ these bounds appear very strong constraints, but, for the models introduced above, they are all easily satisfied because squarks get a large, flavour independent, mass from the supersymmetry breaking mechanism. The mass differences between them are introduced by their Yukawa couplings and thus are of order the quark mass differences and small for the u and c squarks. We will discuss these mechanisms for spontaneous symmetry breakdown in more detail in section (3).

The other rare kaon decay modes also have contributions from the new states, but they provide less stringent bounds than the $K_L - K_S$ systems.[10]

As discussed above a novel feature of supersymmetric models is that they may generate lepton number violating processes through the mixing of the sleptons. The graph of Fig.(1b) give rise to the process $\mu \to e\gamma$. Currently the upper limit on $(\mu \to e\gamma)/\Gamma(\mu \to e\nu\bar{\nu})$ is 2×10^{-10} which leads to the limit

$$\frac{1}{(m_{\tilde{1}}^2 \text{ or } m_{\tilde{B}}^2)} \frac{\Delta m^{2}_{\tilde{1}}}{m_{\tilde{1}}^2} < 0(10^{-7})\text{GeV}^{-2} \tag{33}$$

Once again this condition can be met in the models above, but, because slepton masses are usually less than squark masses, it may be that the rate $\mu \to e\gamma$ is close to the current limit.

1.7 Phenomenology of the minimal SUSY model

We have already discussed some of the novel features of the supersymmetric models, following from their predicted flavour violation. However, the most direct and cleanest test of supersymmetry is to observe directly one of the many new states predicted by theory. In section (1.3) we introduced the multiplets needed to build the basic SU(3) x SU(2) x U(1) [SU(3) x SU(2) x U(1)] x (N = 1 supersymmetry) model. The various Grand Unified versions of the theory all have this low energy structure, with the possible addition of a light singlet field. In fact most of the supersymmetric models that have been considered, whether global or local, have the same low energy spectrum. What differs between these models is the pattern of masses. In particular, different models have as the lightest supersymmetric particle (LSP) the photino, the gravitino or the Higgsinos. As we discussed above in most models there is a multiplicative R parity conserved, where R is +1 for conventional hadrons and -1 for the new supersymmetric states (see section (1.5)). In models with an unbroken R parity, the new states may only be produced in pairs and, once produced, a new supersymmetric state will ultimately decay into the lightest such state which will be stable. Their decay patterns will thus depend sensitively on the identity of the lightest new state. As we have discussed, it is also possible to construct models in which the R parity is broken, allowing for single production of the new states and their decay into convention states only.

For the case of theories with a conserved R parity, the phenomenology for various choices of LSP has been extensively discussed[2,3]. The R_π odd particles have definite couplings to conventional states (cf. section (1.4)) and it is straightforward to discuss their production and decay. For example, squarks and sleptons can be produced via $e^+e^- \to \tilde{q}\tilde{q}, \tilde{\ell}\tilde{\ell}$. The sleptons are expected to decay via

$$\tilde{\ell} \to X + \ell$$

where X may be the photino, gravitino or higgsino depending on their masses. The current experimental bound on the process $e^+e^- \to \ell^+\ell^- + 2X$ places a lower limit of about 20 GeV on the slepton mass and the equivalent processes involving squarks yields $m_{\tilde{q}} > 15$ GeV, the difference being that squarks may decay into gluinos giving a more complicated final state. A full discussion of supersymmetric phenomenology is beyond the scope of this article. Interested readers may find excellent reviews in ref (3) in which the phenomenology of theories with a conserved R parity is discussed. What might we expect in models without an R_π parity[8]? For definiteness, and since it is in any case most interesting, suppose R parity is broken spontaneously through the tan sneutrino acquiring a vev, $\langle\nu_\tau\rangle \neq 0$. In this case, the physical τ^- is a mixture of the original τ^- together with the wino w^- and Higgsino H_2^-. In realistic models, it turns on the τ_L^- mixes predominantly with H_2^- while the τ_R^+ is mainly unmixed. Also the physical ν_τ is a mixture of the current ν_τ and H_2^o and there is a new neutrino-like state ν_N. Remarkably, the forward backward symmetry in $e^+e^- \to \tau^+\tau^-$ and the τ lifetime are unchanged. However, since τ and ν_τ are mixtures of R_π even and R_π odd states, new supersymmetric states may decay singly. For example, squarks (sleptons) may decay into quarks (leptons) plus τ or ν_τ, and the photino can decay into $\nu_\tau(e^+e^-$ or $\mu^+\mu^-$ or $\tau^+\tau^-$ or $q\bar{q}$ or $\nu_e\nu_e$ etc.)

Single production of R_π odd states is also possible. For example, the graphs of Fig. 2 would allow for the single production of gluinos or squarks. The production rate is suppressed by Yukawa couplings, but for a top quark of O(40 GeV) the suppression factor is not small and the production of $q\tilde{q}\nu_\tau$, for example, could be appreciable.

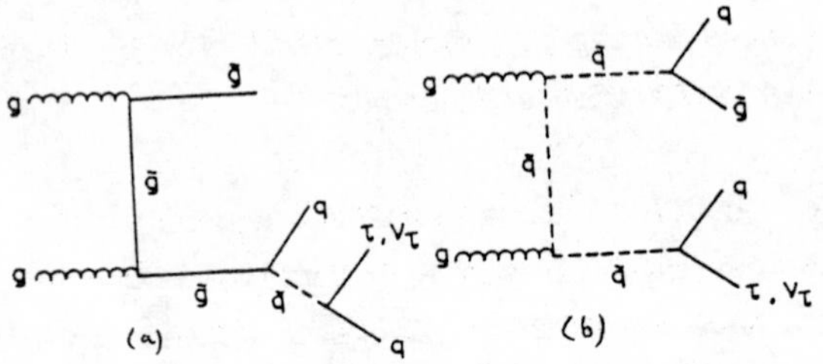

Fig. 2
Typical graphs leading to single production of R_π odd states
(a) gluino production (b) squark production

1.8 Symmetry breaking in supersymmetric models

So far we have discussed the multiplet structure, interactions and phenomenology common to all supersymmetric generalisations of the standard model. We have not yet discussed the sectors of the theory responsible for breaking both the electroweak gauge symmetry, SU(3) x SU(2) x U(1) → SU(3) x $U(1)_{em}$, and the supersymmetry. The former is necessary to give mass to the $W^\pm$, Z bosons. The latter is necessary to split the degeneracy between squarks and quarks, between sleptons and leptons and between gauginos and gauge bosons. It turns out that there are a variety of possibilities giving rise to a profusion of models and we will try to discuss the various possibilities.

Electroweak breaking proceeds via the Higgs scalars, H_1° and H_2° acquiring vevs spontaneously breaking the symmetry in the usual way. We will discuss in section (3.6) in detail how this comes about; for the moment we can assume they do.

Supersymmetry breaking introduces novel features compared to the spontaneous breaking of Lie groups[1]. It is straightforward to calculate the Hamiltonian, H, in terms of the supersymmetry generators Q_α and $Q_{\dot{\alpha}}$, for a globally supersymmetric theory.

$$H = P_o = \frac{1}{4} \sum_{\alpha,\dot{\alpha}=1,2}^{2} (Q_\alpha^{\ 2} + Q_{\dot{\alpha}}^{\ 2}) \tag{34}$$

Thus the energy of a state is positive semidefinite. This immediately implies that a global supersymmetry is spontaneously broken if, and only if, the ground state of the theory has non-zero energy since the broken case requires $Q_{\alpha,\dot{\alpha}}|o> \neq 0$ for some $\alpha,\dot{\alpha}$. From eq.(7), we see H can be non-zero either through the F or the D term acquiring non-zero vevs[14]. The latter requires an Abelian factor in the group(in addition to the U(1) of the standard model) and cannot be Grand Unified in a simple group. For this reason, most models use the F term to break supersymmetry. It is straightforward to construct superpotentials involving new chiral superfields which lead to a non-zero F term. To construct realistic supersymmetric models, it is necessary then to couple these fields and the supersymmetric breaking to the conventional sector of Table 1. Various models have been constructed in which this is done by Yukawa, gauge or gravitational interactions. We will consider detailed models in section 3, but first it is necessary to discuss some of the new features that appear when gravitational effects are included[15, 16, 17] in a supersymmetric theory.

Supersymmetry breaking in locally supersymmetric models. If supersymmetry is made a local symmetry, one automatically obtains gravitational interactions because the supersymmetry algebra includes the Poincare group and the local version of the Poincare group generates Einstein's gravity. The (N = 1) locally supersymmetric gauge theory modifies the Lagrangian given by eq.(7) in the globally supersymmetric case. The potential in the minimally coupled case, may be written as[15]

$$V(\phi,\bar{\phi}) = e^{\sum_i |\phi_i|^2/M^2} \{\sum_j \left|\frac{\partial P}{\partial \phi_j} + \frac{\phi_j^*}{M^2}\right|^2 - \frac{3|P|^2}{M^2}\} + \tfrac{1}{2}\sum_\alpha D_\alpha^* D^\alpha \qquad (35)$$

where ϕ_j are the scalar components of the chiral superfields in the theory, P is the usual superpotential, D the usual D term, and M is related to the Planck mass

$$M = \frac{1}{\sqrt{8\pi}} M_{Planck} = 2.4 \times 10^{18} \text{ GeV}$$

In the limit $M \to \infty$, V reduced to the globally supersymmetric form, eq.(7).

Supersymmetry is broken if the quantity μ is non-zero where

$$\mu^4 \equiv [e^{\sum_i |\phi_i|^2/M^2} \sum_j \left|\frac{\partial P}{\partial \phi_j} + \frac{\phi_j^* P}{M^2}\right|^2 + \tfrac{1}{2}\sum_\alpha D_\alpha^* D^\alpha] \qquad (36)$$

Due to the extra term in eq.(35), and in contrast to the globally supersymmetric case, it is possible for μ to be non-zero while V remains zero. It is usual, in building supergravity models, to exploit this possibility and keep V zero because this corresponds to the desirable case of zero cosmological constant. If this is done the P itself must have non-zero vacuum expectation value. This is related to the gravitino mass

$$m_{3/2} = \frac{\langle P \rangle}{M^2} \exp \left(\tfrac{1}{2}\frac{\sum|\langle\phi_i\rangle|^2}{M^2}\right) \qquad (37)$$

where $\langle P \rangle$ and $\langle \phi_i \rangle$ are the vacuum expectation values of P and ϕ_i.

Now we can ask about symmetry breaking in the gauge nonsinglet sector. The interesting thing is that there are terms in eq.(35) which, at tree level, give a common mass to the scalar fields. Expanding the term $\left|\frac{\partial P}{\partial \phi_j} + \phi_j^* \frac{P}{M^2}\right|$, we find, using eq.(37), the scalar mass term $m_{3/2}|\phi_j|^2$.

From eq.(35) we can see that, for zero cosmological constant, $\sqrt{3}\,\frac{\langle P \rangle}{M} = \mu^2$, where is the supersymmetry breaking scale. Thus $m_{3/2} \approx \frac{\mu^2}{M}$ and gravitational corrections, as anticipated above, give a common (supersymmetry breaking) mass to the scalar fields of magnitude $m_{3/2}$.

To be more specific in giving the predictions of locally super-symmetric N=1 theories, we need to specify the supersymmetry breaking mechanism. We will discuss the simplest class of model in which supersymmetry is broken in the gauge singlet sector and is only communicated to the gauge non - singlet sector by gravitational effects[17].

The choice of a suitable supersymmetry breaking potential is rather arbitrary, particularly in the heavy sector. One simple form that has been analysed in detail is the superpotential

$$P = p(Z) + g(y) \tag{37}$$

where p(Z) involves only singlet fields, Z, and g(y) involves the gauge nonsinglet fields of the standard model. Let us assume that the form of p and g are chosen so that, at the minimum of the potential, some fields acquire a vacuum of the form

$$\begin{aligned} \langle Z_i \rangle &= b_i M \\ \left\langle \frac{\partial p}{\partial Z_i} \right\rangle &= a_i \mu M \\ \langle p \rangle &= M^2 \\ \langle y_a \rangle &= 0 \end{aligned} \tag{39}$$

The condition the cosmological constant vanish at the minimum is

$$\sum_i |a_i + b_i|^2 = 3 \tag{40}$$

The effective low energy potential is found by keeping those terms non-vanishing in the limit $M \to \infty$

$$V = |\tilde{g}_i|^2 + m_{3/2}{}^2 |y_i|^2 + m_{3/2}{}^2 (\tilde{A}g + \text{h.c.}) + \tfrac{1}{2} D_\alpha D^\alpha \tag{41}$$

where

$$m_{3/2} = \exp(\tfrac{1}{2}\sum_i |b_i|^2)\mu$$

$$\tilde{g} = \exp(\tfrac{1}{2}\sum_i |b_i|^2) g$$

and

$$A = b_i^*(a_i + b_i) \tag{42}$$

The scalar potential of eq.(41) shows that the gravitational corrections couple supersymmetry breaking from the singlet "hidden" sector to the nonsinglet sector in a simple manner, through soft supersymmetry breaking terms of $O(m_{3/2})$. Every scalar field obtains a common mass $m_{3/2}$ and there are further soft terms contributing to the scalar potential proportional to the superpotential of the nonsinglet sector, with all fields replaced by their scalar partners. The matter fermions do not acquire a mass from gravitational corrections. Gauginos can acquire a mass of order $m_{3/2}$ if the gauge kinetic energy is non-minimal[15]. It is not related to the scalar masses and can be much smaller (see section (3.31)).

The simplest example of a suitable superpotential (the Polonyi potential (18)) is to choose for p(Z)

$$p(z) = \lambda Z + \Delta \tag{43}$$

In this case $\left| \frac{\partial P}{\partial Z} + \frac{Z^* P}{M^2} \right|$ is non-zero, so supersymmetry is broken with the parameter A of eq.(42) given by

$$A = 3 - \sqrt{3} \tag{44}$$

To summarise, the supergravity potential contains terms that give mass to all supersymmetric partners of quarks and leptons, and to the Higgs scalars. Gauginos may also have a mass at tree level of $O(m_{3/2})$ or, if for some reason this is absent, they will be generated in radiative order (12), as the constant term in eq.(43) violates the R symmetry forbidding their masses (see section (1.7)).

2. GRAND UNIFICATION AND SUPER-GRAND UNIFICATION

The basic idea of grand unification is that the standard model, based on the gauge group SU(3) x SU(2) x U(1), is part of a larger semisimple gauge group G with a single gauge coupling constant (or, possible, a product of such semisimple groups with a discrete symmetry relating these groups)[19]. G is spontaneously at a scale, M_x, to the standard model giving, to order $\frac{M^2}{M_x^2}$, where M is a typical hadronic mass scale, the standard low energy phenomenology. The best known example is that invented by Georgi and

Glashow with G = SU(5) with the breaking pattern

$$G = SU(5) \underset{M_X}{\rightarrow} SU(3) \times SU(2) \times U(1) \underset{M_W}{\rightarrow} SU(3) \times U(1)_{em} \tag{45}$$

The multiplet structure needed for the simplest version of SU(5) is shown in Fig. 3. It is minimal in the sense that the fewest number of scalar representations are used, although it should be said that there is no symmetry principle requiring this and there are reasons to suppose that the minimal theory is already sick because of the predicted fermion mass spectrum.

PARTICLE CONTENT		SPIN
$A_\mu^{a\,=\,1\ldots 24}$ $(W^{\pm}_{\mu}, Z_\mu, \gamma_\mu, A_\mu^{-a=1\ldots 8}, X_\mu^{\pm 4/3}, Y_\mu^{\pm 1/3})$	$= \underline{24}$	$J = 1$

+ 2 generations

$$\begin{bmatrix} \bar{d}^1 \\ \bar{d}^2 \\ \bar{d}^3 \\ e^- \\ \nu_e \end{bmatrix}_L = \underline{\bar{5}}; \quad \frac{1}{\sqrt{2}} \begin{bmatrix} 0 & \bar{u}^3 & -\bar{u}^2 & u_1 & d_1 \\ -\bar{u}^3 & 0 & \bar{u}^1 & u_2 & d_2 \\ \bar{u}^2 & -\bar{u}^1 & 0 & u_3 & d_3 \\ -u_1 & -u_2 & -u_3 & 0 & e^+ \\ -d_1 & -d_2 & -d_3 & -e^+ & 0 \end{bmatrix} = \underline{10} \qquad J = \tfrac{1}{2}$$

$$\bar{H} = \begin{bmatrix} \phi^{1/3} \\ \phi^{1/3} \\ \phi^{1/3} \\ \phi^- \\ \phi^0 \end{bmatrix} = \underline{\bar{5}}; \quad \Sigma^i_j = \underline{24} \qquad J = 0$$

Fig. 3
Particle content of minimal SU(5) theory.

One problem arises in implementing the breaking schemes such as is given in eq.(45). To inhibit proton decay, the scale M_X is often very large ($O(10^{15}$GeV) in SU(5)), much larger than the electroweak scale $M_W = O(10^2$GeV). Unless one requires unnatural cancellations on the theory, radiative corrections will give $\frac{M_W}{M_X} \geq O(\alpha_G)$, where α_G is the grand unified fine structure constant which is typically $> O(10^{-2})$. This is clearly inconsistent with the required ratio $\frac{M_W}{M_X} \leq O(10^{-13})$ giving rise to the hierarchy problem of GUTs[20]. The best solution is to increase the symmetry of the system, the new symmetry preventing the feed through of the large mass scale, M_X, to the electroweak scale, M_W. The only symmetry which can do this is supersymmetry giving rise to supersymmetric grand unified theories (or SUSY GUTs). The SUSY GUTs constructed so far are based on the direct product structure G^{SUSY} = G x [N extended supersymmetry]. In this section I will discuss the construction and phenomenology of the simplest SUSY GUT.

2.1 A supersymmetric version of SU(5)[21]

To construct an SU(5) x [N=1 supersymmetry] SUSY GUT we need to assign the states of Table 1 to N=1 supermultiplets. Following the reasoning of section (1.3) leads immediately to the multiplet structure of Table 2.

Table 2

Supermultiplets used in SU(5) SUSY GUT. The index a=1,...N_g refers to the family index and α, β are group indices.

Role	Notation	SU(5) Representation Content
Matter	$\psi_a^{\alpha}, \chi_a{}^{\alpha\beta}$	N_g x ($\bar{5}$ + 10)(chiral supermultiplet)
Higgs	H_1^{α}, H_2	($\bar{5}$ + 5)(chiral supermultiplet)
	Σ_{α}^{β}	24 (chiral supermultiplet)
Vector	V^{α}	24 (vector supermultiplet)

The Yukawa couplings needed to give all charged fermions a mass has a superpotential of the form

$$P_{5_M} = \frac{1}{\sqrt{2}} M_{ij}^{(d)} \psi_{i\alpha} \chi^{\alpha\beta} H_{2\beta} - \frac{1}{4} M_{ij}^{(u)} \varepsilon_{\alpha\beta\gamma\delta\rho} \chi_i^{\alpha\beta} \chi_j^{\gamma\delta} H_1^{\rho} \qquad (46)$$

where H_1 and H_2 are (left handed) chiral superfields transforming as 5 and $\bar{5}$, respectively, under SU(5) and i and j are family indices. It is necessary that H_1 and H_2 be distinct chiral supermultiplets, and not hermitian conjugates, for supersymmetry does not allow us to build the superpotential with a chiral superfield and its conjugate. It is also impossible to identify H_2, transforming as a $\bar{5}$, with one of the $\bar{5}$s introduced to describe the quark and lepton sector. In addition to the reasons given in section (1.3) for this, there is an even more pressing one. The colour triplet components of H_2 mediate proton decay and must have very large mass ($\geq O(10^{10}$ GeV)) if the proton is not to decay too quickly. If we identify the doublet components of H_2 with the sneutrino, selectron members of a $\bar{5}$, then the triplet components will be partners of the down antiquarks. Since, as we will shortly show, supermultiplet splitting in a gauge nonsinglet representation is $\lesssim O(1\text{TeV}\sqrt{\frac{\pi}{\alpha}})$, the triplet components will be much lighter than the GUT scale and will mediate proton decay far too fast.

Consequently, to accommodate the necessary Higgs scalars, we must choose two new chiral supermultiplets H_1 and H_2 transforming as a 5 and $\bar{5}$ respectively. We must also ensure that the H_1 and H_2 multiplets are split so that their triplet components are heavy ($\geq 10^{10}$ GeV) while leaving their doublets light, O(1TeV). We postpone a discussion of the mechanisms generating this multiplet splitting until section (3.5). In addition to the multiplets introduced above, it is necessary to add an adjoint chiral supermultiplet, Σ, to accommodate the adjoint of Higgs scalars necessary to break SU(5) to SU(3) x SU(2) x U(1).

The final multiplet choice for our SU(5) x N=1 supersymmetry model is given in Table 2. It will, of course, be necessary to break the supersymmetry, and this requires further fields. We will discuss this later.

2.2 The Hierarchy problem in supersymmetry

The hierarchy problem is the difficulty in keeping a low electroweak scale, M_W when grand unification occurs at a scale $M_X >> M_W$[20]. Although we have not yet discussed the mechanism responsible for electroweak breaking, we can still ask whether a low scale for M_W will be stable against the radiative corrections.

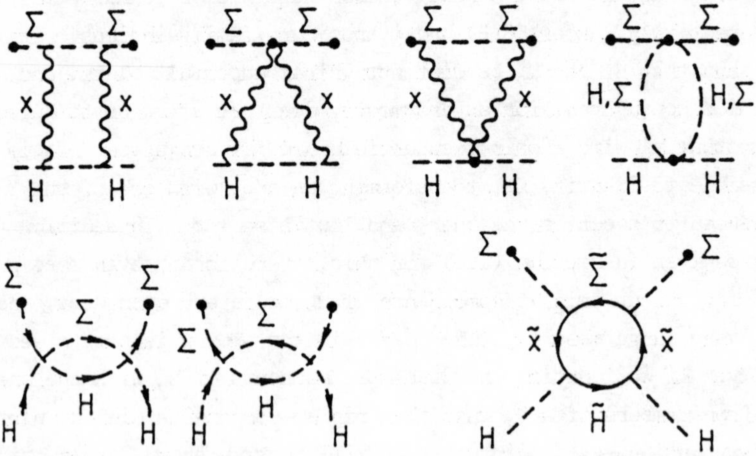

Fig. 4
Graphs contributing to Higgs doublet masses in supersymmetric SU(5).

The dangerous terms are of the form $\lambda H^+H \,\mathrm{Tr}(\Sigma^+\Sigma)$ which generate a large doublet mass. We may avoid complications concerning the renormalisation conditions by performing a finite calculation relating the Higgs mass at two momentum scales

$$m_H^2(p^2) = m_H^2(M_X^2) + [\lambda(p^2) - \lambda(M_X^2)] < \mathrm{Tr}(\Sigma^+\Sigma) > \tag{47}$$

where the graphs of Fig. (4) generate a coupling, λ, which is logarithmically dependent on the momentum scale. Since the vev $<\mathrm{Tr}(\Sigma^+\Sigma)>$, is of order M_X^2 the condition that $m_H^2(1\ \mathrm{TeV}^2)$ should be $< 0(1\ \mathrm{TeV}^2)$ (so M_W is < 100 GeV) can only be achieved if $m_H^2(M_X^2)$ is $0(M_X^2)$and there is a delicate cancellation of the terms on the right hand side. This is the hierarchy problem - why should physics at the microscopic scale be complicated ($m_H(M_X) \simeq 10^{15}$GeV) in order to achieve simplicity at the macroscopic scale ($m_H(1\ \mathrm{TeV}) \simeq 0$)?

The remarkable thing is that λ is zero in supersymmetric theories. This is a manifestation of the nonrenormalisation theorem[(6)] which says terms in the superpotential are not radiatively corrected and which comes about through a cancellation of the graphs involving boson and fermion loops.

Thus supersymmetric theories avoid the hierarchy problem (an improvement on ordinary non-supersymmetric SU(5) for which λ is non-zero, because there are no cancelling fermion contributions).

2.3 Supersymmetry breaking and the hierarchy problem

If supersymmetry were unbroken, there would be scalar leptons and scalar quarks degenerate with the charged leptons and quarks as well as gauginos degenerate with the gauge bosons. No such states have been observed, so supersymmetry must be broken. However, once the degeneracy between such states is broken, the cancellation between the graphs of Fig. 4 is spoilt by the difference in masses in the propagators. Evaluating the graphs of Fig. 4 gives

$$m_H^2(\mu^2) = m_H^2(M_X^2) + \frac{\alpha}{\pi}(m_X^2 - m_X^2) \tag{48}$$

Now, the condition that $m_H(\mu)$ should naturally be less than 0(1 TeV), at $\mu \approx 1$ TeV, puts a bound on the mass splitting with a vector supermultiplet

$$(\tilde{m}_X^2 - m_X^2) < 0(\frac{(1 \text{ TeV})^2\pi}{\alpha}) \tag{49}$$

Similarly, if the mass splitting between supersymmetric partners in a chiral supermultiplet is Δm_H^2, the condition that the contribution from the loops similar to Fig. 4 do not contribute unacceptably to m_H is

$$\Delta m_H^2 < 0(\frac{(1 \text{ TeV})^2\pi}{\beta}) \tag{50}$$

where is the coupling of the chiral supermultiplet to the Higgs doublet sector.

Thus, although supersymmetry can solve the hierarchy problem, it does so only if the masses of the super partners are so low that they are accessible to experimental discovery. For this reason, there has been considerable effort to try to construct viable theories, and to determine the expected spectrum of states and their phenemonology.

2.4 Phenomenology of SUSY-GUTs

Grand unification requires that the three couplihgs constants of the standard model be related to a single, grand unified coupling. In non-supersymmetric SU(5) the relation is

$$g_3 = g_2 = \sqrt{\tfrac{5}{3}}\, g_1 = g_5 \tag{51}$$

The couplings, however, depend on the scale, μ^2, at which they are measured for radiative corrections, such as shown in Fig. 5 cause them to "run".

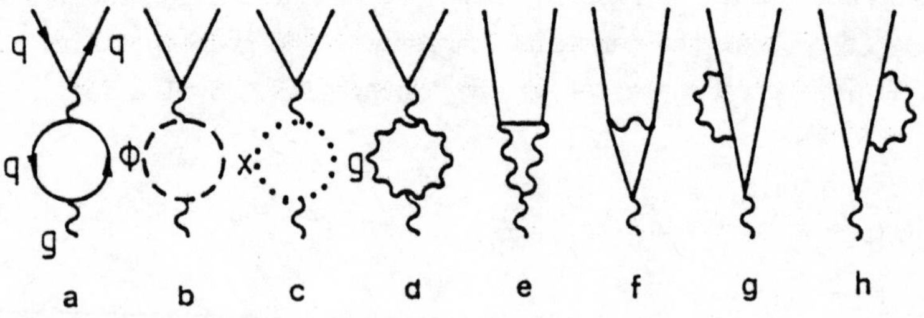

Fig. 5
Diagrams contributing to the running of gauge couplings: solid lines are fermions, wavy lines gauge bosons, dashed lines scalars and dotted lines ghosts.

In leading order the renormalisation group summation gives for the running couplings

$$\alpha_i^{-1}(\mu^2) \equiv \left[\frac{g_i^2(\mu^2)}{4\pi}\right]^{-1} = \alpha_i^{-1}(M_X) + \frac{1}{6\pi} b_i \ln \frac{M_X}{\mu} \tag{52}$$

where $\alpha_i^{-1}(M_X)$ are given by eq. (51) and the b_i are obtained from the graphs of Fig. 5

$$b_3 = 33 - 4N_g \quad ; \quad b_2 = 22 - 4N_g \quad ; \quad b_1 = 4N_g \tag{53}$$

Plotting α_i v/s μ^2 gives the graph of Fig. 6.

Remarkably it is found that the values of $\alpha_i(\mu^2)$ as measured in the laboratory are consistent with eq. (51) provided[22]

$$M_X = 2.4^{+2.8}_{-1.6} \times 10^{14}\ \text{GeV} \quad \text{for} \quad \Lambda_{\overline{MS}} = 0.16^{+0.1}_{-0.08}\ \text{GeV}$$

$$\sin^2\tilde{\theta}_W(80\text{GeV}) = \frac{g_2^2}{g_1^2 + g_2^2}(80\text{GeV}) = 0.212^{-0.004}_{+0.006} \tag{54}$$

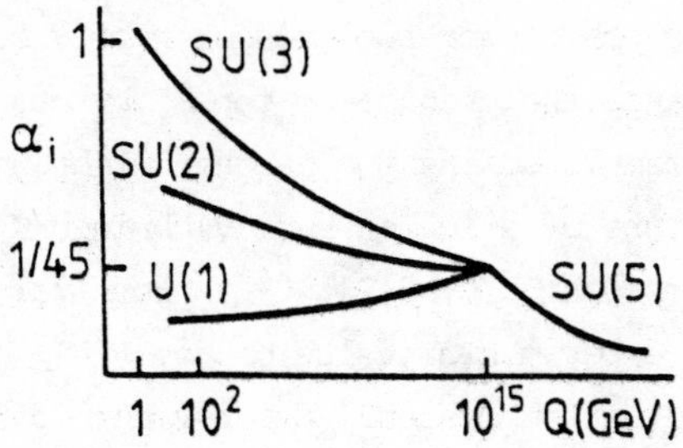

Fig. 6
Behaviour of the (3, 2, 1) couplings with energy

(These results came from a more exact two loop calculation of the running couplings). The weak angle $\sin^2\tilde{\theta}_W$ is evaluated in the $\overline{\text{MS}}$ regularisation scheme (important when calculating beyond the leading order) and for comparison the experimental value for the same quantity is

$$\sin^2\tilde{\theta}_W(80)\Big|_{\text{Expt}} = 0.215 \pm 0.015 \tag{55}$$

in remarkable agreement with SU(5) prediction.

What happens in the SU(5) SUSY-GUT? Because it involves new light supersymmetric particles (see Table 1) it is necessary to recompute the values for the grand unified mass and $\sin^2\theta_W$. The initial ratios of the couplings, eq. (51), are unchanged since we still have the same assignment of states to representations of SU(5). However, the radiative corrections will differ for now, in Fig. 5, we must include loops containing the new states. The values of b_i are easily calculated, including the new light supersymmetric partners of the quarks, leptons and SU(3) x SU(2) x U(1) gauge bosons in the graphs of Fig. 5. This gives [23]

$$b_3 = 27 - 6N_g \quad ; \quad b_2 = 18 - 6N_g \quad ; \quad b_1 = 6N_g \tag{56}$$

Including two loop effects[24] gives the predictions (for $\Lambda_{\overline{\text{MS}}} = 150$ MeV)

$$M_X = 7.7 \times 10^{15} \text{ GeV}$$

$$\sin^2\theta_W = 0.233 \tag{57}$$

The reason M_X has increased is principally because the octet of gluinos reduces b_3 in eq. (56) relative to the SU(5) value in eq. (53) thus slowing the evolution of α_3 and postponing the crossover of α_3 and α_2 (cf. Fig. 6). It is possible to change the prediction for M_X by changing the rate of evolution of α_2; for example adding two further light Higgs SU(2) singlets (mimicing the structure of a lepton family) brings M_X back to the usual SU(5) value and lowers $\sin^2\theta_W$ by 0.215[23]. Both values for $\sin^2\theta_W$ are consistent with the experimental value of eq. (55). SU(5) also makes predictions for quark and lepton masses. Supersymmetric SU(5) does not change these predictions at the scale M_X. At this scale

$$m_b = m_\tau \; ; \; m_s = m_\mu \; ; \; m_d = m_e \tag{58}$$

Radiative corrections change this at low scales and, surprisingly, the predictions are very similar for SUSY SU(5) and ordinary SU(5)[24,25]

$$m_b \approx 3m_\tau \; ; \; m_s \approx 3m_\mu \; ; \; m_d \approx 3m_s \quad \text{at} \quad \mu \approx 1 \text{ GeV} \tag{59}$$

The first relation is in good agreement with experiment but the latter two are not and imply there must be additional structure beyond the minimal SU(5) introduced above.

2.5 Nucleon decay

The most dramatic prediction of SU(5) and most GUTs is that the proton will decay via the graphs of Fig. 7a, through the exchange of X or Y bosons, the new gauge bosons in SU(5) in addition to those in SU(3) x SU(2) x U(1). Evaluation of these graphs gives the dominant contribution

$$\frac{1}{4} L_{\Delta B=1} = \frac{g^2}{8M_X^2} \varepsilon_{ijk} \bar{u}^{c\,k}_L \gamma_\mu u^j_L \bar{e}^+_L \gamma^\mu d^i_L \tag{60}$$

To compute the estimated proton decay lifetime one must estimate the overlap of the quark states appearing in eq. (60) with the hadronic states, the proton, etc. This gives for the dominant decay mode (including radiative corrections)(26) of conventional SU(5)

$$\tau_{p\rightarrow e^+\pi^0} = 5 \times 10^{29 \pm 1.7} \text{ years} \tag{61}$$

a result inconsistent with current limits.

In supersymmetric SU(5) one might think that the proton lifetime will be invisibly long for, from eq. (60), $\tau_p \propto M_X^4$ and M_X in SUSY SU(5) is (cf. eq. (54) and (57)) ≈ 40 times the SU(5) value. However, in SUSY SU(5) there are new processes involving squarks which mediate nucleon decay[(4)(27)]. These are shown in Figs(7b,7c)(plus graphs with d and s interchanged on external states.)

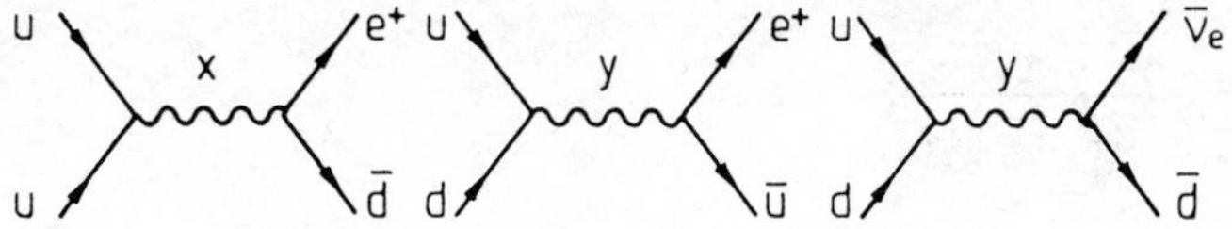

a.

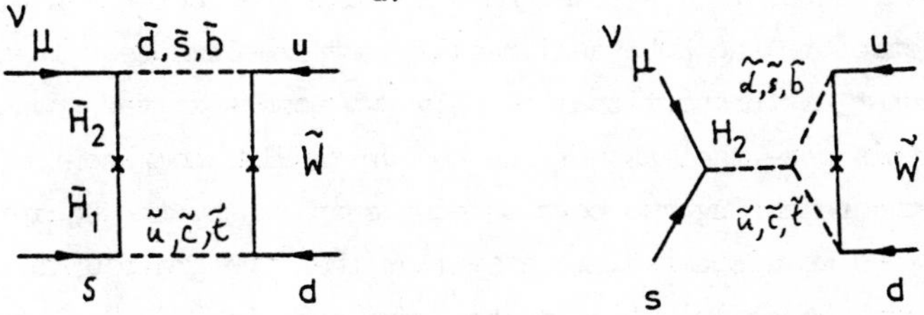

b.

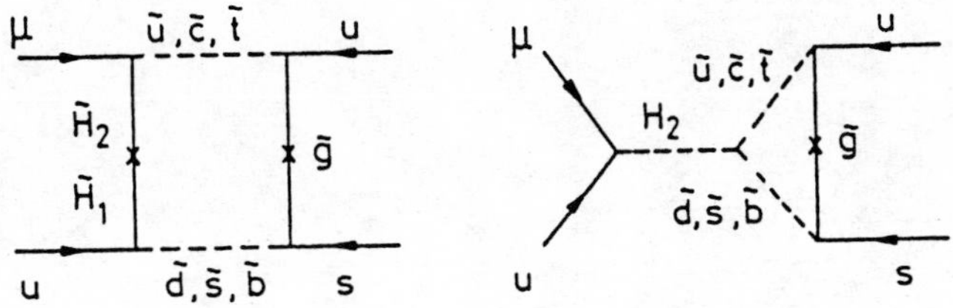

c.

Fig. 7
Graphs contributing to proton decay (a) in GUTs, (b), (c) in SUSY-GUTs. The x denotes a mass insertion.

Consider first the graphs of Fig. (7b). Because the graphs involve Higgsino exchange whose couplings are proportional to masses (cf. eqs. (7) and (8)) the dominant graph involves a strange quark and hence the expected proton decay mode in SUSY SU(5) is $p \to K^+\bar{\nu}$ rather than $p \to e^+\pi^o$ in SU(5). Evaluating these graphs and estimating the quark matrix elements gives[28]

$$\tau^{-1}_{p\to K^+\bar{\nu}} = 4\left|\frac{m_c m_s m_p^2}{2v_1v_2M_{H_X}}\right|^2 b^{o2} \sin^4\theta_c \times 10^{29} \text{ years}^{-1} \tag{62}$$

where

$$b^o = \frac{g_2^2 M_2^{\tilde{W}}}{512\pi^2}\left(\frac{1}{M_{\tilde{q}}^2 \text{ or } M_{\tilde{W}}^2}\right) \tag{63}$$

Here $M_2^{\tilde{W}}$ is the supersymmetry breaking component of the Wino mass which need not be equal to its total mass $M_{\tilde{W}}$, and the factor $\sin^4\theta_c$ is an approximation to the Kobayashi Maskawa matrix elements. In eq. (63) it is clear that t_p is proportional to only two powers of the grand unified mass M_{H_X} (in this case the mass of the colour triplet component of the Higgs field), the remaining two powers being supplied by the squark or Wino mass^2 which is, as we discussed, much smaller than the grand unified mass. For this reason the contribution of the graphs of Fig. (7b, c) are the dominant ones. The actual value of the decay lifetime depends sensitively on the values of the wino and squark masses. For a lifetime $\tau_p \geq 10^{31}$ years consistent with current limits

$$b^o \approx \frac{1}{3} \times 10^{-8} \text{ GeV}^{-1}$$

which implies

$$\frac{M_2^{\tilde{W}}}{(m_{\tilde{q}}^2 \text{ or } M_{\tilde{W}}^2)} < 10^{-5} \text{ GeV}^{-1} \tag{64}$$

which can be satisfied for $M_{\tilde{W}} < m_{\tilde{q}}$ if $M_2^{\tilde{W}} < 10^{-5} m_q^2$, a not unreasonable restriction.

What is the dominant decay mode in SUSY SU(5)? It is easy to check that the μ decay modes in the Wino dressed graphs are suppressed by a factor of

$\frac{m_u}{m_c \sin\theta}$ because the H_1 coupling is proportional to $m^{(u)}$ and μ decay is associated always with an external $Q = \frac{2}{3}$ quark leading to a m_u factor whereas $\bar{\nu}$ decay is associated with an internal $Q = \frac{2}{3}$ quark which can be the charm squark with a m_c factor. The gluino dressed graphs, Fig. 7(c), can contribute to μ decay modes but, if the gluino does not change flavour and the squarks are degenerate the graphs cancel identically. However, the gluino can change flavour (see section 1.6). As a result the graphs of Fig. 7(c) contribute. The dominant contribution to $p \to \mu^+ K^o$ comes from an internal top squark with a factor m in the second of Fig. 7(c). Evaluating this diagram gives the relative rate(29)

$$\frac{\tau_p \text{ (gluino exchange)}}{\tau_p \text{ (wino exchange)}} \approx \left| \frac{m_c M_2^{\tilde{W}} \alpha_2}{m_t \sin^2\theta_c \sin^2\tilde{\theta}_c M_2^{\tilde{g}} g^2 \alpha_3} \right|^2 \tag{65}$$

where we have written $\sin^2\tilde{\theta}_c = (U^+\tilde{U})_{23}$. Model dependent estimated(10, 11) (see section (3.5)) suggest $\sin^2\tilde{\theta}_c \lesssim \sin^2\theta_c$ and $M_2^{\tilde{g}}/M_2^{\tilde{W}} \equiv \alpha_3/\alpha_2$. Using this we find the gluino exchange graphs may be dominant for $m_t \geq 40$ GeV. In this case, Figs. 7(c) and 7(d) suggest the dominant SUSY GUT nucleon decay modes will be $p \to \mu^+ K^o$, νK^+ and $n \to \nu K^o$.

The absolute prediction for the proton lifetime through gluino dressed graphs depends on numerous poorly known parameters and is therefore itself poorly determined. With $M^{(g)} = 100$ GeV, $m_u = 1$ TeV and mixing angles $\tilde{U} = O(U)$ for $m_t = 40$ GeV the proton lifetime is(29)

$$\tau_p = 6.10^{31} \text{ years} \tag{66}$$

This choice of parameters seems quite reasonable within the context of supersymmetric models and the result holds out some hope that proton decay may be visible to the experiments now running.

As we have seen, supersymmetry can generate proton decay through new operators involving squarks and sleptons. These new states can also contribute to many other processes; for example, in section (1.7) we discussed their role in strangeness violation and lepton-flavour violation. More general studies have been made of their role in $n\text{-}\bar{n}$ oscillation, ν masses, etc., and we encourage the interested reader to refer to the extensive literature on the subject(30).

3. MASS SCALES IN SUPERSYMMETRIC GAUGE THEORY MODELS

We turn now to a discussion of specific SUSY-GUT models including the spontaneous breaking mechanisms needed to complete the models introduced above. In order to construct realistic supersymmetric models, it is necessary to introduce supersymmetry breaking to split the new states from their supersymmetric partners. Thus we must couple the supersymmetry breaking mechanism (either F type or D type) (see section (1.9)) to the light sector involving SU(3) x SU(2) x U(1) gauge nonsinglet fields. There are several different ways to do this and each leads to a different type of model.

(i) Couplings via gauge interactions: For D type breaking, the U(1) Abelian gauge group interactions couple the symmetry breaking to the various fields in the theory, and the supersymmetry breaking in a given supermultiplet is dependent on its U(1) charge.

Symmetry breaking models using F type breaking may be built using gauge non-singlet fields in the O'Raifeartaigh potential and then gauge interactions will, in higher order, couple the symmetry breaking to the light sector[31].

(ii) Coupling via Yukawa interactions: F type models with singlet fields, may have additional Yukawa couplings to gauge non-singlet fields. The supersymmetry breaking in a given supermultiplet is then proportional to its effective Yukawa coupling to the singlet fields[32].

(iii) Coupling via gravity[15, 16, 17]: F (or D) type models, with supersymmetry breaking only in the gauge singlet sector, will still induce supersymmetry breaking in the non-singlet sector via gravitational corrections. This provides an attractive alternative to method (ii) as no, ad hoc, Yukawa coupling between the sectors need be introduced. Such gravitationally induced masses are typical of order (E^2 vac/M_{planck}), where M_{planck}(10^{19} GeV) is the Planck mass. For significant effects, the supersymmetry breaking scale, E_{vac}, must be large, or order 10^{11} GeV. Of course, gravitational corrections may also be important in models with supersymmetry breaking in the non-singlet sector.

We will consider three types of model which illustrate the points discussed above.

3.1 F type models with Yukawa coupling[32]

Supersymmetry is broken if the F term is non-zero. This may be achieved by adding to the SU(5) theory discussed above three gauge singlet fields A, B, C with the superpotential (known as the O'Raifeartaigh potential)

$$P = \lambda_1(A^2 - M^2)B + \lambda_2 A^2 C \qquad (67)$$

Using eq. (7) gives

$$V = |\lambda_1(A^2 - M^2)|^2 + |\lambda_2 A^2|^2 + |2\lambda_1 B + 2\lambda_2 C|^2 \, |A|^2 \qquad (68)$$

For no value of A is V zero, so supersymmetry is broken, but only in the A, B, C singlet sector at this stage with $F_C \neq 0$.

To couple supersymmetry breaking to the gauge nonsinglet sector, we add to the superpotential a term

$$P_\phi = \lambda_3 \phi_1{}^a \phi_{2a} \, A + \lambda_4 A M^2 \qquad (69)$$

where ϕ_1 and ϕ_2 transform as 5 and $\bar{5}$ under SU(5). The last term is to break R invariance, defined in section (1.7), which, if exact, forbids gaugino masses. It is easy to check that ϕ_1 and ϕ_2 do not acquire a vacuum expectation value. At tree level ϕ_1 and ϕ_2 do not couple to F_B or F_C, so they do not feel the supersymmetry breaking. However, beyond tree level, terms of the form $[\phi_1{}^+\phi_1 C^+ C]_D$ are generated via the graph of Fig. 8 which gives a scalar, supersymmetry breaking mass

$$m_\phi{}^2 \approx \frac{\lambda_3{}^2}{8\pi^2}\frac{\langle F_C\rangle^2}{\mu^2} = \frac{\lambda_3{}^2}{32\pi^2}\mu^2 \quad (\text{for } \lambda_3 \ll \lambda_1, \lambda_2) \qquad (70)$$

Similarly, there may be contributions to masses of all the other gauge nonsinglet fields in the SU(5) theory occurring through their gauge couplings to ϕ_1 and ϕ_2 and hence to the supersymmetry breaking sector. This gives squarks, sleptons, Higgs scalars and gauginos a mass, just as is required. On the other hand, quarks, leptons and gauge bosons remain massless at this stage because they require SU(2) x U(1) breaking to be non-

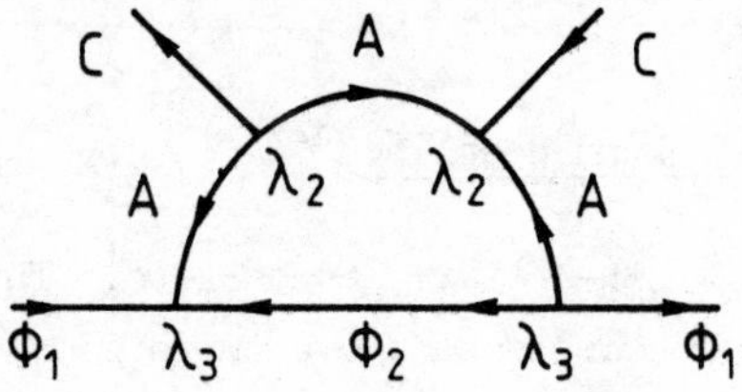

Fig. 8
Supergraph contributing to m_{ϕ_1}

zero. Thus it is naturally explained why we have not yet seen the new supersymmetric states - they are heavy ≃ 0(1 TeV).

3.2 F type model with gauge couplings - the geometric hierarchy(31)

It is straightforward to build a version of the O'Raifeartaigh potential with gauge nonsinglet fields. We introduce the chiral supermultiplets Y and Σ, adjoints under SU(5), together with a singlet field X. The superpotential is

$$P = \lambda_1 \mathrm{Tr}(Y\Sigma^2) + \lambda_2(\mathrm{Tr}\Sigma^2 - \mu^2)X \tag{71}$$

The potential at tree level is

$$V = |F_X|^2 + \sum_{ij} |F_{Y_{ij}}|^2 + \sum_{ij} |F_{\Sigma_{ij}}|^2 + \frac{1}{2}\mathrm{Tr}|D|^2 \tag{72}$$

where

$$D = -g([\Sigma^*,\Sigma] + [Y^*,Y] + \text{quark, lepton and Higgs terms}) \tag{73}$$

F_X, F_Y and F_Σ are given (for a general SU(N)) group by

$$-F_Y^* = \lambda_2(\mathrm{Tr}\Sigma^2 - \mu^2)$$

$$-F^*_{Y_{ij}} = \lambda_1((\Sigma)^2_{ij} - \frac{1}{N}\delta_{ij}(\mathrm{Tr}\Sigma^2)) \tag{74}$$

$$-F^*_{\Sigma_i} = (2\lambda_1 Y\Sigma + 2\lambda_2 X\Sigma)_{ij} - \frac{2}{N}\delta_{ij}\lambda_1(\mathrm{Tr}(Y\Sigma))$$

For SU(6) there is a supersymmetric minimum with

$$<\Sigma> \frac{\mu}{\sqrt{6}} \times \text{Diagonal } (1, 1, 1, -1, -1, -1) \tag{75}$$

For SU(5) it is not possible to achieve the supersymmetric minimum, but the minimum requires $<\Sigma>$ be as close to the identity as possible

$$<\Sigma> = V \times \text{Diagonal } (2, 2, 2, -3, -3) \tag{76}$$

$$\text{with } \quad V = \frac{\lambda_2 \mu}{(\lambda_1^2 + 30\lambda_2^2)^{\frac{1}{2}}} \tag{77}$$

Also at the minimum

$$<Y> = \frac{H_2}{\lambda_1} X_o \times \text{Diagonal } (2, 2, 2, -3, -3) \tag{78}$$

$$<X> = X_o$$

where X_o is undetermined at tree level. SU(5) is uniquely broken in this scheme to SU(3) x SU(2) x U(1).

Radiative corrections to the effective potential give, for large X_o (31)

$$V_{1 \text{ loop}} = \frac{\lambda_1^2 \lambda_2^2 \mu^4}{(\lambda_1^2 + 30\lambda_2^2)} \left| 1 + \frac{\lambda_2^2}{\lambda_2^2 + \frac{\lambda_1}{30}} \left(\frac{29\lambda_1^2 - 50g^2}{80\pi^2}\right) \ell n \frac{X_o}{\mu^2} \right| \tag{79}$$

For $50g^2 < 29\lambda_1^2$ this requires X_o should be larger than μ. The minimum of V can be found by resumming the large logs, using the renormalisation group, and thus X_o is determined in radiative order usually with $X_o >> \mu$.

The symmetry breaking pattern in this model is easy to estimate qualitatively. The Σ field couples through the terms of eq. (72) to the non-vanishing F terms F_Y and F_X. Thus all components of Σ, apart from the hypercharge component which does not couple to these F terms, have supersymmetry breaking masses of order μ. On the other hand the potential V of eq. (72) generates supersymmetric mass terms for the Σ superfield of order $\lambda_2 X_o \simeq M_G$. This may be much larger than μ.

Supersymmetry breaking in other sectors is communicated by radiative corrections. Let us denote by $\tilde{X}$ the combination of chiral superfields X and Y which has non-zero F term. Its fermion component is the goldstino, which in local supersymmetry becomes a component of the gravitino. The goldstino couplings to a multiplet are proportional to the supersymmetry breaking in that multiplet, so, to estimate the supersymmetry breaking in a given supermultiplet, it is sufficient to look for the possible terms generating goldstino couplings to that supermultiplet. There are only two possibilities for a radiatively induced coupling.

$$[\tilde{X}^* \tilde{X} \phi_i \phi_i]_D$$

$$[\tilde{X}^* \tilde{X} \phi_i \phi_j^*]_D \tag{80}$$

where ϕ_i and ϕ_j are general chiral superfields. These can generate scalar masses or order $\langle F_{\tilde{X}} \rangle^2/M^2$ where M is a mass term generated on evaluating the Feynman diagram. As these terms are given by graphs involving the Σ field we know these terms vanish as $M_\Sigma \to \infty$. A careful analysis shows that the mass parameter M^2 must be quadratic in M_Σ and so the scalar masses^2 are of order $\langle F_{\tilde{X}} \rangle^2/M_\Sigma^2$. From eq. (74) we see that this is $0(\mu^4/M_\Sigma^2)$. Thus, multiplets not directly coupled to the $\tilde{X}$ field will obtain a supersymmetry breaking mass of at most $0(\mu^2/M_\Sigma)$. Since M_Σ is to be identified with the Grand Unified mass ($\approx 10^{16}$ GeV), this requires that the supersymmetry breaking scale μ^2 should be very large. Allowing for the fact that coupling constants are also needed, a scale for μ of $0((10^9 - 10^{11})\text{GeV})$ is necessary to generate supersymmetry breaking masses of the order of 10^2 GeV in the light sector. This geometric relation between masses has led to the name geometric hierarchy for this type of model.

Squarks, slepton and Higgs scalars can get mass via gauge boson and gaugino exchanges. They give masses of order $\frac{\alpha}{2\pi}\frac{\mu^2}{M_G}$, where α is the relevant coupling; they are larger for coloured states than for uncoloured states. We note particularly that masses driven by gauge interactions are the same for squarks or sleptons of different families. The corrections to this come from graphs involving quarks and leptons and are proportional to their masses m_i. Evaluation of these graphs gives only a small mass splitting $\delta\mu_i^2$ between squarks and sleptons, in agreement with the bounds of section (1.7).

$$\frac{\delta\mu_i^2}{\mu_i^2} \approx m_i/(\mu^2/M_G) \tag{81}$$

This feature is true also in the F type models discussed above.

3.3 Models with supersymmetry breaking communicated by gravity(15, 16, 17)

A particularly economical class of Grand Unified models have been constructed in which supersymmetry breaking is communicated via gravitational couplings in a manner analagous to that discussed in section (1.8) for the standard model. From eq. (41) we see that gravitational effects couple the supersymmetry breaking in the gauge singlet sector to the gauge non-singlet sector giving a common mass, $m_{3/2}$, at tree level to the scalar fields.

The form of eq. (35) is not the only one allowed by N=1 supergravity(15). The most general form is characterised by a real function $G(Z, Z^*)$ and analytic functions $f_{ab}(Z)$ of the complex chiral superfields Z. G is written in terms of the real "Kahler" function $K(Z,Z^*)$ and the usual analytic superpotential P(z) as

$$G(Z, Z^*) = -3\ell n\, K(Z, Z^*) + \ell n|g(Z)|^2 \tag{82}$$

The kinetic energy of the scalar fields is given by

$$G_i^j \quad \partial_\mu z^i \, \partial^\mu z_j^*$$

$$G_i^j \equiv \frac{\partial}{\partial z^i} \frac{\partial}{\partial z_j^*} G \tag{83}$$

Clearly, for $K(Z, Z^*) = |Z|^2$, $G_i^j = \delta_i^j$ and the usual form of the scalar kinetic energy term is recovered. This "minimal coupling" form for G leads to the expression for the chiral Lagrangian given in section (1.8). However, more general forms are possible and the generalisation of eq. (35) to a general G leads to a scalar potential of the form

$$V = e^G \left(\frac{\delta G}{\delta z_j^*} G^{-1}{}_j^i \frac{\delta G}{\delta z^i} -3\right) \tag{84}$$

In eq. (84) we have not included the D term coming from gauge interactions.

The functions $f_{ab}(Z)$ parameterise possible non-minimal kinetic energy terms for the vector superfields

$$L^{vector}_{kin} = -\tfrac{1}{4}\,\mathrm{Re}(f_{ab}(Z))\, V^a_{mn} V^b_{mn} + \tfrac{1}{4}\,\mathrm{Im}(f_{ab}(Z))\, V^a_{mn} \tilde{V}^b_{mn} - i\,\mathrm{Re}(f_{ab})\, \bar{\lambda}^a \not{D} \lambda^b \tag{85}$$

For $f_{ab} \neq \delta_{ab}$ one gets non-minimal coupling for the vector fields and, in addition new sources of mass for the gauginos which may break the degeneracy between gauge boson and gaugino. For example with $K|Z|^2$ one finds

$$L^F = \bar{\psi}_A M^{AB} \psi_b + 2\bar{\psi}_A M^{Aa} \lambda_a + \bar{\lambda}_a M^{ab} \lambda_b + \text{h.c.} \tag{86}$$

where

$$M^{AB} = \frac{1}{K^3} e^{-g/2} (g^{AB} - \frac{1}{3} g^A g^B) = e^{-G/2} (G,_{AB} - \frac{1}{3} G,_A G,_B)$$

$$M^{Aa} = -\frac{1}{3} i (\mathrm{Ref})^{-1}_{ab} D^b G^A - i (\mathrm{Ref})^{-1/2}_{ab} \partial D^b/\partial z_A +$$

$$\frac{1}{4} i (\mathrm{Ref})^{-1/2}_{ac} (\mathrm{Ref})^{-1}_{bd} (\partial f^{cd}/\partial z^A) D^b$$

$$M^{ab} = -\frac{1}{6} e^{g/2} (\mathrm{Ref})^{-1/2}_{ac} (\mathrm{Ref})^{-1/2}_{bd} D^c D^d - \frac{1}{2} e^{-g/2} g^A \partial f_{ab}/\partial z^A \tag{87}$$

where D^a is the usual D term, and $G,_A \equiv \partial G/\partial Z_A$ etc.

Thus not only can supergravity theories break the symmetry between matter fermions and their scalar partners, giving the scalars a common mass of $O(m_{3/2})$ at tree level, but non-minimally coupled theories can also break the symmetry between gauge bosons and gauginos giving the latter a mass of $O(m_{3/2})$. Even in minimally coupled theories with zero gaugino masses at tree level, gaugino masses will be generated in radiative order as R symmetry, which forbids gaugino masses (cf. eq. (20)), is violated by the constant term in eq. (46).

3.4 SU(5) breaking and multiplet splitting

SU(5) may be broken to SU(3) x SU(2) x U(1) through an adjoint of scalar fields, Σ, acquiring a vev as in eq. (76). An example of a superpotential giving rise to this form is given in eq. (71); other (simpler) forms may easily be constructed[33]. One problem we must tackle is the need to split the Higgs multiplets, H_1 and H_2, so that the colour triplet components should be superheavy, while the doublet components remain light $\leq O(M_W)$. This is necessary to inhibit proton decay, while allowing for SU(2) x U(1) symmetry breaking (see section (2.2)).

The simplest way to achieve the splitting is by introducing the superpotential[21] (I is the unit matrix)

$$P_{5H} = H_1(m'I + \gamma_{24}\Sigma)H_2 \tag{88}$$

This gives terms in the low energy potential of the form

$$V = |H_1(m'I + \gamma_{24}\Sigma)|^2 + |(m'I + \gamma_{24}\Sigma)H_2|^2 \tag{89}$$

From this we find Higgs masses

$$m_{H_D}^2 = m' - m$$

$$m_{H_T}^2 = m' + \frac{2}{3}m \tag{90}$$

where $m = -3\gamma_{24}v$ and subscripts D and T refer to SU(2) doublet and SU(3) colour triplet components. If $m' = m$ then the doublets remain massless while the triplets acquire a mass of order M_X. This certainly achieves the aim of splitting the Higgs multiplets although there is no symmetry giving the relation $m = m'$.

Do radiative corrections spoil the result? We have already noted the property of supersymmetry theories that the parameters in the superpotential are not renormalised[6]. Thus when supersymmetry is exact (as it is is at the scale M_X), there will only be wave function renormalisation affecting the scalar masses. Consequently m_{H_D} remains zero beyond tree level, if $m = m'$, up to supersymmetry violating effects which we will discuss later.

This illustrates the first method for splitting the H and $\bar{H}$ multiplets, namely fixing it by hand at a scale M_X and relying on the non-renormalisation theorems to maintain the splitting at low scales. A second way has been suggested which has the advantage of explaining why $m' = m$ initially. In this method the condition $m = m'$ is necessary to minimise the potential energy[34]. Consider the modified form of eq. (88)

$$P_{5H} = H_1(\gamma_1 ZI + \gamma_{24}\Sigma)H_2 \tag{91}$$

where Z is an SU(5) singlet superfield. If this is the only time Z is mentioned in the superpotential the potential will only involve Z through the terms

$$V = |H_1(\gamma_1 ZI + \gamma_{24}\Sigma)|^2 + |(\gamma_1 ZI + \gamma_{24}\Sigma)H_2|^2 \tag{92}$$

Now, if SU(2) is broken through H_1 and H_2 acquiring vevs (how this happens we will discuss later) minimisation of V with respect to the vev of Z gives

$$\gamma_1\langle Z\rangle - m = 0 \tag{93}$$

Clearly, $\gamma_1\langle Z\rangle$ play the same role as m' in eq. (90) so the above equation requires $m' = m$ as is needed to split the multiplets. For obvious reasons this method has become known as the "sliding singlet" technique.

The sliding singlet scheme is elegant and economical, but works only if the vacuum expectation value of the field Z is determined solely by the terms in eq. (92). This can be easily upset by radiative corrections and it turns out that the solution only works if the scale, μ, of supersymmetry breaking is small ($\ll 10^{11}$ GeV).[35]

In models involving a large supersymmetry breaking scale, μ, such as the Geometric Hierarchy or supergravity models described above, it is necessary to adopt another solution. The most popular method is known as the missing doublet scheme[36]. This idea relies on choosing a new Higgs representation whose SU(3) x SU(2) x U(1) content does not include a SU(2) doublet, SU(3) singlet. As a result its Yukawa couplings to H_1 and H_2 cannot give their doublet components a mass, but can give one to their triplet components. For example, the SU(3) x SU(2) content of a $\underline{50}$ under SU(5) is

$$\underline{50} = (8,2) + (6,3) + (\bar{6},1) + (3,2) + (\bar{3},1) + (1,1) \tag{94}$$

It has, as desired, no (1,2) components. We couple two 50s, θ and $\bar{\theta}$ (together these have no anomalies) to H_1 and H_2 via the superpotential

$$P_{MD} = b\theta\Sigma H_1 + b'\bar{\theta}\Sigma H_2 + \tilde{M}\,\bar{\theta}\theta + F(\Sigma) \tag{95}$$

where Σ is here a 75, needed instead of a 24 to construct the θ,H mixing terms. $F(\Sigma)$ is chosen so that Σ develops a v.e.v. M_X breaking SU(5) to SU(3) x SU(2) x U(1). Then P_{MD} gives mass terms involving H_1 and H_2

$$P_{MD} = \theta_3(bM/g)H_{1_T} + \bar{\theta}_3(cM/g)H_{2_T} + \tilde{M}\tilde{\theta}_3\theta_3 \tag{96}$$

where the subscripts refer to the SU(3) transformation properties. From this it follows that the triplets acquire a mass $\simeq \frac{M^2_X}{\tilde{M}}$ while the doublets remain massless. With $\tilde{M} >> M_X$ the triplets may be much lighter than the vector bosons of the theory.

3.5 Electroweak breaking and the problem of the hierarchy mass scales

One of the main motivations for building N=1 supersymmetric models was to solve the hierarchy problem to allow the electroweak breaking scale to coexist with the grand unified scale. As we have seen, the models introduced above can do this, having light Higgs doublets, H_1 and H_2, which, if they acquire vacuum expectation values, will generate electroweak breaking. It is easy to construct a superpotential which will force H_1 and H_2 to acquire vevs, but only if we are prepared to add a mass parameter of order 300 GeV to the theory. This raises the question of where such a scale should come from. In our theory, including gravity, there are four fundamental mass scales: M_{Planck}, M_X, $m_{3/2}$ and M_W. Obviously it would be aesthetically better to relate these scales, and build a theory with only one fundamental mass scale. It turns out that, in many models, electroweak breaking and M_W may be related either to $m_{3/2}$ (a measure of the supersymmetry breaking scale) or directly to M_{Planck}, and in this section we will concentrate on examples of these models.

At scales much below the Grand Unified mass the scalar potential, involving the light Higgs fields, has the form(2)

$$V(H_1,H_2) = m_1^2|H_1|^2 + m_2^2|H_2|^2 + m_3^2\, H_{1_\alpha} H_{2_\beta}\, \varepsilon^{\alpha\beta} \tag{97}$$

$$+\frac{g_2^2+g_1^2}{8}\left(|H_1|^4+|H_2|^4\right)+\frac{(g_2^2-g_1^2)}{4}|H_1|^2|H_2|^2-\frac{g_2^2}{2}|\varepsilon^{\alpha\beta}H_{1_\alpha}H_{2_\beta}|^2 \qquad (97)$$

where the quartic terms are D terms and we have allowed for possible mass terms. Electroweak breaking occurs if this potential has a minimum away from the origin. For example, this will happen if m_1^2 becomes negative.

Let us consider, first, models of the type of section (3.1), in which gauge nonsinglet scalar fields acquire a symmetry breaking mass through radiative coupling to the symmetry breaking sector. Scalar particles get a contribution to their mass in radiative order of the form

$$m^2 \approx \frac{\alpha}{\pi}(m_F^2 - m_B^2) \qquad (98)$$

where m_F and m_B are the masses of the fermion and boson partners in the virtual loop. If, for example, the exchanged particles are gluons and gauginos the contribution will be positive for the gluons are massless while the gluinos must acquire a supersymmetry breaking mass. The squarks will obtain a larger radiative mass than sleptons or Higgs scalars through their larger gauge couplings. On the other hand, the Higgs scalar H_1 couples via a large Yukawa coupling to the top quark and squark. In this case the contribution of eq. (98) will be negative for the top squark is much heavier than the top quark. If the Yukawa coupling is large enough this term dominates the H_1^2 mass^2, and H_1 will acquire a vev[37]. Minimising eq. (97) shows that H_2 will also acquire a vev along its neutral direction (ie. charge conservation is unbroken). The scale of these vevs is given by the magnitude of m_1^2 which is clearly related to the supersymmetry breaking scale and hence to $m_{3/2}$. It is remarkable that it is only the Higgs multiplets that acquire a negative mass squared, the fact that $\alpha_3 > \alpha_2 > \alpha_1$ ensuring that SU(3) x SU(2) x U(1) is broken to SU(3) x U(1)$_{em}$, for the top squark gets a large positive contribution to its mass^2 from graphs involving gluons and gluinos and thus does not become negative even when the negative contribution from the loops involving the large top Yukawa coupling are included.

In supergravity models the masses in eq. (97) arise at tree level through gravitational coupling to the supersymmetry breaking sector. They are given in eq. (41). If $\tilde{g}$ is cubic in the light fields we have

$$|g_i \pm m_{3/2}y_i|^2 = |g_i|^2 + m_{3/2}^2|y_i|^2 \pm 3m_{3/2}(g + g^*) \qquad (99)$$

Thus V of eq. (41) is positive for A < 3 and is minimised if all light fields have zero vacuum expectation values leaving SU(3) x U(1) unbroken at tree level. For A > 3 there is a lower minimum with non-zero vevs and in this case SU(2) x U(1) may be broken at tree level. As in our first example M_W is related to $m_{3/2}$ in this approach

Unfortunately the global minimum is not the desired one, breaking charge and lepton number. For some values of A the desired minimum (a local one) is stable and so this method of electroweak breaking may be a possibility, but it looks somewhat contrived.

However. even for A < 3, as is the case for the simple Polonyi potential eq. (44), once we include radiative corrections there is the possibility for SU(2) x U(1) breaking[38]. From eq. (41) we see that all scalar masses are equal to ${m_{3/2}}^2$ at tree level. Presumably, this result applies at the Planck scale. Radiative corrections due to the renormalisable gauge and Yukawa interactions will split this equality and, in continuing down to low scales of $O(M_W)$, they contribute terms proportional to $[\ell n\,(M_{Planck}/M_W)]^n$ These terms may be resumed using the renormalisation group analysis, leading to running masses $m_i^2({H_i}^2)$ in eq. (97). For the same reason as was discussed in our first example these radiative corrections may selectively drive the Higgs scalar mass^2 negative triggering SU(2) x U(1) breaking. Minimisation of the potential eq. (97) gives two possibilities. In the first the combination of $m_i^2(H_i^2){H_i}^2$ and ${H_i}^4$ terms lead to a minimisation with $\langle H_i\rangle^2 = O({m_i}^2)$. Since the scale of m_i is set by $m_{3/2}$ this gives a solution with M_W related to $m_{3/2}$. In the second possibility the ${H_i}^4$ terms are anomalously small through a cancellation. The minimum is then close to the point at which ${m_i}^2({H_i}^2)$ goes negative. Since $m_i^2(H_i^2)$ evolves logarithmically from the scale of the Planck mass, this point will be directly related to the Planck mass times an exponentially small factor. This is known as dimensional transmutation[39].

The equality of the squark masses at the Planck scale has immediate implications for the mixing matrices $\tilde{X},\tilde{Y}$ and the related Kobayash-Maskawa matrix $\tilde{U}_{KM}$ in the scalar sector. These matrices were introduced in section (1.5) and $\tilde{X}$, $\tilde{Y}$ were the matrices needed to diagonalise the up and down squark mass matrices respectively. Clearly if, at the Planck scale, the squarks are degenerate $\tilde{X} = \tilde{Y} = \tilde{U} = I$ and there are no mixing angles. However, radiative corrections involving Yukawa couplings split the squark degeneracy at lower scales and will give nontrivial mixings. These corrections are calculable in terms of the Yukawa couplings and so the mixing angles in the scalar sector may be determined in models with a simple Higgs structure having few Yukawa couplings. In the minimal theories of sections (1.2) and (2.1) with Yukawa couplings given by eq. (8) or eq. (46) it is easy to get a qualitative estimate of the mixing angles. In these models the dominant radiative corrections come from the largest Yukawa couplings which are those involving the up quarks. Then the squark mass matrices have the form, in one loop order,

$$m^2_{\tilde{u}} \simeq m^{(u)+}m^{(u)} + b_u m^{(u)+}m^{(u)}$$

$$m^2_{\tilde{d}} \simeq m^{(d)+}m^{(d)} + b_d m^{(u)+}m^{(u)}$$

where $m^{(u)}$, $m^{(d)}$ are the quark mass matirces and b_i is the coefficient determined by evaluating the one loop graph. The dominant contributions to b come from loops with light particular (mass μ) which give logarithmically enhanced contributions $\propto \ln(\frac{M_{Planck}}{\mu})$. The mass matrix $m^2_{\tilde{u}}$ is clearly diagonalised by the rotation that diagonalises $m^{(u)}$, so $V^u_L \simeq V^u_R \simeq I$ as in eq. (28). The right handed down quarks/squarks have no large up-quark Yukawa couplings (cf. eq. (8)) and so for them $b_d = 0$, and their mass matrix is diagonalised by the same relation that diagonalises $m^{(d)}$, so $V^d_R \simeq I$ in eq. (28). The left handed down squarks do couple via the large Yukawa couplings so $b_d \neq 0$ for them. Indeed quantitative estimates for b_d suggest the term $m^{(u)+}m^{(u)}$ is the dominant one for these squarks so they are diagonalised by the rotation that diagonalises the left handed <u>up</u> quarks. Using this and eq. (27) leads to the result $V^d_L \simeq U_{KM}$ and $\tilde{U}_{KM} = I$ as in eq. (28).

Clearly this discussion relies on the fact that there are only one set of light Higgs scalars coupling to up quarks and similarly for down quarks. In the SU(5) theory of section (2.1) this structure leads to relations eq. (59) between the masses of down quarks and leptons which are in conflict with experiment and to remedy this models with a richer Higgs structure have been introduced. In these theories the predictions for the mixing angles change[11]. In particular sizeable values for V_L^u are possible, important for proton decay estimates (cf. eq. (65)).

We have seen that supersymmetric models can have the electroweak breaking scale as a dependent mass scale. However there still remain three independent mass parameters M_{Planck}, M_X and $m_{3/2}$. Can these be related? Since, in supersymmetry models, M_X is usually very close to M_{Planck} they perhaps should not be considered as independent mass scales, and may be related by the mechanism giving rise to the effective N=1 theory valid below the Planck scale. However, the supersymmetry breaking scale μ, or the related gravitino mass $m_{3/2}$, are much smaller and it is clearly desirable to relate it to the other scales. Two mechanisms have been proposed. The first[40] applies to a class of models in which, in the globally supersymmetric limit, supersymmetry is unbroken but in achieving the supersymmetric minimum a field, B, acquires an infinite vev. Including the gravitational strength couplings of eq. (35), the vev of B is limited by the Planck mass and consequently the potential energy is non zero, but suppressed from its natural scale by powers of (M_X/M_{Planck}). As a result the scale of superheavy breaking is very small.

We may illustrate this by a very simple example[40] involving singlet fields X, Y and B with superpotential

$$P = \lambda_1(X\,Y - M^2)B + \lambda_2\,X^3 \tag{100}$$

The globally supersymmetric potential following from this is

$$V = |\lambda_1(XY - M_X^{\,2})|^2 + |\lambda_1\,XB|^2 + |\lambda_1 YB + 3\lambda_2\,X^2|^2 \tag{101}$$

which has a supersymmetric minimum at

$$\begin{aligned} \langle XY\rangle &= M_X^{\,2} \\ \langle X\rangle &= 0 \end{aligned} \tag{102}$$

This minimum requires $\langle Y\rangle = \infty$, a physically unreasonable value for we expect large gravitational corrections of the form Y^{n+4}/M^n. Indeed in the N=1 locally supersymmetric case the potential has the form

$$V = e^{(|X|^2 + |Y|^2 + |B|^2)/M} \{|\lambda_1(XY - M_X{}^2) + B^*P/M^2|^2$$
$$+ |\lambda_1 X B + Y^*P/M^2|^2 + |\lambda_1 YB + 3\lambda_2 X^2 + X^*P/M^2|^2$$
$$- 3|P|^2/M^2\} \qquad (103)$$

For large $\langle Y\rangle$, V has then the form

$$V = e^{Y^2/M^2} \{ \left| - \frac{3}{\langle Y\rangle^4} + \frac{1}{M^2\langle Y\rangle^2} \right|$$
$$- \frac{3}{M^2} \left|\frac{1}{\langle Y\rangle^3}\right|^2\} |\lambda_2 M^6|^2 \qquad (104)$$

This has a minimum with $\langle Y\rangle^2 = 6M^2$. Thus gravitational effects limit the vev of $\langle Y\rangle$ and prevent the supersymmetric minimum eq.(102) being reached. As anticipated, the deviation from this minimum is suppresses by powers of (M_X/M) but the surprise is that for the very simple potential given above at the minimum, each term of eq.(103) is of the order $\lambda_2^2 \frac{M_X^{12}}{M^8}$ giving a supersymmetry breaking scale

$$M_{SUSY} = 0(\sqrt{\lambda_2} \frac{M_X^3}{M^2}) \qquad (105)$$

and gravitino mass

$$m_{3/2} = M \lambda_2 (\frac{M_X}{M})^6 \qquad (106)$$

For a rather moderate value for the initial ratio, $(\frac{M_X}{M})$ of $0(10^{-2.5})$, a gravitino mass of $0(10^2 \text{GeV})$ is easily obtained. Although this model has been built using gauge singlet fields X, Y, B the idea can easily be generalised to gauge non-singlets fields(41) in which case M_X can be related to the GUT breaking scale.

The model presented above is an example of how $m_{3/2}$ may be made much smaller than M_{Planck} through gravitational corrections present at tree level in the N=1 supergravity theory. The reason it works is because without gravitational corrections the potential, eq.(101), is flat in the Y direction at minimum ($m_Y = 0$) and so small gravitational corrections can play an important role in fixing the Y vev. It is possible to invent models(42,43,44,45) in which a singlet field, Y, is massless even in the presence of tree level gravitational corrections. In these models the vev of Y is completely undetermined at tree level as are the supersymmetry breaking scale and $m_{3/2}$ which are related to <Y>. It is only in radiative order that the vev of Y and the supersymmetry breaking scale occur. However, radiative corrections are only logarithmically with scale and one finds $m_{3/2}$ is related to the Planck mass times an exponentially small factor giving rise to a natural hierarchy as desired. The beauty of this approach is that the original flatness of the potential in the Y direction arises as a result of a symmetry which is a general property of many supergravity models rather than, as in our first example, as a consequence of a particular choice of superpotential.

These models have a non-minimal form of some, or all, of the scalar kinetic energy corresponding, in eq. (82), to

$$K = 1 - \frac{|Z|^2}{3} \qquad ; \quad |Z|^2 = \sum_{i=1}^{N} |Z_i|^2 \tag{107}$$

This, in fact, is the natural generalisation of the flat space kinetic energy found minimally substituting covariant derivatives etc.[46] The N scalar fields Z_i have kinetic energies invariant under U(n) rotations and also [43] under non-linear transformations of the coset space $U(n,1)/U(n) \times U(1)$. The metric of eq. (83) is the induced metric of a couplex time-like mass shell of the coset space. In the models with all scalars having K given by eq. (107) the supertrace of the mass matrix vanishes

$$\mathrm{Str}\, M^2 \equiv \sum_{\mathrm{Bosons}} M^2 - \sum_{\mathrm{Fermion}} M^2 = 0 \tag{108}$$

This is in contrast to the minimally coupled models of section (1.8) in which $\mathrm{Str}M^2 = Nm_{3/2}{}^2$ where N is the number of chiral supermultiplets. From this we can see that these models will not generate a common mass of $0(m_{3/2})^2$. Indeed, if we impose the constraints that the potential should vanish at the minimum, ensuring zero cosmological constant, and that supersymmetry is spontaneously broken, then there will always be a massless scalar field at tree level. One can use the non-linear transformations of the coset space to transform all scalar fields to have zero vacuum expectation value. Then the form of the superpotential is uniquely fixed to be

$$g(y, Z) = \mu(1 + \frac{|y|^2}{\sqrt{3}})^3 + 0(Z^2) \tag{109}$$

where y is defined as the direction of nonvanishing gradient of g and Z refers to the remaining scalars. The potential is given by eq. (84) and using the form of eq. (108) one finds[15]

$$V(y, Z = 0) \equiv 0 \tag{110}$$

This means the field y is massless and its vev is undetermined. The gravitino mass is given by

$$m_{3/2} = \mu(\frac{1 + a}{\sqrt{1-|a|^2}}) \qquad |a| < 1 \tag{111}$$

where y has a vev $y = \sqrt{3}\ a$. Thus both the vev of y and the gravitino mass are undetermined at tree level, a consequence of the U(n, 1) invariance of the kinetic energy.

Beyond tree level the degeneracy is lifted and $m_{3/2}$ is determined. In certain models[43,44] the form of the radiative corrections can give scalar masses of the form

$$m^2 = p + q\ \ell n(1 + a) \tag{112}$$

where p and q involve various couplings in the model. The minimum of the potential may favour m^2 negative, and this requires $(1 + a) < e^{-p/q}$. Thus (1 + a)must be exponentially small, giving $m_{3/2}$ in eq. (111) exponentially smaller than the scale μ, which has a natural value of $0(M_{Planck})$. In this way one can build models with a hierarchy of mass scales determined by exponentially small factors whose exponents are related to ratios of the couplings of the theory.

(We may absorb the vev[44] of y using a U(n, 1)/U(n) x U(1) boost

$$\bar{y} = \frac{y - \sqrt{3}a}{1 - \frac{a^* y}{\sqrt{3}}} \quad ; \quad \bar{Z}^i = \frac{Z^i\sqrt{3 - |a|^2}}{1 - \frac{a^* y}{\sqrt{3}}} \tag{113}$$

Then the form of the superpotential in given by eq. (109).)

To summarise, locally supersymmetric theories can solve the problem of mass scales and we can build theories with only one fundamental mass scale M_{Planck} in which both the electroweak breaking scale M_W and the supersymmetry breaking scale $m_{3/2}$ are given in terms of M_{Planck} times naturally small factors.

However, these models do not shed any light on the most severe hierarchy problem of all - that of ensuring the cosmological constant is small. If supersymmetry were exact the cosmological constant in any supersymmetric theory would be zero (cf. the result in the globally supersymmetric case). However, we know that supersymmetry is broken so we expect the cosmological constant to be non-zero at the scale of supersymmetry breaking. Unfortunately this is 10^{50} times too large! Moreover the problem seems to go beyond perturbation theory for the strong interaction chiral phase transition which occurs at several hundred Mev would be expected to give a contribution 10^{45} times too large. It remains to be seen whether these problems can be solved within the framework of spontaneously broken supersymmetric theories, or whether completely new structure is required.

4. SUMMARY

Much progress has been made constructing realistic low-energy supersymmetric models. In the main these have used the minimal N=1 supersymmetry, for only this has the complex representations needed to accomodate the known quarks and leptons without the introduction of mirror fermions. These models have the merit of being able to generate a heirarchy of mass scales in which the electroweak breaking scale is much lower than the Planck scale or the grand-unified breaking scale. They all require a rich structure of new states, supersymmetric partners of the presently known particles and, in order to explain the hierarchy of mass scales, these states must be relatively light, $\lesssim 0$ (1 TeV). However, model calculations show that the natural expectation is that these states be heavier than their conventional partners and that non-observation to date is not an indication that these models are wrong. The new states should, however, be accessible to the next generation of experiments; their production and decay mechanisms are determined and, once the threshold for their production of passed, they should be easily visible.

Models have been built which can relate the weak breaking scale to the supersymmetry breaking scale and that, in turn, to the Planck scale. They need only a single dimensionful constant to specify the theory, and do not require any fine-tuning of couplings to preserve the hierarchy of mass scales. Howeve the supersymmetric models do not shed any light on the cosmological problem of ensuring the cosmological constant be small, in supergravity theories this is achieved by an unnatural adjustment of parameters.

Supersymmetric theories have many beautiful and appealing features which have encouraged theorists to use them to model the low-energy world. In the end, however, we must rely on experiment to decide whether these low-energy models are relevant.

References

(1) For a discussion of superspace, and the rules for forming supersymmetric Lagrangians see:
P. Fayet and S. Ferrara, Phys. Rep. 32C (1977) 259
J. Wess and J. Bagger, Supersymmetry and supergravity, Princeton University, New Jersey (1983)
S.J. Gates, M.T. Grisaru, M. Rocek and W. Siegel, Superspace or One Thousand and One Lessons in Supersymmetry (Benjamin Cummings, 1983)

(2) For other reviews see:
R. Barbieri and S. Ferrara, Surveys in High Energy Physics 4(1983) 33
H.P.A. Nilles, Phys. Rep. 110 (1984) 1
P. Fayet, in Supersymmetry and Supergravity '84, Proceedings of the Trieste Spring School, World Scientific, p.114

(3) The phenomenological aspects of the supersymmetric standard model are covered in H. Haber and G.L. Kane, Phys. Rep. (to be published).
J. Ellis, SUpersymmetric GUTs Lectures presented at the Advanced Study Institute on Quarks, Leptons and Beyond, Munich CERN preprint Ref. TH 3802-CERN: see also Cornell Conference proceedings 1983
P. Fayet, Proceedings of the XVIIthe Rencontre de Moriond, Les Arcs (France 1982) 483

(4) S. Weinberg, Phys. Rev. D26 (1982)

(5) L.J. Hall, M. Suzuki, Nucl. Phys. B231, 419 (1984)
M. Bowick, M. Chase and P. Ramond, Phys. Lett. 128B (1983) 185
I.H. Lee, Phys. Lett. 138B (1984) 121

(6) P. West, Nucl. Phys. B106 (1976) 219
M. Grisaru, M. Rocek and W. Siegal, Nucl. Phys. B159 (1979) 429

(7) For an introduction to R symmetries see also Fayet and Ferrara, ref. 2

(8) G.G. Ross and J.W.F. Valle, Rutherford Appleton Laboratory preprint RAL-84-102
J. Ellis, G. Gelmini, C. Jarlskog, G.G. Ross and J.W.F. Valle, Rutherford Appleton Lab. preprint RAL-84-085

(9) R. Barbieri and R. Gatto, Phys. Lett. 110B (1982) 211
M.J. Duncan, Nucl. Phys. B224 (1983) 289; ibid B221 (1983) 285
M. Suzuki, Phys. Lett. 115B (1982) 40

(10) J. Donoghue, H. Nilles and D. Wyler, Phys. Lett. 135B (1984) 423
A. Bouquet, J. Kaplan and C.A. Savoy, Univ. of Paris preprint PAR LPTHE 8425 (1984)
J.P. Derendinger and C.A. Savoy, Nucl. Phys. B237 (1984) 307

(11) G.G. Ross and S. Wilkinson, Oxford preprint (in preparation)

(12) R. Barbieri, L. Girardello and A. Masiero, Phys. Lett. 127B (1983)429
W. Buchmuller and D. Wyler, Phys. Lett. 121B (1983) 321
R. Arnowitt, A.H. Chamseddine and P. Nath, Phys. Lett. 50 (1983) 232
S. Weinberg, ibid, 50 (1983) 387
D.V. Nanopoulos and M. Srednicki, Phys. Lett. 128B (1983) 61
F. Del Aguila, M. Gavela, J. Grifols and A. Mendez, Phys. Lett. 126B (1983) 71

(13) J. Ellis and D.V. Nanopoulos, Phys. Lett. 110B (1982) 44
T. Inami and C.S. Lim, Nucl. Phys. B207 (1982)

(14) P. Fayet and J. Iliopoulos, Phys. Lett. 51B (1974) 461
L. O'Raifeartaigh, Nucl. Phys. B96 (1975) 331
P. Fayet, Phys. Lett. 58B (1975) 67

(15) E. Cremmer, B. Julia, J. Scherk, P. van Nieuwenhuizen, S. Ferrara and L. Giradello, Phys. Lett. 79B (1978) 231; Nucl. Phys. B147 (1979)105
E. Cremmer, S. Ferrara, L. Giradello and A. van Proven, Phys. Lett. 116B (1982) 231; Nucl. Phys. B212 (1983) 413
J. Bagger and E. Witten, Phys. Lett. 115B (1982) 202

(16) L. Ibanez, Phys. Lett. 118B (1982) 73
R. Barbieri, S. Ferrara and C.A. Savoy, Phys. Lett. 119B (1982) 343
A. Chamseddine, P. Nath, and R. Arnowitt, Phys. Rev. Lett. 49 (1982)970
P. Nath, R. Arnovitt and A.P. Chamseddine, Phys. Lett. 121B (1983)33
J. Ellis, D.V. Nanopoulos and K. Tamvakis, Phys. Lett. 121B (1983)123
L. Hall, J. Lykken and S. Weinberg, Phys. Rev. D27 (1983) 2359

(17) H.P. Nilles, M. Srednicki and D. Wyler, Phys. Lett. 120B (1982) 346

(18) J. Polonyi, Univ. of Budapest report No. KFKI-1977-93 (1977)

(19) J.C. Pati and A. Salam, Phys. Rev. D8 (1973) 1246
H. Georgi and S.L. Glashow, Phys. Rev. Lett. 32(1974) 438

(20) E. Gildener, Phys. Rev. D14 (1976) 1667; Phys. Lett. 92B (1980) 111
E. Gildener and S. Weinberg, Phys. Rev. D15 (1976) 3333
G. 't Hooft, Proceedings of the advanced study institute, Cargese 1979
Eds. G. 't Hooft et al. (Plenum Press NY 1980)

(21) N. Sakai, Z. Phys. (1981) 153
S. Dimopoulos and H. Georgi, Nucl. Phys. B193 (1981) 150

(22) H. Georgi, H.R. Quinn and S. Weinberg, Phys. Rev. Lett. 33 (1974) 451
J. Ellis, M.K. Gaillard, D.V. Nanopoulos and S. Rudaz, Nucl. Phys. B176 (1980) 61
T. Goldman and D.A. Ross, Phys. Lett. 843 (1979) 208
C.H. Llewellyn Smith, G.G. Ross and J. Wheater, Nucl. Phys. B177 (1981) 263
S. Weinberg, Phys. Lett. 91B (1980) 51
I. Hall, Nucl. Phys. B178 (1981) 75
P. Binetruy and T. Schucker, Nucl. Phys. B178 (1981) 293

(23) L. Ibanez and G.G. Ross, Phys. Lett. 105B (1981) 439
J. Ellis, D.V. Nanopoulos and S. Rudaz, Nucl. Phys. B202 (1982) 43

(24) M.B. Einhorn and D.R.T. Jones, Nucl. Phys. B196 (1982) 475
W. Marciano and G. Senjanovich, Phys. Rev. D25 (1982) 3092

(25) D. Nanopoulos and D.A. Ross, Phys. Lett. 118B (1982) 99

(26) For reviews see P. Langacker, Phys. Reports 72 (1981) 185
W. Marciano, BNL preprint 33415 (1983)

(27) N. Sakai and T. Yanagida, Nucl. Phys. B197 (1982) 83
S. Dimopoulos, S. Ruby and F. Wilczek, Phys. Lett. 112B (1982) 133

(28) J. Ellis et al ref.(21)
P. Salati and C. Wallet, Nucl. Phys. B209 (1982) 389
J. Ellis, J.S. Hagelin, D.V. Nanopoulos and K. Tamvakis, Phys. Lett. 124B (1983) 464
S. Chadha and M. Daniel, Nucl. Phys. B229 (1983) 105

(29) S. Chadha, G. Coughlan, M. Daniel and G.G. Ross, Physics Lett. 1984

(30) J. Ellis, ref.(3); G.G. Ross, "Grand Unified Theories", The Benjamin Cummings Publishing Co. (1984); Proceedings of the XVII International Conference on ν physics, Noordkirching (1984)

(31) S. Dimopoulos and S. Raby, Los Alamos report (1982)
E. Witten, Phys. Lett. 105B (1981) 267
J. Polchinski and L. Susskind, Phys. Rev. 26D (1982) 3661

(32) J. Ellis, L.E. Ibañez and G.G. Ross, Phys. Lett. 113B (1982) 283; Nucl. Phys. B221 (1983) 29
L.E. Ibañez and G.G. Ross, Phys. Lett. 110B (1982) 215

(33) See Georgi and Dimopoulos (ref.(20)), also Nilles (ref.(2))

(34) E. Witten, ref.(30)
L.E. Ibañez and G.G. Ross, Phys. Lett. 105B (1981) 439

(35) L. Alvarez-Gaume, J. Polchinski and M. Wise, Nucl. Phys. B221 (1983) 495
H.P. Nilles, M. Svednicki and D. Wyler, Phys. Lett. 124B (1983) 337
A.B. Lahanas, Phys. Lett. 124B (1983) 341
J. Polchinski and L. Susskind, Phys. Rev. D26 (1982) 3661
N. Dragon, N. Marinesen and M.G. Schmidt, Z. Phys. C21 (1984) 383
U. Ellwanger, Phys. Lett. 133B (1983) 187

(36) B. Grinstein, Nucl.Phys. B206 (1982) 387
A. Masiero, D.V. Nanopoulos, K. Tamvakis and T. Yanagida, Phys. Lett. 115B (1982) 380

(37) L.E. Ibañez and G.G. Ross, Phys. Lett. 110B (1982) 215
L. Alvarez- Gaume, M. Claudson and M.B. Wise, Nucl. Phys. B207 (1982) 96
C.R. Nappi and B.A. Ovrut, Phys. Lett. 113B (1982)
K. Inoue, A. Kakuto, H. Komatsu and S. Takeshita, Prog. Theor. Phys. 68 (1982) 927

(38) L.E. Ibanez, Phys. Lett. 118B (1982) 73
J. Ellis, D.V. Nanopoulos and K. Tamvakis, Phys. Lett. 121B (1983)
L. Alvarez-Gaume, J. Polchinski and M.B. Wise, Nucl. Phys. B221 (1983) 495
S.K. Jones and G.G. Ross, Phys. Lett. 135B (1984) 69

(39) S. Coleman and E. Weinberg, Phys. Rev. D7 (1973) 1888

(40) C.J. Oakley and G.G. Ross, Phys. Lett. 125B (1983) 59

(41) L. Ibañez and G.G. Ross, Phys. Lett. 131B (1983) 335
B.A. Ovrut and S. Raby, Phys. Lett. 134B (1984) 51

(42) E. Cremmer, S. Ferrara, C. Kounnas and D.V. Nanopoulos, Phys. Lett. 133B (1983) 61

(43) J. Ellis, A.B. Lahanas, D.V. Nanopoulos and K. Tamvakis, Phys. Lett. 134B (1984) 429
J. Ellis, C. Kounnas and D.V. Nanopoulos, Nucl. Phys. B241 (1984) 406
J. Ellis, C. Kounnas and D.V. Nanopoulos, CERN preprint TH.3824 (1984)

(44) N. Dragon, M.G. Schmidt and U. Ellwanger, CERN preprints TH.3915 (1984) and TH. 3974 (1984)

(45) N. Chang, S. Ouvry and X. Wu, Phys. Rev. Lett. 51 (1983) 327
S. Ouvry, Phys. Lett. 136B (1984) 165
X. Wu, MIT Preprint CTP 1164 (1984)

(46) J. Wess and B. Zumino, Phys. Lett. 79B (1978) 394
B. Ovrut and J. Wess, Phys. Lett. 112B (1982) 346
J. Wess and J. Bagger, Supersymmetry and Supergravity, Princeton University Press (1983)
R. Barbieri, S. Ferrara, D.V. Nanopoulos and K. Stelle, Phys. Lett. 113B (1982) 219

SUPERSYMMETRY AND COSMOLOGY

S RABY

OUTLINE

1. REVIEW OF MINIMAL LOW ENERGY SUPERGRAVITY (MLES) SPECTRUM.
2. DEFINE R-PARITY AND DISCUSS CANDIDATES FOR LIGHTEST SUPER-SYMMETRIC PARTNER (LSP).
3. REVIEW STANDARD "BIG BANG" COSMOLOGY AND RELATIVISTIC BOLTZMANN EQUATION.
4. COSMOLOGICAL BOUNDS FOR THE MASS OF THE LSP.
5. COSMOLOGICAL BOUNDS FOR THE GRAVITINO.

1. SPECTRUM OF MLES

In Table I we list the states in a MLES model (Alvarez-Gaumé, Polchinski and Wise, 1983; Ellis, Hagelin, Nanopoulos and Tamvakis, 1983; Ibáñez and López, 1983; Ibáñez, 1983). On the left-hand side are all the ordinary particles; three families of quarks and leptons, higgs bosons and gauge bosons of $SU_3 \times SU_2 \times U_1$. In addition, we note that supersymmetry requires two sets of Higgs doublets in order to a) give mass to both up and down quarks, and b) to have no weak hyper-charge anomaly. We also have the graviton and the scalar partners of the goldstino. The right-hand side contains all the super-partners. Experimentally, we know that the slepton mass, $m_{\tilde{\ell}^\pm} \gtrsim 20$ GeV (CELLO Collaboration, Behrend et al., 1982; JADE Collaboration, Bartel et al., 1982; MARK J Collaboration, Barber et al., 1981, TASSO Collaboration, Brandelik et al., 1982; MARK II Collaboration, Blocker et al., 1982), and the gluino has a mass, $m_{\tilde{g}} \geq 5$ GeV (Ball et al., 1983). The fact is that the right-hand side has remained quite invisible; although recently some authors have argued that the jet events at CERN

Table I

$q = \begin{pmatrix} u \\ d \end{pmatrix}$ $\ell = \begin{pmatrix} \nu \\ e \end{pmatrix}$	$\tilde{q} = \begin{pmatrix} \tilde{u} \\ \tilde{d} \end{pmatrix}$ $\tilde{\ell} = \begin{pmatrix} \tilde{\nu} \\ \tilde{e} \end{pmatrix}$
$\bar{u}$ $\bar{e}$	$\tilde{\bar{u}}$ $\tilde{\bar{e}}$
$\bar{d}$	$\tilde{\bar{d}}$
quarks and leptons	**squarks and sleptons**
scalar partner of goldstino -graviton	**G goldstino -gravitino**
$H_1 = \begin{pmatrix} H^+ \\ H^0 \end{pmatrix}$ $\bar{H}_2 = \begin{pmatrix} \bar{H}^- \\ \bar{H}^0 \end{pmatrix}$	$\tilde{H}_1 = \begin{pmatrix} \tilde{H}^+ \\ \tilde{H}^0 \end{pmatrix}$ $\tilde{\bar{H}}_2 = \begin{pmatrix} \tilde{\bar{H}}^- \\ \tilde{\bar{H}}^0 \end{pmatrix}$
Higgs bosons	**Higgsinos**
$g, \gamma, Z_0, W^{\pm}$	$\tilde{g}, \tilde{W}_3, \tilde{B}, \tilde{W}^{\pm}$
gauge bosons	**guaginos**

may indicate squarks or gluinos with mass of order 40 GeV (Ellis and Kowalski, 1984a; 1984b).

2. R-PARITY AND THE LSP

Farrar and Fayet (1978a; 1978b) have defined a discrete symmetry, called R-parity, which distinguishes the ordinary particles from their super-partners. Consider a chiral superfield, $\phi(x,\theta)$. R-parity is defined by the transformation

$$\phi'(x,\theta) = \eta\phi(x,-\theta) \quad , \tag{II.1}$$

which for the components ϕ, ψ, F gives

$$\phi'(x) = \eta\phi(x) \quad ,$$
$$\psi'(x) = -\eta\psi(x) \quad ,$$
$$F'(x) = \eta F(x) \quad . \tag{II.2}$$

Thus the R-charge of a particle with spin s is

$$(-1)^{2s}\eta \quad . \tag{II.3}$$

For gauge fields $V(x,\theta,\bar{\theta})$, we have

$$V'(x,\theta,\bar{\theta}) = V(x,-\theta,-\bar{\theta}) \quad , \tag{II.4}$$

which for the components χ, A_μ, D gives

$$\chi'(x) = -\chi(x) \quad ,$$
$$A'_\mu(x) = A_\mu(x) \quad ,$$
$$D'(x) = D(x) \quad . \tag{II.5}$$

The Lagrangian for the MLES model is given (schematically) by the expression

$$\mathcal{L}_{MLES} \sim \int d^4\theta\phi_i^*(x,\theta)e^{-gV(x,\theta,\bar{\theta})}\phi_i(x,\theta)(1 + m_G^2\theta^2\bar{\theta}^2)$$
$$+ \{ \int d^2\theta[mH_1\bar{H}_2 + gH_1Q\bar{U} + \bar{g}\bar{H}_2Q\bar{D} + \bar{g}'\bar{H}_2L\bar{E}](1 + m_G\theta^2) + h.c.\}$$
$$+ \{ \int d^2\theta W^\alpha W_\alpha(1 + m_G\theta^2) + h.c.\} \quad , \tag{II.6}$$

where i labels the chiral superfield Q, $\bar{U}$, etc. and $W_\alpha \equiv \bar{D}^2[e^{gV}D_\alpha e^{-gV}]$ is the gauge field strength. $\mathcal{L}_{MLES}$ is invariant under the R-parity transformation with $\eta = -1$ for the fields $Q,\bar{U},\bar{D},L,\bar{E}$; $\eta = +1$ otherwise. Note that, in Table I, all fields on the left-hand side (right-hand side) are R even (odd). Thus, the MLES model has an exact R-parity. It was noted by Farrar and Weinberg (1983) that the generator of the R-parity defined above is equivalent to the operator

$$R \equiv (-1)^{3(B-L)}(-1)^F \quad , \tag{II.7}$$

where B is baryon number, L is lepton number, and F is fermion number. This observation makes it clear that if (B-L) and F are conserved, then R-parity is an exact symmetry.

In the following text, we shall assume that R-parity is conserved. We shall not consider the possibility of explicitly breaking R-parity with operators of the form

$$\int d^2\theta H_1 L \quad , \tag{II.8}$$

or spontaneously broken R-parity via a non-vanishing $\tilde{\nu}$ expectation value. If R-parity is conserved, then the Lightest Supersymmetric Partner (LSP) is absolutely stable. We are then obliged to ascertain the present expected abundance of the LSP (using standard early universe assumptions) and see if it is consistent with observations. In particular, we shall obtain limits on the mass of the LSP by requiring that its energy density ρ_{LSP} is less than the cosmological critical energy density ρ_c. Before we do this, however, let us discuss the possible candidates for the LSP.

What are the likely candidates for the LSP? We shall only consider electrically neutral, color singlet states. The reason is that a) since they have less interactions, they are probably lighter (for example, if gluinos and photinos are degenerate at $M_{p\ell}$, then one loop renormalization group corrections give the ratio $m_{\tilde{g}}:m_{\tilde{\gamma}} = \alpha_3:\alpha$ at the weak scale); and b) if they are charged, then they have already been ruled out. Wolfram (1979) has calculated the expected abundance of stable heavy leptons (n_L) or hadrons (n_H). He finds

$$n_L \sim 10^{-5} n_b \left(\frac{m_L}{1\ \text{GeV}}\right) ,$$

$$n_H \sim 10^{-10} n_b \left(\frac{m_H}{1\ \text{GeV}}\right) , \tag{II.9}$$

where n_b is the observed baryon number density and $m_L(m_H)$ is the mass of the heavy lepton (hadron). A search has been made for such stable particles (Smith and Bennett, 1979) with the result that for a mass $m < 350$ GeV, the observed density is less than $10^{-21} n_b$. Hence, a charged LSP is not consistent with observation.

We shall henceforth consider the following LSP candidates:

1. photino-Higgsino,
2. sneutrino (scalar neutrino),
3. gravitino. (II.10)

3. REVIEW STANDARD "BIG BANG" COSMOLOGY

We shall use units such that $\hbar = c = k = 1$. Newton's constant G_N defines

$M_{p\ell} \equiv G_N^{-1/2} = 1.9 \times 10^{19}$ GeV, or sometimes $M \equiv \frac{M_{p\ell}}{\sqrt{8\pi}} = 2.4 \times 10^{18}$ GeV.

We assume space-time is described, on large scales, by the Friedmann-Robertson-Walker metric

$$ds^2 = -dt^2 + a^2(t)\left[\frac{dr^2}{1-kr^2} + r^2 d\Omega^2\right], \qquad \text{(III.1)}$$

where $d\Omega^2 \equiv d\theta^2 + \sin^2\theta\, d\phi^2$, and the parameter k determines the spatial topology of the universe: k = +1, closed; k = 0, open and flat; k = -1, open and curved. a is the cosmological scale parameter. Recall that the FRW metric is uniquely determined by requiring space to be both homogeneous and isotropic.

The dynamical equations of cosmology are given by:

Einstein's equation

$$G_{\mu\nu} = \frac{1}{M^2}T_{\mu\nu} \quad , \quad T^{\mu\nu}{}_{;\nu} = 0 \quad , \qquad \text{(III.2)}$$

which implies $\left(\dot{a} \equiv \frac{da}{dt} \quad , \quad \dot{\rho} \equiv \frac{d\rho}{dt}\right)$

$$H^2(t) \equiv \left(\frac{\dot{a}}{a}\right)^2 = \frac{\rho}{3M^2} - \frac{k}{a^2} \quad , \qquad \text{(III.3a)}$$

$$\dot{\rho} = -3(\rho + p)H \quad , \qquad \text{(III.3b)}$$

where

$$T_{\mu\nu} \equiv (\rho + p)u_\mu u_\nu + pg_{\mu\nu} \quad ,$$

$$u^\mu \equiv \frac{dx^\mu}{d\tau} = (1,\vec{0}) \quad . \tag{III.4}$$

ρ,p are the energy, pressure densities, respectively, for a perfect fluid, u^μ is the fluid velocity field 4-vector, and H(t) is the Hubble expansion rate. The geodesic equations of motion for a particle (or fluid element) is given by

$$\frac{du^\gamma}{d\tau} + \Gamma^\gamma_{\mu\nu}u^\mu u^\nu = 0 \quad . \tag{III.5}$$

In a FRW universe, we find

$$\frac{du^\gamma}{d\tau} = 0 \quad \text{for } u^\gamma = (1,\vec{0}) \quad . \tag{III.6}$$

Thus, the FRW coordinate position r,θ,ϕ in Eq. (III.1) describes a coordinate frame which is co-moving with the fluid elements. The distance between two such elements is then increasing with the scale factor a(t).

The dynamical equations (III.3) must be supplemented with an equation of state, $p = p(\rho)$. We shall consider three forms:

1) radiation;

$$p_r = \frac{1}{3}\rho_r \quad , \tag{III.7}$$

(using III.3b implies)

$$\rho_r \sim \frac{1}{a^4} \quad .$$

2) non-interacting conserved massive particles, i.e., matter;

$$p_m = 0 \quad , \tag{III.8}$$

(using III.3b implies)

$$\rho_m \sim \frac{1}{a^3} \quad .$$

3) cosmological constant;

$$p_\Lambda = -\rho_\Lambda \, . \tag{III.9}$$

(using III.3b implies)

$$\rho_\Lambda \equiv \text{constant} \quad .$$

The cosmological constant Λ is typically defined by

$$\Lambda \equiv \frac{\rho_\Lambda}{3M^2} \, . \tag{III.10}$$

We now have a complete system of equations (III.3,7-9).

Let us now discuss some observed properties of the present universe which shall be useful later.

1. The present value of the Hubble expansion parameter is

$$H_0 = h100\text{km/s/Mpc} \quad , \tag{III.11}$$

with $1/2 \leq h \leq 1$. (1pc = 3.26 light years, 1Mpc $\simeq 3 \times 10^{24}$cm). H_0 determines the observed red shift, $Z \equiv \frac{\lambda_{observed} - \lambda_{emitted}}{\lambda_{emitted}} \left(= \frac{a(t_0)}{a(t_e)} - 1 \right)$ of distant stars using the relation

$$d = H_0^{-1}\left[Z + \frac{1}{2}(1 - q_0)Z^2 + 0(Z^3)\right] \quad , \tag{III.12}$$

where d is the present distance to the star and q_0 is the present value of the decelleration parameter $\left[q_0 \equiv - \frac{\ddot{a}a}{\dot{a}^2} \bigg|_{today} \leq 1 \right]$.

2. We define a critical enery density

$$\rho_c(t) \equiv 3H^2(t)M^2 \quad . \tag{III.13}$$

Using (III.3a) we see that for

$$\rho \begin{Bmatrix} > \\ = \\ < \end{Bmatrix} \rho_c \quad , \; k = \begin{Bmatrix} 1 \\ 0 \\ -1 \end{Bmatrix} \quad . \tag{III.14}$$

Hence, ρ_c is the boundary between an open or closed universe. Using H_0 (III.11) we find today

$$\rho_c^0 \simeq 2 \times 10^{-29} g/cm^3 h^2$$

$$\simeq (3 \times 10^{-3} eV)^4 h^2 \quad . \tag{III.15}$$

3. The observed energy density in 2.7°k black body radiation is

$$\rho_r^0 = 4.5 \times 10^{-34} g/cm^3 \quad . \tag{III.16}$$

Cosmologists typically define the ratio, for matter of type i,

$$\Omega_i \equiv \frac{\rho_i}{\rho_c} \quad . \tag{III.17}$$

Thus

$$\Omega_r^0 \equiv 2.25 \times 10^{-5} h^{-2} \quad . \tag{III.18}$$

For luminous matter ("baryons") a typical value is

$$\Omega_b^0 \simeq .02 \quad . \tag{III.19}$$

However, the dominant form of energy in the universe is not luminous. This "dark matter" contributes an amount

$$\Omega_{dm}^o \simeq .2 \quad . \tag{III.20}$$

Note that the energy in the universe is apparently matter dominated today. (III.19, 20) are subject to considerable debate (for a review see Blumenthal, Faber, Primack and Rees, 1984). Note that for $p_0 \ll \rho_0$, the decelleration parameter satisfies

$$q_0 \equiv \frac{1}{2} \frac{\rho_{total}^0}{\rho_c^0} \equiv \frac{1}{2} \Omega_{total}^0 \lesssim 1 \quad , \tag{III.21}$$

which implies

$$\Omega^0_{total} \leq 2 \quad . \tag{III.22}$$

This is the fundamental cosmological constraint which shall be imposed on model building. As a first example, if we require

$$\rho^0_\Lambda < \rho^0_c \quad , \tag{III.23}$$

we find

$$\Lambda < 10^{-120} M^2 \quad . \tag{III.24}$$

Hence the cosmological constant is one of the most finely-tuned parameters in all physics.

4. The ratio of the number of baryons to photons is

$$\left.\frac{n_b}{n_\gamma}\right|_0 \simeq 10^{-9} \quad , \tag{III.25}$$

where $n^0_\gamma \simeq 400/\mathrm{cm}^3$, $n^0_b \simeq \rho^0_b/1\ \mathrm{GeV}$.

As a result, the thermodynamic variables, pressure, and entropy are radiation dominated.

5. Since ρ is matter dominated, we have, using (III.3,8)

$$\left(\frac{\dot{a}}{a}\right)^2 \sim \frac{\Omega^0_m \rho^0_c}{3M^2}\left(\frac{a_0}{a}\right)^3 \quad , \tag{III.26}$$

where a_0 is the present value of $a(t)$, and we neglected k. The solution is

$$a(t) \sim t^{2/3} \quad . \tag{III.27}$$

Thus, $H_0 = 2/(3t_0)$, and the age of the universe t_0 is determined by the Hubble constant

$$t_0 \sim H_0^{-1} \sim 10^{10} \mathrm{years} \quad . \tag{III.28}$$

This calculation is valid for $\rho_m > \rho_r$ and $a_0/a > \Omega_0^{-1} - 1$. The second constraint has probably been satisfied until perhaps very recently.

Let us now determine how long the universe has been matter dominated. We define the time, t_{eq}, and the scale factor, a_{eq}, such that ρ_m equals ρ_r. We find

$$\frac{a_0}{a_{eq}} \sim 10^4 \quad , \text{ and } t_{eq} \sim 10^4 \text{ years} \quad . \tag{III.29}$$

Hence, the universe has been matter dominated throughout most of its history.

Relevant statistical mechanics (Kolb and Wolfram, 1980)

For a particle in thermal equilibrium, the phase space density $f(\vec{p})$ is given by

$$f(\vec{p}) \equiv \frac{dN}{d^3\vec{p}d^3\vec{x}} = \frac{N_h}{(2\pi)^3} \frac{1}{\exp\left(\frac{E-\mu}{T}\right) + \theta} \quad , \tag{III.30}$$

where $\theta = \begin{Bmatrix} +1 & \text{fermions} \\ -1 & \text{bosons} \end{Bmatrix}$, N_h is the number of helicity states,

$E \equiv \sqrt{\vec{p}^2 + m^2}$, T is the temperature and μ is the chemical potential.

The number density n and energy density ρ are then given by

$$\begin{aligned} n &= \int d^3\vec{p} f(\vec{p}) \quad , \\ \rho &= \int d^3\vec{p} E(\vec{p}) f(\vec{p}) \equiv n\langle E\rangle \quad , \end{aligned} \tag{III.31}$$

where brackets, <>, denotes thermal average.

For T >> m, we obtain the number and energy density appropriate for "radiation".

$$n_r = \frac{g'\zeta(3)T^3}{\pi^2} \quad ,$$

$$T \gg m \tag{III.32}$$

$$\rho_r = g\,\frac{\pi^2}{30}\,T^4 \quad .$$

$g' = N_{hB} + 3/4\, N_{hF}$, $g = N_{hB} + 7/8\, N_{hF}$ are weighted sums of boson (N_{hB}) and fermion (N_{hF}) helicities and $\zeta(3) = \sum_{i=1}^{3} i^{-3} \sim 1.2$. Note that the thermal averaged photon energy is

$$\langle E\gamma\rangle \simeq 2.6T \quad . \tag{III.33}$$

For $T \ll m$, we obtain densities appropriate to non-relativistic "matter".

$$n_m = N_h \left(\frac{mT}{2\pi}\right)^{3/2} e^{-m/T} \quad ,$$

$$\begin{matrix} T \ll m \\ \mu = 0 \end{matrix} \tag{III.34}$$

$$\rho_m = mn_m + \frac{3}{2} n_m T \quad , \; p_m = n_m T \quad ,$$

where p_m is the pressure density. As noted previously, $p_m/p_r|_0 \sim 10^{-9}$. Comparing Eqs. (III.7) and (III.32) we see that

$$T \sim \frac{1}{a} \quad , \tag{III.35}$$

for radiation in thermal equilibrium. For matter, on the other hand, $T \sim 1/a$ only if $\mu \neq 0$, or it is out of thermal equilibrium (see III.8 and III.34). In thermal equilibrium, n_m is Boltzmann suppressed ($e^{-m/T}$) and thus decreases much faster than $1/a$. It is not always possible, however, for the reaction rates, which govern the approach to equilibrium, to keep up with the expansion and the system goes out of equilibrium. Let us, therefore, briefly discuss the approach to thermal equilibrium.

Consider two species of particles, x (which is out of equilibrium), and r (in thermal equilibrium). We want to know how x approaches equilibrium (Kolb and Wolfrum, 1980).

The relativistic version of Boltzmann's equation is given by

$$\frac{dn_x}{dt} + 3\frac{\dot{a}}{a}n_x = \Lambda^{x}_{r_1r_2}[f_{r_1}f_{r_2}(1-\theta_x f_x)|\mathcal{M}(r_1r_2\to x)|^2$$

$$- f_x(1-\theta_r f_{r_1})(1-\theta_r f_{r_2})|\mathcal{M}(x\to r_1r_2)|^2]$$

$$+ 2\Lambda^{34}_{12}[f_{r_1}f_{r_2}(1-\theta_x f_{x_3})(1-\theta_x f_{x_4})|\mathcal{M}(r_1r_2\to x_3x_4)|^2$$

$$- f_{x_1}f_{x_2}(1-\theta_r f_{r_3})(1-\theta_r f_{r_4})|\mathcal{M}(x_1x_2\to r_3r_4)|^2] \quad , \tag{III.36}$$

where the phase space factor Λ is defined by

$$\Lambda^{b_1b_2}_{a_1a_2} \equiv \int \frac{d^4p_{a_1}}{(2\pi)^3}\frac{d^4p_{a_2}}{(2\pi)^3}\frac{d^4p_{b_1}}{(2\pi)^3}\frac{d^4p_{b_2}}{(2\pi)^3}\,\delta^4(p^2_{a_i} - m^2_{a_i})$$

$$\times\ \delta^4(p^2_{b_i} - m_{b_i})(2\pi)^4\delta^4(\sum_i p_{a_i} - \sum_i p_{b_i}) \quad ,$$

$\theta_{r,x}$ is defined in Eq. (III.30) and $\mathcal{M}$ is the scattering amplitude for the indicated process. The four terms in Eq. (III.36) represent the four processes in fig. 1, respectively. If we assume that the interactions are both CPT and CP invariant, use the fact that f_r is an equilibrium distribution and assume that the gas is non-degenerate, we obtain

$$\frac{dn_x}{dt} + 3\frac{\dot{a}}{a}n_x = (n^{eq}_x - n_x)\langle\Gamma_{x\to r_1r_2}\rangle$$

$$+ 2[(n^{eq}_x)^2 - (n_x)^2]\langle v\sigma_{xx\to rr}\rangle \quad . \tag{III.37}$$

n^{eq}_x is the thermal equilibrium number density. Note that equilibrium is achieved <u>if</u> $H \equiv \dot{a}/a \ll \langle\Gamma\rangle$ or $\langle\sigma v n\rangle$, i.e., the expansion rate is less than a typical reaction rate. Otherwise we find $n_x \sim 1/a^3$, as in Eq. (III.8).

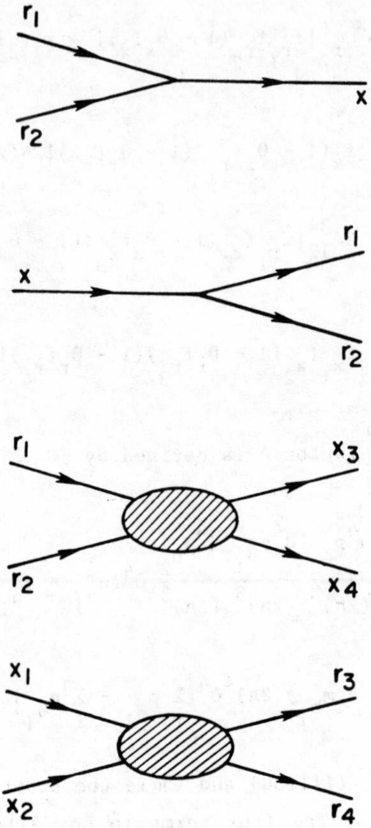

Fig. I

We now have the necessary tools to review the standard picture of the early universe. In Tables II and III we have listed the major events. We shall assume that all particles are in thermal equilibrium at the Planck temperature. As the universe expands it goes through several phase transitions listed in Table III. There may be an inflationary phase. Typically during this transition the universe cools adiabatically and then reheats to approximately the initial transition temperature. We reset our clocks at that time and then continue as before. We shall not discuss inflation more than to say that in such a transition a huge amount of entropy, of order 10^{84}, is produced.

By the time the universe gets down to a temperature of order 100 MeV, particles with mass greater than this temperature are severely Boltzmann suppressed or are out of equilibrium. At about 100 MeV,

Table II

t(s)	T(GeV)	comments
10^{-44}	10^{19}	"initial conditions" assume all particles in thermal equilibrium.
10^{-4}	10^{-1}	$\mu^+\mu^-$ annihilate: $e^{-m_\mu/T}$
10^{-2}	10^{-2}	ν_μ decouple: $\langle \sigma v n \rangle < H$ $\frac{n_n}{n_n + n_p} : \frac{1}{2} \rightarrow \frac{1}{6} : e^{-(m_n - m_p)/T}$
	2×10^{-3}	ν_e decouple : $\langle \sigma v n \rangle < H$
4	$1/2 \times 10^{-3}$	e^+e^- annihilate : $e^{-m_e/T}$ $(4(7/8) + 2)\, T_\nu^3 = 2T_\gamma^3$ (entropy conservation) $\rightarrow T_\nu = (4/11)^{1/3} T_\gamma$
100	10^{-4}	Helium synthesis $p + n \leftrightarrow d + \gamma$ $d + d \leftrightarrow H_e^3 + n \leftrightarrow H^3 + p$ $H^3 + d \leftrightarrow H_e^4 + n$ all free neutrons depleted $\rightarrow H_e^4$ thermal bath includes $p, H_e^4, d, \gamma, \nu, e$ ratio $d/H_e^4 \sim 10^{-5}$
$10^{12} = 3 \times 10^4 y$	$10^{-9} \Omega_m h^2$	$T = T_{eq}$, $\rho_m = \rho_r \simeq 1.68\, \rho_\gamma$
	$1/3 \times 10^{-9}$	$T = T_{recombination}$, $p + e^- \rightarrow H$ photons decouple

muons begin to annihilate (Weinberg, 1972) (see Table II). Neutrinos will soon decouple, leaving an asymptotic ratio of neutrons to neutrons plus protons of about 1/6. This will be the initial ratio for the subsequent process of helium synthesis, (at the temperature, $T_{He} \sim .1$ MeV), where all the remaining neutrons will be incorporated into helium nuclei. At a temperature $T \sim 1/2$ MeV, electrons annihilate and heat up the photons but not the neutrinos, since they have

Table III

t(s)	T(GeV)	comments
10^{-38}	10^{16}	GUT scale
10^{-30}	10^{12}	inflation
		reset clocks
10^{-28}	10^{11}	SUSY breaking scale ↑
10^{-18}	10^{6}	SUSY breaking scale ↓
10^{-10}	10^{2}	weak breaking
		$g \gtrsim 100$
		quark-gluon plasma ↑
10^{-5}	1/3	QCD phase transition
		↓ Hadrons

already decoupled. Using entropy conservation we find that neutrinos are thus colder than photons,

$$T_\nu = \left(\frac{4}{11}\right)^{1/3} T_\gamma \quad . \tag{III.38}$$

Throughout the epoch from $T_{p\ell}$ to $T_{eq} \sim \Omega_m h^2$ eV, the universe is radiation dominated (i.e., the first 3×10^4 years). Thereafter, the energy density is matter dominated. At about $T \sim 1/3$ eV, free protons and electrons combine and form neutral Hydrogen. This is the so-called recombination temperature. Once this occurs, photons decouple. These photons continue to red-shift for the next $\sim 10^{10}$ years. They are presently observed as the 2.7°k, microwave background radiation.

4. COSMOLOGICAL BOUNDS ON PARTICLE MASSES

We are now ready to obtain the bounds, on particle masses, coming from cosmology. Before we discuss the bounds on the possible candidates for the LSP (see II.10), we shall first warm up by discussing the standard cosmological bounds on the "neutrino" mass.

Neutrinos

"Neutrinos" are stable, weakly interacting, neutral fermions. We want to calculate the present energy density in neutrinos, ρ_ν^0. We shall require $\rho_\nu^0 < \rho_c^0(\Omega_\nu^0 < 1)$. At early times, neutrinos are in thermal equilibrium with the thermal bath. If they are light enough so as to decouple when they are still relativistic, their annihilation rate is given by

$$\langle\sigma v n\rangle \sim G_F^2 T^5 \quad . \tag{IV.1}$$

Assuming the universe is radiation dominated at the decoupling temperature, the Hubble expansion rate is given by

$$H \sim \frac{T^2}{M} \quad . \tag{IV.2}$$

The decoupling temperature, T_d, is then obtained by requiring (see III.37)

$$\langle\sigma v n\rangle = H \quad . \tag{IV.3}$$

We obtain

$$T_d = 1 \text{ MeV} \quad . \tag{IV.4}$$

At $T \sim 1$ MeV, the universe is radiation dominated, hence (IV.2) is valid. Moreover, if we require $m_\nu < 1$ MeV then the approximation (IV.1) is also valid. When neutrinos decouple we have

$$n_\nu = \frac{3}{4} n_\gamma \quad . \tag{IV.5}$$

After electron annihilation at $T \sim 1/3$ MeV, we have

$$n_\nu = \frac{3}{4} \cdot \frac{4}{11} n_\gamma \quad . \tag{IV.6}$$

This relation remains valid to the present, so that today

$$\rho_\nu^0 = (\Sigma m_\nu) n_\nu^0 = (\Sigma m_\nu)\frac{3}{11} n_\gamma^0 . \tag{IV.7}$$

(Σm_ν represents the sum over all neutrino species.) Using $\rho_\gamma^0 = (2.6T_0)n_\gamma^0$, Eq. (III.33), we obtain

$$\rho_\nu^0 = (\Sigma m_\nu)\frac{3}{11}\frac{\rho_\gamma^0}{(2.6T_0)} \leq \rho_c^0 ,$$

or

$$(\Sigma m_\nu) \leq 100 \text{ eV} h^2 \text{ (Cowsik and McClelland, 1972 and 1973).} \tag{IV.8}$$

Before we continue with heavy neutrinos, we note that the discussion of light gravitinos is very similar to that of light neutrinos. The only difference is that G_F is replaced by Newton's constant, G_N in (IV.1). As a result the decoupling temperature for light gravitinos is of order 100 GeV. There are many more states in thermal equilibrium at 100 GeV($g \gtrsim 100$) and thus the effective gravitino temperature is much lower than that of light neutrinos. Pagels and Primack (1982) obtain

$$m_{\tilde{G}} \leq 1 \text{ KeV} , \tag{IV.9}$$

or

$$\Lambda_{ss} \leq 10^6 \text{ GeV}$$

(since $m_{\tilde{G}} \simeq \Lambda_{ss}^2/M$). This is the origin of our lower limit for the supersymmetry breaking scale in Table III.

We now consider neutrinos with mass greater than 1 MeV. (We use the symbol N_0 to denote heavy neutrinos). From the previous analysis, we conclude neutrinos (N_0) are non-relativistic when they decouple. We expect a Boltzmann suppression for heavy neutrinos of order $\exp(-m_{N_0}/T_d)$ and thus for some value of $m_{N_0} > T_d$ we should find a rea-reasonable neutrino energy density. We must however reevaluate the decoupling temperature, using the correct annihilation rate for a non-relativistic process.

Following Lee and Weinberg (1977), we use

$$\langle\sigma v\rangle \sim G_F^2\, m_{N_0}^2\, \frac{N_A}{2\pi} \quad , \tag{IV.10}$$

for processes of the type indicated in fig. 2. (N_A is the number of final states in the annihilation of N_0.) The rate equation for N_0 is given by (see Eq. III.37):

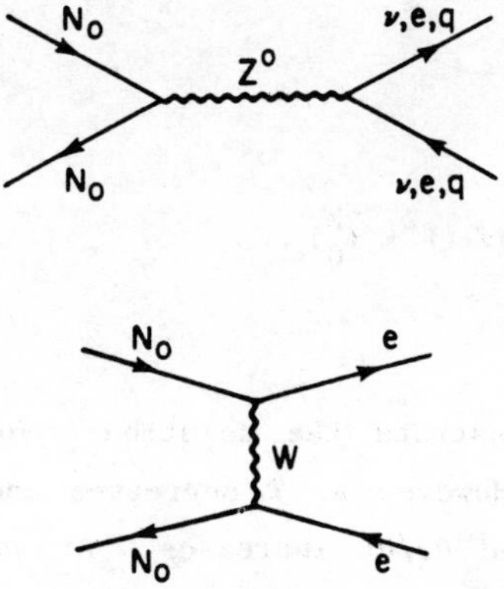

Fig. 2

$$\frac{dn}{dt} + 3\,\frac{\dot{a}}{a}\,n = \left(n_{eq}^2 - n^2\right)\langle\sigma v\rangle \quad , \tag{IV.11}$$

where n is the number of N_0 per unit volume. Using the relations

$$aT = \text{constant} \quad , \tag{III.35}$$

$$\frac{\dot{a}}{a} = -\frac{\dot{T}}{T} = \left(\frac{\rho}{3M^2}\right)^{1/2} \quad , \tag{III.3a}$$

with

$$\rho = g\,\frac{\pi^2}{30}\,T^4 \quad , \tag{III.32}$$

and defining the new parameters

$$x \equiv \frac{T}{m_{N_0}} \quad , \quad f \equiv \frac{n}{T^3} \quad ,$$

$$f_0 \equiv \frac{n_{eq}}{T^3} = \begin{cases} g' \dfrac{\zeta(3)}{\pi^2} & x >> 1 \quad , \quad \text{(III.32)} \\ 2(2\pi x)^{-3/2}\, e^{-1/x} & x \leq 1 \quad , \quad \text{(III.34)} \end{cases} \tag{IV.12}$$

we obtain

$$\frac{df}{dx} = \left(\frac{3M^2 m^2_{N_0}}{g\pi^2/30}\right)^{1/2} <\sigma v>(f^2 - f_0^2) \quad . \tag{IV.13}$$

In fig. 3 we illustrate the solution. For $x >> 1$, $f \sim f_0 \sim$ constant; $df/dx \sim 0$. However as T decreases and $x \leq 1$, f_0 becomes Boltzmann suppressed and df/dx increases. At some temperature, T_f, the effective density f can no longer keep up with f_0. At this point N_0 decouples or freezes out of thermal equilibrium. This occurs approximately at the temperature, T_f, defined by the relation

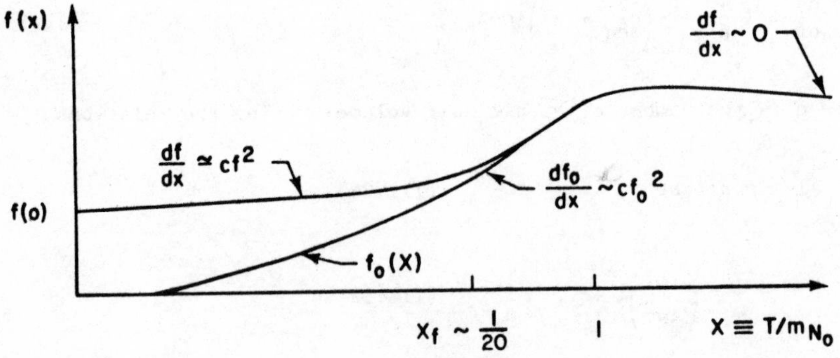

Fig. 3

$$\frac{df_0}{dx} = cf_0^2 \quad , \tag{IV.14}$$

$$\text{with } c \equiv \left(\frac{3M^2 m^2_{N_0}}{g\pi^2/30}\right)^{1/2} <\sigma v> \quad .$$

We find

$$\frac{T_f}{m_{N_0}} \equiv x_f \sim \frac{1}{\ell nc} \sim \frac{1}{20} \quad . \tag{IV.15}$$

For $x \ll x_f$, f_0 can be neglected in (IV.13) and we obtain the analytic result

$$f(0) \sim \left[cx_f + \frac{1}{f(x_f)} \right]^{-1} \sim \frac{20}{c} \quad . \tag{IV.16}$$

Let us now evaluate the energy density, ρ_{N_0}. We have

$$\rho^0_{N_0} = m_{N_0} n(T_0) \simeq m_{N_0} f(0) T_0^3 < \rho^0_c \quad , \tag{IV.17}$$

or using (IV.16),

$$\left(\frac{\pi^2}{15} T_0\right)^{-1} \frac{\rho^0_\gamma}{\rho^0_c} < \left[m_{N_0} f(0) \right]^{-1} \sim \frac{1}{20} \left(\frac{3M^2}{g\pi^2/30} \right)^{1/2} \langle\sigma v\rangle \quad . \tag{IV.18}$$

Since $\langle\sigma v\rangle$ is proportional to $m^2_{N_0}$ (IV.10), we obtain for $N_A \sim 14$, $g \sim 4.5$

$$m_{N_0} \geq 2 \text{ GeV} \quad . \tag{IV.19}$$

Photinos-Higgsinos

The previous calculations are a paradigm for the following discussion. The only changes will be in the value of the annihilation rates, which greatly depends on the particle physics details.

In a MLES Lagrangian, the four neutral fermions $\tilde{W}_3$, $\tilde{B}$, $\tilde{H}^0$, $\tilde{\bar{H}}^0$ (see Table I) have a mass matrix of the form

$$\begin{array}{c} \\ \tilde{W}_3 \\ \tilde{B} \\ \tilde{H}^0 \\ \tilde{\bar{H}}^0 \end{array} \begin{array}{c} \begin{array}{cccc} \tilde{W}_3 & \tilde{B} & \tilde{H}^0 & \tilde{\bar{H}}^0 \end{array} \\ \left(\begin{array}{cc|cc} M_2 & 0 & \frac{-g_2 v_1}{\sqrt{2}} & \frac{g_2 v_2}{\sqrt{2}} \\ 0 & \frac{5}{3}\frac{\alpha_1}{\alpha_2} M_2 & \frac{g_1 v_1}{\sqrt{2}} & \frac{-g_1 v_2}{\sqrt{2}} \\ \hline \frac{-g_2 v_1}{\sqrt{2}} & \frac{g_1 v_1}{\sqrt{2}} & 0 & \varepsilon \\ \frac{g_2 v_2}{\sqrt{2}} & \frac{-g_1 v_2}{\sqrt{2}} & \varepsilon & 0 \end{array}\right) \end{array} \qquad \text{(IV.20)}$$

ε, M_2, M_1 are defined by the Lagrangian terms

$$\mathcal{L} \supset \varepsilon \tilde{H}_1 \tilde{\bar{H}}_2 - M_2 \tilde{W}_a \tilde{W}_a - M_1 \tilde{B}\,\tilde{B} \quad , \qquad \text{(IV.21)}$$

where, as a result of one-loop renormalization group analysis,

$$M_1 = \frac{5}{3}\frac{\alpha_1}{\alpha_2} M_2 \text{ at } M_W \qquad \text{(IV.22)}$$

if $M_1 = M_2$ at $M_{p\ell}$. $v_1 = \langle H_0 \rangle$ and $v_2 = \langle \bar{H}_0 \rangle$.

Diagonalizing (IV.20) we find the four eigenstates

$$\tilde{Z}_i = \alpha_i \tilde{W}_3 + \beta_i \tilde{B} + \gamma_i \tilde{H}_0 + \delta_i \tilde{\bar{H}}_0$$
$$i = 1, \ldots, 4 \text{ where } \alpha_i, \beta_i, \gamma_i, \delta_i \qquad \text{(IV.23)}$$

are the mixing angles.

We shall consider the lightest $\tilde{Z}_i$ (with $i = i^*$):

1)a <u>Higgsino</u>: if it is predominantly a higgsino ($\gamma_i^*, \delta_i^* \gg \alpha_i^*, \beta_i^*$) and hence couples with Yukawa couplings or,

2)a <u>Photino</u>: if it is predominantly a gaugino ($\alpha_i^*, \beta_i^* \gg \gamma_i^*, \delta_i^*$) and hence couples with gauge couplings.

We note that in the limit $\varepsilon, M_2 \to 0$, there are two light states:

$$1. \quad \tilde{S}_0 \equiv \frac{v_2 \tilde{H}_0 + v_1 \tilde{\bar{H}}_0}{v} \quad , \quad v \equiv \sqrt{v_1^2 + v_2^2} \quad , \qquad \text{(IV.24)}$$

with mass

$$m_{\tilde{S}_0} \sim \frac{2 v_1 v_2}{v} \varepsilon \quad ,$$

and

$$2. \quad \tilde{\gamma} \equiv \frac{g_1 \tilde{W}_3 + g_2 \tilde{B}}{\sqrt{g_1^2 + g_2^2}} \qquad \text{(IV.25)}$$

with mass

$$m_{\tilde{\gamma}} \sim \frac{8}{3} \frac{g_1^2}{g_1^2 + g_2^2} M_2 \quad .$$

<u>Higgsino</u>

Higgsino annihilation proceeds via the graphs of fig. 4. The annihilation rate for a relativistic higgsino is of the form

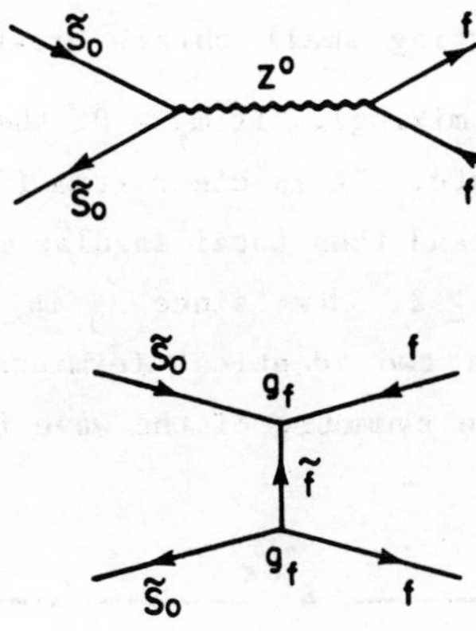

Fig. 4

$$\langle \sigma v \rangle \sim G_F^2 T^2 \quad . \tag{IV.26}$$

If $m_{\tilde{S}_0} < 1$ MeV then, like light neutrinos, Higgsinos are relativistic when they decouple. We thus obtain

$$m_{\tilde{S}_0} < 100 \text{ eV}h^2 \quad . \tag{IV.27}$$

For $m_{\tilde{S}_0} > 1$ MeV, Higgsinos are non-relativistic at decoupling. However, the results are not identical to those of Lee-Weinberg, since higgsinos (and photinos) are Majorana fermions. As a result there is a p-wave suppression in their annihilation rate (Goldberg, 1983). We find

$$\langle \sigma v \rangle \sim G_F^2 \left[\langle v_{rel}^2 \rangle + m_f^2 \right] \quad , \tag{IV.28}$$

unlike the rate for N_0 annihilation Eq. (IV.10). m_f is the mass of the fermion in the final state, and v_{rel} is the relative velocity of the two higgsinos. (Note, IV.28 is valid for $v_1 \neq v_2$. For $v_1 = v_2$, Z_0 exchange is suppressed and the dominant contribution comes from the Yukawa terms, which are typically smaller.)

Before we continue let us briefly illustrate the reason for the p-wave suppression. Consider the graphs of fig. 4. The vertices conserve chirality (neglecting small chiral breaking corrections of order $m_f/m_{\tilde{G}}$ coming from $\tilde{f}\tilde{\bar{f}}$ mixing). If $m_f = 0$, then helicity is also conserved for the final state. It is clear from fig. 5 that the final state must have helicity 1 and thus total angular momentum, $J_{final} \geq 1$. We then conclude $J_{initial} \geq 1$. Now since $\tilde{S}_0$ is a Majorana fermion, the initial state contains two identical fermions, and must thus be odd under interchange. The symmetry of the wave function of the initial state is given by

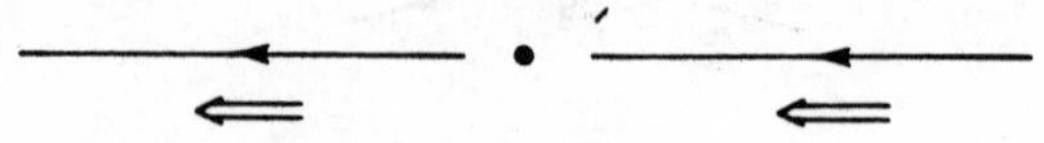

Fig. 5

$$(-1)^{\ell+s+1} \quad , \tag{IV.29}$$

where ℓ is the orbital angular momentum and s is the spin. If we assume $\ell = 0$ (s-wave), then $J_{initial} \geq 1$ implies $s = 1$, which is even. Thus we conclude that the initial state must have $\ell \geq 1$ and hence at least a p-wave suppression. If $m_f \neq 0$, there may then be an s-wave component proportional to m_f^2.

Using the exact cross-sections, Ellis et al. (1983), find

$$\rho^0_{\tilde{S}_0} < \rho^0_c \tag{IV.30}$$

if

$$m_{\tilde{S}_0} \geq m_b \sim 4.5 \text{ GeV} \tag{IV.31}$$

for $v_1 \neq v_2$, and

$$m_{\tilde{S}_0} \geq m_t \sim 30 \text{ GeV} \tag{IV.32}$$

for $v_1 = v_2$ (see note following IV.28).

<u>Photinos</u>

Photino annihilation proceeds via the graph of fig. 6. The annihilation rate is of the form

Fig.6

$$\langle\sigma v\rangle \sim G_F^2\left(\frac{M_W}{m_{\tilde{f}}}\right)^4 \left[v_{rel}^2 + m_f^2\right] \tag{IV.33}$$

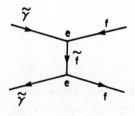

The limits in this case depend on both m_f and $m_{\tilde{f}}$. Ellis et al. (1983), find for

$$\rho^0_{\tilde{\gamma}} < \rho^0_c$$

$$m_{\tilde{\gamma}} \geq \frac{1}{2} \quad ; \quad 1.8 \quad ; \quad 5 \text{ GeV} \quad , \tag{IV.34}$$

for $m_{\tilde{f}} \sim 20$; 40 ; 100 GeV, respectively.

As $m_{\tilde{f}}$ increases, m_f must also increase in order for $\langle\sigma v\rangle$ to remain large enough (see IV.18), and thus $m_{\tilde{\gamma}}$ correspondingly increases. They also find for

$$\rho^0_{\tilde{\gamma}} < .1\ \rho^0_c$$

$$m_{\tilde{\gamma}} \geq 1.8 \quad ; \quad 3 \quad ; \quad 15 \text{ GeV} \quad , \tag{IV.35}$$

for $m_{\tilde{f}} \sim 20$; 40; 100 GeV, respectively. We have discussed only higgsinos or photinos while the lightest $\tilde{Z}_i$ may not be identifiable as such. Ellis et al., find however, that in a more general treatment, the results are dominated by these two special limits.

Sneutrinos (Hagelin, Kane and Raby, 1984; Ibáñez, 1984)

Sneutrinos annihilate via the processes of fig. 7. In fact, fig. 7a gives the dominant contribution:

$$\langle\sigma v\rangle \sim G_F \frac{M_W^2}{m_{\tilde{Z}}} \quad . \tag{IV.36}$$

This is because, this is the only process with no p-wave suppression. The annihilation rate (IV.36) is typically large (superweak), except in the limit $M_2,\varepsilon \to 0$. In this limit, the eigenstates $\tilde{Z}_i$ (Eq. III.23) are $\tilde{S}_0$, $\tilde{\gamma}$, and a dirac fermion. $\tilde{S}_0$ and $\tilde{\gamma}$ do not couple to neutrinos, as in fig. 7a. Moreover, the process of fig. 7a is proportional to the Majorana mass of $\tilde{Z}_i$ and thus the dirac fermion also doesn't contribute. We find, for M_2 and ε sufficiently large, no cosmological bound on $m_{\tilde{\nu}}$.

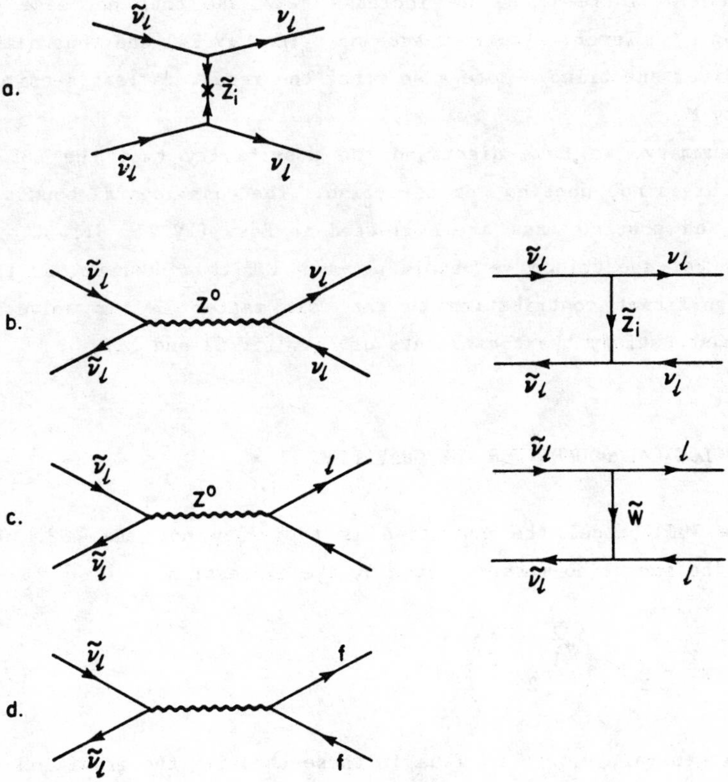

Fig. 7

We have, however, assumed throughout that the sneutrino is the LSP. Thus $m_{\tilde{\nu}} < m_{\tilde{S}_0}$, $m_{\tilde{\gamma}}$. We find that if we require

$$\rho^0_{\tilde{\nu}} \sim \rho^0_c \quad , \tag{IV.37}$$

then

$$m_{\tilde{\nu}} < m_{\tilde{S}_0} \lesssim 2 \text{ GeV} \quad ,$$

or for

$$\rho^0_{\tilde{\nu}} \sim .1 \ \rho^0_c \quad , \tag{IV.38}$$

then

$$m_{\tilde{\nu}} < m_{\tilde{S}_0} \lesssim 10 \text{ GeV} \quad .$$

Note, when we increase ε, we increase $\langle\sigma v\rangle$ and thus decrease $\rho^0_{\tilde{\nu}}$. Increasing ε, however, also increases $m_{\tilde{S}_0}$ (Eq. IV.24) and thus allows for a heavier sneutrino. Note also, that the result is less sensitive to varying M_2.

In summary, we have discussed the possibility that the LSP is either a higgsino, photino, or sneutrino. The cosmological bounds on higgsino and photino mass are collected in Eqs. (IV.27, 31, 32, 34, and 35). For sneutrinos we obtain no mass limits. However, if they are a significant contribution to the dark matter in our universe, then $m_{\tilde{\nu}}$ must satisfy the constraints of Eqs. (IV.37 and 38).

5. COSMOLOGICAL BOUNDS FOR THE GRAVITINO

In a MLES model the gravitino is typically not the LSP. For example, the sneutrino mass is given by the expression

$$m^2_{\tilde{\nu}} = m^2_{\tilde{G}} - \frac{1}{2} M^2_Z \left(\frac{v_1^2 - v_2^2}{v^2} \right), \tag{V.1}$$

where, in general, $v_1 \geq v_2$. Thus in these theories the gravitino can decay via processes illustrated in fig. 8. The decay rate is of order (Weinberg, 1982)

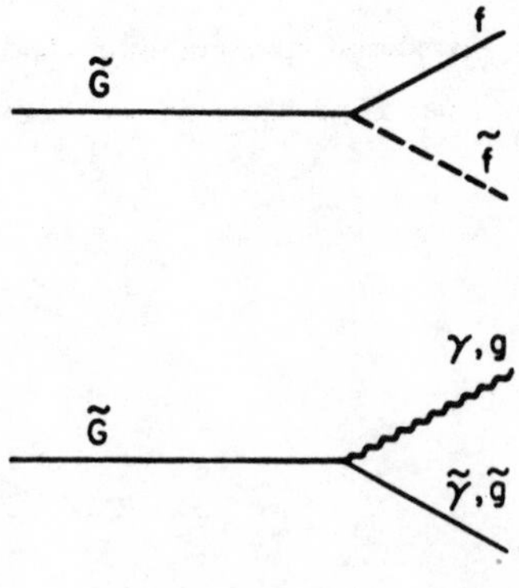

Fig. 8

$$\Gamma_{\tilde{G}} \simeq \frac{m_{\tilde{G}}^3}{M_{p\ell}^2} \quad . \tag{V.2}$$

Gravitinos can be produced singly, in scattering processes with lighter states, as in fig. 9. The production rate is, typically, of order (Krauss, 1983; Weinberg, unpublished; Ellis et al, 1983; Ellis, Linde, Nanopoulos, 1982; Khlopov and Linde, 1984)

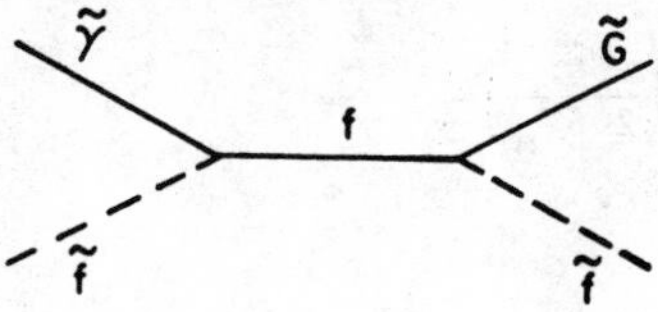

Fig. 9

$$\Gamma_{prod} \simeq \frac{\alpha N T^3}{M_{p\ell}^2} \equiv \langle \sigma v n_{\tilde{f}} \rangle \quad , \tag{V.3}$$

where N is the number of squarks and sleptons in the thermal bath.

Gravitinos decouple just below $M_{p\ell}$. If we require

$$\Gamma_{prod} = H \sim \frac{g^{1/2} T^2}{M_{p\ell}} \quad , \tag{V.4}$$

we find the decoupling temperature

$$T_d \sim \frac{g^{1/2}}{\alpha N} M_{p\ell} \quad . \tag{V.5}$$

If, for the moment, we neglect $\Gamma_{\tilde{G}}$, then

$$n_G a^3 = \text{constant} \tag{V.6}$$

for $T < T_d$. This can lead to problems, as we shall now show.

We assume that the initial gravitino number density is given by

$$n_{\tilde{G}}\Big|_{T_{p\ell}} = n_\gamma\Big|_{T_{p\ell}} \quad . \tag{V.7}$$

If there is no subsequent entropy production (which would increase n_γ without affecting $n_{\tilde{G}}$) and we take $\Gamma_{\tilde{G}} \equiv 0$, then today we obtain

$$n^0_{\tilde{G}} = n^0_\gamma \; . \tag{V.8}$$

Hence

$$\Omega_{\tilde{G}} \equiv \frac{\rho^0_{\tilde{G}}}{\rho^0_c} = \frac{m_{\tilde{G}} n^0_{\tilde{G}}}{\rho^0_c} = \left(\frac{m_{\tilde{G}}}{2.6T_0}\right)\Omega^0_\gamma \quad ,$$

$$\Omega_{\tilde{G}} = 4 \times 10^9 h^{-2}\left(\frac{m_{\tilde{G}}}{10^2 \text{ GeV}}\right) . \tag{V.9}$$

This is much too large. Of course, we have neglected $\Gamma_{\tilde{G}}$; perhaps $\Gamma_{\tilde{G}} \neq 0$ can solve the problem (Weinberg, 1982). Otherwise we must have $n^0_{\tilde{G}}/n^0_\gamma \leq .25 \times 10^{-9} h^2 \; (m_{\tilde{G}}/10^2 \text{GeV})^{-1}$.

We define the parameters Y_G, Y'_G by the expressions

$$Y_G = \frac{n_{\tilde{G}}}{n_r} \quad , \tag{V.10}$$

or

$$n_{\tilde{G}} = Y_G \, \frac{g'\zeta(3)}{\pi^2} \, T^3 \equiv Y'_G T^3 \quad .$$

We shall obtain the following two possible solutions: $\Gamma_{\tilde{G}} \neq 0$ and

1) $Y'_G \lesssim 10^{-9}$, at the decay temperature T_D, or

$$\text{2)} \quad 10^{-9} < Y'_G \lesssim 10^{-4} \text{ at } T_D \text{ and } m_{\tilde{G}} \gtrsim 10^4 \text{ GeV}. \tag{V.11}$$

Scenario 1:

Consider the possibility that

$$T_D > T_{eq} \quad , \tag{V.12}$$

i.e., the universe is radiation dominated when gravitinos decay. The decay time t_D is given in terms of T_D by the expression

$$\Gamma_{\tilde{G}}^{-1} = t_D \sim H_D^{-1} \;, \tag{V.13}$$

where

$$H_D^2 = \frac{g_D\,(\pi^2/30)T_D^4}{3M^2} \;,$$

and

$$\Gamma_{\tilde{G}} = \frac{N}{2\pi}\,\frac{m_{\tilde{G}}^3}{M_{p\ell}^2} \;.$$

We obtain

$$T_D = \left(\frac{N}{g_D^{1/2}}\right)^{1/2} 10^{-7}\,\mathrm{GeV}\left(\frac{m_{\tilde{G}}}{10^2\ \mathrm{GeV}}\right)^{3/2} \;. \tag{V.14}$$

Now using (V.12), i.e.,

$$\rho_{\tilde{G}}\Big|_{T_D} < \rho_r\Big|_{T_D} \;, \tag{V.15}$$

or

$$m_{\tilde{G}} Y'_G < g_D\,\frac{\pi^2}{30}\,T_D \;,$$

we obtain

$$Y'_G < 10^{-9}\left(\frac{m_{\tilde{G}}}{10^2\ \mathrm{GeV}}\right)^{1/2} \;. \tag{V.16}$$

Scenario 1 is acceptable.

1) The universe is radiation dominated at Helium synthesis, since $T_{H_e}4 \sim 10^{-4}$ GeV > T_D (V.14). Thus Helium synthesis is unchanged.

2) $\rho_{\tilde{G}}/\rho_r|_{T_D} < 1$ implies that when gravitinos decay, there is neglibible reheating of the radiation.

3) Gravitinos eventually decay into the LSP($\tilde{\gamma}$, $\tilde{S}_0$, or $\tilde{\nu}$) with

$$n_{\tilde{G}}\Big|_{T_D} = n_{LSP}\Big|_{T_D} \tag{V.17}$$

as a result of R-parity. However, since $n_{\tilde{G}}\Big|_{T_D} \le 10^{-9} n_\gamma\Big|_{T_D}$ (V.16), there is not a problem with too many of the LSP. We have

$$\Omega^0_{LSP} \le 10^{-9}\left(\frac{M_{LSP}}{2.6T_0}\right)\Omega^0_\gamma \quad , \tag{V.18}$$

which is safe [compare with (V.9)]

Scenario 2:

Consider:

$$T_D < T_{eq} \quad , \tag{V.19}$$

i.e., the universe is matter dominated when gravitinos decay. This implies

$$Y'_{\tilde{G}} > 10^{-9}\left(\frac{m_{\tilde{G}}}{10^2 \text{ GeV}}\right)^{1/2} \quad . \tag{V.20}$$

Since $\rho_{\tilde{G}}\Big|_{T_D} > \rho_r\Big|_{T_D}$, the energy which is released when the gravitinos decay will reheat the radiation. We define T'_D to be the reheat temperature for the radiation; $T'_D > T_D$. There are now three relevant possibilities:

$$\begin{aligned} &1)\quad T_D > T_{H_e^4} \sim 10^{-4} \text{ GeV}, \\ &2)\quad T'_D > T_{H_e^4} > T_D \text{ or}, \\ &3)\quad T_{H_e^4} > T'_D > T_D \quad . \end{aligned} \tag{V.21}$$

Case 3 can be ruled out immediately. Since, in this case, the universe reheats after Helium synthesis, the ratio of baryons to photons at the time of Helium synthesis is larger than can be inferred from present observations. When a larger value, $\rho_b/\rho_r\Big|_{T_{H_e^4}}$, is put into the standard calculations, one obtains more Helium and less Deuterium as output. This is unacceptable since the standard result already agrees with the lower limit for the ratio of abundances, $D/H_e^4 \gtrsim 10^{-5}$.

Consider case 1; $T_D > T_{H_e^4}$.

We have

$$H_D^2 = \frac{m_G Y'_G T_D^3}{3M^2} = \Gamma_{\tilde{G}}^2 = \left(\frac{N}{2\pi}\frac{m_{\tilde{G}}^3}{M_{p\ell}^2}\right)^2 ,$$

or

$$T_D = \left(\frac{3N^2}{(16\pi^2)^2}\right)^{1/3} \frac{m_{\tilde{G}}^{5/3}}{M^{2/3}} Y'_G{}^{-1/3} \quad . \tag{V.22}$$

Thus

$$Y'_G < 2N^2 \times 10^{-19}\left(\frac{m_{\tilde{G}}}{10^2 \text{ GeV}}\right)^5 \quad . \tag{V.23}$$

But recall $Y'_G > 10^{-9}\left(\frac{m_{\tilde{G}}}{10^2 \text{GeV}}\right)^{1/2}$, Eq. (V.20). These two equations (V.20) and (V.23) are inconsistent unless

$$m_{\tilde{G}} \geq 10^4 \text{ GeV} \quad . \tag{V.24}$$

Consider case 2; $T'_D > T_{H_e^4} > T_D$.

Using energy conservation, we obtain an expression for T'_D:

$$\rho_{\tilde{G}}\Big|_{T_D} = g(T'_D)\frac{\pi^2}{15}(T'_D)^4 \quad . \tag{V.25}$$

Using

$$\rho_{\tilde{G}}\Big|_{T_D} = m_{\tilde{G}}\, Y'_G T_D^3 \quad , \tag{V.26}$$

and Eq. (V.22) we obtain

$$m_{\tilde{G}} = (16\pi^2)^{1/3} \left(\frac{\pi^2}{45N^2}\right)^{1/6} g(T_D')^{1/6} M^{1/3} (T_D')^{2/3} \quad . \tag{V.27}$$

Using $T_D' > T_{H_e^4}$, we find

$$m_{\tilde{G}} > \left(\frac{g(T_D')}{N^2}\right)^{1/6} \times 1.2 \times 10^4 \text{ GeV} \tag{V.28}$$

for $T_{H_e^4} \sim .1$ MeV, or

$$m_{\tilde{G}} > \left(\frac{g(T_D')}{N^2}\right)^{1/6} \times 1.9 \times 10^4 \text{ GeV} \tag{V.29}$$

for $T_{H_e^4} \sim .2$ MeV.

In either case (1 or 2) we thus find

$$m_{\tilde{G}} \geq 10^4 \text{ GeV} \quad . \tag{V.30}$$

We now show that there is an upper limit on Y_G' (Krauss, 1983; Weinberg, unpublished; Ellis et al, 1983; Ellis, Linde and Nonopoulos, 1982; Khlopov and Linde, 1984). In the standard scenario for baryogenesis, in grand unified theories, the ratio $n_b/n_\gamma \sim 10^{-9}$ is generated via the baryon violating decays of heavy leptoquark gauge bosons or color triplet Higgs bosons. In either case the value of n_b/n_γ which can be obtained is typically no greater than about 10^{-6}. The production of entropy, after the epoch of baryogenesis, will cause the ratio n_b/n_γ to decrease. We must therefore limit the amount of entropy production at low temperatures, i.e., T less than $\sim 10^{12}$ GeV.

We define $\Delta = s_f/s_i$, where s_i is the entropy before gravitino decay and s_f is the entropy after. Then

$$\Delta = \left(\frac{T'_D}{T_D}\right)^3 = \left(\frac{M}{m_{\tilde{G}}}\right)^{1/2} Y'_G \left(\frac{(16\pi^2)^2}{3N^2}\right)^{1/4} \left(\frac{15}{g\pi^2}\right)^{3/4} . \tag{V.31}$$

If we require $\Delta \leq 10^3$, we obtain an upper limit on Y'_G, i.e.,

$$Y'_G \leq 1.6 \times 10^{-4} (N^2 g^3)^{1/4} \left(\frac{m_{\tilde{G}}}{10^4 \text{ GeV}}\right)^{1/2} \left(\frac{\Delta}{10^3}\right) . \tag{V.32}$$

Finally since $Y'_G > 10^{-9}$, we must also worry about the decay products of the gravitino.

In conclusion, scenario 1, i.e., $Y'_G \lesssim 10^{-9}$ at T_D is apparently the easiest one to live with. It is not difficult imagining mechanisms which might produce nine orders of magnitude of entropy, prior to baryogenesis, which would give $Y'_G \leq 10^{-9}$.

To complete our discussion of gravitinos, we should mention the limit $Y'_G \leq 10^{-9}$ places on inflation scenarios (Krauss, 1983; Weinberg, unpublished; Ellis et al, 1983; Ellis, Linde and Nanopoulos, 1982; Khlopov and Linde, 1984). In order to solve the cosmological horizon and flatness problems, we need a mechanism to create $\sim 10^{84}$ orders of magnitude of entropy, prior to baryogenesis (Guth, 1981; Linde, 1982; Albrecht and Steinhardt, 1982). In standard scenarios the universe inflates exponentially and adiabatically, and then reheats to, approximately, the temperature just prior to inflation. During inflation the gravition number density vanishes as $1/a^3$. We now show, however, that the reheat temperature, T_R, can not be too high; otherwise, one runs into the problem of recreating the gravitinos (Krauss, 1983; Weinberg, unpublished; Ellis et al., 1983; Ellis, Linde and Nanopoulos, 1982; Khlopov and Linde, 1984).

Consider the process of fig. 9 and the production rate (V.3). The rate equation for producing gravitinos is

$$\frac{dn_{\tilde{G}}}{dt} + 3\frac{\dot{a}}{a} n_{\tilde{G}} = \Gamma_{prod} n_r ,$$

$$\frac{dn_r}{dt} + 4\frac{\dot{a}}{a} n_r = 0 . \tag{V.33}$$

(valid for $Y_G = n_{\tilde{G}}/n_r << 1$), or for Y_g,

$$\dot{Y}_G = HY_G + \Gamma_{prod} \quad . \tag{V.34}$$

The steady state solution, evaluated at T_r is

$$Y_G = -\frac{\Gamma_{prod}}{H_R} \sim -\alpha\sqrt{N}\frac{T_R}{M_{p\ell}} \tag{V.35}$$

where $H_R \sim \sqrt{N}\, T_R^2/M_{p\ell}$. Finally at T_D, we have

$$Y_G\Big|_{T_D} = \frac{2}{N}\, Y_G\Big|_{T_R} \sim \frac{2\alpha}{\sqrt{N}}\frac{T_R}{M_{p\ell}} \quad . \tag{V.36}$$

The factor 2/N takes into account the annihilation of N-2 degrees of freedom between T_R and T_D. If we now require $Y_G \lesssim 10^{-9}$, we find

$$T_R \lesssim 10^{-12}\text{GeV}\left(\frac{Y_G}{10^{-9}}\right) \quad . \tag{V.37}$$

This is the upper limit on the reheat temperature in an inflationary universe scenario.

We have discussed cosmological mass bounds for supersymmetric partners. The candidate lightest supersymmetric partners were higgsinos, photinos, and sneutrinos. Gravitinos were also discussed since they are also an important contribution to the energy density of the universe. We have demanded that $\rho_{LSP} \lesssim \rho_c$ today. If the LSP does in fact satisfy, $\rho^0_{LSP} \sim \rho^0_c$, then it is also an excellent candidate for the so-called dark matter which is apparently the dominant form of energy. Any one of the LSP candidates, with mass of order a few GeV or greater, would fall into the category of "cold" dark matter (Blumenthal, et al, 1984; Primack and Blumenthal, 1983a, 1983b). One particular property of "cold" dark matter is that it clusters on all scales as seems to be the case observationally (see Table IV). It may also be able to explain the large scale voids which have been observed.

Table IV

Dark matter[a]

Evidence	scale $M[M_{\odot} = 2 \times 10^{33} g]$	M/M_{lum}
large clusters	10^{15}	$8.4^{+7.0}_{-1.0}$
small spiral dominated groups	2×10^{13}	14.2^{+36}_{-6}
whole milky way	10^{12}	14
dwarf spheroidal galaxies	10^{6-8}	12

[a]Blumenthal, et al., 1984; Primack and Blumenthal, 1983a, 1983b.

References

Albrecht, A. and Steinhardt, P. J. 1982, Phys. Rev. Lett. 48, 1220

Alvarez-Gaumé, L., Polchinski, J., and Wise, M. B. 1983, Nucl. Phys. B221, 495

Ball R. C., et al. 1983, in Proceedings of the 1983 International Europhysics Conference on High Energy Physics, Brighton, p. 318, published by Rutherford Appleton Laboratory.

Blumenthal, G. R., Faber, A. M., Primack, J. R., and Rees, M. J. 1984, SLAC preprint, SLAC-PUB-3307

CELLO Collaboration, Behrend, H., et al. 1982, Phys. Lett. 114B, 287

Cowsik, R. and McClelland, J. 1972, Phys. Rev. Lett. 29 669

Cowsik, R. and McClelland, J. 1973, Ap. J. 180, 7

Ellis, J., Hagelin, J., Nanopoulos, D. V., and Tamvakis, K. 1983, Phys. Lett. 125B, 275

Ellis, J. and Kowalski, H. 1984a, CERN preprint, Ref. Th. 3843

Ellis, J. and Kowalski, H. 1984b, DESY preprint, DESY 84-045

Ellis, J., Hagelin, J. S., Nanopovlos, D. V., Olive, K. and Srednicki, M. 1983, Nucl. Phys. B.

Ellis, J. A., Linde, D. and Nanopoulos, D. V. 1982, Phys. Lett. 118B, 59

Farrar, G. R. and Fayet, P. 1978a, Phys. Lett. 76B, 575

Farrar, G. R. and Fayet, P. 1978b, Phys. Lett. 79B, 442

Farrar, G. R. and Weinberg, S. 1983, Phys. Rev. D27, 2732

Goldberg, H. 1983, Phys. Rev. Lett. 50, 1419

Guth, A. 1981, Phys. Rev. D23, 347

Hagelin, J. S., Kane, G. L. and Raby, S. 1984, Nucl. Phys. B241, 638

Ibáñez, L. E. 1983, Nucl. Phys. B218, 514

Ibáñez, L. E. 1984, FTUAM preprint 83-28

Ibáñez, L. E. and López, C. 1983, CERN preprint, Ref. Th. 3650
JADE Collaboration, Bartel, W., et al. 1982, Phys. Lett. 114B, 211
Khlopov, M. Yu. and Linde, A. D. 1984, Phys. Lett 138B, 265
Kolb, E. W. and Wolfram, S. 1980, Nucl. Phys. B172, 224
Krauss, L. M. 1983, Nucl. Phys. B227, 556
Lee, B. W. and Weinberg, S. 1977, Phys. Rev. Lett. 39, 165
Linde, A. 1982, Phys. Lett. 108B, 389
MARK J Collaboration, Barber, D. P., et al. 1981, Phys. Rev. Lett. 45, 1904
MARK II Collaboration, Blocker, C. A., et al. 1982, Phys. Rev. Lett. 49, 517
Pagels, H. and Primack, J. 1982, Phys. Rev. Lett. 48, 223
Primack, J. R. and Blumenthal, G. R. 1983a, UCSC-TH-162-83
Primack, J. R. and Blumenthal, G. R. 1983b, UCSC-TH-164-83
Smith, P. F. and Bennett, J. R. J. 1979, Nucl. Phys. B149, 525
Tasso Collaboration, Brandelik, R., et al. 1982, Phys. Lett. 117B, 365
Weinberg, S. 1972, Gravitation and Cosmology, John Wiley and Sons, Inc.
Weinberg, S. 1982, Phys. Rev. Lett., 48, 1303
Weinberg (unpublished)
Wolfram, S. 1979, Phys. Lett. 82B, 65

SUPERSYMMETRY AND MATHEMATICS

L ALVAREZ-GAUMÉ

One of the most interesting aspects of supersymmetry, which has received a good deal of attention in recent years is that the supersymmetric algebra provides a new way of formulating some classic problems in topology and differential geometry. The usefulness of supersymmetry to address some mathematical question was already realized in the mid-seventies [1] in connection with the computation of the fermionic zero modes in the presence of instanton backgrounds; and also in showing that the spectra of various operators appearing in various one loop instanton determinants are identical [2]. These results which were originally derived for the gauge theory case found also interesting applications in the study of gravitational instantons and the euclidean formulation of quantum gravity [3]. The reason making supersymmetry useful in this context, is that it provides a set of transformation rules which mix fields of different spin. Thus under certain circumstances, knowing the spectrum of a particular operator acting on bosons allows one to compute the spectrum of the Dirac or Rarita-Schwinger operator, and vice-versa, by simply applying the supersymmetry transformation rules to the eigenfunctions of the operators involved.

More recently, we have learned that supersymmetric quantum mechanics leads to rather simple presentations and proofs [4] of the Atiyah-Singer index theorem [5] and Morse theory [6]. In this note we will try to exhibit why supersymmetry can be useful to understand some topological questions. Further details can be found in the literature (see, for instance [7]).

The interplay between topology and supersymmetric quantum mechanics stems from a very interesting analysis of E. Witten [8] of the necessary conditions for supersymmetry to be broken in a supersymmetric field theory. One of the aspects which makes supersymmetry qualitatively different from ordinary symmetries, is that its breaking can be studied at finite volume. The breaking of supersymmetry is related to the vanishing of the vacuum energy of the field theory. More precisely, the order parameter which decides whether supersymmetry is broken, is the vacuum energy density. Recall that in supersymmetric theories, the supercharge Q provides a "square root" of the hamiltonian $H = Q^2$. The theory will have unbroken supersymmetry if the vacuum is annihilated by Q, $Q|0\rangle = 0$, and thus $H|0\rangle = 0$. This in itself does not help very much, because unless we know how to exactly solve the theory, it is in general very difficult to decide, using standard methods, that the vacuum energy is not equal to zero. It is at this point that Witten's insight [8] becomes crucial. In order to analyze the question further, notice that the spectrum of a supersymmetric theory is such that for any bosonic states $|E,b\rangle$ with energy $E \neq 0$, there is always a fermionic state $|E,f\rangle$ with the same energy, obtained from $|E,b\rangle$ by acting with Q: $|E,f\rangle = Q|E,b\rangle/\sqrt{E}$ and vice versa. If $E \neq 0$, then Q cannot annihilate $|E\rangle$, because $Q^2|E\rangle = H|E\rangle = E|E\rangle \neq 0$ i.e., non-zero energy states provide at least two dimensional representations of supersymmetry with equal number of bosons and fermions per energy level. Things are quite different if we look at the (possibly empty) set of zero energy states $|\Omega\rangle$. Since $|\Omega\rangle$ has zero energy, it is annihilated by the supercharge $Q|\Omega\rangle = 0$, so that zero energy states provide one-dimensional irreducible representations of supersymmetry, and there is no need for them to be paired. Thus if we compute the trace of $(-1)^F$ (where F is fermionic number) or its regularized for:

$$\Delta \equiv \mathrm{Tr}(-1)^F e^{-\beta H} \tag{1}$$

we find that the contributions to Δ from non-zero energy bosonic states are exactly cancelled by their fermionic partners, and only the zero energy states remain:

$$\Delta = n_B(E=0) - n_F(E=0) \tag{2}$$

where $n_B(E=0)$ ($n_F(E=0)$) counts the number of zero energy bosonic (fermionic) states in the spectrum. In order to avoid some subtle difficulties with the continuous spectrum, let us assume that the spectrum of the hamiltonian H in a box is discrete. In this case Δ is independent of β, and also of any ultraviolet regulator (as long as it respects supersymmetry). Thus if we find that $\Delta \neq 0$, the theory contains zero energy states, and supersymmetry is unbroken. Computing Δ is far easier than computing the vacuum energy directly, because Δ is a topological invariant of the full quantum theory. If we change H continuously, then the states of the spectrum will also change continuously. It could happen that as we change H a bosonic state ends up with zero energy, thus changing $n_B(E=0)$ by one unit. By continuity, however, all along the deformation of H, the bosonic state is accompanied by a fermionic partner and we also end up increasing $n_F(E=0)$ by one unit so that Δ does not change. Similarly, under some continuous deformation of H, one of the zero energy states may acquire a non-zero energy. Again by continuity some other state of zero energy and opposity fermion number must leave the zero energy sector, leading to no change in Δ. More precisely, notice that $(-1)^F$ is a good quantum number, so that we can split the Hilbert space $\mathscr{H} = \mathscr{H}_B \oplus \mathscr{H}_F$, $\mathscr{H}_B(\mathscr{H}_F)$ = bosonic (fermionic) states. With respect to this splitting, the supercharge Q is off diagonal:

$$Q = \begin{bmatrix} 0 & S \\ S^+ & 0 \end{bmatrix} \tag{3}$$

then:

$$n_B(E=0) = \dim \mathrm{Ker}\, S$$
$$n_F(E=0) = \dim \mathrm{Ker}\, S^+. \tag{4}$$

Ker S is the space of solutions of $S|\Psi\rangle = 0$, and Ker S^+ is the space of solutions of $S^+|\Psi\rangle = 0$. Consequently:

$$\Delta = \mathrm{index}\, S \tag{5}$$

which is indeed invariant under smooth deformations. In order to check whether supersymmetry is broken, what one can do is to change the parameters of the theory continuously up to a point where we can make reliable approximations to determine the spectrum of the theory. In this case we can compute Δ. The topological invariance of Δ then guarantees that Δ will have the same numerical value for the original theory. For ordinary field theories, S is an operator acting on wavefunctional, and it is not easy to calculate Δ for theories which may have phenomenological interest, for example, for supersymmetric chiral gauge theories there is no known systematic procedure to compute Δ (for some progress in this direction see [9]).

If we restrict our attention to field theories in 0+1-dimensions, i.e., ordinary elliptic operator, we can use standard quantum mechanical methods to compute ind S through evaluating (1). The question is therefore whether we can construct supersymmetric quantum mechanical systems whose supercharge is related to those operators appearing naturally in index theory like the Dirac operator with or without gauge fields, the Dolbeault operator, etc. The answer to this question is yes, and we will explain now how this can be shown.

Given an arbitrary quantum mechanical system whose dynamics is described by the hamiltonian H, we say that H admits N supersymmetries if there exist N pairs of operators S^i, S^{+i} $i = 1,N$, such that:

$$\{S^i,S^{+j}\} = 2\delta^{ij}H$$
$$\{S^i,S^j\} = \{S^{+i},S^{+j}\} = 0 \qquad (6)$$

although this is not the most general supersymmetric algebra we could write down, it will be sufficient for our purposes. For any of the pairs S^i, S^{i+}, we have $(S+S^+)^2 = H$. Since H is generically a second order elliptic operators in quantum mechanics, S and S^+ will be first order linear operators. Before we described the quantum mechanical systems which lead to the index theorem for the Dirac operator, etc., let us quickly review some elementary results of differential geometry and the theory of characteristic classes so that at the end we can recognize the answer in proper mathematical terms.

Throughout this note (M,g) will denote an n-dimensional compact manifold without boundary endowed with a Riemannian metric g_{ij}. Introducing vielbein one-forms:

$$ds^2 = \delta_{ab}e^a \otimes e^b$$
$$e^a = e^a{}_\mu(x)dx^\mu \qquad (7)$$

The vielbeins define an orthonormal frame at each point of the manifold. Note that once the vielbein is introduced, we have invariance under local Lorentz transformations $e^a \to L^a{}_b(x)e^b$, where L is a local SO(n) rotation. If we want to consider spinors on M_n we are forced to introduce orthonormal frames from the beginning (GL(n) which is the group whose representations define tensors on M does not admit spinor representations). Given the one-forms e^a, the spin connection (a matrix valued one-form with values in the Lie algebra of SO(n)) and all the standard formuli of local Riemannian geometry follow from the torsion free condition and the definition of the curvature two form $\Omega^a{}_b$ [10]:

$$de^a + \omega^a{}_b e^b = 0 \qquad (8.6a)$$

$$\Omega^a{}_b = d\omega^a{}_b + \omega^a{}_c\omega^c{}_b \qquad (8.6b)$$

From (8.a) we can uniquely compute $\omega^a{}_b$ in terms of the coefficients $b^a{}_{cd}$ appearing in:

$$de^a = \frac{1}{2} b^a{}_{cd} e^c e^d . \tag{9}$$

In (8,9) d represents the exterior derivative, and the standard wedge product between forms is implicitly understood. As a consequence of $d^2 = 0$, we can immediately derive the Bianchi identities from (8.a,b):

$$\Omega^a{}_b e^b = 0$$

$$d\Omega^a{}_b + \omega^a{}_c \omega^c{}_b - \Omega^a{}_c \omega^c{}_b \equiv D\Omega^a{}_b = 0. \tag{10}$$

This is entirely analogous to the gauge theory case where the basic object is the connection one form A taking values on the Lie algebra of some gauge group G, and the curvature $F = dA+A^2$. Under local frame or local gauge transformations, ω and A transform respectively as:

$$\omega \to L^{-1}(\omega+d)L, \quad A \to g^{-1}(A+d)g \tag{11}$$

making the analogy between gauge theories and Riemannian geometry complete. If we want to deal not only with local but also global questions, we have to understand a bit better the global structure of M, and what is the information required to set up orthonormal frames or gauge fields on M (i.e. the orthonormal frame bundle or a principal G-bundle over M). The global structure of M is specified by providing a covering of M by patches $\{U_i\}$ and a set of transition functions f_{ij} on the overlaps $U_i \cap U_j$ which tell us how U_i and U_j are patched together. Any gauge field A (which could be also the spin connection in what follows) is defined separately on each patch A_i, with the requirement that A_i, A_j in overlapping patches $U_i \cap U_j \neq \emptyset$ be related by a gauge transformation $A_i = g_{ij}^{-1}(A_j+d)g_{ij}$. All of the topological information is thereby encoded in the transition functions g_{ij}. (If $M_n = S^n$, the n-dimensional sphere, then we only used one transition function to glue together the northern and southern hemisphere. The transition function g is thus a map from the equatorial S^{n-1} into the gauge group G, and the topological nontriviality of this set-up is determined by the homotopy class of $\Pi_{n-1}G$ generated by G). One general way of measuring the topological non-triviality of a vector bundle is in terms of characteristics classes. Loosely speaking, a characteristic class is a local differential form on M constructed in terms of Ω and F, and constructed so that its integrals over M or appropriate submanifolds are sensitive to the existence of non-trivial transition functions. These classes are generically closed but not exact forms. Using Poincaré's lemma, each of these forms α can be written locally as $d\phi_i$ for each patch U_i, hence by Stoke's theorem the integral of α will only depend on the transition functions. There is a general procedure for constructing characteristic classes [11] which we now outline. Let Ω stand for either the curvature two-form or the gauge field strength. If $P(\alpha)$ is an invariant polynomial under gauge transformation, i.e.,

$$P(g^{-1}\alpha g) = P(\alpha). \tag{12}$$

Then the polynomial $P(\Omega)$ is a closed form whose integrals over M (or compact submanifolds of M) are independent of the connection used (as long as it is on the same topological class). That $P(\Omega)$ is closed follows from the Bianchi identity. To make things simpler, we can always write $P(\Omega)$ in terms of monomials of the form $Pm(\Omega) = \mathrm{Tr}\,\Omega^m$. If each P_m is closed, then combinations of products of them will also be closed:

$$dPm = m\,\mathrm{Tr}\,d\Omega\,\Omega^{m-1} = m\,\mathrm{Tr}\,(d\Omega + [\omega,\Omega])^{m-1} = 0 \tag{13}$$

by the Bianchi identity (10). Now let ω_1, ω_2 be two connections in the same topological class; and let ω_t be an interpolation between the two

$$\omega_{t=0} = \omega_1$$

$$\omega_{t=1} = \omega_2.$$

Let $P_m(t) = \mathrm{Tr}\,\Omega_t^m$. Then

$$\frac{\partial}{\partial t} P_m(t) = m\ \mathrm{Tr}\ \frac{\partial\Omega_t}{\partial t}\,\Omega_t^{m-1} = m\ \mathrm{Tr}\ D_t\,\frac{\partial\omega_t}{\partial t}\,\Omega_t^{m-1}$$

$$= m\,d\,\mathrm{Tr}\,\frac{\partial\omega_t}{\partial t}\,\Omega_t^{m-1} \tag{14}$$

where we have used the Bianchi identity, and used the covariant derivative Dt with respect to the connection ω_t; since $\partial\omega_t/\partial t$ is a one-form, note that

$$D_t\,\frac{\partial\omega_t}{\partial t} = d\,\frac{\partial\omega_t}{\partial t} + \omega_t\,\frac{\partial\omega_t}{\partial t} + \frac{\partial\omega_t}{\partial t}\,\omega \tag{15}$$

integrating (14) from 0 to 1 in t we get:

$$P(\omega_2) - P(\omega_1) = d\,m\int_0^1 dt\,\mathrm{Tr}\,\frac{\partial\omega_t}{\partial t}\,\Omega_t^{m-1}. \tag{16}$$

Thus if we integrate (16) over a 2m-dimensional compact submanifold of M we get:

$$\int_{N_{2m}\subset M} P_m(Q_2) = \int_{N_{2m}\subset M} P_m(\Omega_1) \tag{17}$$

with a little extra work one can also show that (17) only depends on the homology class of N_{2m}. If the gauge group is U(N), (or if we work on

Kähler manifolds which have U(N) holonomy) any characteristic polynomial can be written in terms of Chern classes defined as follows:

$$c(\Omega) \equiv \det\left(1 + \frac{i\Omega}{2\pi}\right) = 1 + c_j(\Omega) + c_2(\Omega) + \dots \qquad (18)$$

if we formally diagonalize the matrix two-forms Ω, and let x_i be the formal eigenvalues of $i\Omega/2\pi$. Then:

$$c_1(\Omega) = \sum_i x_i$$

$$c_k(\Omega) = \sum_{i,<\cdots<i_k} x_{i_1}\cdots x_{i_k} \qquad (19)$$

a useful combination of Chern classes is known as the Chern character, defined by

$$ch(\Omega) = \mathrm{Tr}\ e^{i\Omega/2\pi}. \qquad (20)$$

If Ω turns out to have block diagonal form $\Omega = \Omega_1 + \Omega_2, [\Omega_1,\Omega_2] = 0$, then

$$c(\Omega_1+\Omega_2) = c(\Omega_1)c(\Omega_2)$$

$$ch(\Omega_1+\Omega_2) = ch(\Omega_1) + ch(\Omega_2) \qquad (21)$$

When the group is SO(n) as in ordinary Riemannian geometry, then Ω is an antisymmetric matrix of two-forms, and it can be formally skew-diagonalized:

$$\Omega \sim \begin{bmatrix} 0 & x_1 & & & & & \\ -x_1 & 0 & & & & & \\ & & 0 & x_2 & & & \\ & & -x_2 & & & & \\ & & & & \cdot & & \\ & & & & & \cdot & \\ & & & & & & \cdot \end{bmatrix} \qquad (22)$$

The analog of the c_i's are called the Pontrjagin classes $P_i(\Omega)$ defined by:

$$P(\Omega) = \det\left(1 + \frac{\Omega}{2\pi}\right) = 1 + p_1 + p_2 + \dots$$

$$p_1 = Z_i\, x_i^2$$

$$P_k = \sum_{i,<\cdots<i_k} x_{i_1}^2\cdots x_{i_k}^2. \qquad (23)$$

What the index theorem does for us, is to compute the index of some operator S in terms of an integral over M of a characteristic polynomial constructed in terms of Chern classes. Pontjagin classes or both. What we want to do now, is to show that supersymmetry provides a rather simple way of computing the index of certain operators using rather simple quantum mechanical considerations. Since the most interesting operators appearing in physical applications is the Dirac operator, we will first derive the index theorem in this case. The strategy we follow is to first find a supersymmetric quantum mechanical system whose supercharge is the Dirac operator, and we evaluate (1) by means of a Feynmann path integral in the $\beta \to 0$ limit. The supersymmetric lagrangian we have to consider is just an $N = 1/2$ supersymmetric σ-model in 0+1 dimensions:

$$L = \frac{1}{2} g_{ij}(\phi) \frac{d\phi^i}{dt} \frac{d\phi^j}{dt} + \frac{i}{2} g_{ij}(\phi)\Psi^i \frac{D}{dt} \Psi^j \tag{24}$$

$$\frac{D}{dt} \Psi^i = \frac{d\Psi^i}{dt} + \Gamma^i_{jk} \frac{d\phi^j}{dt} \Psi^k .$$

The $\phi^i(t)$'s transform like coordinates under reparameterization of the manifold M, and the Ψ^i's are real anticommuting fermion fields transforming like vectors under reparametrization of M and Γ is the Christoffel symbol. The lagrangian (24) is invariant under the supersymmetry transformation:

$$\delta\phi^i = i\,\varepsilon\Psi^i$$

$$\delta\Psi^i = - \frac{d\phi^i}{dt} \varepsilon . \tag{25}$$

In order to make the arguments more clear, let us choose a different set of fermionic variables. Since the manifold is Riemannian, we can refer all geometrical objects to orthonormal frames. If $e^a{}_i(\phi)$ represents the vielbein field, let $\Psi^a \equiv e^a{}_i(\phi)\Psi^i$. Thus

$$L = \frac{1}{2} g_{ij}(\phi) \frac{d\phi^i}{dt} \frac{d\phi^j}{dt} + \frac{i}{2} \delta_{ab}\Psi^a \frac{D}{dt} \Psi^b$$

$$\frac{D}{dt} \Psi^a = \frac{d\Psi^a}{dt} + \omega^a{}_{bi} \frac{d\phi^i}{dt} \Psi^b . \tag{26}$$

$\omega^a{}_{bi}$ is the spin connection, so that $\omega_{abi} = -\omega_{bai}$, and for completeness we recall that

$$\omega^a{}_{bi} = -(e^{-1})^j{}_b \nabla_i e^a_j$$

$$= -(e^{-1})^j{}_b (\partial_i e^a_j - \Gamma^k_{ij} e^a_k) \tag{27}$$

upon canonical quantization of (26), the fermions satisfy the canonical anticommutation relations:

$$\{\psi^a,\psi^b\} = \delta^{ab} \tag{28}$$

i.e. the fermions generate a Clifford algebra on M_{2n}, and can be represented by $\gamma^a/\sqrt{2}$, where the γ^a's are the standard Dirac matrices. In order to define fermion number $(-1)^F$, notice that we must require the manifold M to be even dimensional. This is because fermion number can be defined if we can construct an operator which anticommutes with all the fermionic operators ψ^i. In odd dimensions, however, there is no matrix which anticommutes with all the γ-matrices. In even dimensions we can always construct the analog of γ_5, $\bar{\Gamma} = i^n\gamma^1\cdots\gamma^{2n}$, $\bar{\Gamma}^2 = 1$. Since $\bar{\Gamma}$ is defined in terms of the totally antisymmetric tensor $\varepsilon_{a_1\cdots a_{2n}}$, $\bar{\Gamma}$ can be defined only if the manifold is orientable. Thus in order to make sense of the quantum mechanics determined by (26), we have to require that M be at least even dimensional and orientable. In a moment we will also see that the manifold must also be a spin manifold. Using Noether's theorem, the supercharge can be derived to be

$$S = \psi^i g_{ij}(\phi)\frac{d\phi^i}{dt} \tag{29}$$

since the canonically conjugate momentum to ϕ^i is:

$$P_i = g_{ij}(\phi)\left[\frac{d\phi^i}{dt} + \frac{i}{4}\,\omega_{ab,i}[\psi^a,\psi^b]\right] \tag{30}$$

we find that after canonical quantization the supercharge S becomes

$$S = -\frac{i\gamma^i}{\sqrt{2}}\,\frac{\partial}{\partial\phi^i} + \frac{1}{8}\,\omega_{ab,i}[\gamma^a,\gamma^b] = -\frac{i}{\sqrt{2}}\,\not{D}$$

$$\gamma^a = e^a{}_i(\phi)\gamma^i \tag{31}$$

which is the Dirac operator. The wave functions of this system are spinors on the manifold M, and here we need to require that M be a spin manifold in order to be able to define the wave functions consistently. If the manifold does not admit a spin structure, then the Hilbert space of the theory is empty. The index of the Dirac operator is thus:

$$\begin{aligned}\text{ind}\,\not{D} &= \mathrm{Tr}(-1)^F e^{-\beta H}\\ &= \mathrm{Tr}\,\bar{\Gamma}\, e^{-\beta(i\not{D})^2/2}.\end{aligned} \tag{32}$$

We can write (32) as a path integral for the Wick rotated version of the lagrangian (25). Notice that (32) is just the partition function for the density matrix $(-1)^F \exp-\beta H$. Using standard methods in quantum mechanics (see, for example, [7]), one can show that

$$\text{ind}\ i\not{D} = \int_{PBC} [d\phi(t)][d\psi(t)] \exp - S_E(\phi,\psi) \tag{33}$$

where PBC stands for periodic boundary conditions. Due to the presence of $(-1)^F$ both the bosons and fermions have to be integrated over with periodic boundary conditions: $\phi^i(\beta) = \phi^i(0)$; $\Psi^i(\beta) = \Psi^i(0)$. $S_E(\phi,\psi)$ represents the Wick rotated version of (26). Since (32) is independent of β, we can do our computations in the limit $\beta \to 0$ or "high temperature" limit. In this limit the functional integral is dominated by the constant configuration. Thus we can evaluate (32) by expanding $\phi(t)$ and $\Psi(t)$ around the constant configurations (which in this case are exact solutions to the classical equations of motion with zero action). Expanding the action around the constant configurations (ϕ_0,ψ_0), we find the leading contribution to (33) by computing the expansion to second order. This reduces the evaluation of (33) to the computation of one-dimensional determinants. The most convenient way of doing the expansion around (ϕ_0,ψ_0) is by using Riemannian normal coordinates. These are defined as follows. Let ϕ_0^i be a constant configuration, and let $\phi^i(t)$ be a configuration in a neighborhood of ϕ_0^i. Except for exceptional points, one can always find a geodesic which joins ϕ_0^i and ϕ^i. The geodesic is completely defined once we specify the tangent vector ξ^i to the geoderic at ϕ_0. Normalizing ξ^i so that ϕ^i is reached along the geodesic after a unit of proper time has elapsed, we can get ϕ^i as a function of ϕ_0^i and ξ^i by simply solving the geodesic equation via a Taylor series:

$$\phi^i(t) = \phi_0^i + \xi^i - \frac{1}{2}\Gamma^i{}_{jk}(\phi_0)\xi^j\xi^k + \cdots \tag{34}$$

if we now implement this coordinate transformation on (26) one obtains after some algebra (see [4,7] for details) that to second order:

$$\begin{aligned} L^{(2)} &= \frac{1}{2}\delta_{ab}\dot{\xi}^a\dot{\xi}^b + \frac{i}{4}R_{abcd}\Psi_0^c\Psi_0^d\xi^a\dot{\xi}^b \\ &\quad + \frac{i}{2}\delta_{ab}\eta^a\frac{d}{dt}\eta^b \\ \xi^a(t) &= e^a{}_i(\phi_0)\xi^i \end{aligned} \tag{35}$$

and R_{abcd} is the Riemann tensor at the point ϕ_0^i referred to orthonormal frames. In order to avoid overcounting, the constant pieces of ξ and η has been removed. Then (33) becomes:

$$\text{ind } i\not{D} = \int \frac{(d\phi_0)(d\psi_0)}{(2\pi)^n} \left[\frac{\det' - \frac{d^2}{dt^2}\delta_{ab}}{\det'\left(-\frac{d^2}{dt^2}{}_{ab} + M_{ab}\frac{d}{dt}\right)} \right]^{1/2} \tag{36}$$

since the Ψ_0's are real anticommuting variables, they provide a representation for the one-forms on the manifold M. $(d\phi_0)$ gives the invariant volume element on M, and the fact of $(2\pi)^{-n}$ is due to Feynmann's normalization of the path integral. If we formally skew diagonalize M_{ab}:

$$M_{ab} \sim \begin{bmatrix} 0 & x_1 & & & \\ -x_1 & 0 & & & \\ & & \ddots & & \\ & & & 0 & x_n \\ & & & -x_n & 0 \end{bmatrix} \tag{37}$$

we can compute (36) in a plane wave basis:

$$\text{index } i\not{D} = \int \frac{(d\,\text{Vol})(d\psi_0)}{(2\pi)^n} \; \frac{1}{\prod_{i=1}^{n}\prod_{n=1}^{\infty}\left(1 + \frac{(x_i/2)^2}{\pi^2 n^2}\right)}$$

$$= \int \frac{d(\text{Vol})}{(2\pi)^n} \; d\Psi_0^1 \cdots d\Psi_0^{2n} \prod_{i=1}^{n} \frac{x_i/2}{\sinh x_i/2} \tag{38}$$

Equation (38) can be interpreted as follows: Think of Ψ_0^a as one-forms. Then M_{ab} is the curvature two-form, and the x_i's are the formal curvature eigenvalues defined in (22). Hence the integrand of (38) is a polynomial in Pontrjagin classes of various degrees. The Grassmann integration over the Ψ_0's simply projects out the form in the integrand whose degree is equal to the dimension of M. The characteristic polynomial:

$$\hat{A}(M) = \prod_{i=1}^{n} \frac{x_i/4\pi}{\sinh x_i/4\pi} \tag{39}$$

is known as the A-roof or Dirac genus. What we have shown is that

$$\text{ind } i\not{D} = \int_M [\hat{A}(M)]_{\text{Vol}} \tag{40}$$

where the subscript vol. means that we only have to consider the term in the integrand proportional to the volume of the manifold. Rather than proceding to derive some other interesting cases of the general index theorem, let us now consider the case of the Dirac operator in the presence of an arbitrary gauge field with gauge group G. This case is in fact equivalent to the general case, and all other index theorems can be derived from by an appropriate choice of G and the representation of G that appears explicitly in the lagrangian to be derived momentarily. Geometrically we consider an arbitrary vector bundle V over M, with connection 1-form A taking values on the Lie algebra of the bundle group G. The eigenvalue problem for the Dirac equation in this case is:

$$i\gamma^i(\partial_i + 1/2\ \omega_{iab}\sigma^{ab} + A_i^\alpha T^\alpha)_{AB}\Psi_B = \lambda\Psi_A$$
$$\sigma^{ab} = \frac{1}{4}[\gamma^a,\gamma^b]. \tag{41}$$

$(T^\alpha)_{AB}$, A,B = 1,2...,dim T, α =1, dim G are the antihermitian generators of G in the representation carried by the vector bundle V. In order to find the one-dimensional analog of (41), we introduce for each index A = 1,2,,..,dim T a pair of fermionic creation and annihilation operators C_A, C_A^* satisfying the usual canonical anticommutation relations:

$$\{C_A, C_B^*\} = \delta_{AB}, \quad \{C_A, C_B\} = 0. \tag{42}$$

In the Hilbert space generated by the C's, we can consider states of the form

$$|\Psi\rangle = \sum_A \Psi_A(\phi)C_A^*|0\rangle. \tag{43}$$

Then (41) can be rewritten as

$$i\gamma^i(\partial_i + 1/2\ \omega_{iab}\sigma^{ab} + A_i^\alpha C^* T^\alpha C)|\Psi\rangle = \lambda|\Psi\rangle \tag{44}$$

a trivial feature of (41) or (44) is that if $|\Psi,\lambda\rangle$ is an eigenfunction of the Dirac operator with eigenvalue λ, then $\bar{\Gamma}|\Psi,\lambda\rangle$ is an eigenfunction with eigenvalue $-\lambda$(if $\lambda \neq 0$). This is a consequence of the fact that $i\not{D}$ contains an odd number of γ-matrices in even dimensions and that $\{\bar{\Gamma},\not{D}\} = 0$. Now we can generalize (25) in order to obtain a lagrangian whose hamiltonian is the square of the Dirac operator (41). It is given by:

$$L = \frac{1}{2} g_{ij}(\phi)\frac{d\phi^i}{dt}\frac{d\phi^j}{dt} + \frac{i}{2}\psi^a\frac{D}{dt}\psi^b\delta_{ab} + iC_A^*\left(\frac{d}{dt}C_A - A_i^\alpha(\phi)\frac{d\phi^i}{dt}T_{AB}^\alpha C_B\right)$$
$$- \frac{1}{2}\psi^a\psi^b F_{ab}^\alpha C_A^* T_{AB}^\alpha C_B \tag{45}$$

where F_{ab}^α is the gauge field strength associated to A_i^α. Now the index theorem follows from

$$\mathrm{Tr}\ \bar{\Gamma}\ e^{-\beta(i\not{D})^2} = n^{E=0}(\bar{\Gamma}=+1) - n^{E=0}(\bar{\Gamma}=-1). \tag{46}$$

Next, we need to write a functional integral representation for (46) using the lagrangian (45). If one considers the space of states of c-fermions, one notices that only the set of one-particle states carries the representation T^{α}. All other particle states transform as tensor products of the representation T. Since we want to compute the index theorem for spinors in the representation T, we have to restrict the functional integral representation of (46) to include only one-particle states of the C-fermions.

$$\text{ind}\, \not{D} = \lim_{\beta\to 0} \int_{\text{A.B.C.}}^{1} (dC^* dC) \int_{\text{P.B.C.}} (d\phi d\Psi) \exp - \int_0^{\beta} L_E(t)dt. \qquad (47)$$

The prime indicates that one has to extract the result corresponding to one-particle states. In the small β-limit, we again have to expand around the constant configurations (ϕ_0, ψ_0, c=0) to second order. Note that A.B.C. (antiperiodic boundary conditions) for the C's are required by the fact that the trace (46) contains $(-1)^F$ (fermion number for the 4 fermions), but not fermion number for the c-fermions. Since there are no constant configurations for the C's, the last two terms in (45) are already second order in small fluctuations, and we can evaluate their coefficients at the constant configurations (ϕ_0, ψ_0). This makes it particularly easy to evaluate the trace over c-fermions. The result is

$$\text{ind}\, i\not{D} = (2\pi)^{-n/2} \int (d\phi_0 d\psi_0)\, \text{Tr}\exp[1/2\, i F_{ab}(\phi_0)\psi_0^a\psi_0^b] \cdot$$
$$\cdot \int (du d\eta) \exp - \int_0^{\beta} dt \left\{ \frac{1}{2}\delta_{ab} \frac{du^a}{dt}\frac{du^b}{dt} + \frac{i}{4} R_{abcd}\psi_0^c\psi_0^d u^a \frac{du^b}{dt} + \frac{i}{2}\eta^a \frac{d}{dt}\eta^a \right\} \qquad (48)$$

and the computation is reduced to the case without gauge fields. If we recall the definition of the Chern-character (19) for some vector bundle V with gauge field strength two-form F, our final answer can be written as:

$$\text{ind}(i\not{D}) = \int_{-M} [\text{ch}(F)\hat{A}(M)]_{\text{Vol}} \qquad (49)$$

where the subscript vol. has the meaning explained after Eq. (40). It is not difficult to understand why (49) is enough to compute many other index theorem. The wave functions associated to the quantum mechanics of (45) are spinors with an extra index Ψ_A. Since any tensor representation of SO(2n) can be obtained in tensor products of spinor representations, but choosing the bundle V appropriately we can generate any representation we please. For example, if we want to consider the index of the Rarita-Schwinger operator, we simply have to define the vector bundle V as the tangent bundle over M, and then F is the standard Riemann curvature, etc. Similar methods can be used to obtain the generalized fixed point formulae of Atiyah and Bott [10] for the index theorem; the character valued index theorem of Atiyah and Hirzebruch which has been used recently [13] to analyze the issue of chiral fermions in Kaluza-Klein theories, etc. For more examples and details, the reader is referred to [7].

So far supersymmetry has been used toderive results in mathematics that were already known. It is hoped that in the future new mathematical results can be obtained by techniques similar to those presented here. In any case, the interplay between supersymmetry and Differential Topology has been a rather interesting one, and there is no reason to believe that this interplay will not provide some nice surprises in years to come.

References

[1] B. Zumino, Phys. Lett. 69B (1977) 369.

[2] A. D'Adda and P. DiVecchia, Phys. Lett. 73B (1978) 162.

[3] See S. W. Hawking in "General Relativity: An Einstein Survey", Eds. S. W. Hawking and W. Israel, Cambridge University Press (1979).

[4] L. Alvarez-Gaumé, Comm. Math. Phys. 90 (1983) 161; J. Phys. A16 (1983) 4177; D. Friedan and P. Windey, Nucl. Phys. B235 (1984) 395; E. Getzler, Comm. Math. Phys. 92 (1983) 163; B. Zumino, *Proceedings of the Second Shelter Island Conference* (June 1983), LBL-17972; B. Zumino and J. Mañes, LBL preprint (1985).

[5] M. F. Atiyah and J. M. Singer, Ann. of Math. 87 (1968) 485; 87 (1968) 546; 93 (1971) 119; 93 (1971) 139. M. F. Atiyah and G. B. Segal, Ann. of Math. 87 (1968) 531.

[6] E. Witten, J. Diff. Geom. 17 (1982) 661.

[7] L. Alvarez-Gaumé on "Supersymmetry", *Proceedings of the 1984 NATO School at Bonn*, Eds. K. Dietz, R. Flume, G. V. Gehlen, and V. Rittenberg (1984).

[8] E. Witten, Nucl. Phys. B202 (1982) 253.

[9] I. Affleck, M. Dine, and N. Seiberg, Nucl. Phys. B256 (1985) 557.

[10] A. Lichnerowicz, "General Theory of Connections and the Holonomy Group", Noordhoof Ed. (1976).

[11] S. S. Chern, "Complex Manifolds Without Potential Theory", Springer-Verlag (1984).

INDEX